Annals of Mathematics Studies

Number 168

The Structure of Affine Buildings

Richard M. Weiss

PRINCETON UNIVERSITY PRESS
PRINCETON AND OXFORD
2009

Published by Princeton University Press, 41 William Street, Princeton, New Jersey 08540

In the United Kingdom: Princeton University Press, 6 Oxford Street, Woodstock, Oxfordshire 0X20 1TW

Library of Congress Cataloging-in-Publication Data

Weiss, Richard M. (Richard Mark), 1946–
The structure of affine buildings / Richard M. Weiss.
p. cm.
Includes bibliographical references and index.
ISBN: 978-0-691-13659-2 (cloth : acid-free paper)
ISBN: 978-0-691-13881-7 (paper : acid-free paper)
1. Buildings (Group theory) 2. Moufang loops. 3. Automorphisms. 4. Affine algebraic groups. I. Title.
QA174.2 .W454 2008
512′.2–dc22 2008062106

British Library Cataloging-in-Publication Data is available

This book has been composed in LaTeX

The publisher would like to acknowledge the author of this volume for providing the camera-ready copy from which this book was printed.

Printed on acid-free paper. ∞

press.princeton.edu

Printed in the United States of America

10 9 8 7 6 5 4 3 2 1

Contents

Preface

The main goal of this book is to present the complete proof of the classification of Bruhat-Tits buildings. By Bruhat-Tits building we mean an affine building whose building at infinity satisfies the Moufang property.[1] The proof we give is distilled from the contents of the first of the series of fundamental articles [6]–[10] by Jacques Tits and François Bruhat and the article [35] Tits wrote for the proceedings of a conference held in Como in 1984. A secondary goal of this book is to provide the reader with a detailed approach to this remarkable literature.

Bruhat-Tits buildings arise in the study of algebraic groups over a field with a discrete valuation.[2] More precisely, there is a unique Bruhat-Tits building Δ in our sense of the term associated to each pair (G, F), where F is a field complete with respect to a discrete valuation and G is an absolutely simple algebraic group of F-rank ℓ at least 2.[3] The classification of Bruhat-Tits buildings says, essentially, that all Bruhat-Tits buildings arise either in this way or by some small variation on this theme.

The apartments of an affine building Δ are Euclidean spaces of a fixed dimension ℓ tiled by the fundamental domains of the action of an affine Coxeter group. These fundamental domains are the *chambers* of Δ. The building Δ itself can be thought of as a set of apartments amalgamated according to certain rules so that, in particular, every chamber is a chamber in many apartments.

Attached to each affine building is a "building at infinity" that is related to the celestial sphere of a Euclidean space. The building at infinity is a spherical building whose rank is the dimension ℓ of the apartments. As indicated above, we say that an affine building is Bruhat-Tits if $\ell \geq 2$ and its building at infinity satisfies the Moufang property. The *Moufang property* means, roughly, that the automorphism group of the building at infinity is rich enough to contain a special system of subgroups called a *root datum*.

The notion of a root datum is fundamental in the theory of buildings. In particular, a spherical building satisfying the Moufang property is uniquely

---

[1]This means, in particular, that the rank ℓ of the building at infinity is at least 2. All buildings are assumed to be thick and irreducible in the following discussion.

[2]The basic reference for the connection between algebraic groups defined over a field with a discrete valuation and affine buildings is the article [33] Tits wrote for the proceedings of a conference held in Corvallis in 1977.

[3]In the literature, the term "Bruhat-Tits building" is often used to denote only the affine buildings arising from such pairs (G, F) but including those where G is of F-rank 1.

determined by its root datum.

(For the affine building associated to a field F complete with respect to a discrete valuation and an absolutely simple group G of positive F-rank ℓ, the building at infinity is just the spherical building associated with G and F. This building always satisfies the Moufang property (assuming $\ell \geq 2$), its apartments correspond to the maximal F-split tori of G and its root datum consists, essentially, of the root groups associated with a fixed maximal F-split torus.)

All spherical buildings of rank $\ell \geq 3$ as well as all their irreducible residues of rank at least 2 satisfy the Moufang property.[4] Thus if the apartments are assumed to have dimension ℓ at least 3, then an affine building and a Bruhat-Tits building are the same thing.

The description of spherical buildings in terms of root data turns out to be exactly the point of view needed to carry out the classification of Bruhat-Tits buildings. In *The Structure of Spherical Buildings* [37] we gave an introduction to the theory of spherical buildings including the proof of Theorem 4.1.2 in [32] and its most important consequences. The present monograph is intended as a sequel to [37].

We assume that the reader has some familiarity with the basic facts about Coxeter groups, buildings and the Moufang property contained in [37].[5] The most important of these facts are summarized here in Appendix A (Chapter 29). In Chapter 2 we assume some standard facts about root systems which can be found in [3] or [17]. In Appendix B (Chapter 30), we summarize the classification of Moufang spherical buildings of rank $\ell \geq 2$ (as carried out in [32] and [36]) in terms of root data. This summary plays a central role starting in Chapter 16.

Here is a brief overview of the organization of this monograph. In Chapters 1–2 and 4–6, the basic properties and substructures of the apartments of an affine building are presented. We introduce affine buildings themselves and construct the building at infinity in Chapters 7–8. In Chapter 9 we pause to examine the case of an affine building of rank 1.

An affine building of rank 1 is simply a tree without vertices of valency 1 or 2 together with a set of subgraphs of valency 2. These are the apartments, which are to be thought of as Euclidean spaces of dimension 1 tiled by the intervals between successive integers. One purpose of Chapter 9 is to prove an uncanny lemma (9.24) that draws a connection between trees and fields

[4] This is a corollary of the fundamental result Theorem 4.1.2 of [32]. In [36], spherical buildings of rank $\ell = 2$, also known as generalized polygons, *assumed* to satisfy the Moufang condition (hence: Moufang polygons) were classified. In Chapter 40 of [36] the classification of Moufang polygons is used to give a revised proof of Tits's famous classification [32] of spherical buildings of rank $\ell \geq 3$. Once it is established that such a building as well as all its irreducible rank 2 residues satisfy the Moufang property (this is 11.6 and 11.8 in [37]), it remains only to examine how the root data of Moufang polygons (i.e. root data of rank 2) can be assembled to form the root datum of a spherical building of higher rank.

[5] The essential facts needed are all contained in Chapters 1–3, Chapters 7–9 and Chapter 11 of [37].

with a discrete valuation. This is the first hint of the subtle connection between Euclidean geometry and number theory that pervades the theory of affine buildings.[6]

In Chapters 10 and 11 we return to the study of an affine building Δ of arbitrary rank and construct two whole forests of trees. These forests are uniquely determined by a family of functions defined on certain substructures of the building at infinity. In Chapter 12 we show that the affine building Δ is uniquely determined by its building at infinity together with this "tree structure."

At this point we invoke (for the first time) the assumption that the building at infinity is Moufang. This means that the building at infinity is uniquely determined by its root datum. We show in Chapter 13 how the tree structure of Δ is uniquely determined by something called a valuation of this root datum (a notion previously introduced in Chapter 3), and in Chapter 14 we show that every root datum with valuation arises from an affine building in this manner. This yields the conclusion that Bruhat-Tits buildings are classified by root data with valuation.[7]

It is here that we invoke the classification of spherical buildings satisfying the Moufang condition. This means that we know all possible root data. Each root datum, with some mild exceptions, is determined by algebraic data defined over a field K.[8] We show that a valuation of a given root datum is uniquely determined by a discrete valuation of this field K (in the number-theoretical sense of the term). We are thus left with the problem of determining for each root datum necessary and sufficient conditions for a discrete valuation of the field K to extend to a valuation of the root datum. This problem is reduced to a problem in rank 2 in Chapters 15–16 and solved for each family of Moufang polygons in Chapters 19–25.

The building at infinity of a Bruhat-Tits building depends not only on the building but also on the choice of a "system of apartments." It is thus necessary to determine when two root data with valuation correspond to the same affine building but with perhaps two different systems of apartments. We do this in Chapters 17–26 by studying "completions." A good part of Chapters 17–26 is also devoted to determining the structure of the residues of a Bruhat-Tits building.

In Chapter 27 we summarize the conclusions of the classification in the form of a list of the different families of Bruhat-Tits buildings together with references to a few of their principal features. In Chapter 28 we describe in more detail the classification of *locally finite* Bruhat-Tits buildings and correlate the results with the tables in [33].

---

[6]Euclid himself would have taken great delight in this connection between his two favorite topics!

[7]See 14.54 for the exact statement of this result.

[8]More accurately, K is, according to the case in question, a field, a skew field or an octonion division algebra. For those spherical buildings arising from the F-points of an absolutely simple algebraic group (and are thus defined over F in the sense of algebraic groups), F is the either the center of K or the intersection of the center of K with the set of fixed points of an involution of K.

For the most part we have followed the proof of the classification of Bruhat-Tits buildings as it is carried out in [6] and [35] with only minor modifications where we thought we could show something more plainly. Our proof of 12.3 is different from the proof given in [35]. There is also a greater emphasis here on the role of root data in rank 2 (which is inevitable since, after all, the classification of Moufang polygons was not available when [35] was written). In particular, the notion of a *partial valuation* in Chapter 15 as well as many of the results about exceptional quadrangles in Chapters 21–22 and many of the explicit calculations of residues and completions in Chapters 18–26 are, we believe, new.

In parts of this book we were influenced greatly by the book on buildings by Mark Ronan [25]. We refer the reader also to the excellent books by Kenneth Brown [4] and Paul Garrett [14] as well as the "new edition" [1] of [4] by Peter Abramenko and Kenneth Brown.

This book was begun while the author was a guest of the University of Würzburg and supported by a Research Prize from the Humboldt Foundation. Subsequent portions of this book were written while the author was a guest of the University of Birmingham, the University of Bielefeld and the Korean Institute for Advanced Study. Some of this work was also supported by a Senior Faculty Research Fellowship from Tufts University. The author would like to express his gratitude to all these institutions. It is also a pleasure to thank Tom De Medts, Theo Grundhöfer, Petra Hitzelberger, Linus Kramer, Andreas Mars, George McNinch, Bernhard Mühlherr, Holger Petersson, Andrei Rapinchuk, Nils Rosehr and especially Rebecca Waldecker for their assistance with the preparation of various parts of this manuscript.

Finally, the author would like to acknowledge his great debt to Jacques Tits with the profoundest regard and esteem.

The Structure of Affine Buildings

Chapter One

Affine Coxeter Diagrams

By the results summarized in Chapter VI, Section 4.3, of [3], affine Coxeter groups can be characterized as groups generated by reflections of an affine space (by which is meant a Euclidean space without a fixed coordinate system or even a fixed origin). In an effort to get to affine buildings as quickly as possible, however, we will define them more succinctly, but with less motivation, simply as the Coxeter groups associated with the following Coxeter diagrams (and direct products of such groups):

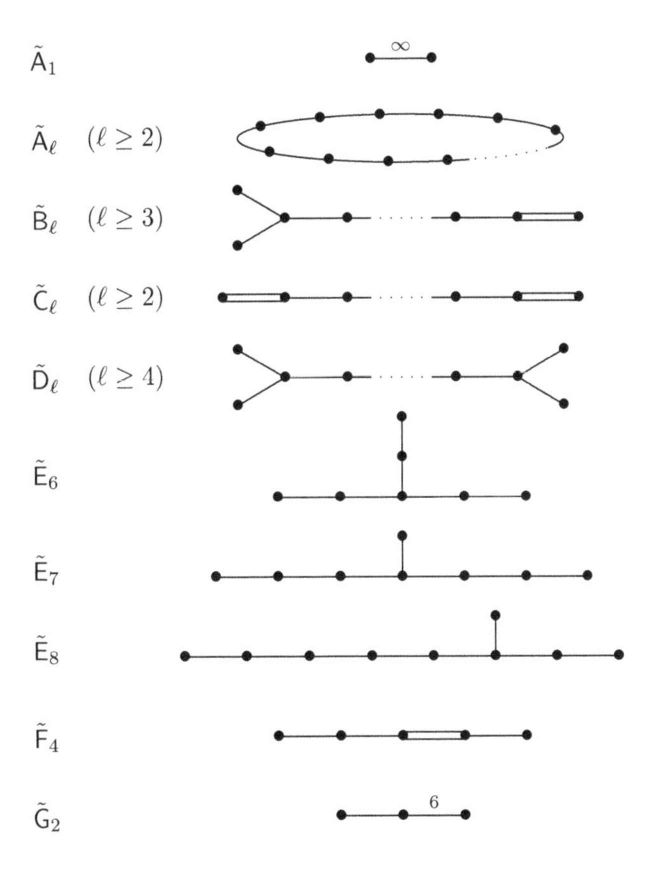

Figure 1.1 The Irreducible Affine Coxeter Diagrams

Figure 1.2 Some Small Affine Coxeter Diagrams

Each Coxeter diagram[1] in Figure 1.1 has a name of the form $\tilde{X}_\ell$, where

$$X = A, B, C, D, E, F \text{ or } G$$

and ℓ is 1 less than the number of vertices. In each case, X_ℓ is the standard name of a spherical diagram with ℓ vertices that can be found in Figure 1.3. [2] Moreover, the spherical diagram X_ℓ can be obtained from $\tilde{X}_\ell$ by deleting a single vertex (and the edge or double edge connected to it).

Following the usual convention, we assign the diagram $\tilde{C}_2$ the alternative name $\tilde{B}_2$. In Figure 1.2, we have reproduced a few of the smaller affine diagrams.

Definition 1.1. A vertex of $\tilde{X}_\ell$ is called *special* if its deletion yields the diagram X_ℓ.

Thus, for example, every vertex of $\tilde{A}_\ell$ is special, the special vertices of $\tilde{B}_\ell$ are the two farthest to the left in Figure 1.1 and the special vertices of $\tilde{C}_\ell$ are those at the two ends. Note that in each case, the special vertices of $\tilde{X}_\ell$ form a single orbit under the action of $\mathrm{Aut}(\tilde{X}_\ell)$ (the group of automorphisms of the underlying graph of $\tilde{X}_\ell$ that preserve the edge labels).

The names A_ℓ, B_ℓ, etc. are also the standard names of irreducible root systems. We explain this connection in Chapter 2.

Affine buildings are frequently called *Euclidean buildings* in the literature. These two terms are synonymous.

As in [37], we regard buildings exclusively as chamber systems throughout this book. We assume that the reader is familiar with the basic facts about

[1] A Coxeter diagram is a graph whose edges are labeled by elements of the set

$$\{m \in \mathbb{N} \mid m \geq 3\} \cup \{\infty\}.$$

The Coxeter diagrams in Figures 1.1–1.3 are drawn using the usual conventions that an edge with label 3 is drawn simply as an edge with no label and an edge with label 4 is drawn as a double edge (also with no label).

[2] A Coxeter diagram is *spherical* if the associated Coxeter group W_Π (defined in 29.5) is finite. By [13], the only irreducible spherical Coxeter diagrams that do *not* appear in Figure 1.3 are the diagrams

$$\overset{m}{\bullet\!\!-\!\!\!-\!\!\!-\!\!\bullet}$$

(sometimes called $I(m)$) for $m = 5$ and $m > 6$ and the diagrams H_3 and H_4.

Coxeter chamber systems and buildings—especially with properties of their roots, residues and projection maps—as covered in Chapters 1–5, 7–9 and 11 of [37]. We have summarized the most basic definitions and results from [37] in Appendix A (Chapter 29). We will also make frequent reference as we go along to the exact results in [37] and Appendix A that we require.

* * *

An affine building is a building whose Coxeter diagram is affine. The apartments of a building with Coxeter diagram Π are all isomorphic to the Coxeter chamber system Σ_Π (as defined in 29.4). We begin our study of affine buildings, therefore, by studying properties of the Coxeter chamber systems Σ_Π for Π a connected affine diagram.[3]

In this chapter we describe three fundamental properties of these Coxeter chamber systems. These properties are stated in 1.3, 1.9 and 1.12 below. In Chapter 2 we will prove that these properties do, in fact, hold. To do this, we will rely on properties of root systems and thus on a representation of Σ_Π as a system of "alcoves" in a Euclidean space. In Chapters 3, we introduce the notion of a "root datum with valuation" which depends on this representation in an essential way. The notion of a root datum with valuation will not be required, however, until Chapter 13. Thus the reader who is impatient to see what affine buildings look like and who is willing to regard the three fundamental properties of affine Coxeter chamber systems as axioms can jump from the end of this first chapter straight to Chapter 4.[4]

For the rest of this chapter, we suppose that Π is one of the affine diagrams in Figure 1.1 and let $\Sigma = \Sigma_\Pi$ denote the Coxeter chamber system of type Π as defined in 29.4.

Definition 1.2. Two roots α and α' of Σ are *parallel* if $\alpha \subset \alpha'$ or $\alpha' \subset \alpha$.[5]

Here is the first of our fundamental properties:

Proposition 1.3. *For each root α there is a bijection $i \mapsto \alpha_i$ from $\mathbb{Z}$ to the set of roots of Σ parallel to α such that $\alpha_0 = \alpha$ and $\alpha_i \subset \alpha_{i+1}$ for each i.*

Proof. See 2.46. □

Corollary 1.4. *Parallelism is an equivalence relation on the set of roots of Σ.*

---

[3] By 7.32 in [37], an arbitrary building Δ of finite rank is a direct product (in an appropriate sense) of buildings $\Delta_1, \dots, \Delta_k$ of type $\Pi_1, \dots, \Pi_k$, where $\Pi_1, \dots, \Pi_k$ are the connected components of the Coxeter diagram of Δ.

[4] Our intention is not, however, to elevate these three properties to the level of axioms, but rather to make clear and explicit the use we are making of the representation of Σ_Π in Euclidean space. Even in those chapters where we are not using it (i.e. Chapters 4–12), this representation remains, of course, an important source of intuition.

[5] The symbol $\subset$ is always to be interpreted in this book as allowing equality.

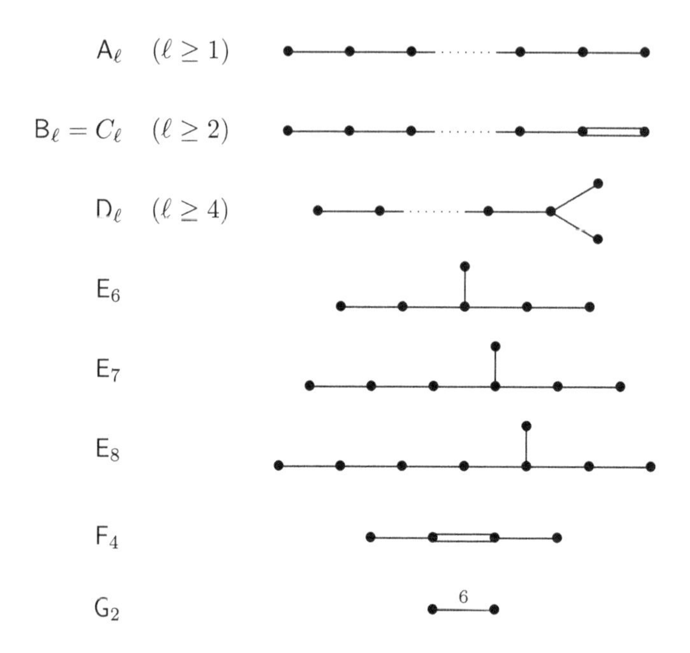

Figure 1.3 Some Spherical Coxeter Diagrams

Proof. This holds by 1.3. □

Notation 1.5. The parallel class of a root α will be denoted by $[\alpha]$.

Definition 1.6. A *gem* of Σ is a residue of type $I\backslash\{o\}$, where I denotes the vertex set of Π and o is a special vertex of Π as defined in 1.1. A gem of type $I\backslash\{o\}$ for some special vertex o is called *o-special.*

Proposition 1.7. *All proper residues of Σ are finite and all gems are, in fact, Coxeter chamber systems of type X_ℓ (up to isomorphism). In particular, all gems have the same number of chambers.*

Proof. This holds by 29.7.iii (and 5.17 of [37]). □

Definition 1.8. Let $\mathrm{Aut}(\Sigma)$ be as defined in 29.2 and let $T = T_\Pi$ denote the set of elements of $\mathrm{Aut}(\Sigma)$ that map each root of Σ to a parallel root. The elements of T are called *translations* and two objects in the same orbit of T are called *translates* of each other. By 1.4, T is a normal subgroup of $\mathrm{Aut}(\Sigma)$.

Here is our second fundamental property:

Proposition 1.9. *The group T of translations acts transitively on the set of gems of Σ.*

Proof. See 2.46. □

The set $[R, d]$ defined next will play a central role in this book.

Notation 1.10. Let R be a gem and let d be a chamber of R. We denote by $[R, d]$ the set of roots α of Σ that contain d but not some chamber of R adjacent to d.

Let R, d and $[R, d]$ be as in 1.10. As observed in 1.7, R is a Coxeter chamber system of type X_ℓ. This implies that there are exactly ℓ chambers in R adjacent to d. Let e be one of them. By 29.6.i, there is a unique root in $[R, d]$ that does not contain e. Thus $|[R, d]| = \ell$.

Definition 1.11. We will say that a root α *cuts* a residue E of Σ if both α and $-\alpha$ contain chambers of E. By 29.7.iv, this is equivalent to saying that $\alpha \cap E$ is a root of E.

Here is the last of our three fundamental properties:

Proposition 1.12. *Let R be a gem, let $d \in R$, let $[R, d]$ be the set defined in 1.10, let*

$$\alpha_1, \ldots, \alpha_\ell$$

be the roots in $[R, d]$ and for each $i \in [1, \ell]$,[6] *let α_i' be a root parallel to α_i. Then there exists a unique gem that is cut by α_i' for all $i \in [1, \ell]$. In particular, R is the unique gem cut by all the roots in $[R, d]$.*

Proof. See 2.46. □

* * *

We use the remainder of this chapter to record a number of simple consequences of 1.3, 1.7 and our three fundamental properties.

Proposition 1.13. *A proper residue (in particular, a gem) cannot be cut by two distinct parallel roots.*

Proof. Let E be a proper residue and suppose that α and β are parallel roots both cutting E. By 1.6, E is finite. By 29.7.iv, both $\alpha \cap E$ and $\beta \cap E$ are roots of E. By 29.6.i, therefore,

$$|\alpha \cap E| = |E|/2 = |\beta \cap E|.$$

Since α and β are parallel, it follows that $\alpha \cap E = \beta \cap E$. Hence $\alpha = \beta$ by 29.17. □

Proposition 1.14. *Let α and β be roots such that α is parallel to neither β nor $-\beta$. Then there exists a residue of rank 2 cut by both α and β.*

---

[6] We use interval notation throughout this book to denote intervals of integers. Thus, for example, by $[1, \ell]$, we mean the subset $\{1, 2, \ldots, \ell\}$ of $\mathbb{Z}$. This should not cause any confusion with the notation introduced in 1.5 or 1.10.

Proof. This holds by 29.24. □

Proposition 1.15. *Let α and β be roots such that α is parallel to neither β nor $-\beta$. Then there exist gems that are cut by β and contained in α.*

Proof. Let $i \mapsto \alpha_i$ be as in 1.3 with $\alpha_0 = \alpha$ (so $\alpha_i \subset \alpha_{i+1}$ for all $i \in \mathbb{Z}$) and choose $m \leq -2$. By 1.14, there exists a residue E of rank 2 cut by both α_m and β. Let P be the unique panel of E that is cut by β and contained in α_m. We choose a gem containing P if there is one and call it R.

Suppose that P is not contained in a gem and let o be its type. Then o is the only special vertex of Π and there exists (by inspection of Figure 1.1) a unique vertex i of Π adjacent to o and the label on the edge $\{o, i\}$ is 3. Let D be the residue of type $\{o, i\}$ containing P and let Q be the panel of D opposite P. Then Q is of type i and β cuts Q. Since $P \subset D \cap \alpha_m$, the root α_m either cuts D or contains D. By 1.13, it follows that $Q \subset D \subset \alpha_{m+1}$. Let R be the unique gem containing Q.

Thus whether or not P is contained in a gem, R is a gem cut by β that contains chambers in α_{m+1}. Thus either $R \subset \alpha_{m+1}$ or R is cut by α_{m+1}. By 1.13 again, it follows that $R \subset \alpha_{m+2} \subset \alpha_0 = \alpha$. □

Corollary 1.16. *Every root cuts gems.*

Proof. Let α be a root. Either there exists a root β parallel to neither α nor $-\alpha$, in which case 1.15 applies, or $\ell = 1$. If $\ell = 1$, then the unique panel cut by α is a gem. □

Corollary 1.17. *Let α be a root. Then there exists a gem R and a chamber $d \in R$ such that $\alpha \in [R, d]$ (where $[R, d]$ is as defined in 1.10).*

Proof. This holds by 1.16. □

We next observe that Σ satisfies a version of the parallel postulate:

Proposition 1.18. *For each root α and each gem R of Σ, there is a unique root parallel to α cutting R.*

Proof. Let R be a gem and let α be a root of Σ. By 1.16, there exists a gem R_1 cut by α. By 1.9, there exists a translation g mapping R_1 to R. By 1.8 and 1.11, α^g is a root parallel to α that cuts R. By 1.13, this root is unique. □

Proposition 1.19. *Let R be a gem of Σ. Then the map $\alpha \mapsto [\alpha]$ is a bijection from the set of roots cutting R to the set of parallel classes of roots of Σ.*

Proof. This map is injective by 1.13 and surjective by 1.18. □

Proposition 1.20. *The group* $\mathrm{Aut}(\Sigma)$ *acts faithfully on the set of roots of Σ.*

Proof. Let g be an element of $\mathrm{Aut}(\Sigma)$ that maps each root to itself and let d be a chamber of Σ. For each chamber e adjacent to d, there is (by 29.6.i) a unique root containing d but not e. The intersection of these roots is thus $\{d\}$. Since g maps this intersection to itself, it fixes d. □

Proposition 1.21. *Let R be a gem. The only element of T that maps R to itself is the identity.*

Proof. Let g be an element of T that maps R to itself and let α be a root. By 1.18, there is a unique root α_1 in the parallel class $[\alpha]$ that cuts R. By 1.13, g maps α_1 to itself. By 1.3, it follows that g acts trivially on $[\alpha]$. We conclude that g fixes every root of Σ. By 1.20, therefore, $g = 1$. □

Corollary 1.22. *The group T acts sharply transitively on the set of gems of Σ.*[7]

Proof. This holds by 1.9 and 1.21. □

Definition 1.23. Let R be a gem, let d be a chamber of R, let $[R, d]$ be as in 1.10 and let $\alpha \in [R, d]$. We denote by $T_{R,d,\alpha}$ the subgroup of T consisting of all translations that fix every root in $[R, d]$ *except* α.

Proposition 1.24. *Let R be a gem, let d be a chamber of R and let $T_{R,d,\alpha}$ for each $\alpha \in [R, d]$ be as in 1.23. Then $T_{R,d,\alpha}$ is isomorphic to $\mathbb{Z}$ and acts transitively and faithfully on the parallel class $[\alpha]$ for all $\alpha \in [R, d]$ and*

$$T = \bigoplus_{\alpha\in[R,d]} T_{R,d,\alpha}.$$

Thus, in particular, $T \cong \mathbb{Z}^\ell$.

Proof. Let

$$\alpha_1, \ldots, \alpha_\ell$$

be the roots in $[R, d]$, let

$$X = [\alpha_1] \cup \cdots \cup [\alpha_\ell] \tag{1.25}$$

and let $\alpha_i' \in [\alpha_i]$ for all $i \in [1, \ell]$. Choose $i \in [1, \ell]$. By 1.12, there exists a unique residue R_i' cut by α_i' and by α_j for all $j \in [1, \ell]$ different from i. By 1.9, there exists a translation t mapping R to R_i'. Thus $t(\alpha_i)$ is a root in $[\alpha_i]$ cutting R_i' for each $i \in [1, \ell]$. By 1.13, it follows that $t(\alpha_i) = \alpha_i'$ and $t(\alpha_j) = \alpha_j$ for all $j \in [1, \ell]$ different from i. By 1.23, $t \in T_{R,d,\alpha_i}$. We conclude that T_{R,d,α_i} acts transitively on $[\alpha_i]$.

By 1.3, it follows that the group T_{R,d,α_i} induces a transitive group isomorphic to $\mathbb{Z}$ on $[\alpha_i]$ for each $i \in [1, \ell]$. In particular, the commutator group

$$[T_{R,d,\alpha_i}, T_{R,d,\alpha_j}]$$

acts trivially on the set X defined in 1.25 above, and if t is an arbitrary element of T, then there exist elements $t_i \in T_{R,d,\alpha_i}$ for $i \in [1, \ell]$ such that

[7] In other words, for each ordered pair (R_1, R_2) of gems, there is a *unique* element of T that maps R_1 to R_2.

the product $t^{-1}t_1 \cdots t_\ell$ also acts trivially on the set X. It thus suffices to show that T acts faithfully on X.

Let g be an element of T that acts trivially on X. Thus $g(R)$ is a gem cut by all the roots of $[R, d]$. By 1.12, R is the unique gem with this property. Hence $g(R) = R$. By 1.21, therefore, $g = 1$. □

Corollary 1.26. *Let α be a root. Then T acts transitively on the parallel class $[\alpha]$.*

Proof. This holds by 1.17 and 1.24. □

Definition 1.27. We will call a translation a *σ-translation* if it is a σ-automorphism of Σ for some $\sigma \in \text{Aut}(\Pi)$ (as defined in 29.2). Let H be the image of T under the homomorphism from $\text{Aut}(\Sigma)$ to $\text{Aut}(\Pi)$ that sends a σ-automorphism to σ. Thus $H \cong T/T \cap W$ (since, by 29.4, W can be thought of as the group of special automorphisms of Σ). The elements of H will be called *translational automorphisms* of Π.

Proposition 1.28. *Let H be as defined in 1.27. Then H acts sharply transitively on the set of special vertices of* Π. *In particular, $|T/T \cap W|$ equals the number of special vertices.*

Proof. By 1.24, T is abelian. Therefore H is also abelian. It will suffice, therefore, to show that H acts transitively on the set of special vertices of Π. Let o, o_1 be two special vertices of Σ, let R be an o-special gem and let R_1 be an o_1-special gem. By 1.9, there exists a translation t mapping R to R_1. Then t is a σ-translation from some $\sigma \in H$ mapping o to o_1. □

Remark 1.29. By Section 4.7(XII) and Section 4.8(XII) in Chapter VI of [3], the group H defined in 1.27 is, in fact, cyclic except when $\Pi = \tilde{\mathsf{D}}_\ell$ for ℓ even, in which case $H \cong \mathbb{Z}_2 \times \mathbb{Z}_2$.

Corollary 1.30. *The subgroup $T \cap W$ of* $\text{Aut}^\circ(\Sigma)$ *(as defined in 29.2) acts sharply transitively on the set of gems of a given type.*

Proof. Let R and R_1 be two gems of the same type. By 1.9, there exists an element $t \in T$ mapping R to R_1. By 1.28, the element t acts trivially on Π. Thus $t \in W$. The claim holds, therefore, by 1.22. □

Definition 1.31. Let R be a gem and let x be a chamber of Σ. Then there is a unique gem R_1 of the same type as R containing x. By 1.30, there is a unique $g \in T \cap W$ mapping R_1 to R. We will call x^g the *special translate of x to R*.

Proposition 1.32. *Let R be an o-special gem for some special vertex o of* Π, *let $\{x, y\}$ be a panel of Σ of type $i \neq o$ and let u and v be the special translates of the chambers x and y to R (as defined in 1.31). Then $\{u, v\}$ is a panel of R of type i.*

Proof. This holds by 1.31 since the o-special gem containing x is the same as the o-special gem containing y. □

Proposition 1.33. *Let R be an o-special gem for some special vertex o of Π and let W_R be the stabilizer of R in $W = \text{Aut}^\circ(\Sigma)$. For each chamber $x \in R$, let $\delta(x)$ be the special translate of y to R, where y is the unique chamber that is o-adjacent to x. Then $\delta(x^g) = \delta(x)^g$ for all $g \in W_R$.*

Proof. Let $g \in W_R$ and $x \in R$. Let y be the unique chamber that is o-adjacent to x and let t be the unique element of $T \cap W$ such that $y^t \in R$. Thus $g^{-1}tg$ is an element of $T \cap W$ (since $T \cap W$ is normal in W) that maps y^g to R. Therefore

$$\delta(x^g) = (y^g)^{g^{-1}tg} = y^{tg} = \delta(x)^g.$$

□

Proposition 1.34. *Let α and β be roots such that α is parallel to neither β nor $-\beta$. Then α contains some but not all chambers in $\partial\beta$ (where $\partial\beta$ is the border of β as in 29.40).*

Proof. By 1.14, there exists a residue E of rank 2 cut by both α and β. By 29.7.iv, both $\alpha \cap E$ and $\beta \cap E$ are roots of E, and by 29.17, $\alpha \cap E \neq \beta \cap E$. It follows that $\partial\beta \cap E$ contains two chambers and α contains exactly one of them. □

Proposition 1.35. *Let α be a root. Then every chamber is contained in some root parallel to α.*

Proof. Let d be a chamber and let R be a gem containing d. By 1.18, there is a root α' parallel to α that cuts R. By 1.3, there exists a root α'' that contains α' properly (and is, in particular, parallel to α'). By 1.13, α'' must contain R. □

Definition 1.36. Let α be a root. By 1.3, there exists a unique root $\beta \neq \alpha$ that contains α and is contained in every other root containing α. We will say that β *contains* α *minimally*, alternatively, that α is *contained in* β *maximally*.

Definition 1.37. A *strip* is a set of the form $-\alpha \cap \beta$, where α and β are roots such that β contains α minimally (as defined in 1.36).

Proposition 1.38. *Let α and β be roots such that β contains α minimally (as defined in 1.36) and let d be contained in the strip $-\alpha \cap \beta$. Then β is the convex hull of $\alpha \cup \{d\}$.*

Proof. By 1.36, every root containing α and d contains β. The claim holds, therefore, by 29.20. □

Proposition 1.39. *If β and β' are parallel roots and $\beta' \cap \partial\beta \neq \emptyset$, then $\beta \subset \beta'$ (where $\partial\beta$ is as in 29.40).*

Proof. Let $x \in \beta' \cap \partial\beta$ and let y be the chamber adjacent to x in $-\beta$. If $y \in \beta'$, then $\beta \subset \beta'$ simply because β and β' are parallel. If $y \notin \beta'$, then $\beta = \beta'$ by 29.6.i. □

Proposition 1.40. *Let α and β be roots such that β contains α minimally. Then the convex hull of $\partial\beta$ is the strip $-\alpha \cap \beta$.*

Proof. Let β' be a root containing $\partial\beta$. By 1.34, β' is a parallel either to β or to $-\beta$. By 1.39, $\beta \subset \beta'$ if β' is parallel to β. If β' is parallel to $-\beta$, then $-\alpha \subset \beta'$ simply by 1.36. The claim holds now by 29.20. □

Proposition 1.41. *Let α be a root, let $i \mapsto \alpha_i$ be the map from $\mathbb{Z}$ to $[\alpha]$ described in 1.3 and for each $i \in \mathbb{Z}$, let L_i denote the strip $\alpha_i \backslash \alpha_{i-1}$. Then Σ is the disjoint union of the strips L_i for all $i \in \mathbb{Z}$ and for each $k \in \mathbb{Z}$, the root α_k is the disjoint union of the strips L_i for all $i \leq k$.*

Proof. By 1.35, every chamber of Σ lies in a root parallel to α and in a root parallel to $-\alpha$. This means that for each chamber x in Σ the set of integers i such that $x \in \alpha_i$ is non-empty and bounded below. □

Proposition 1.42. *Let α and β be roots such that β contains α minimally and let P be a panel. Let $\mu(\alpha)$ and $\mu(\beta)$ be the walls of α and β as defined in 29.22. Then a panel P is contained in $\mu(\alpha) \cup \mu(\beta)$ if and only if P contains exactly one chamber in the strip $-\alpha \cap \beta$.*

Proof. Let $L = -\alpha \cap \beta$. Suppose that $P \in \mu(\alpha)$. Then P contains one chamber x in $-\alpha$ and one chamber y in α and (by 29.6.i) α is the unique root containing y but not x. Since $y \in \alpha \subset \beta$ but $\beta \neq \alpha$, it follows that β contains x. Thus $x \in L$ but $y \notin L$. Similarly, if $P \in \mu(\beta)$, then P contains exactly one chamber in L.

Suppose, conversely, that P is a panel that contains a unique chamber in L. Let x be this unique chamber and let y be the chamber in P distinct from x. Then $y \notin L$. Thus $y \in \alpha$ or $y \in -\beta$. In the first case, $P \in \mu(\alpha)$, and in the second case, $P \in \mu(\beta)$. □

Proposition 1.43. *For each $g \in T$, there exists $M \in \mathbb{N}$ such that*

$$\text{dist}(x, x^g) \leq M$$

for all chambers x of Σ.

Proof. Let $g \in T$ and let R be a gem. Since R is finite, there exists a positive integer M such that

$$\text{dist}(x, x^g) \leq M$$

for all $x \in R$. Now suppose that x is an arbitrary chamber of Σ. By 1.9, there exists an element $h \in T$ such that $x^h \in R$. By 1.24, T is abelian. Thus

$$\text{dist}(x^g, x) = \text{dist}(x^{gh}, x^h) = \text{dist}(x^{hg}, x^h) = \text{dist}((x^h)^g, x^h) \leq M.$$

□

The following notion plays a central role in the theory of affine buildings.

Definition 1.44. Let Γ be an arbitrary graph, let S be the vertex set of X and let

$$\text{dist}(U, V) = \inf\{\text{dist}(u, v) \mid u \in U,\ v \in V\}$$

for all subsets U and V of S. We will call two subsets X and Y of S *parallel* if the sets

$$\{\text{dist}(\{x\}, Y) \mid x \in X\} \text{ and } \{\text{dist}(X, \{y\}) \mid y \in Y\}$$

are both bounded.[8]

Proposition 1.45. *Let X be an arbitrary set of chambers of Σ and let $g \in T$. Then X and X^g are parallel as defined in 1.44.*

Proof. This holds by 1.43. □

Note that the parallel relation defined in 1.44 is an equivalence relation. We have already introduced the notion of parallel roots, so we need to observe that the two roots are parallel in the sense of 1.2 if and only if they are parallel in the sense of 1.44: By 1.26 and 1.43, roots that are parallel in the sense of 1.2 are parallel as defined in 1.44. Suppose, conversely, that α and β are roots that are not parallel in the sense of 1.2. We will say that a chamber x is *separated* from α by k roots if there are k roots of Σ containing α that do not contain x. If α is parallel to $-\beta$ in the sense of 1.2, then by 1.3, β contains chambers separated from α by arbitrarily many roots. If α is not parallel to $-\beta$ in the sense of 1.2, then by 1.14, $(-\alpha') \cap \beta \neq \emptyset$ for every root α' containing α, so again there are chambers in β separated from α by arbitrarily many roots. By 29.6.iii, we conclude that β contains chambers arbitrarily far from α (whether or not α is parallel to $-\beta$ in the sense of 1.2). Thus α and β are not parallel in the sense of 1.44.

We say that two walls of Σ (as defined in 29.22) are parallel if their chamber sets are parallel in the sense of 1.44.

Proposition 1.46. *Let α and α' be two roots. Then α is parallel to α' or $-\alpha'$ if and only if the walls $\mu(\alpha)$ and $\mu(\alpha')$ are parallel.*

Proof. Suppose that α is parallel to α'. By 1.26, α' is a translate of α. Hence $\mu(\alpha')$ is a translate of $\mu(\alpha)$. By 1.45, it follows that $\mu(\alpha)$ and $\mu(\alpha')$ are parallel. Since $\mu(\alpha') = \mu(-\alpha')$, the same conclusion holds if we assume that α is parallel to $-\alpha'$.

Suppose, conversely, that α is parallel to neither α' nor $-\alpha'$. Let $i \mapsto \alpha_i$ be the bijection from $\mathbb{Z}$ to the parallel class $[\alpha]$ as in 1.3. By 29.6.i, the chamber set of $\mu(\alpha)$ is contained in α_i for all $i > 0$ and α_i is parallel to neither α' nor $-\alpha'$ for all i. By 1.14, therefore, the chamber set of $\mu(\alpha')$ has a non-trivial intersection with $-\alpha_i$ for all i. Thus for each $N > 0$, there are chambers of $\mu(\alpha')$ not contained in any of the roots α_i for $i \in [1, N]$. By 29.6.iii, it follows that there exist chambers in $\mu(\alpha')$ that are arbitrarily far from $\mu(\alpha)$. Thus $\mu(\alpha)$ and $\mu(\alpha')$ are not parallel. □

Comment 1.47. If we identify Σ with its representation in a Euclidean space V as described in the next chapter, then parallel walls of Σ correspond

[8] "At bounded distance" might arguably be a more suitable name for this property than "parallel"; see, however, 1.47.

(by 2.34) to affine hyperplanes of V that are parallel in the usual sense of the word. This accounts for the use of the word “parallel” in 1.2 and 1.44.

Chapter Two

Root Systems

In this chapter, we show that the Coxeter chamber systems corresponding to the connected affine Coxeter diagrams (Figure 1.1) have the three fundamental properties formulated in 1.3, 1.9 and 1.12. Our proofs rely on standard facts about root systems for which we will cite [3] and [17] as our references. With the representation of spherical and affine Coxeter chamber systems in terms of Weyl chambers and alcoves that we describe in this chapter, we are also setting the stage for Chapter 3.

Notation 2.1. Let V be a Euclidean space. For each non-zero $\alpha \in V$, we set

$$s_\alpha(v) = v - 2\frac{v \cdot \alpha}{\alpha \cdot \alpha}\alpha$$

for all $v \in V$.

The elements s_α for $\alpha \in V^*$ are called *reflections*. Note that $s_\alpha = s_{t\alpha}$ for each $t \in \mathbb{R}^*$ and that each s_α is an isometry of V.[1] Let Φ be a *root system* as defined in Chapter VI, Section 1.1, of [3]. Thus Φ is a spanning set of non-zero vectors in a Euclidean space V such that

(R1) $\Phi \cap \{t\alpha \mid t \in \mathbb{R}\} = \{\alpha, -\alpha\}$ for all $\alpha \in \Phi$;
(R2) $s_\alpha(\Phi) = \Phi$ for all $\alpha \in \Phi$, where s_α is as in 2.1;
(R3) $2(\alpha \cdot \beta)/(\beta \cdot \beta) \in \mathbb{Z}$ for all $\alpha, \beta \in \Phi$.

We assume, too, that Φ is *irreducible*, i.e. that Φ does not have a decomposition as the direct sum of two root systems as defined in Chapter VI, Section 1.2, of [3]. Let ℓ denote the *rank* of Φ, i.e. the dimension of V.

Root systems were introduced and classified by W. Killing in the nineteenth century. According to this classification, Φ is, up to an isometry of V, one of the root systems called X_ℓ for

$$\mathsf{X} = \mathsf{A}, \mathsf{B}, \mathsf{C}, \mathsf{D}, \mathsf{E}, \mathsf{F} \text{ or } \mathsf{G}$$

described in Chapter VI, Sections 4.4–4.13, of [3].

Definition 2.2. For all $\alpha \in \Phi$, we denote by H_α the hyperplane

$$\{v \in V \mid v \cdot \alpha = 0\}.$$

A *Weyl chamber* of Φ is a connected component of

$$V \backslash \bigcup_{\alpha \in \Phi} H_\alpha. \tag{2.3}$$

[1] We will always denote by R^* the set of non-zero elements of a field or vector space R. If G is a group, we will always denote by G^* the set of non-trivial elements of G.

Proposition 2.4. *Let C be a Weyl chamber. Then the following hold:*

(i) *There exist ℓ roots $B(C) := \{\alpha_1, \dots, \alpha_\ell\}$ in Φ such that*

$$C = \{v \in V \mid v \cdot \alpha_i > 0 \text{ for all } i \in [1, \ell]\}.$$

(ii) *The set $B(C)$ is a basis of the vector space V.*

(iii) *The set $B(C)$ is uniquely determined by C.*

Proof. These assertions hold by Theorem 2 in Chapter VI, Section 1.5, of [3]. □

Definition 2.5. Let Φ be an irreducible root system. A *basis* of Φ is a set of the form $B(C)$ for some Weyl chamber C as defined in 2.4.[2] The *walls* of a Weyl chamber C are the hyperplanes H_α for $\alpha \in B(C)$. We denote the set of walls of a Weyl chamber C by $M(C)$.

Definition 2.6. Let Γ_Φ be the graph whose vertices are the Weyl chambers of Φ, where two distinct Weyl chambers C_1 and C_2 are joined by an edge if and only if there is a root α such that $H_\alpha \in M(C_1) \cap M(C_2)$ (i.e. H_α is a wall of both C_1 and C_2 as defined in 2.5) and the reflection s_α interchanges C_1 and C_2.

Definition 2.7. Let $W_\Phi = \langle s_\alpha \mid \alpha \in \Phi \rangle$. The group W_Φ is called the *Weyl group* of Φ.

Let Π be the Coxeter diagram called X_ℓ in Figure 1.3. We are thus now using X_ℓ as the name for both the root system Φ and the Coxeter diagram Π.

Proposition 2.8. *Let Γ_Φ be as in 2.6, let C be a Weyl chamber, let*

$$B(C) = \{\alpha_1, \dots \alpha_\ell\}$$

be as in 2.4.i and let I be the vertex set of Π. Then the following hold:

(i) *There is a bijection ϕ from I to $B(C)$ such that the map $r_i \mapsto s_{\phi(i)}$ from the generating set $\{r_i \mid i \in I\}$ of the Coxeter group W_Π (described in 29.4) to the Weyl group W_Φ extends to an isomorphism π from W_Π to W_Φ. The map ϕ is unique up to an automorphism of Π.*

(ii) *Let π be as in (i). Then the map $g \mapsto \pi(g)(C)$ from the graph underlying the Coxeter chamber system Σ_Π (i.e. Σ_Π without its edge coloring) to the graph Γ_Φ is an isomorphism.*

Proof. Assertion (i) holds by Theorem 2 and Proposition 15 in Chapter VI, Section 1.5, of [3]. Assertion (ii) follows by 2.6 and 29.4. □

Definition 2.9. By 2.8 and 29.30, there is a unique way (up to an automorphism of Π) to color the edges of Γ_Φ so that it becomes a Coxeter chamber system isomorphic to Σ_Π. We denote this Coxeter chamber system by Σ_Φ.

---

[2] A basis of a root system Φ is also called a *fundamental system* or a *simple system* of roots.

Proposition 2.10. *The reflections of Σ_Φ as defined in 29.6 are the reflections s_α for all $\alpha \in \Phi$.*

Proof. By 2.9 and 29.6, the reflections of Σ_Φ are all the elements in W_Φ that interchange two adjacent Weyl chambers, and for each pair of adjacent Weyl chambers, there is a unique reflection interchanging them. The claim holds, therefore, by 2.6. □

Notation 2.11. For each $\alpha \in \Phi$, let K_α denote the set of Weyl chambers contained in the half-space

$$\{v \in V \mid v \cdot \alpha > 0\}.$$

Proposition 2.12. *Let $\alpha \in \Phi$, let C_1 be a Weyl chamber in K_α and let C_2 be a Weyl chamber in $-K_\alpha$. Suppose that C_1 and C_2 are adjacent in Σ_Φ. Then the reflection s_α interchanges C_1 and C_2.*

Proof. By 2.6, there exists a root β such that $H_\beta \in M(C_1) \cap M(C_2)$ and the reflection s_β interchanges C_1 and C_2. By 2.4, it follows that the set $\bar{C}_1 \cap \bar{C}_2$ spans the hyperplane H_β.[3] On the other hand, $\bar{C}_1 \cap \bar{C}_2 \subset H_\alpha$ since $C_1 \in K_\alpha$ and $C_2 \in -K_\alpha$. Hence $H_\alpha = H_\beta$. Therefore $s_\alpha = s_\beta$. □

Proposition 2.13. *Let $\alpha \in \Phi$. The two roots associated with the reflection s_α are K_α and $K_{-\alpha} = -K_\alpha$.*

Proof. By 2.12, a gallery in Σ_Φ from a Weyl chamber in K_α to a Weyl chamber in $-K_\alpha$ must pass through a pair of adjacent Weyl chambers interchanged by s_α. The claim holds, therefore, by the definition of the two roots associated with a reflection given in 29.6. □

Proposition 2.14. *Let C be a Weyl chamber and let $B(C) = \{\alpha_1, \ldots, \alpha_\ell\}$ be as in 2.4 and let ϕ be as in 2.8.i. Then the following hold:*

(i) *Every root of Φ can be mapped by an element of W_Φ to an element of $B(C)$.*
(ii) *Two elements $s_{\phi(i)}$ and $s_{\phi(j)}$ of $B(C)$ are in the same W_Φ-orbit if and only if there is a path from i and j in Π that goes only through edges whose label is 3.*
(iii) *If Π is simply laced,[4] then W_Φ acts transitively on Φ. In particular, all the roots in Φ have the same length.*
(iv) *If Π is not simply laced, then there are two W_Φ-orbits of roots in Φ and the roots of Φ are of two different lengths.*
(v) *Suppose that $\Phi = \mathsf{B}_\ell$ or C_ℓ for some $\ell \geq 3$ (in which case the two Coxeter diagrams B_ℓ and C_ℓ are the same) and let i be a vertex of Π that is contained in an edge with label 3. Then $\phi(i)$ is long if $\Phi = \mathsf{B}_\ell$ and short if $\Phi = \mathsf{C}_\ell$.*

---

[3]For each subset $X \subset V$, we denote by $\bar{X}$ the *closure* of X in V.

[4]A Coxeter diagram is *simply laced* if all its labels are 3. The Coxeter diagrams in Figure 1.1 that are simply laced, for example, are $\tilde{\mathsf{A}}_\ell$ for $\ell \geq 2$, $\tilde{\mathsf{D}}_\ell$ for $\ell \geq 4$, $\tilde{\mathsf{E}}_6$, $\tilde{\mathsf{E}}_7$ and $\tilde{\mathsf{E}}_8$.

Proof. Assertion (i) holds by Proposition 15 in Chapter VI, Section 1.5, of [3]. If i and j are vertices of Π joined by an edge with label 3, then the roots $\alpha_{\phi(i)}$ and $\alpha_{\phi(j)}$ span a root system of type A_2 and the subgroup $\langle s_{\alpha_{\phi(i)}}, s_{\alpha_{\phi(j)}}\rangle$ acts transitively on the set of roots of this root system. If i and j are vertices of Π joined by an edge with label 4 or 6, then the roots $\alpha_{\phi(i)}$ and $\alpha_{\phi(j)}$ span a root system of type B_2 or G_2, the subgroup $\langle s_{\alpha_{\phi(i)}}, s_{\alpha_{\phi(j)}}\rangle$ has two orbits on the set of roots of this root system and $\alpha_{\phi(i)}$ and $\alpha_{\phi(j)}$ have different lengths. Thus (ii)–(iv) hold. Assertion (v), which is really just a definition, holds by paragraph V in Chapter VI, Section 4.5, of [3]. □

The following result will play a crucial role in the next chapter; see 3.12 and 3.14.

Proposition 2.15. *Let Ξ be the set of roots of Σ_Π. Then for each isomorphism π from Σ_Π to Σ_Φ, there is a bijection ι from Ξ to Φ such that*

$$\pi(a) = K_{\iota(a)}$$

for all $a \in \Xi$.

Proof. Let π be an isomorphism from Σ_Π to Σ_Φ. For each $a \in \Xi$, $\pi(a)$ is a root of Σ_Φ. The claim holds, therefore, by 2.10 and 2.13. □

* * *

We turn now to the affine Coxeter group associated with Φ.

Notation 2.16. Let $\tilde{\Phi} = \Phi \times \mathbb{Z}$.

Definition 2.17. Let $(\alpha, k) \in \tilde{\Phi}$ (as defined in 2.16). We denote by $s_{\alpha,k}$ the *affine reflection* of V given by

$$s_{\alpha,k}(v) = s_\alpha(v) + 2k\alpha/(\alpha \cdot \alpha)$$

for all $v \in V$ and by $H_{\alpha,k}$ the *affine hyperplane*

$$\{v \in V \mid v \cdot \alpha = k\}.$$

Note that the affine hyperplane $H_{\alpha,k}$ is the set of fixed points of the affine reflection $s_{\alpha,k}$ for all $(\alpha, k) \in \tilde{\Phi}$.

Notation 2.18. Let $\tilde{W}_\Phi = \langle s_{\alpha,k} \mid (\alpha, k) \in \tilde{\Phi}\rangle$. The group $\tilde{W}_\Phi$ is called the *affine Weyl group* of Φ.

Definition 2.19. An *alcove* of Φ is a connected component of

$$V \backslash \bigcup_{(\alpha,k)\in\tilde{\Phi}} H_{\alpha,k}.$$

Proposition 2.20. *The following hold:*

(i) *For each alcove d of Φ, there exists a unique set of $\ell + 1$ elements*

$$B(d) := \{(\alpha_1, k_1), \dots, (\alpha_{\ell+1}, k_{\ell+1})\}$$

of pairs in $\tilde{\Phi}$ such that

$$d = \{v \in V \mid v \cdot \alpha_i < k_i \text{ for all } i \in [1, \ell + 1]\}.$$

(ii) *Let C be a Weyl chamber, let*

$$B(C) := \{\alpha_1, \ldots, \alpha_\ell\}$$

be as in 2.4.i and let $\tilde{\alpha}$ be the highest root of Φ with respect to $B(C)$ as defined in Proposition 25 in Chapter VI, Section 1.8, of [3]. Then there is an alcove d such that

$$B(d) = \{(-\alpha_1, 0), \ldots, (-\alpha_\ell, 0), (\tilde{\alpha}, 1)\}.$$

(iii) *If $(\alpha, k) \in B(d)$ for some alcove d, then the set*

$$\{u - v \mid u, v \in \bar{d} \cap H_{\alpha,k}\}$$

spans the hyperplane H_α.

Proof. Assertions (i) and (ii) are proved in Section 4.3 of [17]. By the Proposition in this same section, the affine Weyl group $\tilde{W}_\Phi$ acts transitively on the set of alcoves. Thus to prove (iii), it suffices to check the claim for the alcove d in (ii). □

Definition 2.21. Let d be an alcove. A *wall* of d is an affine hyperplane $H_{\alpha,k}$ for some pair (α, k) in $B(d)$, where $B(d)$ is as defined in 2.20.i. We denote the set of walls of d by $M(d)$.

Definition 2.22. Let $\tilde{\Gamma}_\Phi$ be the graph whose vertices are the alcoves of Φ, where two alcoves d_1 and d_2 are joined by an edge if and only if there is a pair $(\alpha, k) \in \tilde{\Phi}$ such that the affine hyperplane $H_{\alpha,k}$ is in $M(d_1) \cap M(d_2)$ and the affine reflection $s_{\alpha,k}$ interchanges d_1 and d_2.

For each $u \in V$, we denote by t_u the *translation* of V given by

$$t_u(v) = v + u \tag{2.23}$$

for all $v \in V$. Thus, in particular,

$$s_{\alpha,k} = t_{2k\alpha/(\alpha\cdot\alpha)} s_\alpha \tag{2.24}$$

for all $(\alpha, k) \in \tilde{\Phi}$.

Proposition 2.25. *$t_v s_{\alpha,k} t_v^{-1} = s_{\alpha,k+v\cdot\alpha}$ and $s_{\alpha,k} t_v s_{\alpha,k} = t_{s_{\alpha,k}(v)}$ as well as $t_v(H_{\alpha,k}) = H_{\alpha,k+v\cdot\alpha}$ and $s_\beta(H_{\alpha,k}) = H_{s_\beta(\alpha),k}$ for all $(\alpha, k) \in \tilde{\Phi}$, all $\beta \in \Phi$ and all $v \in V$.*

Proof. These identities follow immediately from 2.17 and 2.23. □

Definition 2.26. Let

$$V_\Phi = \{v \in V \mid v \cdot \alpha \in \mathbb{Z} \text{ for all } \alpha \in \Phi\}.$$

A point in V is called *special* (with respect to Φ) if it lies in V_Φ. A point $v \in V$ is thus special if and only if for each $\alpha \in \Phi$, there exists $k \in \mathbb{Z}$ such that $v \in H_{\alpha,k}$. Let

$$T_\Phi = \{t_v \mid v \in V_\Phi\}.$$

The group T_Φ acts transitively on the set V_Φ of special points of V.

Proposition 2.27. *Let $B(C) = \{\alpha_1, \ldots, \alpha_\ell\}$ be as in 2.4.i for some Weyl chamber C and let $v \in V$. Then $v \in V_\Phi$ if and only if $v \cdot \alpha_i \in \mathbb{Z}$ for all $i \in [1, \ell]$.*

Proof. This holds by Theorem 3 in Chapter VI, Section 1.6, of [3]. □

By condition (R3), $2(\alpha \cdot \beta)/(\beta \cdot \beta) \in \mathbb{Z}$ for all α and β in Φ. By 2.24 and 2.25, therefore, the subgroups T_Φ and $\tilde{W}_\Phi$ of $\mathrm{Aut}(V)$ normalize each other, and both groups act on the graph $\tilde{\Gamma}_\Phi$ defined in 2.22. Thus, in particular, $T_\Phi \tilde{W}_\Phi$ is a subgroup of $\mathrm{Aut}(\tilde{\Gamma}_\Phi)$. By 2.24, in fact, $T_\Phi \tilde{W}_\Phi = T_\Phi W_\Phi$.

Now let $\tilde{\Pi}$ be the Coxeter diagram in Figure 1.1 whose name is $\tilde{\mathsf{X}}_\ell$. Thus $\Sigma_{\tilde{\Pi}}$ is one of the Coxeter chamber systems that were the subject of our attention in Chapter 1.

Proposition 2.28. *Let $\tilde{\Gamma}_\Phi$ be as in 2.22, let C be a Weyl chamber, let $B(C)$ be as in 2.4.i, let $\tilde{\alpha}$ and d be as in 2.20.ii and let $\tilde{I}$ be the vertex set of $\tilde{\Pi}$. Then the following hold:*

(i) *There exists a bijection $\tilde{\phi}$ from $\tilde{I}$ to $B(C) \cup \{\tilde{\alpha}\}$ such that $o := \tilde{\phi}^{-1}(\tilde{\alpha})$ is a special vertex of $\tilde{\Pi}$ and the map*

$$r_i \mapsto \begin{cases} s_{\tilde{\phi}(i)} & \text{if } i \neq o, \\ s_{\tilde{\alpha},1} & \text{if } i = o \end{cases}$$

from the generating set $\{r_i \mid i \in \tilde{I}\}$ of the Coxeter group $W_{\tilde{\Pi}}$ to the affine Weyl group $\tilde{W}_\Phi$ extends to an isomorphism $\tilde{\pi}$ from $W_{\tilde{\Pi}}$ to $\tilde{W}_\Phi$. The map $\tilde{\phi}$ is unique up to an automorphism of $\tilde{\Pi}$.

(ii) *Let $\tilde{\pi}$ be as in (i). Then the map $g \mapsto \tilde{\pi}(g)(d)$ from the graph underlying the Coxeter chamber system $\Sigma_{\tilde{\Pi}}$ to the graph $\tilde{\Gamma}_\Phi$ is an isomorphism.*

Proof. Assertion (i) holds by the results in Sections 4.6 and 4.7 in [17]. Assertion (ii) follows by 2.22 and 29.4. □

Definition 2.29. By 2.28 and 29.30, there is a unique way (up to an automorphism of $\tilde{\Pi}$) to color the edges of $\tilde{\Gamma}_\Phi$ so that it becomes a Coxeter chamber system isomorphic to $\Sigma_{\tilde{\Pi}}$. We denote this Coxeter chamber system by $\tilde{\Sigma}_\Phi$.

It is easy to form a picture of Σ_Φ if $\ell \leq 2$. If $\ell = 1$ (so $\mathsf{X}_\ell = \mathsf{A}_1$), then the alcoves are simply the open intervals $(i, i+1)$ for all $i \in \mathbb{Z}$, the special points are the integers and a panel $\{(i-1, i), (i, i+1)\}$ of $\tilde{\Sigma}_\Phi$ is of one type or the other according to the parity of i. If $\ell = 2$ (so $\mathsf{X}_\ell = \mathsf{A}_2$, C_2 or G_2), then V is the Euclidean plane, a region of which is illustrated in Figures 2.1–2.3. In each case, the alcoves are the triangles, straight lines in these figures are the affine hyperplanes $H_{\alpha,k}$, the type of a panel of $\tilde{\Sigma}_\Phi$ (i.e. a pair of alcoves sharing a common boundary segment) is indicated by the "color" (dotted, dashed or solid) of this segment and the special points are those at the intersection of three of these lines if $\mathsf{X}_\ell = \mathsf{A}_2$, four if $\mathsf{X}_\ell = \mathsf{C}_2$, respectively, six if $\mathsf{X}_\ell = \mathsf{G}_2$.

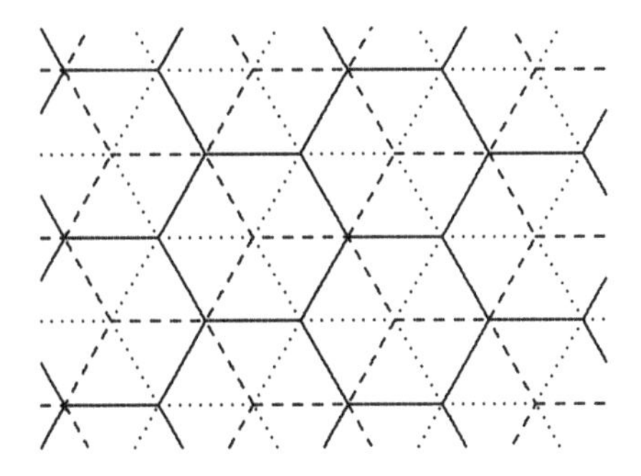

Figure 2.1 Alcoves for $\tilde{\mathsf{A}}_2$

Proposition 2.30. *The reflections of $\tilde{\Sigma}_\Phi$ as defined in 29.6 are the affine reflections $s_{\alpha,k}$ for all pairs $(\alpha, k) \in \tilde{\Phi}$.*

Proof. By 2.29 and 29.6, the reflections of Σ_Φ are all the elements in W_Φ that interchange two adjacent Weyl chambers, and for each pair of adjacent Weyl chambers, there is a unique reflection interchanging them. The claim holds, therefore, by 2.22. □

By 2.18 and 2.29, the group $\tilde{W}_\Phi$ acts transitively on the set of panels of $\tilde{\Sigma}_\Phi$ of a given color. Since T_Φ normalizes $\tilde{W}_\Phi$ and acts on the set of edges of $\tilde{\Gamma}_\Phi$, it follows that T_Φ permutes the orbits of $\tilde{W}_\Phi$ in the set of edges of $\tilde{\Gamma}_\Phi$. Thus T_Φ is a subgroup of $\mathrm{Aut}(\tilde{\Sigma}_\Phi)$, but not one, in general, that preserves colors; see 1.28 and 1.29.

Notation 2.31. For each pair $(\alpha, k) \in \tilde{\Phi}$, let $K_{\alpha,k}$ denote the set of alcoves contained in the affine half-space

$$\{v \in V \mid v \cdot \alpha > k\}.$$

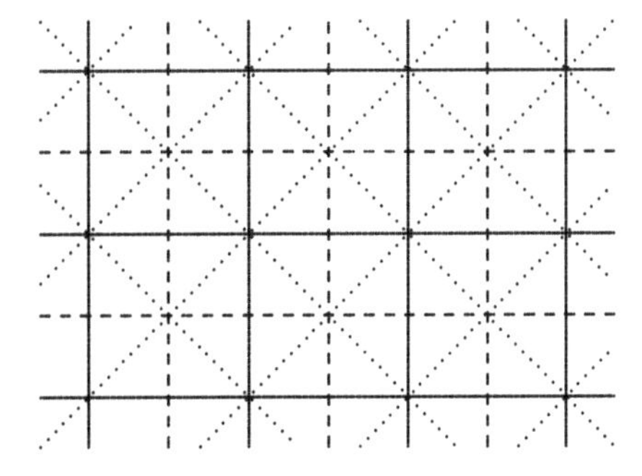

Figure 2.2 Alcoves for $\tilde{\mathsf{B}}_2 = \tilde{\mathsf{C}}_2$

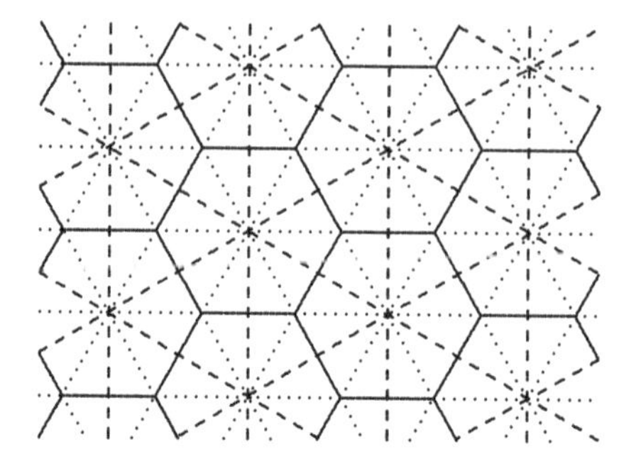

Figure 2.3 Alcoves for $\tilde{G}_2$

Proposition 2.32. *Let $(\alpha, k) \in \tilde{\Phi}$, let d_1 be an alcove in $K_{\alpha,k}$ and let d_2 be an alcove in $K_{-\alpha,-k}$. Suppose that d_1 and d_2 are adjacent in $\tilde{\Sigma}_\Phi$. Then the affine reflection $s_{\alpha,k}$ interchanges d_1 and d_2.*

Proof. By 2.22, there exists a pair $(\beta, l) \in \tilde{\Phi}$ such that the affine hyperplane $H_{\beta,l}$ is a wall of both d_1 and d_2 and the affine reflection $s_{\beta,l}$ interchanges d_1 and d_2. Thus $\bar{d_1} \cap \bar{d_2} \subset H_{\beta,l}$ and (since $s_{\beta,l}$ acts trivially on $H_{\beta,l}$) $\bar{d_1} \cap \bar{d_2} = \bar{d_1} \cap H_{\beta,l}$. By 2.20.iii, it follows that the set

$$\{u - v \mid u, v \in \bar{d_1} \cap \bar{d_2}\}$$

spans H_β. On the other hand, $\bar{d_1} \cap \bar{d_2} \subset H_{\alpha,k}$ since $d_1 \in K_{\alpha,k}$ and $d_2 \in K_{-\alpha,-k}$. It follows that $H_{\alpha,k} = H_{\beta,l}$. Therefore $s_{\alpha,k} = s_{\beta,l}$. □

Proposition 2.33. *Let $(\alpha, k) \in \tilde{\Phi}$. The two roots associated with the reflection $s_{\alpha,k}$ are $K_{\alpha,k}$ and $K_{-\alpha,-k}$.*

Proof. By 2.32, a gallery in $\tilde{\Sigma}_\Phi$ from an alcove in $K_{\alpha,k}$ to an alcove in $K_{-\alpha,-k}$ must pass through a pair of adjacent alcoves interchanged by $s_{\alpha,k}$. The claim holds, therefore, by 29.6. □

Proposition 2.34. *For each pair $(\alpha, k) \in \tilde{\Phi}$, the parallel class of the root $K_{\alpha,k}$ is the set of roots*

$$\{K_{\alpha,m} \mid m \in \mathbb{Z}\}.$$

Proof. Suppose that $K_{\beta,m}$ is parallel to $K_{\alpha,k}$ for some pair $(\beta, m) \in \tilde{\Phi}$. Then α and β are linearly dependent since otherwise the map $x \mapsto (x \cdot \alpha, x \cdot \beta)$ from V to $\mathbb{R}^2$ would be surjective. By condition (R1), therefore, β is either α or $-\alpha$. Since $K_{-\alpha,l}$ is not parallel to $K_{\alpha,k}$ for any $l \in \mathbb{Z}$, we conclude that, in fact, $\alpha = \beta$. □

Definition 2.35. Suppose that Π is *not* simply laced, so the roots of Φ are of two different lengths by 2.14.iv. We will call a root $K_{\alpha,k}$ of $\tilde{\Sigma}_\Phi$ *long* (or *short*) if the root α of Φ is long (or short).

Notation 2.36. Let C and d be as in 2.20.ii. Thus d is the unique alcove in C whose closure contains the origin 0. Let R be the set of alcoves that are in the same W_Φ-orbit as d. Then R is a gem of $\tilde{\Sigma}_\Phi$ (as defined in 1.6) isomorphic to Σ_Φ, the origin 0 is the only point in V contained in the closure of all the alcoves in R, the alcoves in R are the only alcoves whose closure contains 0 and the set of roots that cut R is

$$\{K_{\alpha,0} \mid \alpha \in \Phi\}.$$

Proposition 2.37. *Let R_1 be a gem of $\tilde{\Sigma}_\Phi$ and let $\alpha \in \Phi$. Then there exists a unique $k \in \mathbb{Z}$ such that $K_{\alpha,k}$ cuts R_1.*

Proof. By 1.13 and 2.34, for each $\alpha \in \Phi$ there is at most one $k \in \mathbb{Z}$ such that $K_{\alpha,k}$ cuts R_1. By 1.7, R and R_1 are isomorphic Coxeter chamber systems. In particular, they are each cut by the same number of roots. The claim follows, therefore, from the observation that $K_{\alpha,0}$ cuts R for all $\alpha \in \Phi$. □

Proposition 2.38. *The group T_Φ acts transitively on gems.*

Proof. Let R_1 be a gem, let C be a Weyl chamber, let

$$B := B(C) = \{\alpha_1, \ldots, \alpha_\ell\}$$

be as in 2.4.i and let d and R be as in 2.36. By 2.37, there exist unique integers $k_1, \ldots, k_\ell$ such that K_{α_i,k_i} cuts R_1 for all $i \in [1, \ell]$. By 2.4.ii, the map

$$v \mapsto (\alpha_1 \cdot v, \ldots, \alpha_\ell \cdot v) \tag{2.39}$$

from V to $\mathbb{R}^\ell$ is bijective. By 2.23, therefore, the group T_Φ contains a unique element t mapping $K_{\alpha_i,0}$ to K_{α_i,k_i} for all $i \in [1, \ell]$. Let $d_1 = t(d)$. By 2.25,

$$ts_Bt^{-1} = \{s_{\alpha_i,k_i} \mid i \in [1, \ell]\},$$

where $s_B := \{s_{\alpha_i} \mid i \in [1, \ell]\}$. Since K_{α_i,k_i} cuts R_1, the reflection s_{α_i,k_i} maps R_1 to itself for each $i \in [1, \ell]$. Since $s_{\alpha_i,k_i}(d_1) = ts_{\alpha_i}t^{-1}(d_1) = ts_{\alpha_i}(d)$ and $s_{\alpha_i}(d)$ is a chamber in R adjacent to d for each $i \in [1, \ell]$, it follows that the set of chambers of R_1 adjacent to d_1 is $\{s_{\alpha_i,k_i}(d_1) \mid i \in [1, \ell]\}$. Thus the group

$$M := \langle s_{\alpha_i,k_i} \mid i \in [1, \ell]\rangle$$

maps R_1 to itself and contains elements mapping d_1 to each chamber in R_1 adjacent to d. It follows that $R_1 = M(d_1)$. Thus

$$t(R) = t\langle s_B\rangle(d) = t\langle s_B\rangle t^{-1} \cdot t(d) = M(d_1) = R_1.$$

□

Proposition 2.40. *Let R_1 be a gem. Then the following hold:*

(i) *There is a unique point $x \in V$ contained in the closure of all the alcoves in R_1.*
(ii) *The alcoves in R_1 are the only alcoves whose closure contains x.*

Proof. This holds by 2.36 and 2.38. □

Definition 2.41. Let R_1 be a gem and let x be as in 2.40.i. We call x the *center* of R.

Proposition 2.42. *A point of V is the center of a gem if and only if it is special.*

Proof. By 2.36, 2.38 and 2.41, a point of V is the center of a gem if and only if it is in the same T_Φ-orbit as the origin 0. By 2.26, a point of V is special if and only if it is in the same T_Φ-orbit as the origin 0. □

Proposition 2.43. *The group $T_\Phi W_\Phi$ acts transitively on the set of pairs (R_1, d_1) such that R_1 is a gem of $\tilde{\Sigma}_\Phi$ and d_1 is an alcove in R_1.*

Proof. Let R be as in 2.36. Then W_Φ acts transitively on the set of alcoves in R. The claim holds, therefore, by 2.38. □

Proposition 2.44. *Let $\{\alpha_1, \ldots, \alpha_\ell\}$ be a basis of Φ as defined in 2.5, let R be as in 2.36 and let $k_1, \ldots, k_\ell$ be integers. Then the following hold:*

(i) *There exists a unique point $v \in V$ such that $v \cdot \alpha_i = k_i$ for all $i \in [1, \ell]$, the point v is special (as defined in 2.26), so $t_v \in T_\Phi$, and $t_v(R)$ is a gem cut by K_{α_i,k_i} for all $i \in [1, \ell]$.*
(ii) *$t_v(R)$ is the unique gem cut by K_{α_i,k_i} for all $i \in [1, \ell]$.*

Proof. Since the map in 2.39 is a bijection, there exists a unique point v in V such that $v \cdot \alpha_i = k_i$ for all $i \in [1, \ell]$. By 2.27, $v \in V_\Phi$. Thus $t_v \in T_\Phi$. By 2.31,

$$t_w(K_{\alpha,0}) = K_{\alpha,w\cdot\alpha} \tag{2.45}$$

for all $\alpha \in \Phi$ and all $w \in V_\Phi$. Hence $t_v(R)$ is cut by $t_v(K_{\alpha_i,0}) = K_{\alpha_i,k_i}$ for all $i \in [1, \ell]$. Thus (i) holds. Let R_1 be a second gem cut by K_{α_i,k_i} for all $i \in [1, \ell]$. By 2.38, there exists a special point u such that $t_u(R) = R_1$. By 2.37, $t_u(K_{\alpha_i,0}) = K_{\alpha_i,k_i}$ for all $i \in [1, \ell]$. By 2.45, it follows that $u \cdot \alpha_i = k_i$ for all $i \in [1, \ell]$. Thus $u = v$. Therefore $R_1 = t_v(R)$. Hence (ii) holds. □

We have now reached the main goal of this chapter:

Proposition 2.46. *The assertions in 1.3, 1.9 and 1.12 are all valid.*

Proof. The Coxeter diagram $\tilde{\Pi}$ is one of the diagrams in Figure 1.1. Thus $\tilde{\Pi}$ is playing the role of the Coxeter diagram Π in Chapter 1. By 2.29, it will suffice to show that the assertions 1.3, 1.9 and 1.12 are true for $\tilde{\Sigma}_\Phi$.

By 2.30 and 2.33, all roots of $\tilde{\Sigma}_\Phi$ are of the form $K_{\alpha,k}$. It follows by 2.34 that 1.3 holds.

The group T_Φ maps every root to a parallel root. Thus T_Φ is a subgroup of the group T_Π defined in 1.8. By 2.38, therefore, 1.9 holds. (It follows from 1.22 that, in fact, $T_\Phi = T_\Pi$.)

It remains only to prove 1.12. By 2.43, it suffices to assume that R, C and d are as in 2.36. Let $B(C) = \{\alpha_1, \ldots, \alpha_\ell\}$. By 2.21, the roots containing d but not some alcove in R adjacent to d in $\tilde{\Sigma}_\Phi$ are

$$K_{\alpha_1,0}, \ldots, K_{\alpha_\ell,0}.$$

By 2.34, a root parallel to $K_{\alpha,0}$ for some $\alpha \in \Phi$ is of the form $K_{\alpha,k}$ for some integer k. Thus 1.12 holds by 2.44. □

* * *

We close this chapter with four small observations. The first will be needed in Chapters 25 and 26:

Proposition 2.47. *Suppose that $\Pi = \mathsf{X}_\ell$ is neither simply laced nor $\mathsf{B}_2 = \mathsf{C}_2$. Let J be a subset of the vertex set $\tilde{I}$ of the Coxeter diagram $\tilde{\Pi} = \tilde{\mathsf{X}}_\ell$ of maximal cardinality such that the subdiagram $\tilde{\Pi}_J$ spanned by J is isomorphic to A_k for some k. Let R be a J-residue of $\tilde{\Sigma}_\Phi$. Then every root of $\tilde{\Sigma}_\Phi$ cutting R is long if $\mathsf{X} = \mathsf{B}$, F or G, short if $\mathsf{X} = \mathsf{C}$.*

Proof. Let $\tilde{\alpha}$ and d be as in 2.20.ii. Since $\tilde{W}_\Phi$ acts transitively on the set of residues of a given type, we can assume that $d \in R$. Let $\tilde{\phi}$ be the map in 2.28 that is used (in 2.29) to color the edges of $\tilde{\Sigma}_\Phi$ and let $o = \tilde{\phi}^{-1}(\tilde{\alpha})$. If $\mathsf{X} = \mathsf{F}$ or G, then $o \in J$. If $o \in J$, then R is cut by the root $K_{\tilde{\alpha},1}$. By Proposition 25(iii) in Chapter VI, Section 1.8, of [3], $\tilde{\alpha}$ is long. If $o \notin J$, then $\mathsf{X} = \mathsf{B}$ or C and R is cut by the root $K_{\tilde{\phi}(i)}$ for all $i \in J$. By 2.14.v, these roots are long if $\mathsf{X} = \mathsf{B}$, short if $\mathsf{X} = \mathsf{C}$. By 29.7.iii, we have

$$R \cong \Sigma_{\tilde{\Pi}_J}.$$

Hence by 2.14.iii, the stabilizer of R in $\tilde{W}_\Phi$ acts transitively on the roots of R. It follows that all the roots of $\tilde{\Sigma}_\Phi$ that cut R have the same length. □

Our second observation will be needed in 13.13:

Proposition 2.48. *If Π is not simply laced, then $T_\Phi W_\Phi = \mathrm{Aut}(\tilde{\Sigma}_\Phi)$.*

Proof. Let $g \in \mathrm{Aut}(\tilde{\Sigma}_\Phi)$. Thus g is a σ-automorphism for some $\sigma \in \mathrm{Aut}(\Pi)$ (as defined in 29.2). We claim that $g \in T_\Phi W_\Phi$. By 2.43, we can assume that g fixes a chamber d and a gem R containing d. In particular, σ fixes a special vertex. If Π is not simply laced, it follows (by inspection of Figure 1.1) that σ is the identity map. The only special automorphism of Σ that fixes a chamber is the identity map. □

Our third observation will be needed in the investigation of gems in Chapters 19–24:

Proposition 2.49. *Suppose that $\Phi = \mathsf{B}_\ell$ for some $\ell \geq 3$ or C_ℓ for some $\ell \geq 2$, let C be a Weyl chamber of Φ, let $x_1, \ldots, x_\ell$ be the vertices of the Coxeter diagram B_ℓ going from left to right in Figure 1.3, let ϕ be as in 2.8.i and let $\alpha_i = \phi(x_i)$ for all $i \in [1, \ell]$.*[5] *Furthermore:*

[5]Note that by (iv) and (v) of 2.14, α_ℓ is the unique short root in $B(C) = \{\alpha_1, \ldots, \alpha_\ell\}$ if $\Phi = \mathsf{B}_\ell$ and α_ℓ is the unique long root in $B(C)$ if $\Phi = \mathsf{C}_\ell$.

(i) *If* $\Phi = \mathsf{B}_\ell$ *for some* $\ell \geq 3$, *let* v *be the unique special point in* V *such that* $v \cdot \alpha_i = 0$ *for all* $i \in [2, \ell]$ *and* $v \cdot \alpha_1 = -1$.
(ii) *If* $\Phi = \mathsf{C}_\ell$ *for some* $\ell \geq 2$, *let* v *be the unique special point in* V *such that* $v \cdot \alpha_i = 0$ *for all* $i \in [1, \ell - 1]$ *and* $v \cdot \alpha_\ell = -1$.

Then (in both cases) the gem of $\tilde{\Sigma}_\Phi$ *with center* v *is not of the same type as the gem whose center is the origin.*

Proof. Let R be the gem whose center is the origin, let R_1 be the gem with center v and let $\tilde{\alpha}$ be the highest root of Φ with respect to $B(C)$ (as in 2.20.ii). Since $v \neq 0$, we have $R_1 \neq R$. If we are in case (i), then $v = -\varepsilon_1$ and hence

$$v \cdot \tilde{\alpha} = -\varepsilon_1 \cdot (\varepsilon_1 + \varepsilon_2) = -1,$$

where ε_1 and ε_2 are as in Plate II in [3]. If we are in case (ii), then

$$v = -(\varepsilon_1 + \varepsilon_2 + \cdots + \varepsilon_\ell)/2$$

and hence

$$v \cdot \tilde{\alpha} = v \cdot 2\varepsilon_1 = -1,$$

where $\varepsilon_1, \ldots, \varepsilon_\ell$ are as in Plate III in [3]. Thus in both cases $-v$ is contained in the closure of the alcove d described in 2.20.ii. Scalar multiplication by -1 maps Φ to itself and hence permutes the set of alcoves. Thus $-d$ is an alcove whose closure contains v. By 2.40.ii, therefore, $-d \in R_1$. The alcove $-d$ also lies in R and R is different from R_1. The claim follows since a residue of a given type containing a given chamber is uniquely determined. □

Our last observation will be needed in Chapters 19–25 when we apply 3.41.i:

Proposition 2.50. *Let* $\Phi = \mathsf{B}_2$ *or* G_2 *and let* $\alpha, \beta \in \Phi$. *Then the following hold:*

(i) *If* $\Phi = \mathsf{B}_2$ *and* α *and* β *are at an angle of 135 degrees, then*

$$2(\alpha \cdot \beta)/(\beta \cdot \beta) = \begin{cases} -2 & \text{if } \alpha \text{ is long and} \\ -1 & \text{if } \alpha \text{ is short.} \end{cases}$$

(ii) *If* $\Phi = \mathsf{G}_2$ *and* α *and* β *are at an angle of 150 degrees, then*

$$2(\alpha \cdot \beta)/(\beta \cdot \beta) = \begin{cases} -3 & \text{if } \alpha \text{ is long and} \\ -1 & \text{if } \alpha \text{ is short.} \end{cases}$$

Proof. See Plate X in [3]. □

Note that the hypothesis about the angles in 2.50 is equivalent to the assumption that $\{\alpha, \beta\}$ is a basis of Φ as defined in 2.5.

Chapter Three

Root Data with Valuation

In this chapter, we introduce the notion of a root datum with valuation. We will not need this notion until Chapter 13, but we prefer to introduce it now while the representation of spherical and affine Coxeter chamber systems in terms of root systems, Weyl chambers and alcoves as described in Chapter 2 is still fresh in the reader's mind.

Throughout this chapter, we let $\Pi = \mathsf{X}_\ell$ be one of the spherical diagrams in Figure 1.3 with $\ell \geq 2$ (i.e. not A_1), we let I denote the vertex set of Π, we let Δ be an irreducible spherical building of type Π satisfying the Moufang property as defined in 29.15 and we fix an apartment Σ of Δ. Roots of Σ will be denoted by lowercase Latin letters.

Definition 3.1. Let a and b be two roots of Σ that are not opposite each other. By 29.25, we can choose a residue E of rank 2 of Σ cut by both a and b. By 29.7.iv, both $E \cap a$ and $E \cap b$ are roots of E. There is thus a unique minimal gallery $(x_0, \ldots, x_s)$ in $E \cap \Sigma$ such that a contains x_1 but not x_0 and b contains x_s but not x_{s-1}. For each $i \in [1, s]$, there exists (by 29.6.i) a unique root of Σ containing x_i but not x_{i-1}; we denote this root by a_i. Thus $a_1 = a$, $a_s = b$ and $a_1, a_2, \ldots, a_s$ are precisely the roots of Σ containing $a \cap b \cap E$ but not all of E. Let

$$[a, b] = (a_1, a_2, \ldots, a_s)$$

and

$$(a, b) = (a_2, \ldots, a_{s-1}).$$

These ordered sets will be called the *closed* (respectively, the *open*) *interval from a to b*. By the next result, they are independent of the choice of E.

For each root a of Σ, we denote by U_a the corresponding *root group* of Δ as defined in 29.15.

Proposition 3.2. *Let a and b be two roots of Σ that are not opposite each other. Let E and*

$$[a, b] = (a_1, a_2, \ldots, a_s)$$

be as in 3.1. Then the following hold:

(i) *The roots $a_1, a_2, \ldots, a_s$ are precisely the roots of Σ that contain $a \cap b$.*
(ii) *The set $[a, b]$ and its ordering from a to b are both independent of the choice of E.*

(iii) $[U_1, U_s] \subset U_2U_3 \cdots U_{s-1}$ *if* $s \geq 3$, *where* U_i *denotes the root group* U_{a_i} *for all* $i \in [1, s]$, *and* $[U_1, U_s] = 1$ *if* $s = 2$.

(iv) *Every element in the group* $\langle U_1, U_2, \ldots, U_s \rangle$ *can be written uniquely in the form* $u_1u_2 \cdots u_s$ *with* $u_i \in U_i$ *for all* $i \in [1, s]$.

Proof. Let $(x_0, x_1, \ldots, x_s)$ be as described in 3.1 and let w be an arbitrary chamber of Σ contained in $a \cap b$. By 29.7.i,

$$\text{dist}(x_0, z) + \text{dist}(z, w) = \text{dist}(x_0, w), \tag{3.3}$$ [1]

where $z := \text{proj}_E w$. By 29.16, $z \in a \cap b \cap E$. By 29.7.ii, E is a convex subset of Σ. It follows that x_i is nearer to z than x_{i-1} for each $i \in [1, s]$. By 3.3, therefore, w is nearer to x_i than to x_{i-1} for each $i \in [1, s]$. Hence by 29.6.i, $w \in a_i$ for all $i \in [1, s]$. We conclude that $a \cap b \subset a_i$ for all $i \in [1, s]$.

Now let c be an arbitrary root of Σ containing $a \cap b$ and let w be a chamber in $a \cap b \cap E$. Let x be the chamber opposite w in $E \cap \Sigma$ and let u be the chamber opposite x in Σ. Since $a \cap E$ and $b \cap E$ are both roots of E containing w, neither contains x. Hence neither a nor b contains x. By 29.9.i, it follows that both a and b contain u. Since c contains $a \cap b$, it also contains u. Hence by 29.9.i again, c does not contain x. Thus c contains $a \cap b \cap E$ but not all of E. Therefore $c = a_i$ for some $i \in [1, s]$. Thus (i) holds.

By (i), the unordered set $\{a_1, \ldots, a_s\}$ depends only on a and b and not on the choice of E. From (i), we also deduce that for each $i \in [1, s-1]$, the cardinality $|[a_i, a_s]|$ depends only on a_i and a_s and not on the choice of E. Since for each $i \in [1, s-1]$, a_i is the unique element of $\{a_1, a_2, \ldots, a_s\}$ such that $|[a_i, a_s]| = s + 1 - i$, its position as the i-th entry in the s-tuple $[a_1, a_s]$ is independent of the choice of E. Thus (ii) holds.

By 11.27 of [37], the group

$$\langle U_i \mid i \in [1, s] \rangle$$

acts faithfully on the residue E. By 5.5 and 5.6 of [36], therefore, (iii) and (iv) hold. □

We now introduce the notion of a root datum. There is some ambiguity about this notion in the literature and, in fact, the version of this notion we require is simpler than the version given in Section 6.1 of [6].

Definition 3.4. Let $G^\dagger$ be the subgroup of $\text{Aut}(\Delta)$ generated by all the root groups U_a for all roots a of Δ,[2] let Ξ denote the set of roots of Σ and let ξ denote the map from Ξ to the set of subgroups of $G^\dagger$ given by $\xi(a) = U_a$ for each $a \in \Xi$. We will call the pair $(G^\dagger, \xi)$ the *root datum* of Δ *based at* Σ.[3] By 29.15.iii, $G^\dagger$ acts transitively on the set of apartments of Δ. Thus

---

[1] By 29.13.iii, Σ is a convex subset of Δ. Thus the distance between two chambers of Σ is the same whether it is measured in Σ or in Δ.

[2] When Δ is the building associated with an algebraic group, the group $G^\dagger$ is often called G^+. In the use of the name $G^\dagger$ we are following [34] and [36].

[3] Thus the root datum of Δ consists of the group $G^\dagger$, the subgroup U_a for all $a \in \Xi$ and the map $a \mapsto U_a$ we are calling ξ. We could equally well have denoted the root datum of Δ by

$$(G^\dagger, (U_a)_{a \in \Xi}).$$

$(G^\dagger, \xi)$ is unique (i.e. does not depend on the choice of Σ) up to conjugation by an element of $G^\dagger$.

The following terminology will be used frequently.

Notation 3.5. Let d be a chamber of Σ. For each x in the vertex set I of Π, let e_x be the unique chamber of Σ that is x-adjacent to d. By 29.6.i, there exists for each $x \in I$ a unique root of Σ that contains d but not e_x; we denote this root by a_x. Let

$$\Xi_d^\circ = \{a_x \mid x \in I\}$$

and let

$$\Xi_d = \{b \mid b \in [a_x, a_y] \text{ for some edge } \{x, y\} \text{ of } \Pi\}.$$

It follows from 29.6.i that $e_y \in a_x$ and hence $a_x \neq a_y$ whenever x and y are distinct elements of I. Therefore the map

$$x \mapsto a_x$$

from I to Ξ_d° is a bijection. Note also that if $x, y \in I$ are distinct and E is the $\{x, y\}$-residue of Σ containing d, then $a_x \cap a_y \cap E = \{d\}$ and hence (by 3.1) the interval $[a_x, a_y]$ contains precisely those roots of Σ that contain d but not all of E. Thus if x and y are adjacent vertices of Π and n is the label on the edge $\{x, y\}$, then $|E| = 2n$ and $|[a_x, a_y]| = n$.[4]

The next result says that a Moufang spherical building is uniquely determined by its root datum, in fact, by just a small subset of its root datum.[5]

Theorem 3.6. *Let $\hat{\Delta}$ be a second building having the same Coxeter diagram Π and also satisfying the Moufang condition and let $(\hat{G}^\dagger, \hat{\xi})$ be the root group datum of $\hat{\Delta}$ based at an apartment $\hat{\Sigma}$ (as defined in 3.4). Let $d \in \Sigma$ and let Ξ_d° and Ξ_d be as in 3.5. Suppose that π is a σ-isomorphism from Σ to $\hat{\Sigma}$ for some automorphism σ of Π.[6] Suppose, too, that for each edge $e = \{x, y\}$ of Π, π_e is an isomorphism from the group*

$$U_+^{(e)} := \langle U_b \mid b \in [a_x, a_y] \rangle$$

to the group

$$\hat{U}_+^{(\sigma(e))} := \langle \hat{U}_{\pi(b)} \mid b \in [a_x, a_y] \rangle$$

mapping the root group U_b for each $b \in [a_x, a_y]$ to the root group $\hat{U}_{\pi(b)}$ in $\hat{G}^\dagger$ corresponding to the root $\pi(b)$. Suppose, too, that for all pairs of edges

---

[4]If $x, y \in I$ are distinct but not adjacent, then $|E| = 4$ and $[a_x, a_y]$ contains just a_x and a_y.

[5]The result 3.6 is fundamental in the theory of spherical buildings. In particular, it is the basis of our description of the classification of Moufang spherical buildings in 30.12–30.14.

[6]Thus π maps the interval $[a_x, a_y]$ to the interval $[\hat{a}_{\sigma(x)}, \hat{a}_{\sigma(y)}]$ for all edges $\{x, y\}$ of Π, where $\hat{a}_{\sigma(x)}$ and $\hat{a}_{\sigma(y)}$ are as defined in 3.5 with $\hat{\Sigma}$ and $\pi(d)$ in place of Σ and d.

$e_1 = \{x_1, y_1\}$ and $e_2 = \{x_2, y_2\}$ of Π sharing a common vertex $x := x_1 = x_2$ (so $a_x \in [a_{x_i}, a_{y_i}]$ for both $i = 1$ and 2), the restriction of π_{e_1} to U_{a_x} equals the restriction of π_{e_2} to U_{a_x}. Then there exists a σ-isomorphism ϕ from Δ to $\hat{\Delta}$ extending π that induces the map π_e from $U_+^{(e)}$ to $\hat{U}_+^{(\sigma(e))}$ for each edge e of Π.

Proof. This is proved in 40.17 of [36]. □

Notation 3.7. Let $a \in \Xi$. By 29.6.i, there is a unique reflection of Σ that interchanges the root a with its opposite $-a$ in Σ. We denote this reflection by s_a.

Definition 3.8. Let a be a root of Σ and let $-a$ denote the root opposite a in Σ. By 11.22 of [37], for each $u \in U_a^*$, there exists a unique element in

$$U_{-a}^* u U_{-a}^*$$

that interchanges a and $-a$. We will denote this element here by $m_\Sigma(u)$.[7] Since $m_\Sigma(u)^{-1} \in U_{-a}^* u^{-1} U_{-a}^*$ also interchanges a and $-a$, we have

$$m_\Sigma(u^{-1}) = m_\Sigma(u)^{-1} \tag{3.9}$$

for all $u \in U_a^*$.

Remark 3.10. Let a and $-a$ be opposite roots of Σ, let $g \in m_\Sigma(U_a^*)$ and let P be a panel containing one chamber in a and one in $-a$. Since U_a acts trivially on a and U_{-a} acts trivially on $-a$, the element g maps P to itself. Since g interchanges a and $-a$, it interchanges, in particular, the two chambers in $P \cap \Sigma$. By 29.6, it follows that g induces the reflection s_a on Σ (as defined in 3.7). In particular, g is a special automorphism of Δ and g^2 acts trivially on Σ (but is not, in general, the identity).

Proposition 3.11. *Let $d \in \Sigma$, let Ξ_d° be as in 3.5, let a and b be distinct elements of Π_d°, let*

$$[a, b] = (a_1, a_2, \ldots, a_s)$$

be as in 3.1, let $U_i = U_{a_i}$ for all $i \in [1, s]$, let $m_1 \in m_\Sigma(U_1^*)$ and let $m_s \in m_\Sigma(U_s^*)$. Suppose that $s > 2$. Then the following hold:

(i) *$U_i^{m_1} = U_{s+2-i}$ for all $i \in [2, s]$.*
(ii) *$U_i^{m_s} = U_{s-i}$ for all $i \in [1, s-1]$.*

Proof. We have $a = a_x$ and $b = a_y$ for some edge $\{x, y\}$ of Π. Let E be the $\{x, y\}$-residue of Σ containing d. By 29.7.iv, $a_i \cap E$ is a root of E for each $i \in [1, s]$. By 3.10, m_1 maps E to itself and interchanges $a_1 \cap E$ with the opposite root of E. It follows that

$$(a_i \cap E)^{m_1} = a_{s+2-i} \cap E$$

[7] By 29.56, a is the set of chambers in Σ fixed by an element $u \in U_a^*$, so this notation is unambiguous.

for all $i \in [2, s]$. By 29.19, therefore,

$$a_i^{m_1} = a_{s+2-i}$$

for all $i \in [2, s]$. Thus (i) holds; (ii) holds by a similar argument. □

The following definition is fundamental:

Definition 3.12. Let Ξ be the set of roots of the apartment Σ. A *root map* of Σ is a bijection ι from Ξ to some root system Φ (which we call the *target* of ι) such that

$$\pi(a) = K_{\iota(a)} \tag{3.13}$$

for some isomorphism π from Σ to Σ_Φ and for all $a \in \Xi$, where Σ_Φ is as in 2.9 and $K_{\iota(a)}$ is as in 2.11. Note that by 1.20, the isomorphism π from Σ to Σ_Φ is uniquely determined by ι. We will say that root maps ι and $\hat{\iota}$ are *equivalent* if they have the same target Φ and there is an isometry σ of the ambient Euclidean space V of Φ mapping Φ to itself such that $\hat{\iota} = \sigma \circ \iota$.

Proposition 3.14. *The following hold (with all the notation of 3.12):*

(i) *Root maps of Σ exist.*
(ii) *If Π is simply laced, then all root maps of Σ are equivalent.*
(iii) *If $\Pi = \mathsf{B}_\ell = \mathsf{C}_\ell$ for some $\ell \geq 3$, then there are two equivalence classes of root maps of Σ, one with target B_ℓ and the other with target C_ℓ.*
(iv) *If $\Pi = \mathsf{B}_2 = \mathsf{C}_2$, F_4 or G_2, then there are two equivalence classes of root maps of Σ, both with target the unique root system having the same name as Π.*
(v) *If ι and $\hat{\iota}$ are inequivalent root maps of Σ, then for all $a \in \Xi$, $\iota(a)$ is long if and only if $\hat{\iota}(a)$ is short.*

Proof. By assumption, the Coxeter diagram Π is one of the diagrams X_ℓ with $\ell \geq 2$ in Figure 1.3. Let Φ denote the root system called X_ℓ. Thus Φ is unique except when the Coxeter diagram Π is $\mathsf{B}_\ell = \mathsf{C}_\ell$ for $\ell \geq 3$, in which case Φ can be either of the two distinct root systems B_ℓ or C_ℓ. By 2.15, there exists for each Π a root map with target Φ. Thus, in particular, (i) holds.

Suppose that ι and $\hat{\iota}$ are two root maps of Σ with target Φ. Then there exists a unique automorphism ϕ of Σ such that $\hat{\iota}(a) = \iota(\phi(a))$ for all $a \in \Xi$. Let σ be the automorphism of Π induced by ϕ and let Dyn_Φ denote the Dynkin diagram of Φ. We can think of Dyn_Φ as Π with the addition of an arrow on the multiple edge of Π if there is one, so that $\text{Aut}(\text{Dyn}_\Phi)$ is a subgroup of $\text{Aut}(\Pi)$ of index 1 or 2. By the Corollary in Chapter VI, Section 4.2, of [3], ι and $\hat{\iota}$ are equivalent if and only if $\sigma \in \text{Aut}(\text{Dyn}_\Phi)$. Thus the number of equivalence classes of root maps equals the index of $\text{Aut}(\text{Dyn}_\Phi)$ in $\text{Aut}(\Pi)$. Hence (ii)–(iv) hold.

To prove (v), it suffices by (ii)–(iv) (and 3.12) to produce two root maps ι and $\hat{\iota}$ of Σ such that for each $a \in \Xi$,

$$\iota(a) \text{ is long if and only if } \hat{\iota}(a) \text{ is short} \tag{3.15}$$

under the assumption that Π is not simply laced. Suppose first that $\Pi = \mathsf{F}_4$ or G_2. By 29.29, there exists an automorphism ϕ of Σ that induces the non-trivial automorphism on Π. By 2.14.iv, the Coxeter group W_Π has two orbits in Ξ, one consisting of the long roots and one consisting of the short roots. By 29.27, ϕ interchanges the two W_Π-orbits in Ξ. Let ι be a root map of Σ and let $\hat{\iota} := \iota \circ \phi$. Then $\hat{\iota}$ is also a root map of Σ (with the same target) and the pair $\iota, \hat{\iota}$ satisfies 3.15.

Suppose that $\Pi = \mathsf{B}_\ell$ or C_ℓ. By (V) in Chapter VI, Section 5, of [3], the root systems B_ℓ and C_ℓ can be embedded in the same ambient space in such a way that the map $\alpha \mapsto 2\alpha/(\alpha \cdot \alpha)$ (which we denote by π) is a bijection from B_ℓ to C_ℓ. Thus if ι is a root map of Σ with target B_ℓ, then $\pi \circ \iota$ is a root map of Σ with target C_ℓ (and vice versa) and again the pair $\iota, \hat{\iota}$ satisfies 3.15. Thus (v) holds. □

The identifications described in the following paragraph are important:

Conventions 3.16. Suppose that a root map ι of Σ with target Φ (as defined in 3.12) is given and let π be the unique isomorphism from Σ to Σ_Φ satisfying 3.13. We will always identify the set Ξ of roots of Σ with the root system Φ via the map ι[8] and we will always identify the apartment Σ with the Coxeter chamber system Σ_Φ (defined in 2.9) via the isomorphism π. In the presence of a root map ι with target Φ, it will thus make sense to treat the elements of Ξ as vectors in the ambient space V of Φ and to think of the chambers of Σ as Weyl chambers of Φ. In particular, it will make sense to write linear combinations of elements of Ξ (as in 3.19), to form dot products of vectors in V with elements of Ξ (as in 3.22), to think of the reflection s_a of Σ (for a given root $a \in \Xi$) defined in 3.7 and the reflection s_a of Σ_Φ defined in 2.1 as the same thing and to think of a root a of Σ not only as an element of Φ but also as the root K_a of Σ_Φ.

Remark 3.17. Let $d \in \Sigma$, let x and y be two vertices of Π, let $a = a_x$ and $b = a_y$ be as defined in 3.5 and let s_a and s_b be as defined in 3.7. By 29.4, $[s_a, s_b] = 1$ if and only if the vertices x and y are not adjacent in Π. Now suppose that ι is a root map of Σ with target Φ. By 3.16, we can identify s_a and s_b with the reflections defined in 2.1. Thus $[s_a, s_b] = 1$ if and only if the roots a and b are perpendicular elements of V. Thus the vertices x and y of Π are adjacent if and only if the roots a and b are not perpendicular.

Remark 3.18. For each $d \in \Sigma$, the set Ξ_d° (as defined in 3.5) is a basis of Φ (by 2.5, 2.8 and 3.16) and every basis of Φ is of the form Ξ_d° for some $d \in \Sigma$.

Proposition 3.19. *Let ι be a root map of Σ with target Φ, let a and b be two roots of Σ such that $a \neq \pm b$ and let $c \in (a, b)$, where the open interval (a, b) is as defined in 3.1. Then there exist unique positive real numbers p_c and q_c such that*

$$c = p_c a + q_c b.$$

[8] In particular, the roots of Φ (which are vectors in the ambient Euclidean space V of Φ) will be denoted by lowercase Latin letters.

Proof. There exist $k, l \in \mathbb{R}$ and $v \in \langle a, b\rangle^\perp$ such that $c = ka + lb + v$. By 3.2.i and 3.16, $K_a \cap K_b \subset K_c$. Choose $x \in K_a \cap K_b$. Thus $x \in K_c$, so $x \cdot c > 0$. Let $t \in \mathbb{R}$. The vector $x + tv$ is also in $K_a \cap K_b$, hence $(x + tv) \cdot c > 0$ and therefore

$$c \cdot x > -tv \cdot c = -tv \cdot v.$$

Since t is arbitrary, it follows that $v = 0$. Thus $c = ka + lb$. By condition (R1) on page 13, a and b are linearly independent (since $a \neq \pm b$ by hypothesis) and k and l are both non-zero (since c is equal to neither a nor b by hypothesis). We also have

$$k(x \cdot a) + l(x \cdot b) = x \cdot c > 0 \tag{3.20}$$

for all $x \in K_a \cap K_b$. Since a and b are linearly independent, we can choose $x \in K_a \cap K_b$ so that $|l|x \cdot b < |k|x \cdot a$. By 3.20, it follows that $k > 0$. Similarly, $l > 0$. □

We come now to the principal definition in this chapter.

Definition 3.21. Let ι be a root map of Σ with target Φ as defined in 3.12 and let $(G^\dagger, \xi)$ be the root datum of Δ based at Σ as defined in 3.4. A *valuation* of the root datum $(G^\dagger, \xi)$ of Δ based at Σ is a family of surjective[9] maps $\phi_a : U_a^* \to \mathbb{Z}$, one for each root a in Ξ, satisfying the following conditions:

(V1) For each $a \in \Xi$,

$$U_{a,k} := \{u \in U_a \mid \phi_a(u) \geq k\}$$

is a subgroup of U_a for each $k \in \mathbb{Z}$, where we assign to $\phi_a(1)$ the value ∞, so that $1 \in U_{a,k}$ for all k.

(V2) For all $a, b \in \Xi$ such that $a \neq \pm b$,

$$[U_{a,k}, U_{b,l}] \subset \prod_{c \in (a,b)} U_{c, p_c k + q_c l}$$

for all $k, l \in \mathbb{R}$, where p_c and q_c are as in 3.19.[10]

(V3) For all $a, b \in \Xi$ and $u \in U_a^*$, there exists $t \in \mathbb{Z}$ (depending on a, b and u) such that

$$\phi_{s_a(b)}(x^{m_\Sigma(u)}) = \phi_b(x) + t$$

for all $x \in U_b^*$ (where s_a is as defined in 3.7 and $m_\Sigma(u)$ is as defined in 3.8[11]).[12]

---

[9]We will sometimes refer to the surjectivity of ϕ_a for each $a \in \Xi$ as condition (V0).

[10]The terms in the product on the right-hand side in (V2) are to be understood to be in the same order as the roots in the interval (a, b). In other words, condition (V2) is a refinement of the inclusions in 3.2.iii.

[11]Thus $m_\Sigma(u)$ induces the reflection s_a on Σ by 3.10.

[12]In this book we use the convention that $a^b = b^{-1}ab$ whenever a, b are elements of a group. Thus $x^{m_\Sigma(u)} = m_\Sigma(u)^{-1} x m_\Sigma(u)$. This is the same convention used in [36]. From now on, we will also use the corresponding convention that permutations act from the right. (We used this convention without comment in Chapter 1 but not in Chapter 2.) Thus, in fact, we should really write, for example, b^{s_a} in place of $s_a(b)$ in (V3). We nevertheless write (and will continue to write) $s_a(b)$ rather than b^{s_a} simply because it looks more natural. We assure the reader that this inconsistency will not cause any confusion, and in the one place where difficulties could arise (Chapter 14) we actually use expressions like b^{s_a}; see, for example, the proof of 14.4.

(V4) $\phi_{-a}(x^{m_\Sigma(u)}) = \phi_a(x) - 2\phi_a(u)$ for all $a \in \Xi$ and all $u, x \in U_a^*$. In other words, $t = -2\phi_a(u)$ in (V3) if $a = b$.

We emphasize that the notion of a valuation

$$\phi = \{\phi_a \mid a \in \Xi\}^{13}$$

of the root datum $(G^\dagger, \xi)$ given in 3.21 depends on the choice of the root map ι (and the identifications described in 3.16). When we want to underline this dependence, we will say that ϕ is a valuation of the root datum $(G^\dagger, \xi)$ *with respect to* ι. In fact, the notion of a valuation ϕ of a root datum $(G^\dagger, \xi)$ depends on ι only for the coefficients p_c appearing in condition (V2). These coefficients depend, in turn, only on the equivalence class of ι.[14]

Proposition 3.22. *Let $(G^\dagger, \xi)$ be the root datum of Δ based at Σ (as defined in 3.4), let $\phi := \{\phi_a \mid a \in \Xi\}$ be a valuation of $(G^\dagger, \xi)$ with respect to a root map ι with target Φ as defined in 3.21, let v be a special point in the ambient Euclidean space V of Φ as defined in 2.26 and let*

$$\psi_a = \phi_a + v \cdot a$$

for each $a \in \Xi$ (which we have identified with Φ). Then $\psi := \{\psi_a \mid a \in \Xi\}$ is also a valuation of $(G^\dagger, \xi)$ with respect to ι.

Proof. By 2.26, ψ_a and ϕ_a (for a given root $a \in \Xi$) differ merely by an integer, from which it follows that $\psi_a(U_a^*) = \mathbb{Z}$ for all $a \in \Xi$ and that ψ satisfies conditions (V0) and (V1) of 3.21. The group $U_{a,k}$ defined in condition (V1) with respect to ϕ is the same as the group $U_{a,k-v\cdot a}$ defined with respect to ψ for each $a \in \Xi$ and each $k \in \mathbb{Z}$. If a and b are elements of Ξ such that $a \neq \pm b$, $c \in (a, b)$ and p_c and q_c are as in 3.19, then

$$p_c(v \cdot a) + q_c(v \cdot b) = v \cdot c.$$

It follows that ψ satisfies condition (V2). Simply substituting $\psi - v$ for ϕ, we see that ψ satisfies both conditions (V3), with t replaced by

$$t + v \cdot (s_a(b) - b),$$

and (V4). □

Definition 3.23. Two valuations $\phi = \{\phi_a \mid a \in \Xi\}$ and $\psi = \{\psi_a \mid a \in \Xi\}$ of $(G^\dagger, \xi)$ will be called *equipollent* if there exists a v in the ambient Euclidean space V of Φ (which must necessarily be a special point[15]) such that

$$\psi_a = \phi_a + v \cdot a \tag{3.24}$$

for each $a \in \Xi$. We will write

$$\psi = \phi + v$$

to indicate that 3.24 holds for each $a \in \Phi$.

---

[13] By 3.16, we can equally well write

$$\phi = \{\phi_a \mid a \in \Phi\}.$$

[14] It is a consequence of the observations in 18.32 that if ϕ is a valuation of a root datum $(G^\dagger, \xi)$ with respect to two root maps ι and $\hat{\iota}$, then ι and $\hat{\iota}$ are equivalent.

[15] By condition (V0) in 3.21, $v \cdot a \in \mathbb{Z}$ for each $a \in \Xi$. By 2.26, this means that v is special.

Proposition 3.25. *Let ψ be a valuation of $(G^\dagger, \xi)$, let $u \in U_a^*$ for some $a \in \Phi$ and let v, v' be the elements of U_{-a}^* such that*

$$m_\Sigma(u) = vuv',$$

where $m_\Sigma(u)$ is as in 3.8. Then

$$\psi_{-a}(v) = \psi_{-a}(v') = -\psi_a(u).$$

Proof. By 11.24 of [37],

$$m_\Sigma(u) = m_\Sigma(v) = m_\Sigma(v'). \tag{3.26}$$

By 11.23 in [37], we have

$$m_\Sigma(v^{m_\Sigma(u)}) = m_\Sigma(v)^{m_\Sigma(u)}.$$

By 3.26, it follows that

$$m_\Sigma(v^{m_\Sigma(u)}) = m_\Sigma(u). \tag{3.27}$$

Both $v^{m_\Sigma(u)}$ and u are contained in U_a^*. By 3.27 and condition (V4), it follows that

$$\psi_a(u) = \psi_a(v^{m_\Sigma(u)}). \tag{3.28}$$

Therefore

$$\begin{aligned} \psi_a(u) &= \psi_a(v^{m_\Sigma(u)}) && \text{by } 3.28 \\ &= \psi_a(v^{m_\Sigma(v)}) && \text{by } 3.26 \\ &= \psi_{-a}(v) - 2\psi_{-a}(v) = -\psi_{-a}(v) && \text{by (V4)}. \end{aligned}$$

Since $m_\Sigma(v) = m_\Sigma(v')$ (by 3.26), it follows by condition (V4) (again with $-a$ in place of a) that $\psi_{-a}(v) = \psi_{-a}(v')$. □

Proposition 3.29. *Let ψ be a valuation of $(G^\dagger, \xi)$ as defined in 3.21. Then $\psi_a(u) = \psi_a(u^{-1})$ for all $a \in \Phi$ and all $u \in U_a^*$.*

Proof. Choose $a \in \Phi$ and $u \in U_a^*$ and let $k = \max\{\psi_a(u), \psi_a(u^{-1})\}$. Since the set $U_{a,k}$ defined in condition (V1) of 3.21 is closed under inverses, both $\psi(u)$ and $\psi_a(u^{-1})$ equal k. □

Definition 3.30. Let $\psi := \{\psi_a \mid a \in \Phi\}$ be a valuation of $(G^\dagger, \xi)$, let $u \in U_b^*$ for some $b \in \Phi$ and let $m_u = m_\Sigma(u)$ (as defined in 3.8). For each $a \in \Phi$ we let $\psi_a^{m_u}$ be the map from U_a^* to $\mathbb{Z}$ given by

$$\psi_a^{m_u}(x) := \psi_{s_b(a)}(x^{m_u})$$

for all $x \in U_a^*$, where s_b is as in 3.7. Let

$$\psi^{m_u} := \{\psi_a^{m_u} \mid a \in \Phi\}.$$

In the next result, we show that ψ^{m_u} is a valuation of $(G^\dagger, \xi)$ that is equipollent to ψ.

Proposition 3.31. *Let* $\psi := \{\psi_b \mid b \in \Phi\}$ *be a valuation of* $(G^\dagger, \xi)$ *with respect to a root map* ι. *Let* $u \in U_a^*$ *for some* $a \in \Phi$, *let* $m_u = m_\Sigma(u)$, *let*

$$v = -2\psi_a(u)a/(a \cdot a)$$

and let ψ^{m_u} *be as defined in 3.30. Then* ψ^{m_u} *is a valuation of* $(G^\dagger, \xi)$ *with respect to* ι *and*

$$\psi^{m_u} = \psi + v,$$

where $\psi + v$ *is as defined in 3.23.*

Proof. Choose $b \in \Phi$ and $x \in U_b^*$. Let $k = \psi_a(u)$ and let $l = \psi_b(x)$. Suppose first that $a \neq \pm b$ and let W denote the subspace of V spanned by a and b. Thus $W \cap \Phi$ is a root system of rank 2. There is a natural ordering (clockwise or counterclockwise)

$$(a_0, a_1, a_2, \ldots, a_{2m-1}) \tag{3.32}$$

of the roots in $W \cap \Phi$ such that $2m = |W \cap \Phi|$, $a_0 = a$, $a_m = -a$ and $a_i = b$ for some $i \in [1, m-1]$). Let $\Psi = \{a_1, a_2, \ldots, a_{m-1}\}$. By 2.1 and condition (R2) on page 13 (as well as 3.16), the reflection s_a maps Ψ to itself. In particular, $s_a(b) \in \Psi$.

We observe, too, that if $c, d \in W \cap \Phi$ and $c \neq \pm d$, then by 3.19, the roots in the interval $[c, d]$ are precisely the roots of $W \cap \Phi$ between c and d in the ordering displayed in 3.32, where the subscripts are to be read modulo $2m$. Thus if $\epsilon = +$ or $-$, then

$$[c, d] \subset \{\epsilon a\} \cup \Psi \text{ whenever } c, d \in \{\epsilon a\} \cup \Psi. \tag{3.33}$$

For each $c \in W \cap \Phi$, let

$$h(c) := p_c k + q_c l,$$

where p_c and q_c are the unique real numbers (not necessarily positive) such that $c = p_c a + q_c b$. Thus, in particular, $h(b) = l$. Choose $c, d \in W \cap \Phi$ such that $c \neq \pm d$ and let $e \in (c, d)$. By 3.19, there exist positive real numbers P and Q such that $e = Pc + Qd$. Then

$$e = P(p_c a + q_c b) + Q(p_d a + p_d b)$$

and therefore $h(e) = P(p_c k + q_c l) + Q(p_d k + p_d l) = Ph(c) + Qh(d)$. By condition (V2) (in 3.21), it follows that

$$[U_{c,h(c)}, U_{d,h(d)}] \subset \prod_{e \in (c,d)} U_{e,h(e)}. \tag{3.34}$$

Now let

$$U' := \prod_{c \in \Psi} U_{c,h(c)}.$$

No two roots of Ψ are opposite each other. By condition (V1), 3.33 and 3.34, therefore, U' is a subgroup of $G^\dagger$. By 3.2.iv (with a_1 and a_{s-1} in place of a and b), every element of U' has a unique decomposition into a product of the form

$$\prod_{c\in\Psi} u_c,$$

where $u_c \in U_{c,h(c)}$ for all $c \in \Psi$. Hence

$$U_c \cap U' = U_{c,h(c)} \tag{3.35}$$

for all $c \in \Psi$ (including $c = s_a(b)$). In particular,

$$U_b \cap U' = U_{b,l}.$$

Since $s_a(b) = b - 2(a \cdot b)a/(a \cdot a)$, we have

$$h(s_a(b)) = l - 2(a \cdot b)k/(a \cdot a). \tag{3.36}$$

By 3.8, there exist $v, v' \in U_{-a}^*$ such that

$$m_u = vuv'. \tag{3.37}$$

By 3.25, $\psi_{-a}(v) = \psi_{-a}(v') = -k$. Moreover, $h(a) = k$ and $h(-a) = -k$. Thus by 3.33 and 3.34, we have

$$[U_{a,k}, U_{c,h(c)}] \subset \prod_{e\in(a,c)} U_{e,h(e)} \subset U'$$

and

$$[U_{-a,-k}, U_{c,h(c)}] \subset \prod_{e\in(-a,c)} U_{e,h(e)} \subset U'$$

for all $c \in \Psi$. It follows that U' is normalized by both $U_{a,k}$ and $U_{-a,-k}$. Hence U' is normalized by u, v and v' and thus also by their product $m_u = vuv'$. Therefore

$$\begin{aligned} x^{m_u} \in U_{b,l}^{m_u} &= (U_b \cap U')^{m_u} \\ &= U_b^{m_u} \cap U' \\ &= U_{s_a(b)} \cap U' = U_{s_a(b),h(s_a(b))} \end{aligned}$$

by two applications of 3.35 and hence

$$h(s_a(b)) \le \psi_{s_a(b)}(x^{m_u}).$$

By 3.36, we conclude that

$$\psi_b(x) - 2(a \cdot b)k/(a \cdot a) \le \psi_{s_a(b)}(x^{m_u}). \tag{3.38}$$

Now let $c = s_a(b)$ and let $y = x^{m_u}$. Substituting c for b, y for x and $w := u^{-1}$ for u in the previous argument, we obtain

$$\psi_c(y) - 2(a \cdot c)\psi_a(w)/(a \cdot a) \le \psi_b(y^{m_w})$$

in place of 3.38. By 3.9, $m_w = m_u^{-1}$, so $y^{m_w} = x$. By 3.29, $\psi_a(w) = k$. Since also $a \cdot c = -s_a(a) \cdot s_a(b) = -a \cdot b$, it follows that

$$\psi_c(y) + 2(a \cdot b)k/(a \cdot a) \le \psi_b(x).$$

Hence equality holds in 3.38. By 3.30, we conclude that

$$(\psi^{m_u})_b(x) = \psi_b(x) - 2(a\cdot b)k/(a\cdot a) \tag{3.39}$$

whenever $a \neq \pm b$.

If $a = b$, then

$$\begin{aligned}(\psi^{m_u})_a(x) &= \psi_{-a}(x^{m_u}) && \text{by 3.30}\\ &= \psi_a(x) - 2\psi_a(u) && \text{by (V4)}\\ &= \psi_a(x) - 2(a\cdot a)k/(a\cdot a)\end{aligned}$$

and thus 3.39 holds also in this case. Suppose, finally, that $a = -b$ and let $v \in U_b^*$ be as in 3.37. By 3.26, $m_v = m_u$. Therefore

$$\begin{aligned}(\psi^{m_u})_b(x) &= \psi_a(x^{m_u}) && \text{by 3.30}\\ &= \psi_a(x^{m_v})\\ &= \psi_b(x) - 2\psi_b(v) && \text{by (V4)}\\ &= \psi_b(x) - 2(a\cdot b)\psi_a(u)/(a,a) && \text{by 3.25,}\end{aligned}$$

so 3.39 holds also in this last case. □

Corollary 3.40. *Let ψ, u, m_u and v be as in 3.31. Then*

$$\psi^{m_u^{-1}} = \psi + v.$$

Proof. Let a be as in 3.31. By 3.9 and 3.29, we have $m_u^{-1} = m_{u^{-1}}$ and $v = -2\psi_a(u^{-1})a/(a\cdot a)$. The claim holds, therefore, by 3.31 (with u^{-1} in place of u). □

The following result is of central importance. In particular, the formula in 3.41.i will be applied many times in Chapters 19–25.

Theorem 3.41. *Let ι be a root map of Σ and let ϕ and ψ be two valuations of the root datum $(G^\dagger, \xi)$ with respect to ι. Then the following hold:*

(i) *$\psi_a(x^{m_\Sigma(u)}) = \psi_{s_b(a)}(x) + 2\psi_b(u)a\cdot b/(b\cdot b)$ for all $a, b \in \Phi$, all $x \in U_{s_b(a)}^*$ and all $u \in U_b^*$, where s_b is as in 2.1.*

(ii) *If $\phi_a = \psi_a$ for all roots a in a basis of Φ (as defined in 2.5), then $\phi = \psi$.*

(iii) *If $\phi_a = \psi_a$ for a single root $a \in \Phi$, then there exists $v \in V$ perpendicular to a such that*

$$\psi = \phi + v$$

(so, in particular, ψ and ϕ are equipollent as defined in 3.23).

Proof. Choose distinct roots $a, b \in \Phi$ and let $c = s_b(a)$. Thus

$$b\cdot c = s_b(b)\cdot s_b(c) = -b\cdot a = -a\cdot b. \tag{3.42}$$

Choose $u \in U_b^*$ and $x \in U_c^*$ and let $m = m_\Sigma(u)$. Then

$$\begin{aligned}\psi_a(x^m) &= \psi_c^m(x) && \text{by 3.30}\\ &= \psi_c(x) - 2\psi_b(u)b\cdot c/(b\cdot b) && \text{by 3.31}\\ &= \psi_c(x) + 2\psi_b(u)a\cdot b/(b\cdot b) && \text{by 3.42.}\end{aligned}$$

Thus (i) holds. From (i), we see that ψ_c is uniquely determined by ψ_a and ψ_b. Let

$$B = \{a_1, a_2, \ldots, a_\ell\}$$

be a basis of Φ. By the Theorem and Corollary in Section 1.5 of [17], every root of Φ is the image of a root in B under a product of reflections of the form s_b for $b \in B$. It follows that (ii) holds.

The Coxeter diagram Π is connected. By 3.17, it follows that the ordering of the elements $a_1, \ldots, a_\ell$ of the basis B can be chosen so that $a_{i-1} \cdot a_i \neq 0$ for all $i \in [2, \ell]$. We now assume that $\phi_a = \psi_a$ for some $a \in \Phi$. By the Corollary in Section 1.5 of [17] again, we can assume that $a = a_i$ for some $i \in [1, \ell]$. If we regard x as fixed and u as a variable, then it follows from (i) that ψ_b is uniquely determined by ψ_a up to a constant if $a \cdot b \neq 0$. Therefore there exist integers k_j (with $k_i = 0$) such that

$$\phi_{a_j} = \psi_{a_j} + k_j$$

for all $j \in [1, \ell]$. Let v be the unique element of V such that $v \cdot a_j = k_j$ for all $j \in [1, \ell]$. By 2.27, $v \cdot a \in \mathbb{Z}$ for all $a \in \Phi$. Let $\psi' = \psi + v$ as defined in 3.23. By 3.22, ψ' is a valuation of $(G^\dagger, \xi)$ with respect to ι. By (ii), we conclude that $\phi = \psi'$. Thus (iii) holds. □

By a small variation of the argument at the end of the last proof, we have also the following result:

Proposition 3.43. *Let $\phi = \{\phi_a \mid a \in \Xi\}$ be a valuation of $(G^\dagger, \xi)$ with respect to a root map ι of Σ with target Φ, let d be a chamber of Σ, let $x \mapsto k_x$ be a map from the vertex set I of Π to $\mathbb{Z}$ and let $x \mapsto a_x$ be the map from I to Ξ_d° described in 3.5. Then there exists a unique valuation $\psi = \{\psi_a \mid a \in \Xi\}$ of $(G^\dagger, \xi)$ with respect to ι that is equipollent to ϕ such that*

$$\psi_{a_x} = \phi_{a_x} + k_x$$

for all $x \in I$.

Proof. Let $B = \{a_x \mid x \in I\}$. By 3.18, B is a basis of Φ. Thus by 2.4.ii, B is also a basis of the ambient space V of Φ. There is thus a unique point $v \in V$ such that $v \cdot a_x = k_x$ for all $x \in I$. By 2.27, v is special. The claim holds, therefore, by 3.22 and 3.23. □

Proposition 3.44. *Let ϕ be a valuation of $(G^\dagger, \xi)$, let $a \in \Xi$, let $u \in U_a^*$ and suppose that $\phi_a(u) = 0$. Then for all $b \in \Xi$, the integer t in condition (V3) is zero.*

Proof. Setting a in place of b and $s_a(b)$ in place of a (and thus $s_a(s_a(b)) = b$ in place of $s_b(a)$) in 3.41.i, we have

$$\phi_{s_a(b)}(x^{m_\Sigma(u)}) = \phi_b(x)$$

for all $x \in U_b^*$. □

Corollary 3.45. *Let ϕ and ψ be two valuations of $(G^\dagger, \xi)$ with respect to the same root map ι and suppose that for some $a \in \Phi$ and some $M \in \mathbb{Z}$,*

$$\phi_a(u) = \psi_a(u) + M$$

for all $u \in U_a^*$. Then ϕ and ψ are equipollent.

Proof. By 3.23, there is a valuation ψ' of $(G^\dagger, \xi)$ with respect to ι equipollent to ψ such that

$$\phi_a(u) = \psi'_a(u)$$

for all $u \in U_a^*$. By 3.41.iii, it follows that ϕ and ψ are equipollent. □

Valuations of root data will not reappear until Chapter 13. From that point on, however, all the results of this chapter will play a central role.[16]

[16]See, in particular, the observations in 13.13 and 13.16. These remarks should throw more light on the crucial notion of a root map introduced in 3.12.

Chapter Four

Sectors

We return now to the study of affine Coxeter chamber systems where we left off at the end of Chapter 1.[1] Our goal in this and the next chapter is to introduce sectors and faces. These are fundamental structures in the theory of affine buildings.

As in Chapter 1, we assume that Π is one of the affine Coxeter diagrams $\tilde{\mathsf{X}}_\ell$ in Figure 1.1 and that $\Sigma = \Sigma_\Pi$ is the corresponding Coxeter chamber system as defined in 29.4. Let $W = W_\Pi$ be the Coxeter group associated with Π (as defined in 29.5) which we identify with $\mathrm{Aut}^\circ(\Sigma)$ as described in 1.14. Let $T = T_\Pi$ be the group of translations of Σ (as defined in 1.8).

Definition 4.1. A *sector* of Σ is a set of chambers of the form

$$\bigcap_{\alpha \in [R,d]} \alpha,$$

where R is a gem, d is a chamber of R and $[R, d]$ is as in 1.10. The sector determined by a pair (R, d) will be denoted by $\sigma(R, d)$.

Proposition 4.2. *Let R be a gem. Then $\sigma(R, d) \cap R = \{d\}$ for all $d \in R$.*

Proof. Let $d \in R$. By 29.6.ii and 29.7.ii, both $S := \sigma(R, d)$ and R are convex. Hence $S \cap R$ is convex. By 1.10, no chamber of R adjacent to d is in S. It follows that $S \cap R = \{d\}$. □

In the next result, we refer to the projection map defined in 29.7.i.

Proposition 4.3. *Let R be a gem. Then*

$$\sigma(R, d) = \{u \in \Sigma \mid \mathrm{proj}_R\, u = d\}$$

for all $d \in R$.

Proof. Let $d \in R$, let $x \in \Sigma$ and $S = \sigma(R, d)$. By 29.16, $x \in S$ if and only if $\mathrm{proj}_R\, x \in S$. By 4.2, therefore, $x \in S$ if and only if $\mathrm{proj}_R\, x = d$. □

Corollary 4.4. *Suppose that $\sigma(R, d) = \sigma(R', d')$ for gems R and R' and chambers $d \in R$ and $d' \in R'$. Then $R = R'$ and $d = d'$.*

Proof. Let $S = \sigma(R, d)$ and let $e \in R$ be a chamber adjacent to d. By 29.6.i, there is a unique root α containing d but not e. Since $d \in S$ but $e \notin S = \sigma(R', d')$, there must be a root β in $[R', d']$ such that $d \in \beta$ but

---

[1] The results of Chapters 2 and 3 will not be required until Chapter 13.

$e \notin \beta$. Therefore $\beta = \alpha$ (by the uniqueness of α). Hence $[R, d] \subset [R', d']$. By symmetry, these two sets are, in fact, equal. By 1.12, it follows that $R = R'$. By 4.3, it then follows that $d = d'$. □

Definition 4.5. Let $S = \sigma(R, d)$ for some gem R and some $d \in R$. We will call R the *terminus* of S and d the *apex* of S.

Proposition 4.6. *Let S and S' be two sectors that are translates of each other (as defined in 1.8). Then $S \cap S'$ is a translate of S (and thus $S \cap S'$ is, in particular, also a sector).*

Proof. Let R be the terminus and let $d \in R$ be the apex of S. We have $S' = S^g$ for some $g \in T$. Then $S' = \sigma(R', d')$, where $R' = R^g$ and $d' = d^g$. Let $[R, d] = \{\alpha_1, \ldots, \alpha_\ell\}$. By 1.2, we have

$$S \cap S' = \bigcap_{i=1}^{\ell} \beta_i,$$

where for each $i \in [1, \ell]$, $\beta_i = \alpha_i \cap \alpha_i^g$ (so β_i equals α_i or α_i^g). By 1.24, there exists an element $h \in T$ such that $\alpha_i^h = \beta_i$ for all $i \in [1, \ell]$. Then $S \cap S' = S^h$. □

Proposition 4.7. *Let S and S' be two sectors such that $S' \subset S$. Then S and S' are translates of each other.*

Proof. Let R and R' be the termini and let d and d' be the apices of S and S'. By 1.9, there exists an element $g \in T$ such that $R = (R')^g$. Let $e = (d')^g$ and $S'' = (S')^g$. By 4.6, $S' \cap S''$ is a sector. In particular, $S' \cap S'' \neq \emptyset$. Since $S' \subset S$, it follows that $S \cap S'' \neq \emptyset$. The sectors S and S'' have the same terminus. By 4.3, it follows that they are equal. □

Corollary 4.8. *Let S' and S'' be two subsectors of a sector S. Then $S' \cap S''$ is a subsector and S, S', S'' and $S' \cap S''$ are all translates of each other.*

Proof. This holds by 4.6 and 4.7. □

Corollary 4.9. *Let S and S' be two sectors. Then S and S' are translates of each other if and only if $S \cap S'$ is a subsector.*

Proof. This holds by 4.6 and 4.7. □

Proposition 4.10. *Let S be a sector with terminus R and let S' be a subsector of S. Then S' contains a subsector whose terminus has the same type as R.*

Proof. By 4.7, there exists a translation g mapping S to S'. Thus

$$S \supset S' = S^g \supset S^{g^2} \supset S^{g^3} \supset \cdots .$$

By 1.28, $T/T \cap W$ is a finite group. Thus for some $m > 0$, $g^m \in W = \mathrm{Aut}^\circ(\Sigma)$, i.e. g^m is type-preserving. □

Proposition 4.11. *Let R be a gem, let α be a root cutting R and suppose that $d \in \alpha \cap R$. Then $\sigma(R,d) \subset \alpha$.*

Proof. By 29.16, $\text{proj}_R x \in -\alpha$ for all $x \in -\alpha$. In particular, $\text{proj}_R x \neq d$ for all $x \in -\alpha$. By 4.2, therefore, $\sigma(R,d) \subset \alpha$. □

Proposition 4.12. *Let R be a gem, let $d \in R$, let α be a root cutting R and let $S = \sigma(R,d)$. Then the following hold:*

(i) *If $d \in \alpha$, then every root parallel to α contains a subsector of S.*
(ii) *If $d \notin \alpha$, then no root parallel to α contains a subsector of S.*

Proof. Suppose first that $d \in \alpha$. Then $S \subset \alpha$ by 4.11. Let $\alpha' \in [\alpha]$ (where $[\alpha]$ is as defined in 1.5). By 1.26, $\alpha' = \alpha^g$ for some $g \in T$. Thus $S^g \subset \alpha'$. By 4.6, $S \cap S^g$ is a subsector of S contained in α'. Thus (i) holds.

Now suppose that $d \notin \alpha$. By 4.11, $S \subset -\alpha$. Suppose that α' is a root parallel to α. Then the root $-\alpha'$ is parallel to $-\alpha$. By (i), therefore, $-\alpha'$ contains a subsector of S. By 4.8, two subsectors of S have a non-empty intersection. It follows that α' cannot also contain a subsector of S. □

Corollary 4.13. *Let S be a sector and α a root. Then α or $-\alpha$ (but not both) contains a subsector of S.*

Proof. Let R be the terminus of S. By 1.18, there is a unique root parallel to α cutting R. The claim holds, therefore, by 4.12. □

Proposition 4.14. *Let R be a gem, let α be a root cutting R and let $d \in \alpha \cap R$. Suppose that α' is a root parallel to α that contains d. Then α' contains $\sigma(R,d)$.*

Proof. By 4.11, $\sigma(R,d) \subset \alpha$. By 1.2, $\alpha \subset \alpha'$ or $\alpha' \subset \alpha$. By 1.13, it follows that, in fact, $\alpha \subset \alpha'$ (since $d \in \alpha'$). □

Proposition 4.15. *Let S be a sector. Then there exists a subsector of S whose terminus is contained in S.*

Proof. Let R be the terminus, let d be the apex of S and let

$$[R,d] = \{\alpha_1, \ldots, \alpha_\ell\}.$$

For each $i \in [1,\ell]$, let α_i' be a root contained properly in α_i. By 1.24, there is a unique translation g mapping each α_i to α_i' for each $i \in [1,\ell]$. Let $R' = R^g$ and $S' = S^g$. Then S' is a subsector of S whose terminus is R'. By 1.13, $R' \subset \alpha_i$ for each $i \in [1,\ell]$. Thus $R' \subset S$. □

Corollary 4.16. *Every sector contains gems.*

Proof. This holds by 4.15. □

Proposition 4.17. *Let S' be a subsector of a sector S and let d be the apex of S. Then every root of Σ containing S' and d contains S.*

Proof. Let α be a root containing S' and d and let R denote the terminus of S. By 1.18, there is a root α' parallel to α that cuts R. By 4.12.ii, we have $d \in \alpha'$ (since otherwise α would not be allowed to contain the subsector S' of S). Thus $S \subset \alpha$ by 4.14. □

Corollary 4.18. *Let S' be a subsector of a sector S and let d denote the apex of S. Then S is the convex hull of $S' \cup \{d\}$.*

Proof. By 4.1 and 4.17, S is the intersection of all the roots of Σ that contain $S' \cup \{d\}$. The claim holds, therefore, by 29.20. □

Proposition 4.19. *Let S' be a subsector of a sector S and let d denote the apex of S. Then for each chamber $x \in S$, there exists a subsector S'' of S' (depending on x) such that for each $u \in S''$, x lies on a minimal gallery from d to u.*

Proof. Let R denote the terminus of S and let α be a root containing d but not S. By 1.18, there exists a root $\alpha' \in [\alpha]$ that cuts R. By 4.14, $d \notin \alpha'$. By 4.12.i, therefore, $-\alpha$ contains a subsector of S. By 4.8, it follows that $-\alpha$ contains a subsector of S'.

Now choose a chamber $x \in S$ and let $k = \text{dist}(x, d)$. By 29.6.iii, there are exactly k roots $\alpha_1, \ldots, \alpha_k$ of Σ that contain d but not x. By the conclusion of the previous paragraph, there exist subsectors $S_1, \ldots, S_k$ of S' such that $S_i \subset -\alpha_i$ for all $i \in [1, k]$. Let

$$S'' = S_1 \cap \cdots \cap S_k.$$

By 4.8, S'' is a subsector of S. Let $u \in S''$. Suppose β is a root containing both d and u. Since $u \in S'' \subset -\alpha_i$ for each $i \in [1, k]$, we have

$$\beta \notin \{\alpha_1, \ldots, \alpha_k\}.$$

By the choice of $\alpha_1, \ldots, \alpha_k$, it follows that $x \in \beta$. By 29.6.iv, therefore, x lies on a minimal gallery from d to u. □

By 1.7, the gems of Σ are finite. It thus makes sense to talk about *opposite* chambers in a gem (as defined in 29.9).

Proposition 4.20. *Let R be a gem, let d and e be opposite chambers of R and let $u \in \sigma(R, d)$ and $v \in \sigma(R, e)$. Then there exists a minimal gallery from u to v that passes through d and e. Equivalently,*

$$\text{dist}(u, v) = \text{dist}(u, d) + \text{dist}(d, e) + \text{dist}(e, v).$$

Proof. Let α be a root of Σ containing u and v. By 1.18, there exists a root α' parallel to α that cuts R. By 29.7.iv, $\alpha' \cap R$ is a root of R. Hence $|\alpha' \cap \{e, d\}| = 1$ by 29.9.i. Suppose that $d \in \alpha'$ and $e \in -\alpha'$. Since $v \in \alpha$, we have $\sigma(R, e) \not\subset -\alpha$. By 4.14, it follows that $e \notin -\alpha$, i.e. $e \in \alpha$. Hence $\alpha \not\subset \alpha'$. By 1.2, we thus have $\alpha' \subset \alpha$, so also $d \in \alpha$. By symmetry, this same conclusion (i.e. both d and e lie in α) holds under the assumption that $d \in -\alpha'$ and $e \in \alpha'$. We conclude that every root containing u and v also contains d and e. The claim holds, therefore, by 29.6.iv. □

Proposition 4.21. *Let R be a gem, let $\{d, e\}$ be a panel of R, let α be the unique root of Σ containing d but not e, let $S = \sigma(R, d)$ and let $S' = \sigma(R, e)$. Then $S \cup S'$ is the intersection of all the roots contained in*

$$[R, d] \cup [R, e] \backslash \{\alpha, -\alpha\}.$$

In particular, $S \cup S'$ is convex.

Proof. If $\beta \in [R, d] \backslash \{\alpha\}$, then $e \in \beta$ (by the uniqueness of α) and hence $S' \subset \beta$ by 4.11. Similarly, $S \subset \beta$ for all $\beta \in [R, e] \backslash \{-\alpha\}$. By 4.1, therefore, $S \cup S'$ is the intersection of all the roots contained in $[R, d] \cup [R, e] \backslash \{\alpha, -\alpha\}$. By 29.6.ii, it follows that $S \cup S'$ is convex. □

Proposition 4.22. *Let S and S' be two sectors. Then S is a translate of S' if and only if S and S' are parallel as defined in 1.44.*

Proof. If S' is a translate of S, then by 1.45, S and S' are parallel. Now suppose that S and S' are not translates of each other. We want to show that S and S' are not parallel. By 1.9 and 1.45, we can assume that S and S' have the same terminus but distinct apices d and d'. We can thus choose a root $\alpha \in [R, d]$ that does not contain d'. Let $i \mapsto \alpha_i$ be the map from $\mathbb{Z}$ to $[\alpha]$ as in 1.3. Then

$$S \subset \alpha = \alpha_0 \subset \alpha_i$$

for all $i \geq 0$ and $d' \in -\alpha_0$. By 4.12.i, $-\alpha_i \cap S' \neq \emptyset$ for all i. By 29.6.iii, it follows that S' contains chambers that are arbitrarily far from S. □

Corollary 4.23. *Let S and S' be two sectors. Then S and S' are parallel if and only if $S \cap S'$ is a subsector.*

Proof. This holds by 4.9 and 4.22. □

Proposition 4.24. *Let R be a gem and S a sector. Then S is parallel to a unique sector whose terminus is R.*

Proof. By 1.9 and 4.22, S is parallel to a sector S' whose terminus is R. If S'' is a second sector with terminus R that is parallel to S, then S'' is a translate of S' by 4.22, hence $S' \cap S'' \neq \emptyset$ by 4.6 and thus $S'' = S'$ by 4.3. □

Proposition 4.25. *Let R be a gem, let d and e be opposite chambers of R, let $S = \sigma(R, d)$, let $S' = \sigma(R, e)$ and let S_1 and S_1' be subsectors of S and S'. Then the convex hull of $S_1 \cup S_1'$ is Σ.*

Proof. Suppose that α is a root containing $S_1 \cup S_1'$. By 1.18, there exists a root α' parallel to α that cuts R. By 4.12.ii, α' contains both d and e. By 29.7.iv, however, $R \cap \alpha'$ is a root of R, but by 29.9.i, no root of R can contain two opposite chambers of R. We conclude that there are no roots containing $S_1 \cup S_1'$. The claim holds, therefore, by 29.20. □

Proposition 4.26. *Let α be a root and let $X \neq \Sigma$ be a convex subset containing a subsector of every sector contained in α. Then X is a root parallel to α.*

Proof. Let β be a root containing X. By 1.16, we can choose a gem R cut by α. Let β' be the unique root parallel to β that cuts R. By 4.11, $\sigma(R, d) \subset \alpha$ for all $d \in R \cap \alpha$. By 4.12.ii, therefore, $R \cap \alpha \subset R \cap \beta'$. Since $R \cap \alpha$ and $R \cap \beta'$ are roots of R (by 29.7.iv) and R is finite, we have

$$|R \cap \alpha| = |R|/2 = |R \cap \beta'|$$

by 29.18. Thus, in fact, $R \cap \alpha = R \cap \beta'$. By 29.17, therefore, $\alpha = \beta'$. Thus β is parallel to α. By 29.20, X is the intersection of all the roots containing X. By 1.3, we conclude that X is a root parallel to α. $\square$

Chapter Five

Faces

Roughly speaking, a face is the border between two "adjacent" sectors. Faces play a central role in the construction of the building at infinity, as we will see in Chapter 8.

We continue to denote by Π one of the affine Coxeter diagrams $\tilde{\mathsf{X}}_\ell$ in Figure 1.1 and by Σ the corresponding Coxeter chamber system.

Definition 5.1. Let R be a gem, let $P = \{d, e\}$ be a panel contained in R, let $S = \sigma(R, d)$ and let α be the unique root of Σ containing d but not e. Let $\mu(\alpha)$ be the wall of α (as defined in 29.22), so $|Q \cap \alpha| = 1$ and hence $|Q \cap S| \leq 1$ for each panel $Q \in \mu(\alpha)$. Let

$$\mu(R, P) = \{Q \in \mu(\alpha) \mid Q \cap S \neq \emptyset\}.$$

Thus, in particular, $P \in \mu(R, P)$. The set $\mu(R, P)$ will be called the *P-face* of S or, equivalently, the α-face of S. The set of chambers contained in some panel in $\mu(R, P)$ will be called the *chamber set* of $\mu(R, P)$.

Definition 5.2. Let R be a gem and let $d \in R$. A *face* of a sector $S = \sigma(R, d)$ is a P-face of S for some panel $P \subset R$ containing d. A set of panels will be called a *face* if it is the face of some sector.

Proposition 5.3. *Let R be a gem and let P be a panel of R. Then the face $\mu(R, P)$ is the set of panels Q of Σ such that $\text{proj}_R Q := \{\text{proj}_R u \mid u \in Q\}$ equals P.*

Proof. This holds by 4.3 and 5.1. □

What we are calling a face should be called more precisely a *sector-face* or (as in [25]) a *sector-panel*. We have decided, nevertheless, to use the shorter, unhyphenated term and assure the reader that we do not use the term "face" in any sense other than that described in 5.2 in this book.

Proposition 5.4. *Let $S = \sigma(R, d)$ and $S' = \sigma(R', d')$ be two sectors and suppose that S' is a translate of S and that $[R, d] \cap [R', d']$ contains a root α. Let f be the α-face of S, let f' be the α-face of S' and let $\hat{S} = S \cap S'$. Then the following hold:*

(i) *There is a translation h mapping S to $\hat{S}$; in particular, $\hat{S}$ is a sector.*
(ii) *$\alpha \in [\hat{R}, \hat{d}]$, where $\hat{R}$ and $\hat{d}$ are the terminus and apex of $\hat{S}$.*
(iii) *$f^h = f \cap f'$.*
(iv) *$f \cap f'$ is the α-face of $\hat{S}$.*

Proof. Let g be a translation mapping S to S'. By 4.4, g maps R to R' and d to d'. Thus α and α^g are both contained in $[R,d]^g = [R',d']$. Hence $\alpha^g = \alpha$ by 1.13. Let $[R,d] = \{\alpha_1, \ldots, \alpha_\ell\}$, where $\alpha_1 = \alpha$. Then

$$\hat{S}' = \bigcap_{i=1}^{\ell} \beta_i,$$

where $\beta_i = \alpha_i \cap \alpha_i^g$ (so β_i equals α_i or α_i^g) for each $i \in [1,\ell]$. In particular, $\beta_1 = \alpha$. By 1.24, there exists a translation h mapping α_i to β_i for each $i \in [1,\ell]$. Thus $\alpha^h = \alpha$, so $\alpha \in [R^h, d^h]$, and $S^h = \hat{S}$, so R^h and d^h are the terminus and apex of $\hat{S}$ and f^h is the α-face of $\hat{S}$. In particular, (i) and (ii) hold. Every panel in $\mu(\alpha)$ contains exactly one chamber in α. Since S, S' and $\hat{S}$ are all contained in α, it follows that a panel in $\mu(\alpha)$ contains a chamber in $\hat{S}$ if and only if it contains a chamber in S and a chamber in S'. In other words, the α-face of $\hat{S}$ is $f \cap f'$. Thus (iii) and (iv) hold. □

Proposition 5.5. *Let f and f' be comural faces that are translates of each other.*[1] *Then $f \cap f'$ is a translate of f and of f'.*

Proof. Let $f = \mu(R,P)$, let g be a translation mapping f to f', let $R' = R^g$ and let $P' = P^g$. Then $f' = \mu(R',P')$. There exists a root α such that f and f' are both contained in the wall $M := \mu(\alpha)$. In particular, M contains both P and P'. Hence R' is cut by both α and α^g. By 1.13, therefore, $\alpha = \alpha^g$. Hence

$$\alpha \in [R,d] \cap [R',d'],$$

where d is the unique chamber in $P \cap \alpha$ and $d' = d^g$ is the unique chamber in $P' \cap \alpha$. Hence by 5.4.iii, $f \cap f'$ is a translate of f. □

Proposition 5.6. *Let R be a gem, let $P = \{d,e\}$ be a panel contained in R, let α be a root cutting R and containing both d and e and let α' be a root parallel to α. Then α' contains subsectors of $S := \sigma(R,d)$ and $S_1 := \sigma(R,e)$ sharing a face contained in $f := \mu(R,P)$.*

Proof. Let β denote the unique root containing d but not e. Then $\alpha \neq \pm\beta$ since $d, e \in \alpha$. By 1.13, it follows that α is parallel to neither β nor $-\beta$. By 1.15, therefore, there exists a gem R' cut by β and contained in α'. By 1.9, there exists a translation g mapping R to R'. Thus β and β^g both cut R'. Hence $\beta = \beta^g$ by 1.13, so $\beta \in [R,d] \cap [R,d]^g$. By two applications of 5.4.iv, therefore, $f \cap f^g$ is the β-face of $\hat{S} := S \cap S^g$ and the $-\beta$-face of $\hat{S}_1 := S_1 \cap S_1^g$. By 4.11, α contains both S and S_1. Therefore α^g contains both S^g and S_1^g. Since α^g cuts R' and α' contains R', it follows that α' contains α^g. Thus α' contains both $\hat{S}$ and $\hat{S}_1$. □

Notation 5.7. If u is a chamber and f is a face, we will write $u \in f$ if u is contained in the chamber set of f.

---

[1] We say that two faces are *comural* if they are contained in the same wall. Note that by 29.6.i, every panel is contained in a unique wall. Thus, in particular, the wall containing a face is also unique.

Proposition 5.8. *Let R and R' be two gems, let P and P' be panels contained in R and R' and let $f = \mu(R, P)$ and $f' = \mu(R', P')$ be the corresponding faces as defined in 5.2. Then the following hold:*

(i) *The faces f and f' are parallel (as defined in 1.44) if and only if the unique translation that maps R to R' maps P to P'.*
(ii) *If f and f' are not parallel, then for each $N \in \mathbb{N}$ there exist subfaces $\hat{f}$ of f and $\hat{f}'$ of f' such that $\text{dist}(u, f') \geq N$ and $\text{dist}(f, v) \geq N$ for all $u \in \hat{f}$ and all $v \in \hat{f}'$. In particular, both*

$$\{\text{dist}(u, f') \mid u \in f\}$$

and

$$\{\text{dist}(f, v) \mid v \in f'\}$$

are unbounded.

Proof. Suppose that $R = R'$ but $P \neq P'$. Let $P = \{d, e\}$ and $P' = \{d', e'\}$. Relabeling d' and e' if necessary, we can assume that there exists a root α such that $e' \in \alpha$ and $P \subset -\alpha$. Let β be the unique root containing d but not e. Then $\alpha \neq \pm\beta$. Let α_i for $i \in \mathbb{Z}$ be as in 1.3. By 4.11,

$$\sigma(R, e') \subset \alpha = \alpha_0 \subset \alpha_1 \subset \alpha_2 \subset \cdots.$$

Let $N > 0$. By 5.6, there exists a subface f_N of $\mu(R, P)$ whose chamber set is contained in $-\alpha_N$. By 29.6.iii, $\text{dist}(u, v) \geq N + 1$ for all $u \in f_N$ and all $v \in \sigma(R, e')$ and hence $\text{dist}(u, f') \geq N$ for all $u \in f_N$. By symmetry, there exists a subsector f'_N of f' such that $\text{dist}(f, v) \geq N$ for all $v \in f'_N$. Both (i) and (ii) follow now by 1.45. □

Corollary 5.9. *Let R be a gem, let P and Q be panels of R and suppose that the faces $\mu(R, P)$ and $\mu(R, Q)$ are parallel. Then $P = Q$.*

Proof. This holds by 1.21 and 5.8.i. □

Corollary 5.10. *Two faces are parallel if and only if they are translates of each other.*

Proof. This holds by 1.45 and 5.8.i. □

Proposition 5.11. *Let R be a gem, let $P = \{d, e\}$ be a panel contained in R, let $S = \sigma(R, d)$, let $S' = \sigma(R, e)$ and let $f = \mu(R, P)$. Then every sector having a face parallel to f is parallel to either S or S'.*

Proof. Let S_1 be a sector having a face f_1 parallel to f. By 1.9, we can choose a translation g mapping the terminus of S_1 to R. By 1.43 and 5.9, $f_1^g = f$. Thus $S_1^g = S$ or S'. □

Proposition 5.12. *Let f and f' be parallel comural faces. Then $f \cap f'$ is a face that is a translate of both f and f'.*

Proof. This holds by 5.5 and 5.10. □

Proposition 5.13. *Suppose that f' is a subface of a face f. Then f is parallel to f'.*

Proof. Let $f = \mu(R, P)$ and let $f' = \mu(R', P')$. Let g be the unique translation that maps R' to R, let $P'' = (P')^g$ and let $f_1 = \mu(R, P') = (f')^g$. By 1.43, there exists $N \in \mathbb{N}$ such that $\text{dist}(x^g, x) \leq N$ for all chambers x. Thus $\text{dist}(x, f) \leq N$ for all chambers x in the chamber set of f_1. By 5.8.ii, it follows that f is parallel to f_1. By 1.45, f_1 is parallel to f'. Hence f is parallel to f'. □

Proposition 5.14. *Let f and f' be two faces. Then $f \cap f'$ is a face if and only if $f \cap f'$ contains a face.*

Proof. Suppose that f_1 is a face contained in $f \cap f'$. By 5.13, both f and f' are parallel to f_1. Let M be the unique wall containing f and let M' be the unique wall containing f'. Since both M and M' contain f_1, we must have $M = M'$. By 5.12, therefore, $f \cap f'$ is a face. □

Proposition 5.15. *Let f' and f'' be subfaces of a face f. Then $f' \cap f''$ is also a subface.*

Proof. By 5.13, both f' and f'' are parallel to f. Thus f' is parallel to f''. Since f' and f'' are both contained in f, they are contained in the unique wall containing f. The claim holds, therefore, by 5.12. □

Proposition 5.16. *Let M and M' be walls containing faces f and f' and suppose that f is parallel to f'. Then M and M' are also parallel.*

Proof. By 5.10, $f' = f^g$ for some translate g. Thus M^g is a wall containing f'. Since the wall containing a face is unique, we have $M' = M^g$. The claim holds, therefore, by 1.45. □

Proposition 5.17. *Let R be a gem, let $\{d, e\}$ be a panel contained in R, let S be a subsector of $\sigma(R, d)$ and let S' be a subsector of $\sigma(R, e)$. Then the convex hull of $S \cup S'$ contains subsectors of $\sigma(R, d)$ and $\sigma(R, e)$ sharing a common face that is contained in $\mu(R, P)$.*

Proof. Let α be a root containing $S \cup S'$. By 1.18, there exists a root α' parallel to α that cuts R. By 4.12.ii, α' contains both d and e. By 5.6, therefore, α contains subsectors of $\sigma(R, d)$ and $\sigma(R, e)$ sharing a common face that is contained in $\mu(R, P)$. If $S_1, \ldots, S_m$ and $S'_1, \ldots, S'_m$ are subsectors of $\sigma(R, d)$ and $\sigma(R, e)$ such that S_i and S'_i share a common face contained in $\mu(R, P)$, then by 4.8 and 5.15, the intersections

$$S_1 \cap \cdots \cap S_m$$

and

$$S'_1 \cap \cdots \cap S'_m$$

are also subsectors of $\sigma(R, d)$ and $\sigma(R, e)$ sharing a common face contained in $\mu(R, P)$. By 1.18, there are only finitely many parallel classes of roots. The claim follows, therefore, by 29.20. □

Definition 5.18. Let R be a gem, let $P = \{d, e\}$ be a panel contained in R, let α be the unique root of Σ containing d but not e and let $S = \sigma(R, d)$. We set

$$\partial_P S = S \cap \partial\alpha,$$

where $\partial\alpha$ is as defined in 29.40. Equivalently, $\partial_P S$ is the intersection of S with the chamber set of $\mu(\alpha)$. This set will be called the *P-border* or *α-border* of S and we will refer to it also as $\partial_\alpha S$. Note that the chamber set of the face $\mu(R, P)$ is the disjoint union of the P-border $\partial_P S$ of S and the P-border $\partial_P S'$ of S', where $S' = \sigma(R, e)$.

Proposition 5.19. *Let R be a gem, let P be a panel in R, let d, e be the two chambers in P, let $f = \mu(R, P)$, let $S = \sigma(R, d)$ and let α be the unique root containing d but not e. Let $i \mapsto \alpha_i$ be as in 1.3 (so $\alpha_0 = \alpha$). Then the wall $\mu(\alpha_i)$ contains a face contained in S (i.e. whose chamber set is contained in S) that is parallel to f if and only if $i < 0$.*

Proof. By 1.24, there exists a translation g that maps α_i to α_{i-1} for all i and fixes all the roots in $[R, d]$ different from α. Let $S_i = S^{g^i}$ for all i. By 4.1, $S_i \supset S_{i+1}$ for all i. Let $u \in f \cap S$ and let v be the unique chamber in $-\alpha$ adjacent to u. Let $i < 0$ and let $h = g^i$. Then $u^h \in S_i \subset S$. Since $\alpha^h \notin [R, d]$, it follows that $v^h \in S$ as well. We conclude that f^h is a face contained in S and contained in $\mu(\alpha)^h = \mu(\alpha_{-i})$. By 1.45, f^h is parallel to f.

Suppose, conversely, that $i \geq 0$. Then $S \subset \alpha \subset \alpha_i$. Thus each panel in $\mu(\alpha_i)$ contains a chamber in $-\alpha$. Since $S \subset \alpha$, we conclude that $\mu(\alpha_i)$ does not contain any panels that are contained in S. □

Proposition 5.20. *Let R be a gem, let P be a panel in R, let $f = \mu(R, P)$, let d, e be the two chambers in P, let $S = \sigma(R, d)$ and let $S' = \sigma(R, e)$. Then every face parallel to f contains a subface contained in S, S' or f.*

Proof. Let f_1 be a face parallel to f and let M_1 be the unique wall containing f_1. By 5.16, M_1 is parallel to the unique wall M containing f. Let α be the unique root containing d but not e. By 1.46, there exists a root α_1 parallel to α such that $M = \mu(\alpha)$ and $M_1 = \mu(\alpha_1)$. If $\alpha = \alpha_1$, then f and f_1 are both contained in M and hence (by 5.12) $f \cap f_1$ is a subface of f_1 contained in f. Suppose from now on that $\alpha \neq \alpha_1$. If α_1 is properly contained in α, then by 5.19, there is a face f_1' parallel to f that is contained in M_1 and in S. If $-\alpha_1$ is properly contained in $-\alpha$, then again by 5.19 (this time with α replaced by $-\alpha$), there is a face f_1' parallel to f that is contained in S'. Thus (in both cases) f_1' are f_1 parallel and comural. By another application of 5.12, it follows that $f_1' \cap f_1$ is a subface of f_1. Since $f_1' \cap f_1$ is contained in f_1, it is contained in S or S'. □

Proposition 5.21. *Let S be a sector with terminus R and apex d, let $\alpha \in [R, d]$ and let β be the unique root containing $-\alpha$ minimally (as defined in 1.36). Then $\beta \cap S$ is the convex hull of the α-border $\partial_\alpha S$ of S (as defined in 5.18).*

Proof. Let $U = \beta \cap S$, $X = \partial_\alpha S$ and let $x \in X$. If y is the unique chamber in $-\alpha$ adjacent to x, then $-\alpha$ is the unique root containing y but not x; since $-\alpha$ is a proper subset of β, it follows that $x \in \beta$. Thus $X \subset U$.

Now let α_0 be an arbitrary root of Σ containing X. By 1.18, there is a root α_0' parallel to α_0 that cuts R. If $d \in \alpha_0'$, then $S \subset \alpha_0$ by 4.14 (since $d \in X$) and hence $U \subset \alpha_0$.

Suppose that $d \notin \alpha_0'$. Let e be the unique chamber of $R \cap -\alpha$ adjacent to d. Since $X \subset \alpha_0$, we have $e \in \alpha_0'$ by 5.6. Hence $\alpha_0' = -\alpha$ since $-\alpha$ is the unique root containing e but not d. Thus α_0 is a root parallel to $-\alpha$ that contains d. Hence $\beta \subset \alpha_0$. By the conclusion of the previous paragraph, we thus have $U \subset \alpha_0$ whether or not $d \in \alpha_0'$. By 4.1, we conclude that $U = \beta \cap S$ is the intersection of all the roots of Σ that contain X. By 29.20, therefore, U is the convex hull of X. □

Proposition 5.22. *Let S be a sector, let $S_\mathcal{F}$ denote the set of faces of S and for each $f \in S_\mathcal{F}$, let $\hat{f}$ be a face parallel to f. Then there exists a unique sector S' parallel to S such that for each face f' of S', there exists $f \in S_\mathcal{F}$ such that $f' \cap \hat{f}$ is a face.*

Proof. Let R and d be the terminus and apex of S and let

$$[R, d] = \{\alpha_1, \dots, \alpha_\ell\}.$$

For each face f, let M_f denote the unique wall containing f. By 5.1, there is a bijection $\alpha_i \mapsto f_i$ from $[R, d]$ to $S_\mathcal{F}$ such that $f_i \subset \mu(\alpha_i)$ for all $i \in [1, \ell]$. Thus for each $i \in [1, \ell]$, $\hat{f}_i$ is by hypothesis a face parallel to f_i. For each $i \in [1, \ell]$, there is a unique pair of opposite roots $\pm\alpha_i'$ such that

$$\mu(\alpha_i') = \mu(-\alpha_i') = M_{\hat{f}_i}.$$

By 4.13, we can label these roots so that α_i' contains a subsector of S for each $i \in [1, \ell]$. By 5.16, the wall M_{f_i} is parallel to the wall $M_{\hat{f}_i}$ for each $i \in [1, \ell]$. By 1.46, it follows that for each $i \in [1, \ell]$, the root α_i is parallel to either α_i' or $-\alpha_i'$. Hence by 4.12.ii, α_i' is, in fact, parallel to α_i for all $i \in [1, \ell]$. Thus by 1.24, there exists a unique translation g mapping α_i to α_i' for all $i \in [1, \ell]$. Let $S' = S^g$. Then

$$(S')_\mathcal{F} = \{f_i^g \mid i \in [1, \ell]\}$$

and by 4.22, S' is parallel to S. For each $i \in [1, \ell]$, f_i^g is a face contained in

$$\mu(\alpha_i)^g = \mu(\alpha_i') = M_{\hat{f}_i}$$

and parallel to f_i (by 1.45) and hence to $\hat{f}_i$. By 5.12, it follows that $f_i^g \cap \hat{f}_i$ is a face for all $i \in [1, \ell]$.

Suppose that S'' is another sector parallel to S such that for each face $f'' \in (S'')_\mathcal{F}$, $f'' \cap \hat{f}_j$ is a face for some $j \in [1, \ell]$. By 4.22 again, $S'' = S^h$ for some translation h. Thus

$$(S'')_\mathcal{F} = \{f_i^h \mid i \in [1, \ell]\}.$$

Let $i \in [1, \ell]$. Then $f_i^h \cap \hat{f}_j$ is a face for some $j \in [1, \ell]$. By 1.45 and 5.13, therefore, f_i is parallel to $\hat{f}_j$. By 5.9, two faces of the same sector are never parallel. Hence $i = j$. Thus, in particular, $f_i^h \cap f_j \neq \emptyset$, so f_i^h and $\hat{f}_i$ are comural. Therefore $\mu(\alpha_i^h) = \mu(\alpha_i^g)$. By 1.46, therefore, α_i^h equals α_i^g or $-\alpha_i^g$. Since $S'' \subset \alpha_i^h$, the root α_i^h contains a subsector of S (by 4.23). Thus α_i^h and α_i^g both contain subsectors of S. By 4.13, therefore, $\alpha_i^g = \alpha_i^h$. By the uniqueness of g, it follows that $h = g$. Hence $S'' = S'$. □

Proposition 5.23. *Let R be a gem, let d and e be opposite chambers of R, let $S = \sigma(R, d)$, let $S_1 = \sigma(R, e)$, let P and Q be panels of R containing d and e, let $f = \mu(R, P)$ and let $f_1 = \mu(R, Q)$ (so f is a face of S and f_1 is a face of S_1). Suppose that f and f_1 contain subfaces contained in a common wall. Then the panels P and Q are opposite in R.*[2]

Proof. Let d' be the chamber in P distinct from d, let e' be the chamber in Q distinct from e and let f' and f_1' be subfaces of f and f_1 that are contained in a wall M. Let M_P be the unique wall containing P and let M_Q be the unique wall containing Q. Since each face lies in a unique wall, it follows that $M = M_P = M_Q$. In particular, both P and Q are contained in M. By 29.6, this means that the unique reflection s interchanging d and d' also interchanges e and e'. Since reflections are special automorphisms, it follows that s maps the residue R to itself. It follows that e' is opposite d' in R. Hence $Q = \text{op}_R(P)$. □

[2] "The panels P and Q are *opposite in R*" means that $Q = \text{op}_R(P)$, where op_R is as in 29.9.iv with Σ replaced by R. (This makes sense since by 29.7.iii, R is a spherical Coxeter chamber system.)

Chapter Six

Gems

In this chapter we give a characterization of affine Coxeter chamber systems in terms of their gems (6.6 below). We will need this result in Chapter 12.

We continue to let Π be one of the affine Coxeter diagrams in Figure 1.1 and to denote by $\Sigma = \Sigma_\Pi$ the corresponding Coxeter chamber system.

Definition 6.1. Let R and R_1 be two gems of Σ. We will say that R and R_1 are *adjacent* if $R \cap R_1 = \emptyset$ and there exist chambers in R adjacent to chambers in R_1.

Note that if R and R_1 are adjacent gems of Σ and $d \in R$ and $e \in R_1$ are adjacent chambers, then both R and R_1 are o-special (as defined in 1.6), where o is the type of the panel $\{d, e\}$. Thus adjacent gems have the same type.

Proposition 6.2. *Let R and R_1 be two gems of Σ. Then R and R_1 are adjacent (as defined in 6.1) if and only if there exists a unique root α of Σ such that $R \subset \alpha$ and $R_1 \subset -\alpha$.*

Proof. Let k be the length of a gallery $(d_0, d_1, \ldots, d_k)$ of minimal length such that $d_0 \in R$ and $d_k \in R_1$. By 6.1, R and R_1 are adjacent if and only if $k = 1$. By 29.28, k is the number of roots containing R_1 whose opposite contains R. □

Notation 6.3. For each gem R of Σ, we denote by Ξ_R the set of roots of Σ cutting R and by Ω_R the set of roots of Σ containing R. For each ordered pair (R, R_1) of adjacent gems of Σ, we denote by R^{R_1} the set of chambers in R that are adjacent to a chamber in R_1.

Proposition 6.4. *Let R and R_1 be adjacent gems of Σ and let*

$$X = \bigcap_{\alpha \in \Xi_R \cap \Omega_{R_1}} \alpha$$

(with the notation in 6.3). Then $X \cap R = R^{R_1}$.

Proof. Let $d \in R^{R_1}$, so d is adjacent to a chamber e of R_1, and let $\alpha \in \Xi_R \cap \Omega_{R_1}$. By 29.28, α contains d (since $R \not\subset -\alpha$). Thus $d \in X$. Hence $R^{R_1} \subset X \cap R$.

Suppose, conversely, that $d \in R \cap X$. Let $e = \text{proj}_{R_1} d$, let $\gamma = (d, d_1, \ldots, e)$ be a minimal gallery from d to e and let α be the unique root containing d_1

but not d. By 29.7.i, every vertex of R_1 is nearer to d_1 than to d. By 29.6.i, therefore, $R_1 \subset \alpha$. Since $d \notin \alpha$ but $d \in R \cap X$, it follows that α does not cut R. Hence $R \subset -\alpha$ and $d_1 \notin R$.

Suppose that $d_1 \notin R_1$, let d_2 be the chamber in γ that follows d_1 and let β be the unique root containing d_2 but not d_1 (so $\beta \neq \alpha$ since $d_1 \in \alpha$). By 29.7.i again, every vertex of R_1 is nearer to d_2 than to d_1. Also d is nearer to d_1 than to d_2 (since γ is minimal). By two more applications of 29.6.i, we thus have $R_1 \subset \beta$ and $d \in -\beta$. Since $d \in R \cap X$, it follows that β does not cut R. Thus $R \subset -\beta$. Thus both α and β contain R_1 and their opposites both contain R. By 6.2, however, there can be only one such root. We conclude that $d_1 \in R_1$. Thus $d \in R^{R_1}$. Hence $X \cap R \subset R^{R_1}$. □

We observe that by 6.4, the set R^{R_1} is convex for all gems R and R_1.

Proposition 6.5. *Let R and R_1 be adjacent gems of Σ and let $d \in R^{R_1}$ and $e \in R_1^R$. Then d is adjacent to e if and only if the set of roots in $\Xi_R \cap \Xi_{R_1}$ containing d is the same as the set of roots in $\Xi_R \cap \Xi_{R_1}$ containing e.*

Proof. Suppose d and e are adjacent and let α be a root in $\Xi_R \cap \Xi_{R_1}$ containing d. By 29.28, the unique root containing d but not e contains all of R and hence does not equal α. Thus $e \in \alpha$. By symmetry, we conclude that the set of roots in $\Xi_R \cap \Xi_{R_1}$ containing d is the same as the set of roots in $\Xi_R \cap \Xi_{R_1}$ containing e.

Now suppose that d and e are not adjacent. Since $d \in R^{R_1}$ and $e \in R_1^R$, the chamber d is adjacent to a chamber e_1 in R_1 and the chamber e is adjacent to a chamber d_1 in R. Then $d_1 = \text{proj}_R\, e$. Therefore there exists a minimal gallery

$$\gamma := (d, d_2, \ldots, d_1, e)$$

from d to e that passes through d_1 (by 29.7.i). Since the gem R is convex (by 29.7.ii), the subgallery $(d, d_2, \ldots, d_1)$ is contained in R. In particular, $d_2 \in R$. Let α be the unique root of Σ that contains d but not d_2. Since the galleries γ and (e_1, d, d_2) are both minimal, we have $e \in -\alpha$ and $e_1 \in \alpha$ (by 29.6.i). Thus α is a root in $\Xi_R \cap \Xi_{R_1}$ that contains d but not e. □

We come now to the main result of this chapter:

Theorem 6.6. *Let Π' be a second Coxeter diagram from Figure 1.1 (distinct or not distinct from Π) and let Σ' denote the corresponding Coxeter chamber system. Let κ be a bijection from the set of gems of Σ to the set of gems of Σ', let ρ be a bijection from the set of roots of Σ to the set of roots of Σ' and for each gem R of Σ, let τ_R be an isomorphism from R to $\kappa(R)$. Suppose that the following conditions are satisfied:*

(i) $\rho(-\alpha) = -\rho(\alpha)$ for all roots α of Σ and if α and β are roots of Σ such that $\alpha \subset \beta$, then

$$\rho(\alpha) \subset \rho(\beta).$$

(ii) For all gems R, all $d \in R$ and all $\alpha \in [R, d]$ (where $[R, d]$ is as defined in 1.10),

$$\rho(\alpha) \in [\kappa(R), \tau_R(d)].$$

Then there exists a unique isomorphism τ from Σ to Σ' whose restriction to each gem R of Σ equals τ_R. Moreover, $\tau(\alpha) = \rho(\alpha)$ for each root α of Σ.

Proof. We will prove 6.6 in a series of steps.

Lemma 6.7. *Let R be a gem of Σ, let d and e be adjacent chambers of R and let α be the unique root of Σ that contains d but not e. Then $\tau_R(d)$ and $\tau_R(e)$ are adjacent and $\rho(\alpha)$ is the unique root of Σ' that contains $\tau_R(d)$ but not $\tau_R(e)$.*

Proof. Since τ_R is an isomorphism, the chambers $\tau_R(d)$ and $\tau_R(e)$ of $\kappa(R)$ are adjacent. We have $\alpha \in [R, d]$ and $-\alpha \in [R, e]$. By (i) and (ii) of 6.6, it follows that

$$\rho(\alpha) \in [\kappa(R), \tau_R(d)]$$

and

$$-\rho(\alpha) = \rho(-\alpha) \in [\kappa(R), \tau_R(e)].$$

Thus $\rho(\alpha)$ is the unique root of Σ' containing $\tau_R(d)$ but not $\tau_R(e)$. □

Lemma 6.8. *Let R be a gem of Σ and let α be a root of Σ. Then the following hold:*

(i) α cuts R if and only if $\rho(\alpha)$ cuts $\kappa(R)$.
(ii) $R \subset \alpha$ if and only if $\kappa(R) \subset \rho(\alpha)$.

Proof. Suppose that α cuts R. Thus there exists a chamber $d \in R$ such that $\alpha \in [R, d]$. By 6.6.ii, we have $\rho(\alpha) \in [\kappa(R), \tau(d)]$. In particular, $\rho(\alpha)$ cuts $\kappa(R)$.

Suppose next that $R \subset \alpha$. By 1.18, there is a unique root α' parallel to α that cuts R. By 1.2, we have $\alpha' \subset \alpha$. By 6.6.i, therefore, $\rho(\alpha') \subset \rho(\alpha)$. By the conclusion of the previous paragraph, $\rho(\alpha')$ cuts $\kappa(R)$. Thus $\kappa(R) \subset \rho(\alpha)$ by 1.13.

If $R \subset -\alpha$, then

$$\kappa(R) \subset \rho(-\alpha) = -\rho(\alpha)$$

by 6.6.i and the conclusion of the previous paragraph. Thus if α does not cut R, in which case either α or $-\alpha$ contains R, then either $\rho(\alpha)$ or $-\rho(\alpha)$ contains $\kappa(R)$ and therefore $\rho(\alpha)$ does not cut $\kappa(R)$. Hence (i) holds. Similarly, if $R \not\subset \alpha$, then $\kappa(R) \not\subset \rho(\alpha)$. Thus (ii) holds too. □

Lemma 6.9. *Let R and R_1 be two gems of Σ. Then R and R_1 are adjacent (as defined in 6.1) if and only if $\kappa(R)$ and $\kappa(R_1)$ are adjacent.*

Proof. Let α be a root of Σ. By 6.6.i and 6.8.ii, $R \subset \alpha$ if and only if $\kappa(R) \subset \rho(\alpha)$ and $R_1 \subset -\alpha$ if and only if $\kappa(R_1) \subset \rho(-\alpha) = -\rho(\alpha)$. The claim holds, therefore, by 6.2 (since ρ is bijective). □

Notation 6.10. For each special vertex o of Π, let τ_o be the map from Σ to Σ' such that for each chamber x of Σ,

$$\tau_o(x) = \tau_R(x),$$

where R is the unique o-special gem of Σ that contains x.

Lemma 6.11. *Let o be a special vertex of Π, let α be a root of Σ and let d be a chamber in Σ. Then $d \in \alpha$ if and only if $\tau_o(d) \in \rho(\alpha)$, where τ_o is as defined in 6.10.*

Proof. Suppose that $d \in \alpha$ and let R be the unique o-special gem of Σ that contains d. If $R \subset \alpha$, then

$$\tau_o(d) = \tau_R(d) \in \kappa(R) \subset \rho(\alpha)$$

by 6.8.ii. Suppose that R cuts α. In this case, we can choose adjacent chambers x and y such that $x \in \alpha \cap R$ and $y \in -\alpha \cap R$. Since $d \in \alpha$, we have $\text{dist}(d, x) < \text{dist}(d, y)$ (by 29.6.i). Since τ_R is an isomorphism and R is convex (by 29.7.ii), it follows that

$$\text{dist}(\tau_o(d), \tau_o(x)) < \text{dist}(\tau_o(d), \tau_o(y)).$$

By 6.7, $\rho(\alpha)$ is the unique root of Σ' that contains $\tau_o(x)$ but not $\tau_o(y)$. By 29.6.i again, it follows that $\tau_o(d) \in \rho(\alpha)$. Thus $\tau_o(d) \in \rho(\alpha)$ whether or not $R \subset \alpha$. Since d is arbitrary, we conclude that $\tau_o(\alpha) \subset \rho(\alpha)$. Since α is arbitrary, it follows that also $\tau_o(-\alpha) \subset \rho(-\alpha) = -\rho(\alpha)$. Thus if $d \notin \alpha$, then $\tau_o(d) \notin \rho(\alpha)$. □

Lemma 6.12. *Let R and R_1 be two gems of Σ and suppose that R and R_1 are adjacent as defined in 6.1. Then*

$$\tau_R(R^{R_1}) = \kappa(R)^{\kappa(R_1)}.$$

Proof. By 6.9, $\kappa(R)$ and $\kappa(R_1)$ are adjacent gems of Σ'. Let

$$X = \bigcap_{\alpha \in \Xi_R \cap \Omega_{R_1}} \alpha$$

and

$$X' = \bigcap_{\alpha' \in \Xi_{\kappa(R)} \cap \Omega_{\kappa(R_1)}} \alpha',$$

where $\Xi_{\kappa(R)}$ and $\Omega_{\kappa(R_1)}$ are defined as in 6.3 with respect to Σ'. By 6.8, ρ maps the set $\Xi_R \cap \Omega_{R_1}$ bijectively to the set $\Xi_{\kappa(R)} \cap \Omega_{\kappa(R_1)}$, so

$$X' = \bigcap_{\alpha \in \Xi_R \cap \Omega_{R_1}} \rho(\alpha).$$

By 6.11, we have $\tau_R(R \cap \alpha) = \kappa(R) \cap \rho(\alpha)$ for every root α of Σ. Therefore

$$\tau_R(X \cap R) = X' \cap \kappa(R). \tag{6.13}$$

By 6.4, we have $R^{R_1} = X \cap R$ and $\kappa(R)^{\kappa(R_1)} = X' \cap \kappa(R)$. By 6.13, therefore, $\tau_R(R^{R_1}) = \kappa(R)^{\kappa(R_1)}$ □

Lemma 6.14. *Let o be a special vertex of Π and let τ_o be as in 6.10. Then τ_o is an isomorphism from Σ to Σ'.*

Proof. Since κ is a bijection, so is τ_o. Let d and e be chambers of Σ and let R and R_1 be the unique o-special gems containing d and e. If $R = R_1$, then d and e are adjacent if and only if $\tau_o(d)$ and $\tau_o(d)$ are adjacent (since τ_R is an isomorphism).

Suppose that $R \neq R_1$. By 6.1 and 6.9, we can assume that R and R_1 are adjacent and that $\kappa(R)$ and $\kappa(R_1)$ are adjacent, and by 6.12, we can assume that $d \in R^{R_1}$ and $e \in R_1^R$. We know (by 6.8) that

$$\rho(\Xi_R \cap \Xi_{R_1}) = \Xi_{\kappa(R)} \cap \Xi_{\kappa(R_1)}.$$

Thus by 6.11, α is a root in $\Xi_R \cap \Xi_{R_1}$ containing d, respectively, e, if and only if $\rho(\alpha)$ is a root in $\Xi_{\kappa(R)} \cap \Xi_{\kappa(R_1)}$ containing $\tau_o(d)$, respectively, $\tau_o(e)$. Therefore

$$\{\alpha \in \Xi_R \cap \Xi_{R_1} \mid d \in \alpha\} = \{\alpha \in \Xi_R \cap \Xi_{R_1} \mid e \in \alpha\}$$

if and only if

$$\{\alpha' \in \Xi_{\kappa(R)} \cap \Xi_{\kappa(R_1)} \mid \tau_o(d) \in \alpha'\} = \{\alpha' \in \Xi_{\kappa(R)} \cap \Xi_{\kappa(R_1)} \mid \tau_o(e) \in \alpha'\}.$$

By 6.5, we conclude that d and e are adjacent in Σ if and only if $\tau_o(d)$ and $\tau_o(e)$ are adjacent also when $R \neq R_1$. □

Proposition 6.15. *The map τ_o defined in 6.10 is independent of the choice of the special vertex o.*

Proof. Let o, o_1 be two special vertices of Π. Let d be a chamber of Σ and let R be the unique o-special gem containing d. By 6.11, we have

$$\tau_{o_1}(\alpha) = \rho(\alpha)$$

for all roots α of Σ. By 6.6.ii, it follows that

$$[\kappa(R), \tau_o(d)] = \rho([R, d]) = \tau_{o_1}([R, d]).$$

Since τ_{o_1} is an isomorphism (by 6.14), we have

$$\tau_{o_1}([R, d]) = [\tau_{o_1}(R), \tau_{o_1}(d)].$$

Thus

$$[\kappa(R), \tau_o(d)] = [\tau_{o_1}(R), \tau_{o_1}(d)].$$

By 1.12, it follows that $\kappa(R) = \tau_{o_1}(R)$ and therefore $\tau_o(d) = \tau_{o_1}(d)$. Since d is arbitrary, we conclude that $\tau_o = \tau_{o_1}$. □

Now choose a special vertex o and let $\tau := \tau_o$ be as defined in 6.10. By 6.14, τ is an isomorphism from Σ to Σ'. By 6.10, τ restricted to an arbitrary o-special gem R agrees with τ_R. By 6.15, it follows that τ restricted to an *arbitrary* gem R agrees with τ_R. By 6.11, $\tau(\alpha) = \rho(\alpha)$ for each root α of Σ.

This concludes the proof of 6.6. □

Chapter Seven

Affine Buildings

We now turn to affine buildings. Our goal in this chapter is to prove 7.24. This result is crucial for the construction (described in Chapter 8) of the building at infinity of an affine building. It was first proved in 2.9.6 of [6].

Let Δ be a thick irreducible affine building. More precisely, we suppose that Δ is a thick building of type Π (as defined in 29.10), where Π is one of the affine Coxeter diagrams in Figure 1.1. Let $W = W_\Pi$ be the Coxeter group associated with Π; we identify W with $\text{Aut}^\circ(\Sigma_\Pi)$ as described in 29.4. Let δ be the Weyl distance function of Δ as described in 29.10.

All the apartments of Δ are isomorphic to the Coxeter chamber system Σ_Π. This observation allows us to apply all our results about Σ_Π from previous chapters to the individual apartments of Δ.

Definition 7.1. A *gem* of Δ is a residue of type $I_o := I\backslash\{o\}$ for some special vertex o of Π (as defined in 1.1), where I is the vertex set of Π. A gem of type I_o for some special vertex o is called *o-special.*

Definition 7.2. We will say that a residue D of Δ is *cut* by an apartment A if $A \cap D \neq \emptyset$. If a residue D of Δ is cut by an apartment A, then, in fact, $D \cap A$ is a residue of A of the same type as D (by 29.13.iv).

In the proof of the next result, we refer to the retraction map; see 8.16 of [37] for the definition.[1]

Proposition 7.3. *Suppose that S is a sector in an apartment A and that S is contained in a second apartment A'. Then there is a special isomorphism from A to A' acting trivially on S. In particular, S is also a sector in A'.*

Proof. Let x be a chamber of S. By 8.19 of [37], $\text{retr}_{A',x}$ is a special isomorphism from A to A'. By 8.17 of [37], this map acts trivially on $A \cap A'$. □

Definition 7.4. A *sector* of Δ is a sector in one of its apartments. Let S be a sector, let A be an apartment containing it, let R be the unique gem of Δ such that $A \cap R$ is the terminus of S in A and let $d \in R \cap A$ denote the apex of S. By 7.3, S is a sector in every apartment that contains it, and the gem R and the apex d are independent of the choice of this apartment.

[1]The retraction map will play a central role in this chapter. We suggest, in fact, that the reader review the basic properties of the retraction map proved in 8.17–8.19 of [37] before proceeding.

From now on, we will refer to R (rather than $R \cap A$ which *does* depend on A) as the *terminus* of S. We will denote S by $\sigma_A(R, d)$. Thus $S = \sigma_{A'}(R, d)$ also for any other apartment A' containing S.

Notation 7.5. Let A be an apartment of Δ, let R be a gem cut by A and let d be a chamber of $A \cap R$. We denote by $[R, d]_A$ the set of roots of A containing d but not some chamber of $R \cap A$ adjacent to d.

Proposition 7.6. *Let u be a chamber and let S be a sector of Δ. Then there exists an apartment containing both u and some subsector of S.*[2]

Proof. Let A be an apartment containing S. By induction with respect to the minimal length of a gallery from u to a chamber of A, it suffices to assume that $u \notin A$ but that u is contained in a panel P containing chambers of A. Let x and y be the two chambers in $P \cap A$ and let α be the unique root of A containing x but not y. By 4.13, we can assume (by interchanging x with y if necessary) that α contains a subsector S' of S. By 29.33, there exists an apartment containing $\alpha \cup \{u\}$. □

Proposition 7.7. *Let S be a sector of an apartment A, let $u \in \Delta$ and let S' and S'' be two subsectors of S such that there is an apartment containing u and S' and an apartment containing u and S''. Then*

$$\mathrm{retr}_{A,x}(u) = \mathrm{retr}_{A,y}(u)$$

for all $x \in S'$ and all $y \in S''$.

Proof. Let $S_1 = S' \cap S''$. By 4.8, S_1 is a subsector of S. Choose chambers $x \in S'$ and $z \in S_1$. It will suffice to show that $\mathrm{retr}_{A,x}(u) = \mathrm{retr}_{A,z}(u)$. Let A' be an apartment containing u and S'. Let $\phi = \mathrm{retr}_{A,x}$. By 8.19 of [37], ϕ is a special isomorphism from A' to A, and by 8.17 of [37], ϕ fixes z. By 8.2 of [37], therefore, $\delta(u, z) = \delta(u^\phi, z)$. By 8.17 of [37], it follows that $u^\phi = \mathrm{retr}_{A,z}(u)$. □

Definition 7.8. Suppose that A is an apartment and that S is a sector in A. For each chamber $u \in \Delta$, let

$$\mathrm{retr}_{A,S}(u) = \mathrm{retr}_{A,x}(u),$$

where x is an arbitrary chamber contained in some subsector of S lying in a common apartment with u. By 7.6, there always exists such chambers x, and by 7.7, $\mathrm{retr}_{A,S}(u)$ is independent of the choice of x. The map $\mathrm{retr}_{A,S}$ will be called the *sector retraction* onto A with respect to S.[3]

Proposition 7.9. *Let A be an apartment, let S be a sector of A and let U be a finite set of chambers of Δ. Then there exists a subsector S_1 of S such that*

$$\mathrm{retr}_{A,S}(u) = \mathrm{retr}_{A,x}(u)$$

for all $x \in S_1$ and all $u \in U$.

---

[2]Alternatively, we might say that u *cohabits* with a subsector of S.

[3]Curiously, this important tool will appear only in this chapter.

Proof. This holds by 4.8 and 7.8. □

Proposition 7.10. *Sector retractions are special homomorphisms (as defined in 29.2).*

Proof. Let A be an apartment and let S be a sector in A. By 8.18 of [37], $\text{retr}_{A,x}$ is a special homomorphism for all $x \in A$. Let u, v be adjacent chambers of Δ. By 7.9, there exists a chamber $x \in S$ such that the restriction of $\text{retr}_{A,S}$ to the set $\{u, v\}$ equals the restriction of $\text{retr}_{A,x}$ to the set $\{u, v\}$. Thus $\text{retr}_{A,S}$ is a special homomorphism. □

Lemma 7.11. *Let A be an apartment of Δ, let S be a sector of A and let S' be another sector of Δ. Then there exists a subsector S'_1 of S' such that* $\text{retr}_{A,S}$ *restricted to S'_1 is an isometry.*[4]

Proof. Let $\rho = \text{retr}_{A,S}$. Let R_0 and d be the terminus and the apex of the sector S, let $R = R_0 \cap A$ (so R is a gem of A) and let d' be the apex of S'. Let o be the special vertex of Π such that the gem R is o-special (as defined in 1.6). Let A' be an apartment containing S' and let ϕ be the unique special isomorphism from A' to A mapping d' to $z := \rho(d')$. Let $x \mapsto \bar{x}$ denote the map that sends a chamber of A to its special translate in R (as defined in 1.31). Let W_R be the stabilizer of R in the Coxeter group W. The group W acts sharply transitively on A and preserves types. It follows that W_R acts sharply transitively on R. For each chamber $x \in S'$, there is thus a unique element $\omega(x)$ of W_R such that

$$\overline{\phi(x)}^{\,\omega(x)} = \overline{\rho(x)}. \tag{7.12}$$

Choose adjacent vertices $x, y \in S'$ and let i denote the type of the panel containing them. By 29.6.i, we can assume that

$$\text{dist}(d', x) < \text{dist}(d', y). \tag{7.13}$$

Let h be the unique element in W_R such that

$$\overline{\phi(x)}^{\,h} = \overline{\phi(y)}. \tag{7.14}$$

Suppose first that $\rho(x) \neq \rho(y)$. By 7.10, $\rho(x)$ and $\rho(y)$ are i-adjacent. By 7.12, we have

$$\overline{\phi(x)}^{\,\omega(x)} = \overline{\rho(x)} \tag{7.15}$$

and

$$\overline{\phi(y)}^{\,\omega(y)} = \overline{\rho(y)}. \tag{7.16}$$

If $i \neq o$, then by 1.32, $\overline{\phi(y)}$ is the unique chamber of R that is i-adjacent to $\overline{\phi(x)}$ and $\overline{\rho(y)}$ is the unique chamber of R that is i-adjacent to $\overline{\rho(x)}$, so 7.15 implies that

$$\overline{\phi(y)}^{\,\omega(x)} = \overline{\rho(y)}. \tag{7.17}$$

---

[4]A map π from a subset X of Δ to Δ is an *isometry* if

$$\delta(x, y) = \delta(x^\pi, y^\pi)$$

for all $x, y \in X$.

Now suppose that $i = o$. By 1.30, there exists unique elements t_1, t_2 in $T \cap W$ such that both $\phi(x)^{t_1}$ and $\rho(x)^{t_2}$ are contained in the gem R. We have

$$\overline{w^{t_1}} = \overline{w^{t_2}} = \bar{w}$$

for every chamber w in A (by 1.31). Thus the chamber $\phi(y)^{t_1}$ is o-adjacent to $\phi(x)^{t_1} = \overline{\phi(x)}$ and the chamber $\rho(y)^{t_2}$ is o-adjacent to $\rho(x)^{t_2} = \overline{\rho(x)}$, and if δ is as in 1.33, then

$$\delta\left(\overline{\phi(x)}\right) = \overline{\phi(y)^{t_1}} = \overline{\phi(y)}$$

and, by 7.15,

$$\delta\left(\overline{\phi(x)}^{\omega(x)}\right) = \delta\left(\overline{\rho(x)}\right) = \overline{\rho(y)^{t_2}} = \overline{\rho(y)}.$$

Hence 7.17 holds by 1.33 with $\omega(x)$ in place of g. Therefore 7.17 holds whether or not $i = o$. By 7.16 and 7.17, it follows that $\omega(x) = \omega(y)$ (since W_R acts sharply transitively on R). We conclude that

$$\text{(7.18)} \qquad \text{if } \rho(x) \neq \rho(y), \text{ then } \operatorname{dist}\left(d, \bar{z}^{\omega(x)}\right) = \operatorname{dist}\left(d, \bar{z}^{\omega(y)}\right).$$

Suppose now that $\rho(x) = \rho(y)$. By 7.12 and 7.14, we have

$$\overline{\phi(x)}^{\omega(x)} = \overline{\phi(y)}^{\omega(y)} = \overline{\phi(x)}^{h\omega(y)}.$$

Thus

$$\text{(7.19)} \qquad \omega(x) = h\omega(y),$$

again because the identity is the only element of W_R fixing an element of R. Let p be the unique chamber of A that is i-adjacent to $\rho(x)$, let P be the unique panel of Δ containing $\rho(x)$ and p and let α denote the unique root of A that contains $\rho(x)$ but not p. Suppose that α contains a subsector of S. By 4.8 and 7.9, there exists a chamber $w \in \alpha \cap S$ such that $\rho(u) = \operatorname{retr}_{A,w}(u)$ for $u = x$ and y. Since $w \in \alpha$, we have

$$\operatorname{dist}(w, \rho(x)) < \operatorname{dist}(w, p)$$

by 29.6.i. By 29.10.v, it follows that $\operatorname{proj}_P w = \rho(x) = \rho(y)$. There are thus no minimal galleries from w to $\rho(x) = \rho(y)$ whose type ends in i. By 8.17 of [37] (and 29.10.i), it follows that there are no minimal galleries from w to x or from w to y whose type ends in i. By 29.10.v again, however, this implies that $x = y = \operatorname{proj}_Q w$, where Q is the i-panel containing x and y. Since x and y are distinct, we conclude that the root α contains no subsector of S.

Let g be the unique element of $g \in T \cap W$ mapping $\rho(x)$ to $\overline{\rho(x)}$ and let $\alpha' = \alpha^g$. By 4.12.i and the conclusion of the previous paragraph, we have

$$\text{(7.20)} \qquad d \notin \alpha'.$$

Suppose that $i = o$. Let R_0' be the unique o-special gem containing $\rho(x)$ and let $R' = R_0' \cap A$. Then $p \notin R'$ and hence $\rho(x) = \operatorname{proj}_{R'} p$. Therefore $R' \subset \alpha$ (by 29.6.i and 29.10.v). This implies, however, that

$$R = (R')^g \subset \alpha^g = \alpha'.$$

By 7.20, we conclude that $i \neq o$. Hence $\bar{p} = p^g$. Therefore the chambers $\overline{\rho(x)}$ and $\bar{p}$ are i-adjacent (since $g \in W$) and α' contains $\overline{\rho(x)}$ but not $\bar{p}$. Thus $(\alpha')^{\omega(x)^{-1}}$ contains the chamber

$$\overline{\rho(x)}^{\omega(x)^{-1}} = \overline{\phi(x)}$$

but not the chamber of A that is i-adjacent to $\overline{\phi(x)}$. This chamber is $\overline{\phi(y)}$. Hence $\alpha' = \beta^{\omega(x)}$ and therefore

$$d^{\omega(x)^{-1}} \in -\beta, \tag{7.21}$$

where β is the unique root of A containing $\overline{\phi(x)}$ but not $\overline{\phi(y)}$. By 4.19 and 7.13, there exists a subsector S'' of S' such that for each $u \in S''$, there exists a minimal gallery from u to d' passing through y and then x. In particular, every chamber in $\phi(S'')$ is nearer to $\phi(y)$ than to $\phi(x)$ and hence (by 29.6.i)

$$\phi(S'') \subset -\beta',$$

where β' is the unique root of A that contains $\phi(x)$ but not $\phi(y)$. The root β' is a translate in A of the root β and the sector $\phi(S')$ (whose apex is z) is a translate of a sector S_0 having apex $\bar{z}$. By 4.6 and 4.8, $S_0 \cap \phi(S'')$ is a subsector of S_0. By 4.12.ii, we conclude that

$$\bar{z} \in -\beta. \tag{7.22}$$

Since $\overline{\phi(x)}$ and $\overline{\phi(y)}$ are i-adjacent, the element h in 7.14 must be the unique reflection in W_R interchanging β and $-\beta$. By 3.18 of [37], 7.21 and 7.22, therefore,

$$\operatorname{dist}\big(d^{\omega(x)^{-1}}, \bar{z}\big) < \operatorname{dist}\big(d^{\omega(x)^{-1}h}, \bar{z}\big).$$

By 7.19, we conclude that

$$\text{if } \rho(x) = \rho(y), \text{ then } \operatorname{dist}\big(d, \bar{z}^{\omega(x)}\big) < \operatorname{dist}\big(d, \bar{z}^{\omega(y)}\big). \tag{7.23}$$

Finally, let

$$\psi(u) = \operatorname{dist}\big(d, \bar{z}^{\omega(u)}\big)$$

for all $u \in S'$. Since $\psi(u) \leq \operatorname{diam}(R) < \infty$ for all $u \in S'$, we can choose $v \in S'$ such that

$$\psi(u) \leq \psi(v)$$

for all $u \in S'$. By 4.19, there exists a subsector S_1' of S' such that for each $u \in S_1'$, there is a minimal gallery from d' to u that passes through v. Suppose that w and u are two adjacent chambers in S_1'. Since one of these two chambers is nearer to d' than the other (by 29.6.i), we can assume without loss of generality that there is a gallery

$$\gamma = (d', \ldots, v, \ldots, w, u)$$

from d' to u passing through *both* v and w. By 4.1 and 29.6.ii, sectors are convex. Thus $\gamma \subset S'$. We conclude that if (x, y) is a successive pair of chambers on γ, then x and y satisfy 7.13. By 7.18 and 7.23 and the choice of v, it follows that $\rho(w) \neq \rho(u)$.

Next suppose that w and u are any two chambers of S_1'. Since the sector S_1' is convex, there is a minimal gallery γ' from w to u contained in S_1'. Let f' be the type of γ'. By 29.10.i, f' is reduced. By the conclusion of the previous paragraph, $\rho(\gamma')$ is a gallery from $\rho(w)$ to $\rho(u)$ which is also of type f'. By 29.10, this means that $\delta(\rho(u), \rho(w)) = r_{f'} = \delta(u, w)$, where $r_{f'}$ is as in 29.8. $\square$

We come now to the main theorem of this chapter.

Theorem 7.24. *Let S and S' be two sectors in Δ. Then there exists an apartment A that contains subsectors of both S and S'.*

Proof. Let A be an apartment containing S and let $\rho = \mathrm{retr}_{A,S}$. By 7.11, there exists a subsector S_1' of S' such that the restriction of ρ to S_1' is an isometry. By 29.13.i, it follows that ρ is a special isomorphism from S_1' to a subsector of some apartment. By 7.3, therefore, $\rho(S_1')$ is, in fact, a subsector of A.

Let z be the apex of S_1'. Let R_0 be the terminus of S, let R_0' be the gem having the same type as R_0 that contains $\rho(z)$, let $R = R_0 \cap A$ and let $R' = R_0' \cap A$. Thus R and R' are both gems of A. By 1.22, there exists a unique translate S_1 of S in A whose terminus is R'. By 4.6, $S \cap S_1$ is a sector. By 7.6, there is an apartment A' of Δ that contains both z and a subsector S_2 of $S \cap S_1$. Suppose that

$$\rho(u) = \mathrm{retr}_{A,c}(u) \tag{7.25}$$

for all $u \in S_1'$ and all $c \in S_2$. Then $\delta(c, \rho(u)) = \delta(c, u)$ for all $u \in S_1'$ and all $c \in S_2$ (by 8.17 in [37]). In particular, if $\rho(u) \in S_2$ for some $u \in S_1'$, then $\delta(\rho(u), u) = \delta(\rho(u), \rho(u)) = 1$ and hence $u = \rho(u) \in S_2$. Since the restriction of ρ to S_1' is injective, it follows that there exists an isometry π from $S_2 \cup \rho(S_1')$ to $S_2 \cup S_1'$ such that $\pi(x) = x$ if $x \in S_2$ and $\pi(\rho(u)) = u$ if $u \in S_1'$. The image of π contains $S_2 \cup S_1'$. By 29.13.i, π extends to a special isomorphism from A to Δ. The image of this extension is an apartment. It thus suffices to show that 7.25 holds.

Choose $u \in S_1'$ and $c \in S_2$. We proceed by induction with respect to $\mathrm{dist}(z, u)$. By 7.8, we know that 7.25 holds for $u = z$ since z and S_2 lie in the apartment A'. Suppose that $\mathrm{dist}(z, u) > 0$, let

$$\gamma = (z, \ldots, w, u)$$

be a minimal gallery from z to u, let $w' = \rho(w)$ and let $u' = \rho(u)$. By induction, we can assume that $\rho(w) = \mathrm{retr}_{A,c}(w)$ and hence

$$\delta(c, w) = \delta(c, w') \tag{7.26}$$

by 8.17 of [37]. Let P be the unique panel of Δ that contains w and u and let i be the type of P. Since the restriction of ρ to S_1' is an isometry, the chambers u' and w' are i-adjacent. Since $\mathrm{retr}_{A,c}$ is a special homomorphism, the chambers $w' = \mathrm{retr}_{A,c}(w)$ and $\mathrm{retr}_{A,c}(u)$ are either equal or i-adjacent;

if they are i-adjacent, then $\text{retr}_{A,c}(u)$ must be equal to u' since u' is the unique chamber of A that is i-adjacent to w'. It thus suffices to show that $\text{retr}_{A,c}(w) \neq \text{retr}_{A,c}(u)$. By 8.17 of [37], this is equivalent to showing that

$$\delta(c, w) \neq \delta(c, u). \tag{7.27}$$

If w or u equals $\text{proj}_P c$, then $\text{dist}(c, w) \neq \text{dist}(c, u)$ and hence 7.27 holds by 7.8 of [37]. We thus assume that

$$w \neq \text{proj}_P c \tag{7.28}$$

and observe that under this assumption it will suffice to show that

$$u = \text{proj}_P c. \tag{7.29}$$

Let $p = \text{proj}_P c$, let γ_1 be a minimal gallery from c to p and let P' be the unique panel containing u' and w'. By 7.28, $w \neq p$. By 29.10.v, therefore, (γ_1, w) is a minimal gallery. Let f be its type. Then f is reduced and $\delta(c, w) = r_f$ (by 29.10.i). Hence $\delta(c, w') = r_f$ by 7.26. There thus exists a minimal gallery γ_2 of type f from c to w' (again by 29.10.i). Since the word f ends in i, the penultimate chamber in γ_2 lies in P'. Thus $w' \neq \text{proj}_{P'} c$. By 29.10.v, it follows that c is not nearer to w' than it is to u'. Hence $c \notin \alpha$, where α is the unique root of A containing w' but not u'.

By 7.26 (and 7.8 of [37]), we have $\text{dist}(c, w) = \text{dist}(c, w')$. By 7.10, it follows that (γ_1, w) is mapped by ρ to a gallery of type f from c to w'. The penultimate chamber in such a gallery must be u', the unique chamber in A that is i-adjacent to w'. Thus

$$\rho(p) = u'. \tag{7.30}$$

The sector S_2 is contained in a sector having terminus R'. We have

$$(\rho(z), \ldots, w', u') = \rho(\gamma).$$

Since the restriction of ρ to S_1' is an isometry, the type of $\rho(\gamma)$ is the same as the type of γ. Since γ is minimal, it follows (again by 29.10.i) that $\rho(\gamma)$ is also minimal. By 4.3, $\rho(z) = \text{proj}_{R'} u'$. Thus for each $x \in R'$, there exists a minimal gallery from x to u' that passes through both $\rho(z)$ and w' (by 29.10.v). It follows that every chamber of R' is nearer to w' than to u'. Hence $R' \subset \alpha$. In particular, α contains the apex of the translate S_1 of S which we denote by q. Let α' be the unique root of A parallel to α that cuts R'. Since $c \in S_2 \subset S_1$ but $c \notin \alpha$, we have $S_1 \not\subset \alpha$. By 4.14, therefore, $q \notin \alpha'$. By 4.12.i, it follows that $-\alpha$ contains a subsector of S_1. Hence by 4.8, $-\alpha$ contains a subsector S_3 of S_2. By 4.8 and 7.9, there exists a chamber e in $S_3 \subset -\alpha$ such that

$$\rho(x) = \text{retr}_{A,e}(x)$$

for both $x = u$ and $x = p$. By 7.30, therefore,

$$\text{retr}_{A,e}(p) = \rho(p) = u' = \rho(u) = \text{retr}_{A,e}(u). \tag{7.31}$$

Since $\text{retr}_{A,e}(u) = u'$ and $\text{retr}_{A,e}$ is a special homomorphism, we have $\text{retr}_{A,e}(P) \subset P'$. Since $\text{retr}_{A,e}$ preserves the distance to e (by 7.8 and 8.17 of

[37]), it must therefore map $\text{proj}_P\, e$, but no other chamber on P, to $\text{proj}_{P'}\, e$. Since $e \in -\alpha$, we have

$$\text{dist}(e, u') < \text{dist}(d, w')$$

by 29.6.i and 29.13.iii. By 29.10.v, therefore, we have $\text{proj}_{P'}\, e = u'$. There is thus a unique chamber in P mapped by $\text{retr}_{A,e}$ to u'. Hence $u = p$ by 7.31. Therefore 7.29 holds. □

* * *

We close this chapter with a brief look at the very simplest affine buildings:

Example 7.32. Suppose that $\Pi = \tilde{\mathsf{A}}_1$. Let Γ denote the graph whose vertices are the panels of Δ, where two panels are joined by an edge of Γ if they contain a common chamber. If two panels of any building have a non-empty intersection, then the two panels must have different colors and the intersection must consist of a single chamber. Since Δ has only two colors, these colors must therefore alternate in every nonrepeating sequence of adjacent panels in Γ. Let $r\colon M_I \to W_\Pi$ be as in 29.8. Since the label on the one edge of $\tilde{\mathsf{A}}_1$ is ∞, the restriction of the map r to the set of reduced words in M_I is a bijection. By 7.7 of [37], every two chambers in Δ are connected by an unique minimal gallery. It follows that Γ has no circuits. In other words, Γ is a tree.

We define a tree to be *thick* if every vertex has at least three neighbors. We are assuming that all buildings considered in this monograph are thick. This means (by 29.2) that each panel of Δ contains at least three chambers. It follows that Γ is thick.

Suppose, conversely, that we start with a thick tree Γ. Trees are connected and bipartite, so there is a unique decomposition of the vertex set of Γ into two subsets V_1 and V_2 such that every edge of Γ contains exactly one vertex from each of these two subsets. Let Δ be the edge set of Γ. We declare that two distinct edges e and f are i-adjacent for $i = 1$ or 2 if their intersection consists of a vertex of V_i. Then Δ is a building of type $\Pi = \tilde{\mathsf{A}}_1$.

The two constructions just described are inverses of each other. Thus, in conclusion, buildings of type $\tilde{\mathsf{A}}_1$ and thick trees are equivalent notions. A tree will be called *thin* if every vertex is adjacent to exactly two vertices. If we think of a thick tree as a building of type $\tilde{\mathsf{A}}_1$, then the apartments are the thin subtrees.[5]

We suggest that the reader verify 7.24 in the case $\Pi = \tilde{\mathsf{A}}_1$ with a few simple diagrams.

---

[5] More precisely, an apartment is the set of edges on a thin subgraph of Γ, but there is no reason to be so precise.

Chapter Eight

The Building at Infinity

In this chapter we show that the "boundary" of an affine building is a building of spherical type. This "building at infinity" plays a central role in the study of affine buildings. The notion of the building at infinity generalizes, as we will see, the notion of the celestial sphere of a Euclidean space. The main results of this chapter are 8.24 and 8.36.

We continue to assume that Π is one of the affine Coxeter diagrams in Figure 1.1 and that Δ is a thick building of type Π.

Definition 8.1. Let A be an apartment of Δ, let R be a gem and let P be a panel that is both contained in R and cut by A (as defined in 7.2). Let $d \in P \cap A$, let $S = \sigma_A(R, d)$ be as in 7.4, let α be the unique root of A such that $\alpha \cap P = \{d\}$ and let $\mu(\alpha)$ be the wall of α as defined in 29.32. Let $\mu_A(R, P)$ be the set of panels Q in the wall $\mu(\alpha)$ such that $|Q \cap S| = 1$. Thus

$$\{Q \cap A \mid Q \in \mu_A(R, P)\}$$

is the $P \cap A$-face of S in A as defined in 5.1. By 7.3, $\mu_A(R, P) = \mu_{A'}(R, P)$ if A' is any other apartment containing the sector S. Thus we can set $\mu(S, P) = \mu_A(R, P)$ and we can call $\mu(S, P)$ simply the P-*face* of S. A *face* is a set of the form $\mu(S, P)$ for some sector S and some panel P containing its apex and contained in its terminus. We call the set of chambers contained in some panel of a face the *chamber set* of the face. As in 29.22 and 5.7, we say that two walls (or two faces) are *parallel* if their chambers sets are parallel in the sense of 1.44. If u is a chamber and f is a face, we will write $u \in f$ if u is contained in the chamber set of f.

The walls and faces of Δ will play a central role in the next several chapters.

Proposition 8.2. *Let S and S' be two sectors of Δ (not necessarily contained in the same apartment). Then S and S' are parallel (as defined in 1.44) if and only if $S \cap S'$ contains a sector.*

Proof. Suppose that $S \cap S'$ contains a subsector S''. By two applications of 4.23, S'' is parallel to S and to S'. Thus S is parallel to S'.

Suppose, conversely, that S is parallel to S'. By 7.24, there exists an apartment A that contains subsectors S_1 of S and S_1' of S'. By 4.23 again, S is parallel to S_1 and S' is parallel to S_1'. Thus S_1 is parallel to S_1'. By one more application of 4.23, it follows that $S_1 \cap S_1'$ is a sector contained in $S \cap S'$. □

Definition 8.3. Suppose $\mathcal{A}$ is an arbitrary set of apartments in Δ. An apartment of Δ will be called an $\mathcal{A}$-apartment if it is contained in $\mathcal{A}$, a sector will be called an $\mathcal{A}$-sector if it is contained in an $\mathcal{A}$-apartment, a root will be called an $\mathcal{A}$-root if it is contained in an $\mathcal{A}$-apartment, a wall will be called an $\mathcal{A}$-wall if it is the wall of an $\mathcal{A}$-root and a face will be called an $\mathcal{A}$-face if it is the face of an $\mathcal{A}$-sector.

The following definition is important:

Definition 8.4. A *system of apartments* of Δ is a set $\mathcal{A}$ of apartments satisfying the following two conditions:

(i) Every two chambers of Δ lie in some $\mathcal{A}$-apartment.
(ii) Every two $\mathcal{A}$-sectors contain subsectors that are contained in a common $\mathcal{A}$-apartment.

Definition 8.5. By 7.24 and 29.13.ii, the set of *all* apartments of Δ is a system of apartments. This system of apartments is called the *complete* system of apartments.

Definition 8.6. Let o be a special vertex of Π and let I_o be as in 7.1. We will call a sector *o-special* if its terminus is o-special.

For the rest of this chapter, we fix a system of apartments $\mathcal{A}$ and a special vertex o of Π and let I and I_o be as in 7.1.

Proposition 8.7. *Every $\mathcal{A}$-sector is parallel to an o-special $\mathcal{A}$-sector.*

Proof. Let S be an $\mathcal{A}$-sector and let A be an apartment in $\mathcal{A}$ containing S. By 1.9, there is a translate S' of S in A whose terminus is o-special. By 4.22, S' is parallel to S. □

Definition 8.8. Let $S = \sigma_A(R, d)$ be an o-special sector and let $i \in I_o$. The *i-face* of S is the P-face $\mu_A(S, P)$, where P is the i-panel containing d.

We now define an edge-colored graph $\Delta_{\mathcal{A}}^\infty$ with index set I_o:

Definition 8.9. Two parallel classes x and x_1 of $\mathcal{A}$-sectors will be called *i-adjacent* for some $i \in I_o$ (and we write $x \sim_i x_1$) if $x \neq x_1$ and for every o-special $\mathcal{A}$-sector $S \in x$ and every o-special $\mathcal{A}$-sector $S_1 \in x_1$, the i-faces of S and S_1 are parallel. Let $\Delta_{\mathcal{A}}^\infty$ be the edge-colored graph with index set I_o whose vertices are the parallel classes of $\mathcal{A}$-sectors, where two parallel classes x and x_1 of $\mathcal{A}$-sectors are joined by an edge of color $i \in I_o$ if and only if they are i-adjacent.

Proposition 8.10. *Let S and S_1 be parallel o-special $\mathcal{A}$-sectors. Then the i-faces of S and S_1 are parallel for each $i \in I_o$.*

Proof. Let A and A_1 be $\mathcal{A}$-apartments containing S and S_1. By 4.10 and 8.2, there is an o-special sector $\hat{S}$ contained in $S \cap S_1$. By 1.30, there is a special translate of $\hat{S}$ in A having the same terminus as S. By 4.3, it follows that $\hat{S}$ is, in fact, a special translate of S in A. Thus for each $i \in I_o$, the i-face of $\hat{S}$ is a translate of the i-face of S. Hence by 1.45, the i-face of $\hat{S}$ is parallel to the i-face of S for each $i \in I_o$. Similarly, $\hat{S}$ is a special translate of S_1 in A_1 and hence the i-face of $\hat{S}$ is parallel to the i-face of S_1 for each $i \in I_o$. □

Corollary 8.11. *Let x, x' be distinct vertices of $\Delta_{\mathcal{A}}^{\infty}$ and let $i \in I_o$. Then $x \sim_i x'$ if and only if for some o-special $\mathcal{A}$-sector $S \in x$ and some o-special $\mathcal{A}$-sector $S' \in x'$, the i-face of S and the i-face of S' are parallel.*

Proof. This holds by 8.9 and 8.10. □

Proposition 8.12. *Let H be the subgroup of* Aut(Π) *defined in 1.27. Let o' be another special vertex of* Π *and let $I_{o'}$ and $\sim_i'$ for all $i \in I_{o'}$ be as in 7.1 and 8.9 with o' in place of o. Then there is a unique element τ of H mapping o to o' such that for all $x, x' \in \Delta_{\mathcal{A}}^{\infty}$ and all $i \in I_o$,*

$$x \sim_i x' \text{ if and only if } x \sim_{\tau(i)}' x'.$$

Proof. Let $A \in \mathcal{A}$. By 1.28, there is a unique element τ of H mapping o to o'. By 1.27, there exists a translation g of A that is also a τ-automorphism of A. Thus g maps o-special gems (of A) to o'-special gems (of A). Hence if S is an o-special sector in A and if f is its i-face for some $i \in I_o$, then S^g is an o'-sector and f^g is its $\tau(i)$-face. By 1.45, every sector S of A is parallel to S^g and every face f of A is parallel to f^g. By 8.11, it follows that if $x \sim_i x'$ for some $x, x' \in \Delta_{\mathcal{A}}^{\infty}$ and for some $i \in I_o$, then $x \sim_{\tau(i)}' x'$. Reversing the role of o and o' in this argument, we conclude that also the converse holds. □

By 8.9, $\Delta_{\mathcal{A}}^{\infty}$ is an edge-colored graph with index set I_o, where two vertices x, x' of $\Delta_{\mathcal{A}}^{\infty}$ are connected by an edge with the color i whenever $x \sim_i x'$. By 8.12, this graph is, up to a relabeling of the colors, independent of the choice of the special vertex o of Π.

Proposition 8.13. *Let $x, x_1 \in \Delta_{\mathcal{A}}^{\infty}$ and suppose that $x \sim_i x_1$ and $x \sim_{i'} x_1$ for $i, i' \in I_o$. Then $i = i'$.*

Proof. By 4.23 and 8.4.ii, there exists an $\mathcal{A}$-apartment A containing sectors $S \in x$ and $S_1 \in x_1$. By 4.10 and 4.23, we can assume that S and S_1 are both o-special. By 1.30, there is a special translate S_0 of S_1 in A such that S and S_0 have the same terminus. By 4.22, S_0 is parallel to S_1 and hence lies in x_1. Thus by 8.9, the i-face of S is parallel to the i-face of S_0 and the i'-face of S is parallel to the i'-face of S_0. By 8.9, however, we also have $x \neq x_1$, so $S \neq S_0$. Therefore S and S_0 have at most one face in common (by 5.1). By 5.9, a face of S is parallel to a face of S_0 if and only if the two faces are equal. Hence $i = i'$. □

Proposition 8.14. *The edge-colored graph $\Delta_{\mathcal{A}}^{\infty}$ with index set I_o described in 8.9 is a chamber system as defined in 29.2.*

Proof. By 8.13, each edge has a unique color. Let $i \in I_o$. We set

$$x \approx_i x_1$$

for $x, x_1 \in \Delta_{\mathcal{A}}^{\infty}$ if either $x = x_1$ or $x \sim_i x_1$. Since the parallel relation defined in 1.44 is an equivalence relation, so is the relation $\approx_i$. Now let x be a vertex of $\Delta_{\mathcal{A}}^{\infty}$. By 8.6, we can choose an o-special $\mathcal{A}$-sector S in x. Let R and d be the terminus and apex of S, let A be an $\mathcal{A}$-apartment containing S and let e be the unique chamber in $R \cap A$ that is i-adjacent to d. Let $S_1 = \sigma_A(R, e)$ and let x_1 be the set of $\mathcal{A}$-sectors parallel to S_1. Then S and S_1 have the same i-face. By 4.3, on the other hand, S and S_1 are disjoint. Therefore $x \neq x_1$ by 8.2. Thus each i-panel of $\Delta_{\mathcal{A}}^{\infty}$ contains at least two vertices. □

In light of 8.14, we will start referring to the vertices of $\Delta_{\mathcal{A}}^{\infty}$ as chambers.

Proposition 8.15. *Let S and S' be parallel $\mathcal{A}$-sectors. Then every face of S is parallel to a unique face of S'.*

Proof. By 8.2, $S \cap S'$ contains a sector S_1. By 1.45 and 4.7, every face of S is parallel to a face of S_1 and every face of S_1 is parallel to a face of S'. By 5.9, no two faces of a sector are parallel. □

Proposition 8.16. *Let S and S' be two o-special $\mathcal{A}$-sectors, let $i, i' \in I_o$, let f be the i-face of S and let f' be the i'-face of S'. Suppose that f and f' are parallel. Then $i = i'$.*

Proof. By 4.10 and 8.4.ii, there is an apartment A in $\mathcal{A}$ containing o-special subsectors S_1 of S and S_1' of S'. Let f_1 be the i-face of S_1 and let f_1' be the i'-face of S_1'. By 4.23, S_1 is parallel to S and S_1' is parallel to S'. By two applications of 8.10, therefore, f_1 is parallel to f and f_1' is parallel to f'. Hence f_1 is parallel to f_1'. By 1.30, there is a special translation g of A mapping the terminus of S_1 to the terminus of S_1'. Since g is special, f^g is the i-face of S_1'. By 1.45, f_1^g is parallel to f_1 and hence to f_1'. Thus by 5.9, $f_1^g = f_1'$. Therefore $i = i'$. □

Definition 8.17. For each $\mathcal{A}$-sector S, let S^{∞} denote the chamber of $\Delta_{\mathcal{A}}^{\infty}$ (i.e. the parallel class of $\mathcal{A}$-sectors) that contains S. For each $\mathcal{A}$-face f, let $[f]$ be the set of chambers of $\Delta_{\mathcal{A}}^{\infty}$ that contain $\mathcal{A}$-sectors that have a face parallel to f. Note that by 8.15, if x is a chamber in $[f]$ for some face f, then *every* $\mathcal{A}$-sector in x has a face parallel to f.

Proposition 8.18. *The panels of the chamber system $\Delta_{\mathcal{A}}^{\infty}$ are the sets $[f]$ for all $\mathcal{A}$-faces f.*

Proof. Let F be an i-panel for some $i \in I_o$ and let $x \in F$. By 8.7, we can choose an o-special $\mathcal{A}$-sector S in x. Let f be the i-face of S. Then $F \subset [f]$ by 8.9 and 8.17. Suppose that x' is a chamber of $\Delta_{\mathcal{A}}^{\infty}$ (i.e. a vertex of $\Delta_{\mathcal{A}}^{\infty}$) that contains an $\mathcal{A}$-sector S' having a face f' parallel to f. By 8.7 and 8.15, we can assume that S' is o-special. By 8.16, it then follows that f' is the i-face of S'. By 8.11, therefore, $x' \in F$. Thus $F = [f]$. Hence every panel is of the form $[f]$ for some face f.

Suppose, conversely, that f is a face of an arbitrary $\mathcal{A}$-sector S and let $x = S^{\infty}$. By 8.7, there is an o-special sector S' that is parallel to S. By 8.15,

S' has a face f' that is parallel to f; moreover, f' is the i-face of S' for some $i \in I_o$. By 8.17, $[f] = [f']$. By the conclusion of the previous paragraph, $[f']$ is the i-panel containing S^∞. Thus every set of the form $[f]$ is a panel. □

We will need the following observation in the proof of 8.24.

Lemma 8.19. *Let $A, A' \in \mathcal{A}$, let R be a gem such that $A \cap R = A' \cap R$, let $f = \mu_A(R, P)$ for some panel $P \subset R$ cut by A, let $f' = \mu_{A'}(R, P')$ for some panel $P' \subset R$ cut by A' and suppose that the faces f and f' are parallel. Then $P = P'$.*

Proof. Let $x \in A \cap A'$, let $\rho = \text{retr}_{A,x}$, let $f_1 = A \cap f$ (i.e. the intersection of A with the chamber set of f), let $f_1' = A' \cap f'$, let $g_1 = (f_1')^\rho$ and let $g = \mu_A(R, P')$. By 8.19 of [37], the restriction of ρ to A' is a special isomorphism from A' to A, and by 8.17 of [37], ρ acts trivially on R. By 5.3, it follows that $g_1 = A \cap g$. Suppose that $P' \neq P$. By 5.9, f and g are not parallel. By 5.8.ii, therefore, for each $N > 0$, there exists $x \in f_1'$ such that

$$\text{dist}(x^\rho, f_1) \geq N. \tag{8.20}$$

By 8.18 of [37], ρ is a special homomorphism from Δ to A. Hence (by 1.18 of [37])

$$\text{dist}(u^\rho, v^\rho) \leq \text{dist}(u, v)$$

for all chambers u, v of Δ. Therefore 8.20 implies that

$$\text{dist}(x, f_1) \geq N$$

since ρ acts trivially on $f_1 \subset A$. Hence f_1 and f_1' are not parallel. This implies, however, that f and f' are not parallel. We conclude that, in fact, $P = P'$. □

Notation 8.21. Let Π_o denote the subdiagram of Π spanned by I_o, i.e. the diagram obtained from Π by deleting o and all edges containing o.

Proposition 8.22. *Let $A \in \mathcal{A}$ and let A^∞ denote the subgraph of $\Delta_{\mathcal{A}}^\infty$ spanned by the set of all chambers of $\Delta_{\mathcal{A}}^\infty$ containing sectors that are contained in A.*[1] *Let R be an o-special gem cut by A. Then A^∞ is a Coxeter chamber system of type Π_o and there is a special isomorphism from A^∞ to R that sends the parallel class $\sigma_A(R, d)^\infty$ to d for each chamber d in R.*

Proof. The set $A \cap R$ is a gem of A. By 4.24, every parallel class in A^∞ contains a unique sector of A having terminus $A \cap R$. There is thus a well-defined map π from A^∞ onto R that sends the parallel class $\sigma_A(R, d)^\infty$ to d for each chamber d in R. By 4.3, $\sigma_A(R, d)$ and $\sigma_A(R, d')$ are disjoint if d and d' are distinct elements of R. By 8.2, it follows that π is injective.

Let $\{d, e\}$ be an i-panel of $A \cap R$. By 8.11, the parallel classes $\sigma_A(R, d)^\infty$ and $\sigma_A(R, e)^\infty$ are i-adjacent vertices of $\Delta_{\mathcal{A}}^\infty$. Suppose, conversely, that the

[1] Let Γ be an arbitrary graph and let X be a subset of the vertex set of Γ. The *subgraph of Γ spanned by X* is the graph whose vertex set is X and whose edge set is the set of all edges of Γ connecting two vertices of X.

sector classes $\sigma_A(R,d)^\infty$ and $\sigma_A(R,e)^\infty$ are i-adjacent for $d,e \in R$ and for some $i \in I_o$. By 8.9, the i-faces of the sectors $\sigma_A(R,d)$ and $\sigma_A(R,e)$ are parallel. By 5.9, it follows that d and e are i-adjacent. Thus π is a special isomorphism. By 1.7, R is a Coxeter chamber system of type Π_o. □

Corollary 8.23. *Let $A \in \mathcal{A}$ and let R be a gem cut by A (not necessarily o-special). Then there is an isomorphism from A^∞ to R (not necessarily special) that sends the parallel class $\sigma_A(R,d)^\infty$ to d for each chamber d in R.*

Proof. Let R' be an o-special gem cut by A. By 1.9, $R' = R^g$ for some translation of A. Then $\sigma_A(R,d)$ and $\sigma_A(R,d)^g$ are parallel for every $d \in R$. By 8.22, therefore, there is an isomorphism from A^∞ to R' that sends $\sigma_A(R,d)^\infty$ to $d^{g^{-1}}$ for every $d \in R$. □

We come now to the main result of this chapter.

Theorem 8.24. *Let $\Delta_{\mathcal{A}}^\infty$ be as defined in 8.9. Then $\Delta_{\mathcal{A}}^\infty$ is a building of type Π_o whose apartments are the subgraphs A^∞ for all $A \in \mathcal{A}$, where Π_o and A^∞ are as in 8.21 and 8.22.*

Proof. By 8.14, $\Delta_{\mathcal{A}}^\infty$ is a chamber system with index set I_o, and by 8.22, A^∞ is isomorphic to Σ_{Π_o} for each $A \in \mathcal{A}$. By 8.4.ii, every two chambers of $\Delta_{\mathcal{A}}^\infty$ are contained in A^∞ for some $A \in \mathcal{A}$. Thus 29.35.i holds. Let A, A' be two apartments in $\mathcal{A}$. By 29.34, there is a special isomorphism π from A to A' that acts trivially on $A \cap A'$. This isomorphism induces a special isomorphism ϕ from A^∞ to $(A')^\infty$. If S is a sector contained in both A and A', then ϕ fixes S^∞. Thus 29.35.ii holds.

Suppose now that there exists a sector S contained in both A and A' and a panel F of $\Delta_{\mathcal{A}}^\infty$ such that $F \cap A^\infty$ and $F \cap (A')^\infty$ are both non-empty. This means that there exist sectors S_1 contained in A and S_1' contained in A' such that S_1^∞ and $(S_1')^\infty$ are both contained in F. By 8.18, S_1 has a face f and S_1' has a face f' such that $F = [f] = [f']$. By 4.15, we can choose a gem R such that $R \cap A = R \cap A' \neq \emptyset$. By 1.9 and 1.45, we can assume that $f = \mu_A(R,P)$ for some panel $P \subset R$ cut by A and $f' = \mu_{A'}(R,P')$ for some panel $P' \subset R$ cut by A'. By 8.19, $P = P'$. Since π acts trivially on $A \cap R$, it follows that the isomorphism π maps f to f'. We have

$$F \cap A^\infty = \{\sigma_A(R,x)^\infty \mid x \in P \cap A\}$$

and

$$F \cap (A')^\infty = \{\sigma_{A'}(R,x')^\infty \mid x' \in P \cap A'\}.$$

Therefore the isomorphism ϕ maps $F \cap A^\infty$ to $F \cap (A')^\infty$ as claimed. As we observed in the previous paragraph, ϕ fixes S^∞. Thus 29.35.iii holds.

By 29.35, we conclude that $\Delta_{\mathcal{A}}^\infty$ is a building of type Π_o. In particular, $\Delta_{\mathcal{A}}^\infty$ is spherical. Let Σ be an arbitrary apartment of $\Delta_{\mathcal{A}}^\infty$. By 29.14.i, we can choose two chambers in Σ that are opposite in $\Delta_{\mathcal{A}}^\infty$. As observed above, there exists $A \in \mathcal{A}$ such that the apartment A^∞ contains these two chambers. By

29.14.ii, we must, in fact, have $A^\infty = \Sigma$. We conclude that every apartment of $\Delta_{\mathcal{A}}^\infty$ is of the form A^∞ for some $A \in \mathcal{A}$. □

Definition 8.25. Let o be a special vertex of Π, let Π_o be as in 8.21 and let $\Delta_{\mathcal{A}}^\infty$ be the edge-colored graph defined in 8.9. As we have just shown in 8.24, $\Delta_{\mathcal{A}}^\infty$ is a building of type Π_o.[2] We call $\Delta_{\mathcal{A}}^\infty$ the *building at infinity* of the pair $(\Delta, \mathcal{A})$.

Remark 8.26. We have $\Pi = \tilde{\mathsf{X}}_\ell$, where X_ℓ is one of the diagrams in Figure 1.3. By 8.24, the building at infinity $\Delta_{\mathcal{A}}^\infty$ is a building of type X_ℓ. In particular, $\Delta_{\mathcal{A}}^\infty$ is irreducible. In 11.3, we will show that $\Delta_{\mathcal{A}}^\infty$ is also thick.

Our principal goal for the rest of this chapter is to prove 8.36 where we describe the panels of $\Delta_{\mathcal{A}}^\infty$ in terms of parallel classes of $\mathcal{A}$-faces of Δ and the walls of $\Delta_{\mathcal{A}}^\infty$ in terms of parallel classes of $\mathcal{A}$-walls of Δ.

Proposition 8.27. *The map $A \mapsto A^\infty$ from $\mathcal{A}$ to the set of all apartments of $\Delta_{\mathcal{A}}^\infty$ is a bijection.*

Proof. By 8.24, this map is surjective. Suppose A, A_1 are two elements of $\mathcal{A}$ such that $A^\infty = A_1^\infty$. Let R be an o-special gem cutting A, let d and e be two opposite chambers of $A \cap R$, let $S = \sigma_A(R, d)$ and let $S' = \sigma_A(R, e)$. Then A_1 contains subsectors S_1 and S_1' of S and S'. By 4.25, the convex hull of $S \cup S_1'$ is A. Since A_1 is also convex, it follows that $A \subset A_1$. By symmetry, we also have $A_1 \subset A$. □

Proposition 8.28. *Let f and f' be $\mathcal{A}$-faces such that $[f] = [f']$ (where $[f]$ and $[f']$ are as defined in 8.17). Then f is parallel to f'.*

Proof. Let S be an $\mathcal{A}$-sector such that f is a face of S. By 8.15 and 8.17, S has a face parallel to f'. Now let A be an $\mathcal{A}$-apartment containing S, let R and d be the terminus and apex of S, let P be the panel containing d such that $f = \mu(S, P)$, let e be the unique chamber in $P \cap A$ distinct from d and let $S' = \mu_A(R, e)$. Then S' is also an $\mathcal{A}$-sector and f is also a face of S'. Hence also S' has a face parallel to f'. By 5.9, the only face of S that is parallel to a face of S' is f. It follows that f is parallel to f'. □

Notation 8.29. For each $\mathcal{A}$-root α of Δ, we set

$$\alpha^\infty = \{S^\infty \mid S \text{ is a sector contained in } \alpha\},$$

and for each $\mathcal{A}$-wall M of Δ, we set

$$[M] = \{[f] \mid f \text{ is a face contained in } M\},$$

where $[f]$ is as in 8.17.

Proposition 8.30. *The following hold:*

(i) *For each $\mathcal{A}$-root α, the set α^∞ defined in 8.29 is a root of $\Delta_{\mathcal{A}}^\infty$.*

---

[2]We recall that by 8.12, the building at infinity does not depend in an essential way on the choice of the special vertex o.

(ii) *For each $A \in \mathcal{A}$, the map $[\alpha] \mapsto \alpha^\infty$ is a well-defined bijection from the set of parallel classes of roots of A to the set of roots of A^∞.*

Proof. Let R be an o-special gem of A and let X be the set of roots of A cutting R. By 8.22 and 29.19, the map

$$\alpha \mapsto \{\sigma_A(R,d)^\infty \mid d \in \alpha \cap R\}$$

is a bijection from X to the set of roots of A^∞. By 4.24 and 4.12, we have

$$\{\sigma_A(R,d)^\infty \mid d \in \alpha \cap R\} = \alpha^\infty$$

for each $\alpha \in X$. Thus (i) holds and the map $\alpha \mapsto \alpha^\infty$ is a bijection from X to the set of roots of A^∞. By 1.19, therefore, (ii) holds. □

Proposition 8.31. *Let α and α' be two roots such that $\alpha^\infty = (\alpha')^\infty$. Then $\alpha \cap \alpha'$ is a subroot of α and α'.*

Proof. Since roots are convex, this holds by 4.26. □

Proposition 8.32. *Let M be an $\mathcal{A}$-wall, let R be a gem, let X be the set of panels of R that are contained in M and suppose that $X \neq \emptyset$. Then*

$$M = \bigcup_{P \in X} \mu_A(R,P).$$

Proof. Let A be an $\mathcal{A}$-apartment containing a root α such that $M = \mu(\alpha)$. Let $Q \in M$, let u be the unique chamber in $\alpha \cap Q$, let v be the other chamber in $A \cap Q$, let $d = \text{proj}_R u$, let $e = \text{proj}_R v$, let $S = \sigma_A(R,d)$ and let $S' = \sigma_A(R,e)$. By 29.21, d and e are adjacent. Let P be the unique panel that contains d and e. By 4.3, $u \in S$, $v \in S'$ and $P \in X$. Thus Q is contained in the face $\mu_A(R,P)$. If, conversely, P is an arbitrary panel in X, then $\mu_A(R,P) \subset \mu(\alpha) = M$ by 8.1. □

Proposition 8.33. *For each $\mathcal{A}$-root α, the set $[\mu(\alpha)]$ defined in 8.29 is the wall of the root α^∞ of $\Delta_{\mathcal{A}}^\infty$ and the map $M \mapsto [M]$ is a surjective map from the set of $\mathcal{A}$-walls of Δ to the set of walls of $\Delta_{\mathcal{A}}^\infty$.*

Proof. Suppose that α is a root of an apartment A in $\mathcal{A}$ (so α^∞ is a root of $\Delta_{\mathcal{A}}^\infty$ by 8.30), let $M = \mu(\alpha)$ and let β be the root opposite α in A. By 8.30, it will suffice to show that

$$[M] = \mu(\alpha^\infty).$$

Let f be a face contained in M. Then $f = \mu(R,P)$ for some panel P in M and some gem R containing P. Let d be the unique chamber in $\alpha \cap P$ and let e be the unique chamber in $\beta \cap P$. By 4.11, $\sigma_A(R,d) \subset \alpha$ and $\sigma_A(R,e) \subset \beta$. Thus $\sigma_A(R,d)^\infty \subset \alpha^\infty \cap [f]$ and $\sigma_A(R,e)^\infty \subset \beta^\infty \cap [f]$ by 8.17 and 8.29. It follows that $[f]$ is a panel of $\Delta_{\mathcal{A}}^\infty$ (by 8.18) that is contained in the wall $\mu(\alpha^\infty)$ (by 29.32). We conclude that $[M] \subset \mu(\alpha^\infty)$.

Suppose, conversely, that F is a panel of $\Delta_{\mathcal{A}}^\infty$ contained in the wall $\mu(\alpha^\infty)$. By 8.29, there exist sectors S contained in α and S_1 contained in β such that $S^\infty \in F \cap \alpha^\infty$ and $S_1^\infty \in F \cap \beta^\infty$. Let R be the terminus of S. Translates of sectors are parallel by 4.22 and $[M] = [\mu(\alpha')]$ for each root α' parallel to α by 1.26. By 1.18 and 1.22, therefore, we can assume that α cuts R (by replacing α by a parallel root) and that R is also the terminus of S_1 (by replacing S_1 by a translate). Since S^∞ and S_1^∞ are both contained in F, they are adjacent chambers of $\Delta_{\mathcal{A}}^\infty$. By 8.23, therefore, the apices d of S and e of S_1 are contained in a panel P. Let $f = \mu(R, P)$. Since $d \in \alpha$ and $e \in \beta$, it follows that P is contained in M. By 8.32, therefore, $f \subset M$. Thus $[f]$ is also a panel of $\Delta_{\mathcal{A}}^\infty$ (by 8.18) that contains both S^∞ and S_1^∞. Hence $[f] = F$. Thus $F \in [M]$. We conclude that $\mu(\alpha^\infty) \subset [M]$. □

Proposition 8.34. *Let f and f' be two $\mathcal{A}$-faces that are not parallel. Then for each $N > 0$ there exists a subface f_N of f such that* $\text{dist}(u, f') \geq N$ *for all $u \in f_N$.*

Proof. Let $N > 0$, let S be a sector contained in an $\mathcal{A}$-apartment A such that f is a face of S and let S' be a sector contained in an $\mathcal{A}$-apartment A' such that f' is a face of S'. By 8.4.ii, there exist subsectors S_1 of S and S_1' of S' and an apartment A_1 containing S_1 and S_1'. By 4.7, $S_1 = S^g$ for some translation g of A and $S_1' = (S')^h$ for some translation h of A'. Let f_1 be the unique face of S_1 such that $f_1 \cap A = (f \cap A)^g$ and let f_1' be the unique face of S_1' such that $f_1' \cap A' = (f' \cap A)^h$. By 1.45, f_1 is parallel to f and f_1' is parallel to f'. Therefore f_1 is not parallel to f'.

Choose $m > 0$. By 5.8.ii, there exists a subface $\hat{f}$ of f_1 such that

$$\text{dist}(u, f_1') \geq m$$

for all $u \in \hat{f}$. Let $\tilde{f}$ be the subface of f such that $(\tilde{f} \cap S)^g = \hat{f} \cap S_1$. By 1.43, there exists M such that $\text{dist}(x, x^g) \leq M$ for all chambers $x \in A$ and $\text{dist}(x, x^h) \leq M$ for all chambers $x \in A'$. Thus $\text{dist}(u, f') \geq m - 2M$ for all $u \in \tilde{f}$. □

Proposition 8.35. *Let M and M' be two $\mathcal{A}$-walls. Then M and M' are parallel if and only if $[M] = [M']$, where $[M]$ and $[M']$ are as defined in 8.29.*

Proof. By 1.16, we can choose a gem R containing panels in M. Let X be as in 8.32, let A be an $\mathcal{A}$-apartment containing a root α such that $M = \mu(\alpha)$ and let $f_P = \mu_A(R, P)$ for all $P \in X$. Then $P \mapsto P \cap A$ is an injective map from X to the set of panels of $R \cap A$. By 1.7, $R \cap A$ is finite. Thus X is also finite.

Suppose now that M and M' are parallel. By 1.44, there exists $N > 0$ such that

$$\text{dist}(M, u') \leq N$$

for all $u' \in M'$. Let f' be an $\mathcal{A}$-face contained in M'. Let X_1 be the set of $P \in X$ such that f' is not parallel to f_P. By 5.15 and 8.34 (and the finiteness of X), there exists $u' \in f'$ such that $\text{dist}(f_P, u') > N$ for all $P \in X_1$. By 8.32, it follows that $X_1 \neq X$. Thus f' is parallel to f_P for some $P \in X$. Therefore $[f'] \in [M]$. Hence $[M'] \subset [M]$. By symmetry, we conclude that, in fact, $[M] = [M']$.

Suppose, conversely, that $[M] = [M']$. Let $P \in X$. By 8.29, there exists a face f' contained in M' such that $[f'] = [f_P]$. By 8.28, therefore, f' is parallel to f_P. This means that there exists $N_P > 0$ such that

$$\text{dist}(u, M') \leq \text{dist}(u, f') \leq N_P$$

for all $u \in f_P$. Let N be the maximum of the N_P for all $P \in X$. Then $\text{dist}(u, M') \leq N$ for all

$$u \in \bigcup_{P \in X} f_P.$$

By 8.32 again, it follows that $\text{dist}(u, M') \leq N$ for all $u \in M$. By symmetry,

$$\{\text{dist}(M, u') \mid u' \in M'\}$$

is also bounded. Therefore M and M' are parallel. □

Theorem 8.36. *For each $\mathcal{A}$-face f, let f^∞ denote the set of $\mathcal{A}$-faces parallel to f and let $[f]$ be as in 8.17. For each $\mathcal{A}$-wall M, let M^∞ denote the set of $\mathcal{A}$-walls parallel to M and let $[M]$ be as in 8.29. Then the following hold:*

(i) *The map $[f] \mapsto f^\infty$ is a well-defined bijection from the set of panels of $\Delta_{\mathcal{A}}^\infty$ to the set of parallel classes of $\mathcal{A}$-faces.*
(ii) *The map $[M] \mapsto M^\infty$ is a well-defined bijection from the set of walls of $\Delta_{\mathcal{A}}^\infty$ to the set of parallel classes of $\mathcal{A}$-walls.*

Proof. The first assertion holds by 8.18 and 8.28 and the second by 8.33 and 8.35. □

Chapter Nine

Trees with Valuation

Affine buildings of rank 1 are the same thing as thick trees, as was explained in 7.32. In this chapter we pause in our investigation of affine buildings of arbitrary rank to introduce an important connection between trees and valuations of fields. The principal results in this chapter are 9.14, 9.24 and 9.30.

Notation 9.1. The apartments of a thick tree are its thin subtrees. We will refer to the sectors of thick trees as *rays*. (This will be helpful later on when we talk about rays in trees and sectors in affine buildings of higher rank at the same time.) Thus a ray is simply a path

$$(x_0, x_1, x_2, \ldots)$$

that goes on indefinitely, and two rays are parallel if their intersection is also a ray. The rays of a single apartment all lie in one of two parallel classes. In fact, in an obvious sense, these two parallel classes can be thought of as the two "ends" of the apartment, and we will refer to parallel classes of rays in thick trees as *ends*, the standard term for this notion.

Notation 9.2. Let Γ be a thick tree. Then for each vertex x and each end a, there exists a unique ray that begins at x and is contained in a. We denote this ray by a_x.[1]

Proposition 9.3. *Let Γ be a thick tree and let a and b be distinct ends of Γ. Then there exists a unique apartment of Γ whose ends are a and b.*

Proof. Let x be an arbitrary vertex of Γ, let a_x and b_x be as in 9.2 and let $(x, \ldots, y, z)$ be the intersection of a_x and b_x. Then the union of the two rays a_x and b_x minus the subpath $(x, \ldots, y)$ is an apartment whose ends are a and b. Since Γ is a tree, this apartment is unique. □

Notation 9.4. We will denote the unique apartment in 9.3 by $[a, b]$.[2]

Definition 9.5. Let Γ be a thick tree and let $\mathcal{A}$ be a subset of the set of apartments of Γ. The pair $(\Gamma, \mathcal{A})$ is a *tree with sap* if $\mathcal{A}$ is a system of apartments (= *sap* ![3]) of Γ considered as a building of type $\tilde{A}_1$. Thus by 8.4, $(\Gamma, \mathcal{A})$ is a tree with sap if the following two conditions hold:

---

[1]There should be no confusion with the notation $x \mapsto a_x$ introduced in 3.5.
[2]There should be no confusion with the interval notation introduced in 3.1.
[3]This clever acronym is due to Mark Ronan.

(i) Every two vertices of Γ lie in some $\mathcal{A}$-apartment.[4]

(ii) For every pair a, b of distinct $\mathcal{A}$-ends[5] the apartment $[a, b]$ defined in 9.4 is contained in $\mathcal{A}$.

The set of $\mathcal{A}$-ends of a thick tree with sap $(\Gamma, \mathcal{A})$ will be denoted by $\Gamma^\infty_{\mathcal{A}}$.

Example 9.6. Let Γ be a thick tree. Then the set of *all* apartments of Γ is a system of apartments. We will refer to this as the *complete* system of apartments of Γ.

Example 9.7. Let Γ be a thick tree and let W be a set of paths (x_0, x_1, x_2) of length 2 in Γ such that for every path (x_0, x_1) of length 1 in Γ, there exists at least one vertex x_2 such that $(x_0, x_1, x_2) \in W$. Let $\mathcal{A}_W$ denote the set of apartments A such that all but finitely many of the paths of length 2 contained in A are contained in W. Then $(\Gamma, \mathcal{A}_W)$ is a tree with sap. For most choices of W, however, the set $\mathcal{A}_W$ is not the complete system of apartments.

Remark 9.8. Note that if X is a set of ends of a thick tree and

$$\mathcal{A} = \{[a, b] \mid a, b \in X,\ a \neq b\},$$

then $\mathcal{A}$ satisfies 9.5.ii automatically.

We now fix a thick tree with sap $(\Gamma, \mathcal{A})$.

Definition 9.9. For each triple a, b, c of distinct $\mathcal{A}$-ends, there is a unique vertex of Γ common to the three apartments $[a, b]$, $[b, c]$ and $[a, c]$. We call this vertex the *junction* of a, b and c and denote it by $\kappa(a, b, c)$. The map κ will be called the *junction map* of the pair $(\Gamma, \mathcal{A})$. For every 4-tuple (a, b, c, d) of distinct $\mathcal{A}$-ends, let

$$\omega(a, b, c, d)$$

denote the unique integer n such that $|n|$ is the distance from $\kappa(a, b, c)$ to $\kappa(a, b, d)$ and n is negative if and only if the ray $a_{\kappa(a,b,d)}$ does not contain the vertex $\kappa(a, b, c)$.[6] The map ω is called the *canonical valuation* of the pair $(\Gamma, \mathcal{A})$.

In Figure 9.1, for example, $\kappa(a, b, c)$ is the vertex labeled x, $\kappa(a, b, d)$ is the vertex labeled y, $\omega(a, b, c, d) = 3$ and $\omega(a, b, d, c) = -3$.

Proposition 9.10. *Every vertex of Γ is the junction of three ends in $\mathcal{A}$.*

Proof. This holds by 9.5.i. □

---

[4]The literal translation of 8.4.i is that every two *edges* of Γ lie in some $\mathcal{A}$-apartment. Since Γ is a tree, however, 9.5.i is an equivalent assertion.

[5]An $\mathcal{A}$-*ray* is a ray contained in an $\mathcal{A}$-apartment. By $\mathcal{A}$-*end*, we mean a parallel class of $\mathcal{A}$-rays.

[6]Alternatively, we could define the *signed length* of a path contained in the apartment $[a, b]$ to be the length of the path with a plus sign if the path goes toward the end b (in the obvious sense) or a minus sign if it goes toward a. The number $\omega(a, b, c, d)$ is then the signed length of the unique path from $\kappa(a, b, c)$ to $\kappa(a, b, d)$.

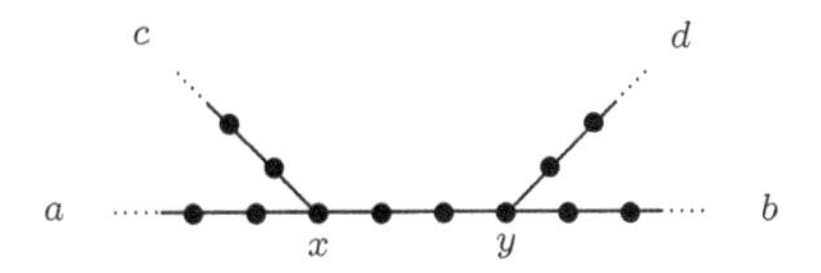

Figure 9.1 The canonical valuation ω

Proposition 9.11. *The following hold:*

(i) $\omega(a,b,c,d) = \omega(c,d,a,b) = -\omega(a,b,d,c)$;
(ii) *if* $\omega(a,b,c,d) = k > 0$, *then* $\omega(a,d,c,b) = k$ *and* $\omega(a,c,b,d) = 0$;
(iii) $\omega(a,b,c,d) + \omega(a,b,d,e) = \omega(a,b,c,e)$; *and*
(iv) *the map* $t \mapsto \omega(a,b,c,t)$ *from* $\Gamma_{\mathcal{A}}^{\infty}\backslash\{a,b,c\}$ *to* $\mathbb{Z}$ *is surjective*

for all a,b,c,d,e *in* $\Gamma_{\mathcal{A}}^{\infty}$ *that are pairwise distinct.*

Proof. The identities (i)–(iii) follow immediately from 9.9 with the help of a few diagrams like the one in Figure 9.1. Let $a,b,c \in \Gamma_{\mathcal{A}}^{\infty}$ be distinct, let $u = \kappa(a,b,c)$ and let $n \in \mathbb{Z}$. If $n \geq 0$, let v be the vertex of $[a,b]$ at distance $|n|$ from u that is contained in the ray b_u, and if $n < 0$, let v be the vertex of $[a,b]$ at distance $|n|$ from u that is not contained in the ray b_u. Since Γ is thick, there exists a vertex w adjacent to v that is not in $[a,b]$ and there exists a vertex z adjacent to w that is not in the ray c_u.[7] By 9.5.i, there is an end $t \in \mathcal{A}$ containing a ray that starts at v and passes through w and z. Hence a,b,c,t are distinct and $\omega(a,b,c,t) = n$. Thus (iv) holds. □

Definition 9.12. Let X be a set. A *projective valuation* on X is a function ω from the set of 4-tuples of distinct elements of X to $\mathbb{Z}$ satisfying (i)–(iv) of 9.11.

Proposition 9.13. *Let* ω *be a projective valuation on a set* X *and let* $a \in X$. *Then* ω *is uniquely determined by its restriction to the set of all 4-tuples of distinct elements of* X *whose first entry is* a.

Proof. Let ω_a denote the restriction of ω to the set of all 4-tuples of distinct elements of X whose first entry is a and let (x,y,z,w) be an arbitrary 4-tuple of distinct elements in X. Then either $a \in \{x,y,z,w\}$ or $a \notin \{x,y,z,w\}$. In the first case, $\omega(x,y,z,w)$ can be calculated from ω_a using 9.11.i, and in the second case, we have

$$\omega(x,y,z,w) = \omega(x,y,z,a) + \omega(x,y,a,w)$$

by 9.11.iii. □

Every projective valuation, it can be shown, is the canonical valuation on the set of ends of a thick tree with sap. We omit the proof since we do not need this fact. Instead, it is the following uniqueness result that will play an important role in Chapter 12. The proof we give is due to Nils Rosehr.

[7] In fact, *every* vertex adjacent to w is not contained in c_u unless $n = 0$.

Theorem 9.14. *Suppose that $(\Gamma, \mathcal{A})$ and $(\hat{\Gamma}, \hat{\mathcal{A}})$ are two thick trees with sap (as defined in 9.5) with canonical valuations ω and $\hat{\omega}$ and let ϕ be a bijection from $\Gamma^\infty_{\mathcal{A}}$ to $\hat{\Gamma}^\infty_{\hat{\mathcal{A}}}$ such that $\hat{\omega} \circ \phi = \omega$. Then there exists a unique isomorphism ρ from Γ to $\hat{\Gamma}$ such that $\rho(x) = \phi(x)$ for all $x \in \Gamma^\infty_{\mathcal{A}}$. (Thus, in particular, ρ maps $\mathcal{A}$ to $\mathcal{A}'$.)*

Proof. Let $X = \Gamma^\infty_{\mathcal{A}}$ and $\hat{X} = \hat{\Gamma}^\infty_{\hat{\mathcal{A}}}$. If $a, b, \ldots$ and $a', b', \ldots$ are elements of X, we denote by $\hat{a}, \hat{b}, \ldots$ and $\hat{a}', \hat{b}', \ldots$ their images under ϕ. Let $\hat{\kappa}$ denote the junction map of $\hat{\Gamma}$.

Choose a vertex x of Γ. By 9.10, there exist distinct ends $a, b, c \in X$ such that $x = \kappa(a, b, c)$. Suppose that a', b', c' is a second triple of distinct ends in X such that $x = \kappa(a', b', c')$. Let $\hat{x} = \kappa(\hat{a}, \hat{b}, \hat{c})$ and $\hat{x}' = \hat{\kappa}(\hat{a}', \hat{b}', \hat{c}')$. We claim that

$$\hat{x} = \hat{x}'. \tag{9.15}$$

After relabeling the ends a', b', c' if necessary, we can assume that for all distinct $u, v \in \{a, b, c\}$, the rays u_x and v'_x do not pass through the same neighbor of x. It follows that for all $u \in \{a, b, c\}$, either $u = u'$ or

$$\omega(u, u', v, w') = 0 \tag{9.16}$$

for all $v, w \in \{a, b, c\} \backslash \{u\}$ such that $v \neq w'$.

Let k be the distance from $\hat{x}$ to $\hat{x}'$ in $\hat{\Gamma}$ and suppose that $k > 0$. Let γ be the unique minimal path in $\hat{\Gamma}$ from $\hat{x}$ to $\hat{x}'$, let $\hat{x}_1$ be the unique neighbor of $\hat{x}$ that lies on γ and let $\hat{x}'_1$ be the unique neighbor of $\hat{x}'$ that lies on γ. We can choose two 2-element subsets $\{u, v\}$ and $\{w, z\}$ of $\{a, b, c\}$ such that the rays $\hat{u}_{\hat{x}}$ and $\hat{v}_{\hat{x}}$ do not pass through $\hat{x}_1$ and the rays $\hat{w}'_{\hat{x}'}$ and $\hat{z}'_{\hat{x}'}$ do not pass through $\hat{x}'_1$. Since $|\{a, b, c\}| = 3$, the sets $\{u, v\}$ and $\{w, z\}$ cannot be disjoint. We can thus assume that $u = z$. Hence $\hat{u}$, $\hat{u}'$, $\hat{v}$ and $\hat{w}'$ are distinct and

$$\hat{\omega}(\hat{u}, \hat{u}', \hat{v}, \hat{w}') = k.$$

By 9.16, however, $\hat{\omega}(\hat{u}, \hat{u}', \hat{v}, \hat{w}') = 0$. With this contradiction, we conclude that 9.15 holds. There is thus a unique map ρ from the vertex set of Γ to the vertex set of $\hat{\Gamma}$ such that

$$\kappa(a, b, c)^\rho = \hat{\kappa}(\hat{a}, \hat{b}, \hat{c})$$

for all triples a, b, c of distinct ends in X.

Since the graph $\hat{\Gamma}$ is thick, the map ρ is surjective (by 9.10). Suppose that x and y are vertices of Γ and let $m = \text{dist}(x, y)$. By 9.5.i and thickness, there exist ends a, b, c, d in X such that $x = \kappa(a, b, c)$, $y = \kappa(a, b, d)$ and

$$\omega(a, b, c, d) = m.$$

Thus

$$\hat{\omega}(\hat{a}, \hat{b}, \hat{c}, \hat{d}) = m.$$

By 9.9, this means that

$$x^\rho = \hat{\kappa}(\hat{a}, \hat{b}, \hat{c})$$

and

$$y^\rho = \hat{\kappa}(\hat{a}, \hat{b}, \hat{d})$$

are at distance m in $\hat{\Gamma}$. Thus ρ preserves distances. It follows that ρ is an isomorphism.

It only remains to show that the map induced by ρ on X coincides with the bijection ϕ. Let $b \in X$. There exist ends a, c, d_i in X such that a, b, c, d_i are distinct and

$$\omega(a, b, c, d_i) = i$$

for all $i \geq 1$. Let $x_i = \kappa(a, b, d_i)$ and $\hat{x}_i = \hat{\kappa}(\hat{a}, \hat{b}, \hat{d}_i) = x_i^\rho$ for all $i \geq 1$. Thus

$$(x_1, x_2, x_3, \ldots)$$

is a ray contained in the end b. Since

$$\hat{\omega}(\hat{a}, \hat{b}, \hat{c}, \hat{d}_i) = i$$

for all $i \geq 1$,

$$(x_1, x_2, x_3, \ldots)^\rho = (\hat{x}_1, \hat{x}_2, \hat{x}_3, \ldots)$$

is a ray contained in $\hat{b}$. It follows that $\rho(b) = \hat{b}$. □

We turn now to valuations of alternative division rings:

Definition 9.17. Let K be a field, a skew field or an octonion division algebra.[8] A *valuation*[9] of K is a surjective map ν from K^* to $\mathbb{Z}$ such that

(i) $\nu(ab) = \nu(a) + \nu(b)$ and
(ii) $\nu(a + b) \geq \min\{\nu(a), \nu(b)\}$

for all $a, b \in K^*$. As is standard, we assign $\nu(0)$ the value ∞. A *uniformizer* is an element $u \in K^*$ such that $\nu(u) = 1$.

To begin, we investigate the consequences of condition 9.17.ii in a more general setting:

Lemma 9.18. *Let U be a group written additively, but which might not be abelian, and let ν be a map from U^* to $\mathbb{Z}$, which we extend to U by setting $\nu(0) = \infty$, such that*

(a) $\nu(a + b) \geq \min\{\nu(a), \nu(b)\}$ *and*
(b) $\nu(-a) = \nu(a)$

[8] By a famous result of Bruck and Kleinfeld, it would be equivalent to say, "Let K be an alternative division ring (as defined in 9.1 of [36])." See, for example, Chapters 9 and 20 of [36] for a proof. Alternative division rings are precisely the algebraic structures that classify Moufang projective planes.

[9] More precisely, this is the definition of a *discrete* valuation *of rank 1* of K. Since these are the only valuations of an alternative division ring that will appear in this monograph, we will almost always omit these extra adjectives. We refer the reader who would like to know more about valuations of skew fields and octonion division algebras to [20], [22] and [39].

for all $a, b \in U^*$. *Let* ∂_ν *be the map from* $U \times U$ *to* $\mathbb{R}$ *defined as follows:*

$$(9.19) \qquad \partial_\nu(x,y) = \begin{cases} 2^{-\nu(x-y)} & \text{if } x \neq y \text{ and} \\ 0 & \text{if } x = y. \end{cases}$$

Then the following hold:

(i) *For all* $a, b \in U^*$,

$$\nu(a+b) = \nu(a)$$

if $\nu(a) < \nu(b)$.

(ii) ∂_ν *is a metric.*

(iii) *If* $(a_k)_{k\geq 0}$ *is a Cauchy sequence in* U *(with respect to* ∂_ν*) that does not converge to zero, then* $\nu(a_k)$ *is constant for* k *sufficiently large.*

Proof. Choose $a, b \in U^*$ such that $\nu(a) < \nu(b)$. By (a), we have $\nu(a+b) \geq \nu(a)$ and

$$\nu(a) = \nu(a+b-b) \geq \min\{\nu(a+b), \nu(-b)\}.$$

Since $\nu(-b) = \nu(b)$ by (b) and $\nu(a) < \nu(b)$, it follows that $\nu(a) \geq \nu(a+b)$. Thus (i) holds. By (b), ∂_ν is symmetric. By (a), ∂_ν satisfies the triangle inequality. Since $\partial_\nu(x,y) \neq 0$ whenever $x \neq y$, it follows that ∂_ν is a metric. Thus (ii) holds.

Let $(a_k)_{k\geq 0}$ be a Cauchy sequence that does not converge to zero. Since (a_k) does not converge to zero, there exists N such that $\nu(a_k) \leq N$ for infinitely many indices k. Since $(a_k)_{k\geq 0}$ is a Cauchy sequence, there exists M such that $\nu(a_l - a_k) > N$ for all $k, l > M$. Choose $k > M$ such that $\nu(a_k) \leq N$. By (i), $\nu(a_l) = \nu(a_k)$ for all $l \geq k$. Thus (iii) holds. □

Notation 9.20. Let U, ν and ∂_ν be as in 9.18. By 9.18.ii, ∂_ν is a metric on U. Let $\hat{U}$ denote the *completion of* U *with respect to* ν, i.e. with respect to this metric. By 9.18.iii, there is a unique extension of ν to a continuous map from $\hat{U}^*$ to $\mathbb{Z}$ (as metric spaces). We usually denote this extension also by ν. We say that U is *complete with respect to* ν if $U = \hat{U}$. In particular, if $U = K$ is as in 9.17, then $\hat{K}$ denotes the completion of K with respect to ν and we denote the unique extension of ν to $\hat{K}$ by the same letter ν.

We will see that in all the cases that interest us, there is a natural extension of the group structure on U to a group structure on $\hat{U}$. For the moment, we observe only that if K is an alternative division ring, then $\hat{K}$ has a unique structure of an alternative division ring such that K is a subring of $\hat{K}$ and addition and multiplication in $\hat{K}$ are continuous with respect to the continuation of ν to $\hat{K}$ defined in 9.20; moreover, this continuation is a valuation of $\hat{K}$.[10]

---

[10]Here is a proof: Suppose that $(t_k)_{k\geq 1}$ and $(u_k)_{k\geq 1}$ are sequences of elements of K that converge to elements t and u in $\hat{K}$. Then $\nu(t_k)$ and $\nu(u_k)$ are bounded from below for k sufficiently large. Since $t_k u_k - t_l u_l = t_k(u_k - u_l) + (t_k - t_l)u_l$ for all k and l, it follows

If U and ν are as in 9.18, then

$$U_k := \{u \in U \mid \nu(u) \geq k\} \tag{9.21}$$

is a subgroup of U for all $k \in \mathbb{Z}$. In all the cases that will interest us, U_{k+1} is, in fact, a normal subgroup of U_k for each k; see 18.20.

Notation 9.22. Let K be an alternative division ring and let ν be a valuation of K in the sense of 9.17. As in 9.21, we set

$$K_m = \{u \in K \mid \nu(u) \geq m\}$$

for each $m \in \mathbb{Z}$. By 9.17.i, K_0 is a subring of K, the set $K_0 \backslash K_1$ is closed under multiplicative inverses and K_1 is a two-sided ideal of K_0. Let $\bar{K}$ denote the quotient ring K_0/K_1. If K is a field, then so is $\bar{K}$, if K is a skew field, then $\bar{K}$ is a field or a skew field and if K is an octonion division algebra, then (by 26.15) $\bar{K}$ is a field, a skew field or an octonion division algebra. In each case, we will call $\bar{K}$ simply the *residue field* of K (with respect to ν). The subring K_0 is called the *ring of integers of K with respect to ν* (or the *valuation ring of ν*). We will denote it by $\mathcal{O}_K$ since the symbol K_0 will have a different meaning in several chapters.

Proposition 9.23. *Let ν be a valuation of an alternative division ring K as defined in 9.17. Then the following hold:*

(i) *$\nu(-b) = \nu(b)$ for all $b \in K^*$ and*
(ii) *$\nu(a + b) = \nu(a)$ for all $a, b \in K^*$ such that $\nu(a) < \nu(b)$.*

Proof. By 9.17.i, we have $\nu(1) = 2\nu(1) = 2\nu(-1)$. Hence $\nu(-1) = \nu(1) = 0$. By 9.17.i again, it follows that (i) holds. Thus by 9.18.i, also (ii) holds. $\square$

We come now to the first main theorem of this chapter. This result explains the use of the word "valuation" in the notion of a canonical (and projective) valuation.

Theorem 9.24. *Let K be an alternative division ring, let ∞ be an additional symbol, let $X := K \cup \{\infty\}$ (the corresponding "projective line"), let H be the permutation group*

$$\langle x \mapsto ax + b \mid a \in K^*, \ b \in K \rangle$$

---

that $(t_k u_k)_{k \geq 1}$ is a Cauchy sequence. We set tu equal to its limit (which depends only on t and u and not on the two sequences). This defines a multiplication in $\hat{K}$ that extends the multiplication in K. Addition is defined similarly. The same argument also shows that addition and multiplication in an arbitrary alternative division ring with valuation are continuous.

It follows from the definition of addition and multiplication that each identity that holds in K (such as the associative law for multiplication) holds in $\hat{K}$, and the extension of ν to $\hat{K}$ defined in 9.20 satisfies conditions (i) and (ii) in 9.17. It therefore remains only to show the existence of multiplicative inverses. Suppose that $t \neq 0$. After deleting finitely many terms, we can assume, by 9.18.iii, that $\nu(t_k)$ is, in fact, a constant (and thus, in particular, $t_k \neq 0$) for all $k \geq 1$. Hence $\nu(t_k^{-1}) = -\nu(t_k)$ is a constant for all $k \geq 1$. Since $t_k^{-1} - t_l^{-1} = t_l^{-1}(t_l - t_k)t_k^{-1}$ if t_k and t_l are both non-zero, it follows that $(t_k^{-1})_{k \geq 1}$ is a Cauchy sequence. Its limit is the multiplicative inverse of t.

on X (which fixes ∞) and let ω be an H-invariant projective valuation on X (as defined in 9.12). Let ν be the map from K^* to $\mathbb{Z}$ sending 1 to 0 given by

$$\nu(x) := \omega(\infty, 0, 1, x) \tag{9.25}$$

for all other $x \in K^*$. Then ν is a valuation of K as defined in 9.17.

Proof. By 9.11.iv, the map ν is surjective. By 9.1.i of [36], $a^{-1} \cdot ab = b$ for all $a, b \in K^*$. Since ω is invariant under the group H, we thus have

$$\begin{aligned}
\nu(ab) &= \omega(\infty, 0, 1, ab) \\
&= \omega(\infty, 0, a^{-1}, b) \\
&= \omega(\infty, 0, a^{-1}, 1) + \omega(\infty, 0, 1, b) \quad \text{by 9.11.iii} \\
&= \nu(a) + \nu(b)
\end{aligned}$$

for all $a, b \in K^*$ such that 1, b and ab are distinct, and thus

$$\nu(a) + \nu(a^{-1}) = \nu(a) + \nu(a^{-1}b) - \nu(b) = \nu(a \cdot a^{-1}b) - \nu(b) = 0$$

for all $a, b \in K^*$ such that 1, $a^{-1}b$ and b are distinct. By 9.11.iv, the set X is infinite. Thus for every $a \in K\backslash\{0, 1\}$, there exists $b \in K\backslash\{0, 1, a\}$. It follows that 9.17.i holds. In particular,

$$\nu(-a) = \nu(a) \tag{9.26}$$

for all $a \in K^*$.

To prove 9.17.ii, we suppose that a, b are elements of K^* such that $a+b \neq 0$ and $\nu(a + b)$ is strictly less than both $\nu(a)$ and $\nu(b)$. Replacing a and b by ac and bc for a suitable element $c \in K^*$, we can assume by 9.17.i that $a \neq 1$, $b \neq 1$ and $a + b \neq 1$. Thus (by the H-invariance of ω)

$$\begin{aligned}
\nu(a + b) &= \omega(\infty, 0, 1, a + b) \\
&= \omega(\infty, 0, 1, a) + \omega(\infty, 0, a, a + b) \quad \text{by 9.11.iii} \\
&= \nu(a) + \omega(\infty, -a, 0, b),
\end{aligned}$$

so $\omega(\infty, -a, 0, b) < 0$. Therefore $\omega(\infty, -a, b, 0) > 0$ by 9.11.i and hence

$$\nu(-b^{-1}a) = \omega(\infty, 0, 1, -b^{-1}a) = \omega(\infty, 0, b, -a) > 0$$

by 9.11.ii. Interchanging a and b in this argument, we conclude that also $\nu(-a^{-1}b) > 0$. By 9.17.i, however, $\nu(-b^{-1}a) + \nu(-a^{-1}b) = \nu(1) = 0$. With this contradiction, we conclude that 9.17.ii also holds. □

The canonical valuation of a tree with sap $(\Gamma, \mathcal{A})$ is sometimes called the *cross ratio* of $(\Gamma, \mathcal{A})$ because of the following observation.

Proposition 9.27. *Let all the hypotheses and notation in 9.24 hold. Then ω can be recovered from ν by the formula*

$$\omega(a, b, c, d) = \nu\big((b - c)^{-1}(a - c) \cdot (a - d)^{-1}(b - d)\big), \tag{9.28}$$

where we allow ourselves to delete factors of the form $u - v$ and $(u - v)^{-1}$ on the right-hand side of 9.28 if u or v equals ∞.

Proof. Using the H-invariance of ω (and 9.26), we have

$$\begin{aligned}\omega(\infty,b,c,d) &= \omega(\infty,0,c-b,d-b)\\ &= \omega\big(\infty,0,1,(c-b)^{-1}(d-b)\big)\\ &= \nu\big((b-c)^{-1}(b-d)\big)\end{aligned} \tag{9.29}$$

for all triples b,c,d of distinct elements of K. Thus

$$\begin{aligned}\omega(a,b,c,d) &= \omega(a,b,c,\infty)+\omega(a,b,\infty,d) && \text{by 9.11.iii}\\ &= \omega(\infty,c,b,a)+\omega(\infty,d,a,b) && \text{by 9.11.i}\\ &= \nu\big((b-c)^{-1}(a-c)\big)+\nu\big((a-d)^{-1}(b-d)\big) && \text{by 9.29}\\ &= \nu\big((b-c)^{-1}(a-c)\cdot(a-d)^{-1}(b-d)\big) && \text{by 9.17.i}\end{aligned}$$

for every 4-tuple (a,b,c,d) of distinct elements of K. Now suppose that (a,b,c,d) is an arbitrary 4-tuple of distinct elements of $K\cup\{\infty\}$. If $b=\infty$, then

$$\begin{aligned}\omega(a,b,c,d) &= -\omega(b,a,c,d) && \text{by 9.11.i}\\ &= -\nu\big((c-a)^{-1}(d-a)\big) && \text{by 9.29}\\ &= \nu\big((a-c)(a-d)^{-1}\big) && \text{by 9.17.i.}\end{aligned}$$

In a similar fashion, we check that 9.28 holds if c or d equals ∞. □

We arrive now at the final result of this chapter.

Theorem 9.30. *Let $(\Gamma,\mathcal{A})$ be a thick tree with sap and let ω be its canonical valuation. Let X denote the set of $\mathcal{A}$-ends of Γ, let ∞ and 0 be two elements of X and let U be a group acting on X which we write additively even though it is not assumed to be abelian. Suppose that the group U fixes ∞ and acts sharply transitively on $X\backslash\{\infty\}$. We identify U with the set $X\backslash\{\infty\}$ via the map $u\mapsto 0^u$. Let H be the permutation group*

$$\{x\mapsto x+b \mid b\in U\}$$

on X (which fixes ∞ and is isomorphic to U) and suppose that ω is invariant under the action of the group H. For each $x\in U^*$, let $\nu_x\colon U^*\to\mathbb{Z}$ be given by

$$\nu_x(w)=\begin{cases}\omega(\infty,0,x,w) & \text{if } w\neq x \text{ and}\\ 0 & \text{if } w=x\end{cases}$$

and let $\partial_x\colon U\times U\to\mathbb{R}$ be given by

$$\partial_x(u,v)=\begin{cases}2^{-\nu_x(u-v)} & \text{if } u\neq v \text{ and}\\ 0 & \text{if } u=v.\end{cases}$$

and suppose that

(a) $\nu_x(w+z)\ge\min\{\nu_x(w),\nu_x(z)\}$ *and*
(b) $\nu_x(-w)=\nu_x(w)$

for all $w, z \in U^*$. *Then the following hold:*

(i) *For all* $x, x' \in U^*$, *the maps* ∂_x *and* $\partial_{x'}$ *are equivalent metrics on* U.
(ii) *Let* $\hat{U}$ *be the completion of* U *with respect to the metric* ∂_x *for some* $x \in U^*$ *(which is independent of the choice of* x *by (i)). Then* $U = \hat{U}$ *if and only if the system of apartments* $\mathcal{A}$ *is complete (in the sense of 8.5).*

Proof. Let x, x' be distinct elements of U^*. By 9.18.ii, ∂_x and $\partial_{x'}$ are metrics on U. We have

$$\begin{aligned}\nu_x(w) - \nu_{x'}(w) &= \omega(\infty, 0, x, w) - \omega(\infty, 0, x', w)\\ &= \omega(\infty, 0, x, w) + \omega(\infty, 0, w, x') \quad \text{by 9.11.i}\end{aligned}$$

and therefore (by 9.11.iii)

$$\nu_x(w) - \nu_{x'}(w) = \omega(\infty, 0, x, x') \tag{9.31}$$

for all $w \in U^* \backslash \{x, x'\}$. In particular, the expression on the left-hand side of 9.31 is independent of the choice of w. Thus (i) holds.

We turn now to (ii). Choose $x \in U^*$ and suppose that y, z are distinct elements of $U \backslash \{0, x\}$ such that $\nu_x(y) = \nu_x(z) = 1$. Then $\nu_x(-z) = 1$ by (b). Since $\nu_x(x) = 0$, it follows that $x \neq -z$ and, by (a), $x \neq y - z$. Therefore

$$\begin{aligned}\nu_x(y - z) &= \omega(\infty, 0, x, y - z)\\ &= \omega(\infty, 0, x, -z) + \omega(\infty, 0, -z, y - z) && \text{by 9.11.iii}\\ &= \omega(\infty, 0, x, z) + \omega(\infty, 0, -z, y - z) && \text{by (b)}\\ &= \omega(\infty, 0, x, z) + \omega(\infty, z, 0, y) && \text{by the } H\text{-invariance of } \omega\\ &= \omega(\infty, z, x, 0) + \omega(\infty, z, 0, y) && \text{by 9.11.ii}\end{aligned}$$

and thus

$$\nu_x(y - z) = \omega(\infty, z, x, y) \tag{9.32}$$

by 9.11.iii.

Now let w be an arbitrary end of Γ (i.e. an end not necessarily in $X = \Gamma_{\mathcal{A}}^{\infty}$) that is distinct from 0 and ∞ and let

$$(u_0, u_1, u_2, \ldots)$$

be the unique ray in the end w starting at the junction $\kappa(\infty, 0, w)$. Let w_0 be the end 0, and for each $k \geq 1$, let w_k be an $\mathcal{A}$-end such that the path $(u_0, u_1, \ldots, u_k)$ extends to a ray contained in w_k and $w_k \neq w_l$ for all $l < k$; such an end exists by 9.5.i. By 9.5.i again, there exists an $\mathcal{A}$-end $x \in U^*$ such that

$$\omega(\infty, 0, x, w_k) = 1 \tag{9.33}$$

for all $k \geq 1$. Thus

$$\omega(\infty, w_k, x, w_l) > \min\{k, l\}$$

for all $k, l \geq 1$. By 9.32, we have

$$\nu_x(w_l - w_k) = \omega(\infty, w_k, x, w_l)$$

for all distinct $k, l \geq 1$. Therefore $(w_k)_{k\geq 1}$ is a Cauchy sequence in U with respect to the metric ∂_x.

Now suppose that $U = \hat{U}$ (i.e. that U is complete with respect to ν_x in the sense of 9.20). Then there exists an element $z \in U$ such that for each N there exists M such that

$$\nu_x(z - w_k) > N \tag{9.34}$$

for all $k > M$. In particular, we have $\nu_x(z) = 1$ by 9.18.i and 9.33. By 9.32,

$$\nu_x(z - w_k) = \omega(\infty, w_k, x, z)$$

for all $k \geq 1$ such that $w_k \neq z$. Hence by 9.34, the $\mathcal{A}$-end z contains the ray

$$(u_0, u_1, u_2, \ldots).$$

Therefore $z = w$. Since $z \in U = X \backslash \{\infty\}$ is an $\mathcal{A}$-end, also w is an $\mathcal{A}$-end. Thus $\mathcal{A}$ is complete.

Suppose, conversely, that $\mathcal{A}$ is complete. Let $(w_k)_{k\geq 1}$ be a Cauchy sequence in U with respect to ν_x for some $x \in U^*$ that does not converge to 0. Our goal is to show that this sequence has a limit in U (with respect to ν_x). We can assume that the terms w_k are all distinct from each other and from 0 and x and, after deleting the first few terms, that $\nu_x(w_k)$ is independent of k (by 9.18.iii). Replacing x if necessary, we can assume by 9.31 that, in fact,

$$\nu_x(w_k) = 1 \tag{9.35}$$

for all $k \geq 1$. Thus $\nu_x(w_l - w_k) = \omega(\infty, w_k, x, w_l)$ by 9.32. Hence for each N there exists M such that

$$\omega(\infty, w_k, x, w_l) > N$$

for all $k, l > M$. It follows that there exists an end z of Γ distinct from ∞, 0, x and (after deleting one of the w_k if necessary) all the w_k's such that for each N there exists M such that

$$\omega(\infty, w_k, x, z) > N$$

for all $k > M$. Since $\mathcal{A}$ is complete, we have $z \in U^*$. By 9.35, we have

$$\nu_x(z) = \omega(\infty, 0, x, z) = 1.$$

Therefore

$$\nu_x(z - w_k) = \omega(\infty, w_k, x, z)$$

for all $k \geq 1$ by 9.32. We conclude that the sequence $(w_k)_{k\geq 1}$ converges to z with respect to ν_x. Thus (ii) holds. □

Chapter Ten

Wall Trees

We return now to our assumptions in Chapter 8: Π is one of the affine Coxeter diagrams in Figure 1.1 with vertex set I, $o \in I$ is a special vertex of Π, Δ is a thick building of type Π, $\mathcal{A}$ is a system of apartments of Δ and $\Delta_{\mathcal{A}}^{\infty}$ is the building at infinity of the pair $(\Delta, \mathcal{A})$ as constructed in 8.9 and 8.25. Our main goal in this chapter is to introduce a family of trees with sap, one for each wall of $\Delta_{\mathcal{A}}^{\infty}$. We also prove (in 10.24) for each gem R of Δ the existence of a canonical epimorphism from $\Delta_{\mathcal{A}}^{\infty}$ to R.

Definition 10.1. We will say that a wall M is *contained* in an apartment A if there exists a root α contained in A such that $M = \mu(\alpha)$. We will say that a wall M is *contained* in a root β if there exists a root α properly contained in β such that $M = \mu(\alpha)$.

Definition 10.2. Let M and M' be two $\mathcal{A}$-walls. We will say that the walls M and M' are *adjacent* if there exists an $\mathcal{A}$-apartment A containing roots α and α' such that $M = \mu(\alpha)$, $M' = \mu(\alpha')$ and α' is contained maximally in α. Note that if α' is contained maximally in α, then $-\alpha$ is contained maximally in $-\alpha'$; adjacency is thus a symmetric relation. Note, too, that adjacent walls are parallel (by 1.46).

Definition 10.3. Let M and M' be adjacent $\mathcal{A}$-walls and let A and α be as in 10.2. We set

$$[M, M'] = -\alpha' \cap \alpha.$$

Thus $[M, M']$ is a strip in A as defined in 1.37.

In 10.6, we show that adjacency and the set $[M, M']$ are independent of the choice of the $\mathcal{A}$-apartment A containing M and M' in 10.2 and 10.3. This will justify referring to $[M, M']$ simply as the *strip between* M *and* M' whenever M and M' are adjacent.

Proposition 10.4. *Let M and M' be adjacent $\mathcal{A}$-walls and let A, α and α' and $[M, M']$ be as in 10.3. Then $[M, M']$ is the convex hull of the border $\partial\alpha$.*

Proof. This holds by 1.40 (and 29.13.iii). □

Proposition 10.5. *Let M and M' be adjacent $\mathcal{A}$-walls and let A, α and α' and $[M, M']$ be as in 10.3. Then for each panel P' in the $\mathcal{A}$-wall M', the set $[M, M']$ is the convex hull of the set*

$$\{\text{proj}_P\, u \mid P \in M \textit{ and } u \in P'\}.$$

Proof. Let $P \in M$, let $P' \in M'$, let x be the unique chamber in $\alpha \cap P$, let x_1 be the other chamber in $A \cap P$, let x' be the unique chamber in $-\alpha' \cap P'$ and let x_1' be the other chamber in $A \cap P'$. Since $x' \in \alpha$, we have

$$\mathrm{dist}(x, x') < \mathrm{dist}(x_1, x')$$

(by 29.6.i and 29.13.iii). There thus exists a minimal gallery

$$\gamma = (x_1, x, \ldots, x')$$

from x_1 to x' that passes through x. Since $x_1 \in -\alpha'$, we have

$$\mathrm{dist}(x_1, x') < \mathrm{dist}(x_1, x_1').$$

Thus the gallery (γ, x_1') is minimal. Let $v \in P' \backslash \{x'\}$. Then the galleries (γ, v) and (γ, x_1') have the same type. Hence (γ, v) is also a minimal gallery (by 29.10.i). It follows that $\mathrm{proj}_P v = x$ for all $v \in P'$ (by 29.10.v). Since $\{x\} = P \cap \partial\alpha$ (by 29.40), we conclude that

$$\{\mathrm{proj}_P u \mid P \in M \text{ and } u \in P'\} = \partial\alpha.$$

The claim holds, therefore, by 10.4. □

Proposition 10.6. *Let M and M' be two $\mathcal{A}$-walls. Then the following hold:*

(i) *The walls M and M' are adjacent as defined in 10.2 if and only if for every $\mathcal{A}$-wall A containing M and M', there exist roots α and α' in A such that $M = \mu(\alpha)$, $M' = \mu(\alpha')$ and α' is contained maximally in α.*
(ii) *If M and M' are adjacent, then the set $[M, M']$ is independent of the choice of A in 10.4.*
(iii) *If M and M' are adjacent, then the set $[M, M']$ is contained in every $\mathcal{A}$-apartment that contains M and M'.*

Proof. Suppose that M and M' are adjacent and let A and A' be two $\mathcal{A}$-apartments that contain them both. By 10.5 and 29.13.iii, both A and A' contain $[M, M']$. All three claims hold, therefore, by 29.34. □

Proposition 10.7. *Let M, M' and M'' be $\mathcal{A}$-walls such that both M' and M'' are adjacent to M (as defined in 10.2) and suppose that*

$$P \cap [M, M'] \cap [M, M''] \neq \emptyset$$

for some $P \in M$. Then $M' = M''$.

Proof. Let A, α and α' be as in 10.2 (with respect to M and M') and let A_1, α_1 and α_1' be as in 10.2 with M'' in place of M'. By 10.3, $P \cap [M, M'] \subset \partial\alpha$ and $P \cap [M, M''] \subset \partial\alpha_1$. By 29.42, therefore, $\partial\alpha = \partial\alpha_1$. By 10.4, it follows that $[M, M'] = [M, M'']$. By 1.42, a panel is contained in M' (respectively, M'') if and only if it contains exactly one chamber in $[M, M']$ (respectively, $[M, M'']$) but is not contained in M. Thus $M' = M''$. □

Proposition 10.8. *Let α be a root of an apartment A in $\mathcal{A}$, let P be a panel in $\mu(\alpha)$, let y be a chamber in P but not in α and let X denote the set of $\mathcal{A}$-apartments containing $\alpha \cup \{y\}$. Then α is a root in every apartment in X, and if $A_1 \in X$ and β is the root of A_1 containing α minimally, then β is contained in all the apartments in X.*

Proof. Let $A_1 \in X$ and let $x \in \alpha$. Then $\text{retr}_{A,x}$ is an isomorphism from A_1 to A acting trivially on α (by 8.17 and 8.18 in [37]). Thus α is a root of A_1. By 1.38, the root of A_1 containing α minimally is the convex hull of $\alpha \cup \{y\}$ and hence lies in every apartment in X (by 29.13.iii). □

Definition 10.9. We call the set of apartments $\mathcal{A}$ *full* if for all choices of A, α, P and y, the set X in 10.8 is always non-empty.

We will show in 10.30 that *every* system of apartments is full. For now, we have only the following observation.

Proposition 10.10. *If $\mathcal{A}$ is the complete system of apartments (as defined in 8.5), then it is full.*

Proof. This holds by 29.33. □

Until we reach 10.30, it will be important to pay careful attention at each step whether $\mathcal{A}$ is assumed to be complete, full or simply arbitrary.

Proposition 10.11. *Suppose that $\mathcal{A}$ is full. Let $M_0, M_1, \ldots, M_k$ be a sequence of $\mathcal{A}$-walls such that M_j is adjacent to M_{j-1} (as defined in 10.2) for all $j \in [1, k]$ and $M_j \neq M_{j-2}$ for all $j \in [2, k]$. Then there exists an apartment A in $\mathcal{A}$ containing roots*

$$\alpha_0, \alpha_1, \ldots, \alpha_k$$

such that $M_j = \mu(\alpha_j)$ for all $j \in [0, k]$ and α_j contains α_{j-1} minimally for all $j \in [1, k]$.

Proof. We proceed by induction with respect to k. If $k = 1$, then (by 10.2) there is nothing to show. Suppose that $k > 1$ and that A_0 is an $\mathcal{A}$-apartment containing roots $\alpha_0, \alpha_1, \ldots, \alpha_{k-1}$ such that $M_j = \mu(\alpha_j)$ for all $j \in [0, k-1]$ and α_j contains α_{j-1} minimally for all $j \in [1, k-1]$. Let $P \in \mu(\alpha_{k-1})$ and let x be the unique chamber in $P \cap \alpha_{k-1}$. By 10.6,

$$[M_{j-1}, M_j] = -\alpha_{j-1} \cap \alpha_j$$

for all $j \in [1, k-1]$. Thus, in particular, $P \cap [M_{k-2}, M_{k-1}] = \{x\}$.

By 10.3, there is a unique chamber y in P that is contained in the strip $[M_{k-1}, M_k]$. By 10.7, we have $x \neq y$ (since $M_{k-2} \neq M_k$). Thus $y \notin \alpha_{k-1}$. Hence by 10.9, there exists an $\mathcal{A}$-apartment A containing $\alpha_{k-1} \cup \{y\}$. Let α_k be the root of A that contains α_{k-1} minimally and let $M^\# = \mu(\alpha_k)$. By 10.3, $[M_{k-1}, M^\#]$ contains y. By 10.7 again, it follows that $M^\# = M_k$. □

Corollary 10.12. *Let $M_0, M_1, \ldots, M_k$ be a sequence of $\mathcal{A}$-walls such that M_j is adjacent to M_{j-1} (as defined in 10.2) for all $j \in [1, k]$ and $M_j \neq M_{j-2}$ for all $j \in [2, k]$. Then $M_k \neq M_0$.*

Proof. We can replace $\mathcal{A}$ by the complete system of apartments. The result holds then by 10.10 and 10.11. □

Definition 10.13. Let m be a parallel class of $\mathcal{A}$-walls. We let T_m denote the graph whose vertex set is the set of walls in m and whose edges are pairs of walls in m that are adjacent as defined in 10.2.

Proposition 10.14. *Let T_m be as in 10.13 for some parallel class m of $\mathcal{A}$-walls. Then T_m is a tree.*

Proof. By 10.12, T_m has no circuits. It will thus suffice to show that T_m is connected. Let M and M' be two walls in m, i.e. two vertices of T_m. Let A and A' be $\mathcal{A}$-apartments containing roots α and α' cutting M and M'. By 8.30, α^∞ and $(\alpha')^\infty$ are roots of the building at infinity $\Delta_{\mathcal{A}}^\infty$. Suppose that they are the same. Then by 8.31, $\alpha \cap \alpha'$ is a root of $A \cap A'$ parallel to both α and α'. It follows that M and M' are in the same connected component of T_m.

Suppose that the roots α^∞ and $(\alpha')^\infty$ are different. Since M and M' are parallel, we have $[M] = [M']$, where $[M]$ and $[M']$ are as defined in 8.29. By 8.35, it follows that α^∞ and $(\alpha')^\infty$ are roots of $\Delta_{\mathcal{A}}^\infty$ having the same wall. Thus by 29.48, the union of these two roots is an apartment of $\Delta_{\mathcal{A}}^\infty$. By 8.27, the apartment is of the form A_0^∞ for a unique apartment A_0 of Δ. Thus there exist opposite roots α_0 and α_0' of A_0 such that $(\alpha_0)^\infty = \alpha^\infty$ and $(\alpha_0')^\infty = (\alpha')^\infty$. Thus by 8.31 again, $\alpha \cap \alpha_0$ is a root of $A_0 \cap A$ parallel to α and $\alpha' \cap \alpha_0'$ is a root of $A_0 \cap A'$ parallel to α'. It follows again that M and M' are in the same connected component of T_m. □

Proposition 10.15. *Suppose that $\mathcal{A}$ is full. Let M be a wall and let $P \in M$. For each $\mathcal{A}$-wall M' adjacent to M, there is a unique chamber $x_{M'}$ in $P \cap [M, M']$ and the map*

$$M' \mapsto x_{M'}$$

is a bijection from the set of $\mathcal{A}$-walls adjacent to M to P.

Proof. By 1.42 and 10.3, $|P \cap [M, M']| = 1$ for each $\mathcal{A}$-wall M' adjacent to M. The map $M' \mapsto x_{M'}$ is injective by 10.7 and surjective by 10.9. □

Corollary 10.16. *Suppose that $\mathcal{A}$ is full and let m be a parallel class of $\mathcal{A}$-walls. Then T_m is a thick tree.*

Proof. By 10.14, T_m is a tree. By 10.15, T_m is thick (since Δ is assumed to be thick). □

Proposition 10.17. *Let R be an arbitrary gem and let S be a sector of Δ. Then S is parallel to a unique sector S_1 whose terminus is R.*[1]

Proof. By 7.6, there exists an apartment A (which might not be in $\mathcal{A}$) containing a subsector of S and a chamber of R. By 4.23, S is parallel to this subsector. By 4.24, therefore, S is parallel to a sector S_1 in A having terminus R. Now suppose that S_2 is a second sector parallel to S whose terminus is R. Let d be the apex of S_1 and let e be the apex of S_2. By 4.3 (and 29.10.v), $d = \text{proj}_R u$ for all $u \in S_1$ and $e = \text{proj}_R u$ for all $u \in S_2$. By 4.23, $S_1 \cap S_2$ is a subsector and thus, in particular, not empty. It follows that $d = e$. By 4.18, therefore, $S_1 = S_2$. □

---

[1]We are not claiming in 10.17 that $S_1 \in \mathcal{A}$ even if $S \in \mathcal{A}$. See, however, 10.27.

Notation 10.18. Let R be a gem. For each sector S, let d_S denote the apex of the sector S_1 described in 10.17. We denote the map $S^\infty \mapsto d_S$ from parallel classes of $\mathcal{A}$-sectors (i.e. chambers of the building at infinity $\Delta_{\mathcal{A}}^\infty$) to R by ψ_R.

Definition 10.19. Let m be a parallel class of $\mathcal{A}$-walls and let T_m be the graph defined in 10.13. Thus by 10.16, T_m is a thick tree if $\mathcal{A}$ is full. For each $A \in \mathcal{A}$, let A_m denote the set of walls in A contained in m, let $m_{\mathcal{A}}$ be the set of all $A \in \mathcal{A}$ such that A_m is non-empty and let

$$\mathcal{A}_m = \{A_m \mid A \in m_{\mathcal{A}}\}.$$

Let $A \in m_{\mathcal{A}}$. By 10.2, the set A_m is the vertex set of a thin subgraph, i.e. of an apartment, of T_m. We can thus talk about rays in A_m: A ray in A_m is a set of the form $\{\mu(\beta) \mid \beta \subset \alpha\}$ (which we will denote by α_m) for some root α of A such that $\mu(\alpha) \in m$. There is thus a canonical correspondence between the two ends of A_m and the two parallel classes of roots α of A such that $\mu(\alpha) \in m$.

Proposition 10.20. *Suppose that $\mathcal{A}$ is full, let m be a parallel class of $\mathcal{A}$-walls and let T_m and $\mathcal{A}_m$ be as in 10.13 and 10.19. Then the pair $(T_m, \mathcal{A}_m)$ is a thick tree with sap as defined in 9.5.*

Proof. By 10.16, T_m is a thick tree. By 10.11, every two vertices of T_m are contained in an element of $\mathcal{A}_m$. Thus 9.5.i holds. Choose apartments A and A' in $m_{\mathcal{A}}$ and roots α of A and α' of A' such that the walls $\mu(\alpha)$ and $\mu(\alpha')$ are both contained in the parallel class m. By 8.30 and 8.33, α^∞ and $(\alpha')^\infty$ are roots of $\Delta_{\mathcal{A}}^\infty$ having walls $[\mu(\alpha)]$ and $[\mu(\alpha')]$. By 8.36, these two walls are the same.

Suppose now that the rays α_m and α'_m of T_m (as defined in 10.19) are not parallel. By 8.31, the roots α^∞ and $(\alpha')^\infty$ of $\Delta_{\mathcal{A}}^\infty$ are distinct. By 8.27 and 29.48, there is a unique apartment $A_1 \in \mathcal{A}$ such that the apartment A_1^∞ is the union of α^∞ and $(\alpha')^\infty$. There thus exist roots α_1 and α_2 in A_1 such that α_2 is opposite α_1 in A_1, $\alpha_1^\infty = \alpha^\infty$ and $(\alpha_2)^\infty = (\alpha')^\infty$. By 8.31 again, $\alpha \cap \alpha_1$ and $\alpha' \cap \alpha_2$ are roots of A'. Therefore $(A_1)_m \in \mathcal{A}_m$ and $(A_1)_m$ contains subrays of α_m and α'_m. Thus 9.5.ii holds. □

Definition 10.21. A *wall tree* of $(\Delta, \mathcal{A})$ is a pair $(T_m, \mathcal{A}_m)$ for some parallel class of $\mathcal{A}$-walls m, where T_m is as in 10.13 and $\mathcal{A}_m$ is as in 10.19.[2]

Proposition 10.22. *Suppose that $\mathcal{A}$ is full and let M be an $\mathcal{A}$-wall. Let α and α' be two $\mathcal{A}$-roots such that $\mu(\alpha) = \mu(\alpha') = M$ but*

$$\alpha \cap \alpha' \cap M = \emptyset. \tag{10.23}$$

Then $\alpha \cup \alpha'$ is an $\mathcal{A}$-apartment.[3]

[2]Thus a wall tree is, in fact, a tree with sap and not simply a tree. In fact, we only know that $(T_m, \mathcal{A}_m)$ is a tree with sap when the system of apartments $\mathcal{A}$ is full (by 10.20); we are thus anticipating 10.30 with this comment.

[3]More precisely, the subgraph of Δ spanned by $\alpha \cup \alpha'$ is an $\mathcal{A}$-apartment.

Proof. Let m be the parallel class of $\mathcal{A}$-walls containing M and let α_m and α'_m be as in 10.19. Then M is a vertex of the tree T_m that is contained in both α_m and α'_m. By 10.15 and 10.23, the vertex of T_m adjacent to M in α_m is distinct from the vertex of T_m adjacent to M in α'_m. By 10.20, therefore, there exists an apartment $A \in \mathcal{A}_m$ such that $\alpha_m \cup \alpha'_m = A_m$. By 1.41 and 10.6.iii, it follows that $\alpha \cup \alpha' = A$. □

Theorem 10.24. *Let R be a gem. Then the map ψ_R from $\Delta_{\mathcal{A}}^\infty$ to R given in 10.18 is a surjective homomorphism (as defined in 29.2).*

Proof. Let $d \in R$. By 8.4.i, there exists an $\mathcal{A}$-apartment A containing d. Let $S = \sigma_A(R, d)$. Then $\psi_R(S^\infty) = d$. Thus ψ_R is surjective.

To prove that ψ_R is a homomorphism, we can assume that $\mathcal{A}$ is complete (and hence full by 10.10). Let x and x' be adjacent chambers of $\Delta_{\mathcal{A}}^\infty$. By 10.17, there exist unique sectors S and S' in x and x' such that the terminus of both S and S' is R. Let d be the apex of S and let e be the apex of S'. Thus $d = \psi_R(x)$ and $e = \psi_R(x')$. By 8.17 and 8.18, S and S' have faces f and f' that are parallel.

Let A be an apartment containing S. There is a unique wall M in A containing f; let $m = M^\infty$. By 8.17 and 8.18, $[f]$ is the panel containing x and x'. By 8.29 and 8.33, $[f] \in [M]$ and $[M]$ is a wall of $\Delta_{\mathcal{A}}^\infty$. By 8.30, 8.33 and 29.47, there is an $\mathcal{A}$-root α such that $[\alpha^\infty] = [M]$ and $x' \in \alpha^\infty$. Let A_1 be an $\mathcal{A}$-apartment containing α and let $M' = \mu(\alpha)$. By 8.35, M' is parallel to M. We have

$$[f] \in [M] = [M'] = \{[f'] \mid f' \text{ is a face contained in } M'\}$$

by 8.29. Hence by 8.28, the wall M' contains a face parallel to f. Since $x' \in [f]$ and $x' \in \alpha^\infty$, there exists a sector S'_1 contained in A_1 that is parallel to S' and has a face f'_1 parallel to f. Let M_1 be the unique wall of A_1 that contains f'_1. By 5.16, M_1 is parallel to M' and hence also to M. We have thus produced an $\mathcal{A}$-apartment A_1 and a wall M_1 contained in A_1 that is parallel to M such that A_1 contains a sector S'_1 parallel to S' having a face f'_1 contained in M_1 and parallel to f. We now assume that A_1 and M_1 are chosen among all such pairs so that the distance k from M to M_1 in the tree T_m is minimal. Let α_1 be the root of A_1 containing S'_1 such that $\mu(\alpha_1) = M_1$.

We suppose first that $M \neq M_1$, i.e. that $k > 0$, and let $(M_1, \hat{M}_1, \ldots, M)$ be the unique path from M_1 to M in T_m. Suppose that $\hat{M}_1$ is contained in some $\mathcal{A}$-apartment $A^\#$ that contains α_1. By 1.26 and 1.46, there is a translation g of $A^\#$ such that $\mu(\alpha_1^g) = \hat{M}_1$. Thus $(S'_1)^g$ is a sector of $A^\#$ having a face $\hat{f}_1$ contained in $\hat{M}_1$. By 1.45, $(S'_1)^g$ is parallel to S' and $\hat{f}_1$ is parallel to f. This contradicts the choice of A_1 and M_1. Hence $\hat{M}_1$ is not contained in any $\mathcal{A}$-apartment containing α_1.

By 10.20, there exists an apartment $A_2 \in m_{\mathcal{A}}$ such that $(A_2)_m$ contains both M_1 and $\hat{M}_1$. There are thus unique roots α_2 and $\hat{\alpha}_2$ in A_2 such that $M_1 = \mu(\alpha_2)$, $\hat{M}_1 = \mu(\hat{\alpha}_2)$ and $\hat{\alpha}_2 \subset \alpha_2$. Since $\alpha_1 \subset A_1$, the wall $\hat{M}_1$ is not contained in A_1 by the conclusion of the previous paragraph. By 10.7 and

29.42, it follows that

$$\alpha_1 \cap \alpha_2 \cap M_1 = \emptyset.$$

By 10.22, therefore, $A^\# := \alpha_1 \cup \alpha_2$ is an $\mathcal{A}$-apartment. This contradicts the conclusion of the previous paragraph since $A^\#$ contains both α_1 and $\hat{M}_1 = \mu(\hat{\alpha}_2)$. We conclude that $M = M_1$.

Since $M = M_1$ and M contains panels that are contained in R, the apartment A_1 contains chambers in R. Let R_1 be the terminus of S_1'. By 1.9, there exists a translation h of A_1 such that $(R_1)^h = R$. By 1.45, $(S_1')^h$ is a sector with terminus R that is parallel to S'. Thus $(S_1')^h = S'$ by the uniqueness of S'. Since $\mu(\alpha_1) = M_1 = M$, both α_1 and α_1^h cut $R \cap A$. By 1.13, it follows that $\alpha_1 = \alpha_1^h$. Therefore S' has a face parallel to f that is contained in M. Since no two faces of a sector are parallel (by 5.9), it follows that $f' \subset M$. Let α be the unique root of A containing S such that $M = \mu(\alpha)$.

Suppose that $\partial\alpha = \partial\alpha_1$ and choose a chamber x in $\partial\alpha$. Then $\text{retr}_{A,x}(S')$ is a sector of A having face f'. Thus f and f' are parallel faces of sectors contained in A having the same terminus. By 5.9, it follows that $f = f'$. Now suppose that $\partial\alpha \neq \partial\alpha_1$. By 29.42, $\partial\alpha$ and $\partial\alpha_1$ are disjoint. Thus by 10.22, $A^\# := \alpha \cup \alpha_1$ is an $\mathcal{A}$-apartment. Therefore, f and f' are parallel faces of sectors contained in $A^\#$ having the same terminus. By 5.9 again, it follows that $f = f'$.

Let P be the unique panel in f contained in R. Since f contains d and f' contains e, it follows that P contains both d and e. Let o' be the unique special vertex such that the gem R is o'-special, let $i \in I_o$ be the type of the panel of $\Delta_{\mathcal{A}}^\infty$ containing x and x', let $j \in I_{o'}$ be the type of the panel P (so f is the j-face of both S and S') and let $\sim_i'$ for all $i \in I_{o'}$ be as defined in 8.9 with o' in place of o. Thus $x \sim_j' x'$. If σ is as in 8.12, then $j = \sigma(i)$. Note that σ depends only on o and o' and not on x or x'. We conclude that if x and x' are i-adjacent chambers of $\Delta_{\mathcal{A}}^\infty$ for some $i \in I_o$ (where o is the special vertex used in the construction of $\Delta_{\mathcal{A}}^\infty$ and thus Π_o is the type of $\Delta_{\mathcal{A}}^\infty$ by 8.24), then $\psi_R(x)$ and $\psi_R(x')$ are both contained in a $\sigma(i)$-panel of R (which is a building of type $\Pi_{o'}$ by 29.10.iii). Thus ψ_R is a homomorphism from $\Delta_{\mathcal{A}}^\infty$ to R. □

Proposition 10.25. *Let R be a gem. Then every panel of R is of the form $\psi_R(F)$ for some panel F of $\Delta_{\mathcal{A}}^\infty$, where ψ_R is as in 10.18.*

Proof. This holds by 10.24 and 29.53. □

Proposition 10.26. *Let R be a gem and let d and e be opposite chambers of R, let A and A' be apartments in $\mathcal{A}$ such that $d \in A$ and $e \in A'$, let $S = \sigma_A(R, d)$ and $S' = \sigma_{A'}(R, e)$. Then there is a unique $\mathcal{A}$-apartment A_1 containing $S \cup S'$ and A_1 is, in fact, the only apartment of Δ (in $\mathcal{A}$ or not) containing subsectors of both S and S'.*

Proof. By 8.4.ii, there is an apartment $A_1 \in \mathcal{A}$ containing subsectors S_1 of S and S_1' of S'. The apartments in $\Delta_{\mathcal{A}}^\infty$ and in R are all isomorphic to Σ_{Π_o}.

By 29.14.i, it follows that $\Delta_{\mathcal{A}}^{\infty}$ and R have the same diameter. Since d and e are opposite in R and the map ψ_R defined in 10.18 is a homomorphism by 10.24, the chambers S^{∞} and $(S')^{\infty}$ are therefore opposite in $\Delta_{\mathcal{A}}^{\infty}$. By 8.27 and 29.14.ii, A_1 is the only apartment in $\mathcal{A}$ containing subsectors of both S and S'. Since $\mathcal{A}$ is an arbitrary system of apartments, we conclude that A_1 is also the only apartment in $\hat{\mathcal{A}}$ containing subsectors of both S and S', where $\hat{\mathcal{A}}$ is the complete system of apartments of Δ.

Choose $u \in S_1$ and $u' \in S_1'$. By 8.4.i, there is an apartment A_2 in $\mathcal{A}$ containing d and e. Let $S_2 = \sigma_{A_2}(R, d)$, let $S_2' = \sigma_{A_2}(R, e)$, let γ be a minimal gallery in S from d to u, let γ' be a minimal gallery in S' from u' to e and let γ_0 be a minimal gallery from d to e. Next let γ_2 be the unique gallery in A_2 starting at d and having the same type as γ and let γ_2' be the unique gallery in A_2 ending at e and having the same type as γ'. Let x be the last chamber of γ_2 and let x' be the last chamber of γ_2'. By 4.3 and 29.10.v, (γ^{-1}, γ_0) is a minimal gallery. Therefore also $(\gamma_2^{-1}, \gamma_0)$ is a minimal gallery (by 29.10.i). It follows $\mathrm{proj}_R\, x = d$. Therefore $x \in S_2$ by another application of 4.3. Similarly, $x' \in S_2'$. By 4.20, we conclude that $(\gamma_2^{-1}, \gamma_0, \gamma_2')$ is a minimal gallery. Therefore also $(\gamma^{-1}, \gamma_0, \gamma')$ is a minimal gallery. Since u' and u, its first and last chambers, both lie in A_1, the whole gallery lies in A_1 (by 29.13.iii). Therefore d and e lie in A_1. By 4.18, it follows that $S \cup S'$ is contained in A_1. □

We can now prove an improved version of 10.17:

Proposition 10.27. *Let R be an arbitrary gem and let S be an $\mathcal{A}$-sector. Then S is parallel to a unique $\mathcal{A}$-sector S_1 whose terminus is R.*

Proof. By 10.17, there exists a unique sector S_1 parallel to S whose terminus is R. We must show that S_1 lies in an apartment in $\mathcal{A}$. Let $S_2 = S \cap S_1$, let d be the apex of S_1 and let e be an arbitrary chamber of R that is opposite d in R. Then S_2 is an $\mathcal{A}$-sector since S is one. By 8.4.i, there exists an $\mathcal{A}$-apartment A containing the chamber e. Let $S' = \sigma_A(R, e)$. By 8.4.ii, there exists an $\mathcal{A}$-apartment A_1 in $\mathcal{A}$ containing subsectors of S' and S_2. By 10.26, the apartment A_1 contains d. By 4.18, therefore, $S_1 \subset A_1$. □

The main idea in the proof of the next result was suggested by Linus Kramer.

Proposition 10.28. *The system of apartments $\mathcal{A}$ is full (as defined in 10.9).*

Proof. Let A, α, P and $y \in P\backslash\alpha$ be as in 10.8 and let x be the unique chamber in $\alpha \cap P$. By 1.16, there is a gem R cut by α. Let Q be a panel cut by α that is contained in R, let $u = \mathrm{proj}_Q\, x$ and let $v = \mathrm{proj}_Q\, y$. By 29.39 and 29.43, $u \in \alpha$ and $v \notin \alpha$. By 29.43, we also have $y = \mathrm{proj}_P\, v$, so there is a minimal gallery from x to v that passes through y (by 29.8.v). Thus if there is an apartment A_1 containing $\alpha \cup \{v\}$, then $y \in A_1$ (since apartments are convex). It thus suffices to assume that $P = Q$, i.e. that P is contained in a gem R.

Let z be the unique chamber in $P \cap A$ distinct from x, let d be the unique chamber opposite z in $R \cap A$ and let e be the unique chamber opposite x in $R \cap A$. By 29.9.v, d and e are adjacent. By 29.9.i, $d \in \alpha$ and $e \notin \alpha$. Thus $d \in \partial\alpha$. Let $S = \sigma_A(R, d)$, $\check{S} = \sigma_A(R, z)$ and $\hat{S} = \sigma_A(R, x)$. Let F be the unique panel of $\Delta_{\mathcal{A}}^{\infty}$ containing the chambers $\check{u} := \check{S}^{\infty}$ and $\hat{u} := \hat{S}^{\infty}$. Thus $\psi_R(\check{u}) = z$ and $\psi_R(\hat{u}) = x$. By 10.25, it follows that $\psi_R(F) = P$. There thus exists a chamber $u' \in F$ such that $\psi_R(u') = y$. By 10.27, there is a unique $\mathcal{A}$-sector S' in the parallel class u' having apex y.

By 29.9.iii, there is a minimal gallery

$$(x_0, \ldots, x_k, z)$$

in A with $x_0 = d$ and $x_k = x$, where $k = \operatorname{diam}(R) - 1$ and thus

$$k = \operatorname{diam}(\Delta_{\mathcal{A}}^{\infty}) - 1. \tag{10.29}$$

Since the gallery

$$\gamma_1 := (x_0, \ldots, x_k, y)$$

has the same type, it is also a minimal gallery (by 29.10.i). Let

$$u_i = \sigma_A(R, x_i)^{\infty}$$

for all $i \in [1, k]$. Thus $u_k = \hat{u}$ and

$$\gamma_2 := (u_0, \ldots, u_k, u')$$

is a gallery in $\Delta_{\mathcal{A}}^{\infty}$ which is mapped by ψ_R to γ_1. By 10.24, it follows that γ_2 is minimal.

By 10.26, there is a unique $A_1 \in \mathcal{A}$ containing $S \cup S'$. In particular, A_1 contains y and d and A_1^{∞} contains u and u'. By 29.13.iii, it follows that γ_2 is contained in A_1^{∞} (since γ_2 is minimal). By 4.11, S and $\hat{S}$ are contained in the root α. Hence u_0 and $u_k = \hat{u}$ are contained in the root α^{∞}. By 10.29, the distance from u_0 to u_k in $\Delta_{\mathcal{A}}^{\infty}$ is $\operatorname{diam}(\Delta_{\mathcal{A}}^{\infty})$. By 29.6.iii and 29.9.ii, α^{∞} is the unique root of A^{∞} that contains u_0 and u_k. By 29.6.iv, therefore, α^{∞} is the convex hull in $\Delta_{\mathcal{A}}^{\infty}$ of $\{u_0, u_k\}$. Thus $\alpha^{\infty} \subset A^{\infty} \cap A_1^{\infty}$. Hence by 4.26, $A \cap A_1$ is a root of A parallel to α. Since $d \in A \cap A_1$ and $d \in \partial\alpha$, it follows that $\alpha \subset A \cap A_1$. Therefore A_1 is an $\mathcal{A}$-apartment containing $\alpha \cup \{y\}$. □

The result in 10.28 is important enough to reformulate as follows:

Theorem 10.30. *Every system of apartments is full.*

Proof. The system of apartments $\mathcal{A}$ is arbitrary. The claim holds, therefore, by 10.28. □

Corollary 10.31. *For each wall m of $\Delta_{\mathcal{A}}^{\infty}$, $(T_m, \mathcal{A}_m)$ is a thick tree with sap.*

Proof. This holds by 10.20 and 10.30. □

Corollary 10.32. *Let M be an $\mathcal{A}$-wall. Then every wall parallel to M is an $\mathcal{A}$-wall.*

Proof. Let $\hat{\mathcal{A}}$ be the complete set of apartments of Δ, let $\hat{m}$ be the wall of $\Delta_{\hat{\mathcal{A}}}$ that contains M and let $m \subset \hat{m}$ be the set of $\mathcal{A}$-walls parallel to M. By 10.15 and 10.30, every wall in $\hat{m}$ adjacent in $T_{\hat{m}}$ to a wall contained in m is contained in m. Since $T_{\hat{m}}$ is connected, it follows that $\hat{m} = m$. □

Corollary 10.33. *Let M be an $\mathcal{A}$-wall and let M' be a wall parallel to M. Then there exists an $\mathcal{A}$-apartment containing both M and M'.*

Proof. By 10.32, M' is an $\mathcal{A}$-wall. The claim holds, therefore, by 10.31. □

Corollary 10.34. *Let M be an $\mathcal{A}$-wall and let α and α' be $\mathcal{A}$-roots cutting M such that*

$$\alpha \cap \alpha' \cap M = \emptyset.$$

Then $\alpha \cup \alpha'$ is an $\mathcal{A}$-apartment.

Proof. This holds by 10.22 and 10.30. □

Corollary 10.35. *Let A be an $\mathcal{A}$-apartment, let α_1 be a root of A and let α_2 be its opposite in A. Then there exists a third $\mathcal{A}$-root α_3 having the same wall as α_1 and α_2 such that $\alpha_i \cup \alpha_j$ is an $\mathcal{A}$-apartment for all pairs i, j of distinct elements of the index set $[1, 3]$.*

Proof. Let P be a panel in $\mu(\alpha_1)$ and let $y \in P \backslash A$. By 10.30, we can choose an $\mathcal{A}$-apartment A_1 containing $\alpha_1 \cup \{y\}$. The root α_1 of A is also a root of A_1; let α_3 be its opposite in A_1. The claim holds now by 29.42 and 10.34. □

Corollary 10.36. *Let A be an $\mathcal{A}$-apartment and let α be a root of A. Then there exists an $\mathcal{A}$-apartment A_1 such that $\alpha = A \cap A_1$.*

Proof. This holds by 10.35. □

Here is an improved version of 7.6:

Proposition 10.37. *Let u be a chamber and let S be an $\mathcal{A}$-sector of Δ. Then there exists an $\mathcal{A}$-sector containing both u and a subsector of S.*

Proof. In light of 10.30, the proof of 7.6 applies verbatim. □

Proposition 10.38. *Let the set of walls of $\Delta_{\mathcal{A}}^{\infty}$ be identified with the set of parallel classes of $\mathcal{A}$-walls of Δ under the map $[M] \mapsto M^{\infty}$ described in 8.36.ii. Let m be a wall of $\Delta_{\mathcal{A}}^{\infty}$ and let ξ_m denote the map from the set of roots of $\Delta_{\mathcal{A}}^{\infty}$ whose wall is m to the set of $\mathcal{A}_m$-ends of the tree T_m given by*

$$\xi_m(\alpha^{\infty}) = (\alpha_m)^{\infty}$$

for each root α of an $\mathcal{A}$-apartment such that $\mu(\alpha) \in m$, where α^{∞} is as in 8.29, the ray α_m is as in 10.19 and $(\alpha_m)^{\infty}$ is the end of T_m containing the ray α_m. Then the map ξ_m is a bijection.

Proof. By 8.30 and 8.33, every root of $\Delta_{\mathcal{A}}^{\infty}$ whose wall equals m is of the form α^{∞} for some root α of Δ whose wall is contained in the parallel class m. The map ξ_m is surjective by the definition of $(T_m, \mathcal{A}_m)$ given in 10.19. Suppose A and A' are two apartments in $\mathcal{A}_m$ and α and α' are roots of A and A' whose walls both lie in m. Suppose, too, that $(\alpha_m)^{\infty} = (\alpha'_m)^{\infty}$. By 1.41 and 10.6.iii, $\alpha \cap \alpha'$ contains a root. This root is parallel to α and to α', so α is parallel to α'. Therefore $\alpha^{\infty} = (\alpha')^{\infty}$. Hence ξ_m is injective. □

Chapter Eleven

Panel Trees

In this chapter, which is a companion to the previous chapter on wall trees, we introduce a second, related family of trees with sap, one for each panel of the building at infinity $\Delta_{\mathcal{A}}^{\infty}$.

We continue to assume that Π is one of the affine Coxeter diagrams in Figure 1.1 with vertex set I, that Δ is a thick building of type Π and that $\mathcal{A}$ is a system of apartments as defined in 8.4.

Conventions 11.1. From now on, we identify the set of panels of $\Delta_{\mathcal{A}}^{\infty}$ with the set of parallel classes of $\mathcal{A}$-faces via the map $[f] \mapsto f^{\infty}$ and we identify the set of $\mathcal{A}$-walls of $\Delta_{\mathcal{A}}^{\infty}$ with the set of parallel classes of walls in Δ via the map $[M] \mapsto M^{\infty}$ as described in 8.36. Thus, in particular, a panel F of $\Delta_{\mathcal{A}}^{\infty}$ is contained in a wall m of $\Delta_{\mathcal{A}}^{\infty}$ if and only if each $\mathcal{A}$-wall of Δ in the parallel class m contains an $\mathcal{A}$-face in the parallel class F.

Proposition 11.2. *Let f be a face contained in an $\mathcal{A}$-wall M of Δ. Then each $\mathcal{A}$-apartment A containing M contains a sector S having f as a face.*

Proof. Let A be an $\mathcal{A}$-apartment containing M. By 10.1, there exists a root α in A such that $M = \mu(\alpha)$. Since f is a face, we can choose an $\mathcal{A}$-apartment A_1 containing a sector S_1 having f as a face. Suppose that $A \cap A_1 \neq \emptyset$ and let $x \in A \cap A_1$. Then the restriction of $\rho := \text{retr}_{A,x}$ to A_1 is a special isomorphism from A_1 to A that acts trivially on $A \cap A_1$ (by 8.17 and 8.18 in [37]). Thus $\rho(P) \subset P$ for each panel $P \in \mu(\alpha)$ and hence $\rho(S_1)$ is a sector of A having f as a face.

Suppose now that $A \cap A_1 = \emptyset$. Choose a panel P in f and let y be a chamber in $P \cap A_1$. By 10.30, there exists an $\mathcal{A}$-apartment A_2 containing $\alpha \cup \{y\}$. Thus $A_1 \cap A_2 \neq \emptyset$, so f is the face of a sector in A_2, and $A \cap A_2 \neq \emptyset$, so f is the face of a sector in A (by two applications of the conclusion of the previous paragraph). $\square$

Proposition 11.3. *The building at infinity $\Delta_{\mathcal{A}}^{\infty}$ is thick.*

Proof. Let f be an $\mathcal{A}$-face, let M be an $\mathcal{A}$-wall of Δ containing f and let F denote the panel $[f]$ of $\Delta_{\mathcal{A}}^{\infty}$ (as defined in 8.17). By 10.35, there exist roots $\alpha_1, \alpha_2, \alpha_3$ such that $\mu(\alpha_i) = M$ for all $i \in [1,3]$ and $\alpha_i \cup \alpha_j$ is an $\mathcal{A}$-apartment for all pairs i, j of distinct elements in $[1,3]$. By 11.2, there exist sectors $S_i \subset \alpha_i$ having f as a face for all $i \in [1,3]$. Since the sectors S_1, S_2, S_3 are pairwise disjoint, the chambers $S_1^{\infty}, S_2^{\infty}$ and S_3^{∞} are pairwise distinct elements of F. By 8.18, we conclude that every panel of $\Delta_{\mathcal{A}}^{\infty}$ contains at least three chambers. $\square$

Proposition 11.4. *Suppose f_1, f_2 and f_3 are three faces such that $f_1 \cap f_2$ and $f_2 \cap f_3$ both contain faces. Then $f_1 \cap f_3$ also contains a face.*

Proof. Let A be an apartment containing a sector S_2 that has f_2 as a face. By 11.2, there are sectors S_1 and S_3 in A such that S_1 has a face $f_1^\#$ contained in $f_1 \cap f_2$ and S_3 has a face $f_3^\#$ contained in $f_2 \cap f_3$. Thus $f_1^\#$ and $f_3^\#$ are both subfaces of f_2 and $f_1^\# \cap f_3^\# \subset f_1 \cap f_3$. By 5.15 (applied to A), $f_1^\# \cap f_3^\#$ is a face. □

Proposition 11.5. *Let f and f' be parallel $\mathcal{A}$-faces both contained in an $\mathcal{A}$-wall. Then $f \cap f'$ is a face.*

Proof. Let M be an $\mathcal{A}$-wall containing f and f' and let A be an $\mathcal{A}$-apartment containing M. By 11.2, f and f' are faces of sectors contained in A. The claim holds, therefore, by 5.12. □

Proposition 11.6. *Let f and f' be $\mathcal{A}$-faces such that $f' \subset f$ contains a face. Then f' is parallel to f.*

Proof. Let A be an $\mathcal{A}$-face containing a sector that has f as a face. By 11.2, f' is also a face of a sector contained in A. The claim holds, therefore, by 5.13. □

Definition 11.7. Two $\mathcal{A}$-faces are called *asymptotic* if their intersection contains a face. By 11.4, this is an equivalence relation. The equivalence classes will be called $\mathcal{A}$-*asymptote classes.* The $\mathcal{A}$-asymptote class containing a face f will be denoted by f°. We will say that an $\mathcal{A}$-asymptote class e is contained in a wall M if there is a face in e that is contained in M. By 11.6, all the $\mathcal{A}$-faces in a given $\mathcal{A}$-asymptote class are parallel to one another. We can thus set $e^\infty = f^\infty$ for each asymptote class e, where f is an arbitrary face in e and f^∞ is as in 8.36. We will say that an $\mathcal{A}$-asymptote class e is parallel to an $\mathcal{A}$-asymptote class e_1 (or to an $\mathcal{A}$-face f) if $e^\infty = e_1^\infty$ (or $e^\infty = f^\infty$).

Definition 11.8. Let f and f' be two $\mathcal{A}$-faces. We will say that f and f' are *adjacent* if they are parallel and there exists an $\mathcal{A}$-apartment A containing subfaces f_1 of f and f_1' of f' such that the unique walls of A containing f_1 and f_1' are adjacent as defined in 10.2. We will say that two $\mathcal{A}$-asymptote classes (as defined in 11.7) are adjacent if they contain adjacent faces (in which case every face in the one asymptote class is adjacent to every face in the other).

Definition 11.9. For each panel F of $\Delta_{\mathcal{A}}^\infty$ (which we are interpreting as a parallel class of $\mathcal{A}$-faces by 11.1), let T_F be the graph whose vertices are the asymptote classes contained in F, where adjacency is defined as in 11.8. For each $A \in \mathcal{A}$, let A_F be the set of asymptote classes in F contained in A (i.e. that contain faces contained in A), let $F_{\mathcal{A}}$ denote the set of $A \in \mathcal{A}$ such that A_F is non-empty and let

$$\mathcal{A}_F = \{A_F \mid A \in F_{\mathcal{A}}\}.$$

Note that if F is a panel and m is a wall of $\Delta_{\mathcal{A}}^{\infty}$, then by 11.1,

$$F \in m \text{ if and only if } m_{\mathcal{A}} \subset F_{\mathcal{A}}. \tag{11.10}$$

Proposition 11.11. *Let F be a panel of $\Delta_{\mathcal{A}}^{\infty}$, let m be a wall of $\Delta_{\mathcal{A}}^{\infty}$ containing F, let $A \in m_{\mathcal{A}}$ (so also $A \in F_{\mathcal{A}}$ by 11.10), let A_m and $m_{\mathcal{A}}$ be as in 10.19 and let A_F and $F_{\mathcal{A}}$ be as in 11.9. Then every $e \in A_F$ is contained in a unique wall M_e of A_m and the map $e \mapsto M_e$ from A_F to A_m is a bijection.*

Proof. Since $F \in m$, each wall in m contains a face in F (by 11.1). Two distinct $\mathcal{A}$-walls in A_m are disjoint and hence do not contain asymptotic faces. We conclude that for each $e \in A_F$, there exists a unique wall $M_e \in A_m$ that contains it, and that the map $e \mapsto M_e$ is surjective. By 11.5, this map is injective. □

Proposition 11.12. *Let F be a panel of $\Delta_{\mathcal{A}}^{\infty}$, let $A \in F_{\mathcal{A}}$ and let m be a wall of $\Delta_{\mathcal{A}}^{\infty}$ containing F. Then there exists a unique apartment A' in $m_{\mathcal{A}}$ (and thus $A' \in F_{\mathcal{A}}$ by 11.10) such that $(A')_F = A_F$.*

Proof. Let S and S' be the distinct sectors of A sharing a face f contained in F (thought of as a parallel class of $\mathcal{A}$-faces). By 8.36.i, the two parallel classes S^{∞} and $(S')^{\infty}$ are both contained in F (thought of as a panel of $\Delta_{\mathcal{A}}^{\infty}$). By 8.27 and 29.49, there exists a unique $\mathcal{A}$-apartment A' containing a wall in m such that $(A')^{\infty}$ contains S^{∞} and $(S')^{\infty}$. The apartment A' contains subsectors of S and S'. Let X be the convex hull of $S \cup S'$. Since apartments are convex, we have $X \subset A \cap A'$. By 5.17, X contains sectors S_1 and S_1' sharing a common face that is contained in f. By 5.20 applied to A and then to A', a face in F is contained in A (respectively, A') if and only if it contains a subface that is contained in S_1, S_1' or f. Thus $A_F = (A')_F$. □

Note that by 11.12, there are many distinct apartments $A \in F_{\mathcal{A}}$ for which the set A_F is the same since in $\Delta_{\mathcal{A}}^{\infty}$ there are many different walls containing a given panel.

Proposition 11.13. *Let m be a wall of $\Delta_{\mathcal{A}}^{\infty}$ and let F be a panel contained in m. Then each asymptote class e contained in F is contained in a unique wall contained in m.*

Proof. Let $f \in F$. By 11.12, there exists a wall $M \in m$ containing the asymptote class f°. By 10.14, distinct walls in m are disjoint. Thus M is unique. □

Proposition 11.14. *Let m be a wall of $\Delta_{\mathcal{A}}^{\infty}$ and let F be a panel contained in m. Let M, M' be two $\mathcal{A}$-walls in m and let e, e' be $\mathcal{A}$-asymptote classes in F contained in M and M'. Then M and M' are adjacent in T_m if and only if e and e' are adjacent in T_F.*

Proof. This holds by 10.2, 11.8 and 11.13. □

Definition 11.15. Let m be a wall of $\Delta_{\mathcal{A}}^{\infty}$ and let F be a panel contained in m. By 11.13, there is a map from the set of $\mathcal{A}$-asymptote classes in F to

m that sends an $\mathcal{A}$-asymptote class to the unique wall in m that contains it. We denote this map by $\phi_{m,F}$.

Proposition 11.16. *Let m be a wall of $\Delta_{\mathcal{A}}^{\infty}$, let F be a panel contained in m and let $\phi_{m,F}$ be as in 11.15. Then $\phi_{m,F}$ is an isomorphism from T_F to T_m that maps $\mathcal{A}_F$ to $\mathcal{A}_m$.*

Proof. By 11.11,

$$\phi_{m,F}(A_F) = A_m \tag{11.17}$$

for each $A \in m_{\mathcal{A}}$. Since every wall in m is contained in an apartment in $m_{\mathcal{A}}$, it follows that $\phi_{m,F}$ is surjective. By 5.12 and 11.2, $\phi_{m,F}$ is injective. By 11.14, therefore, $\phi_{m,F}$ is an isomorphism from T_F to T_m. Let $A \in F_{\mathcal{A}}$ and let A' be as in 11.12. By 11.12 and 11.17, $\phi_{m,F}$ maps $\mathcal{A}_F$ to $\mathcal{A}_m$. □

Corollary 11.18. *For each panel F of $\Delta_{\mathcal{A}}^{\infty}$, $(T_F, \mathcal{A}_F)$ is a thick tree with sap.*

Proof. This holds by 10.31 and 11.16 (since there are walls of $\Delta_{\mathcal{A}}^{\infty}$ containing any given panel). □

Definition 11.19. A *panel tree* is a pair $(T_F, \mathcal{A}_F)$ for some panel F of $\Delta_{\mathcal{A}}^{\infty}$. (Thus a panel tree is, in fact, a tree with sap and not simply a tree.)

Notation 11.20. Let F be a panel of $\Delta_{\mathcal{A}}^{\infty}$ and let S be an $\mathcal{A}$-sector such that the chamber S^{∞} of $\Delta_{\mathcal{A}}^{\infty}$ is contained in F (thought of as a panel of $\Delta_{\mathcal{A}}^{\infty}$). This means that S has a face f contained in F (thought of as a parallel class of faces). Let A be an $\mathcal{A}$-apartment containing S, let α be the unique root of A containing S such that $f \subset \mu(\alpha)$ and let S_F denote the set of asymptote classes that contain faces that are both parallel to f and contained in $\mu(\beta)$ for some root β of A contained in α. Thus $A \in F_{\mathcal{A}}$, S_F is a ray in the apartment A_F of T_F and every $\mathcal{A}_F$-ray of T_F is of this form.

Proposition 11.21. *The ray S_F described in 11.20 is independent of the choice of the $\mathcal{A}$-apartment A.*

Proof. By 5.20, each asymptote class in S_F contains faces contained in S or in f. The claim holds, therefore, by 29.34. □

Proposition 11.22. *Let F be a panel of $\Delta_{\mathcal{A}}^{\infty}$ and let ζ_F denote the map from F to the set of $\mathcal{A}_F$-ends of T_F given by*

$$\zeta_F(S^{\infty}) = (S_F)^{\infty}$$

for all $\mathcal{A}$-sectors S such that $S^{\infty} \in F$, where S_F is the ray of A_F described in 11.9 and $(S_F)^{\infty}$ is the end of T_F containing the ray S_F. Then ζ_F is a bijection.

Proof. Since every $\mathcal{A}_F$-ray of T_F is of the form S_F for some $\mathcal{A}$-sector S such that $S^{\infty} \in F$ (as was observed in 11.20), the map ζ_F is surjective. Suppose that $A' \in F_{\mathcal{A}}$, that S' is a sector in A' having a face in F and that

$(S'_F)^\infty = (S_F)^\infty$. By 8.4.ii, there exists $A_1 \in \mathcal{A}$ containing subsectors S_1 of S and S'_1 of S'. By 4.7, $(S_1)_F$ and $(S'_1)_F$ are subrays of S_F and $(S')_F$. Thus $(S_1)_F$ and $(S'_1)_F$ contain a common subray of the apartment $(A_1)_F$. By 5.11, it follows that the panels S_1 and S'_1 are parallel. Hence $S^\infty = (S')^\infty$. Thus ζ_F is injective. □

Notation 11.23. Let m be a wall of $\Delta_{\mathcal{A}}^\infty$ and let F be a panel contained in m. Let X_m denote the set of roots of $\Delta_{\mathcal{A}}^\infty$ whose wall is m. By 29.47, there is a bijection from F to X_m that sends a chamber u in F to the unique root in X_m that contains u. We denote this map by $\iota_{m,F}$.

Proposition 11.24. *Let m be a wall of $\Delta_{\mathcal{A}}^\infty$ and let F be a panel contained in m. Let ξ_m be as in 10.38, let $\phi_{m,F}$ be as in 11.15, let ζ_F be as in 11.22 and let $\iota_{m,F}$ be as in 11.23. Then $\phi_{m,F}$ induces the map*

$$\xi_m \circ \iota_{m,F} \circ \zeta_F^{-1}$$

from the set of $\mathcal{A}_F$-ends of T_F to the set of $\mathcal{A}_m$-ends of T_m.

Proof. Let p be an $\mathcal{A}_F$-end of T_F. By 11.20, 11.21 and 11.22, there exists an apartment A that is contained in both $F_{\mathcal{A}}$ and $m_{\mathcal{A}}$, a sector S contained in A such that $p = (S_F)^\infty$ and a root α contained in A such that $S \subset \alpha$ and the wall $\mu(\alpha)$ contains the face of S that is contained in both m and F. The map $\phi_{m,F}$ sends $(S_F)^\infty$ to $(\alpha_m)^\infty$ and the map $\iota_{m,F}$ sends S^∞ to α^∞. Thus

$$\begin{aligned}
\xi_m\big(\iota_{m,F}(p)\big) &= \xi_m\big(\iota_{m,F}(S^\infty)\big) \\
&= \xi_m(\alpha^\infty) \\
&= (\alpha_m)^\infty \\
&= \phi_{m,F}\big((S_F)^\infty\big) \\
&= \phi_{m,F}\big(\zeta_F(S^\infty)\big) = \phi_{m,F}\big(\zeta_F(p)\big).
\end{aligned}$$

□

Corollary 11.25. *Let m be a wall of $\Delta_{\mathcal{A}}^\infty$, let F be a panel contained in m and let ω_m and ω_F be the canonical valuations of $(T_m, \mathcal{A}_m)$ and $(T_F, \mathcal{A}_F)$ as defined in 9.9. Let ξ_m be as in 10.38, let ζ_F be as in 11.22 and let $\iota_{m,F}$ be as in 11.23. If the set of roots of $\Delta_{\mathcal{A}}^\infty$ whose wall is m is identified with the set of $\mathcal{A}_m$-ends of T_m via ξ_m and the panel F is identified with the set of $\mathcal{A}_F$-ends of T_F via ζ_F, then*

$$\omega_F = \omega_m \circ \iota_{m,F}.$$

Proof. By 11.16, we have $\omega_F = \omega_m \circ \phi_{m,F}$. The claim holds, therefore, by 11.24. □

Chapter Twelve

Tree-Preserving Isomorphisms

The goal of this chapter is to prove 12.3. This fundamental result gives a necessary and sufficient condition expressed in terms of wall and panel trees for an isomorphism from one building at infinity to another to extend to an isomorphism of affine buildings.[1] In 12.31, we deduce as a corollary that all the elements of all the root groups of the building at infinity extend to automorphisms of the affine building itself.

The main result of this chapter was first proved by Tits in [35]. In fact, Theorem 2 in Section 8 of [35] is more general: It applies to *non-discrete* as well as affine buildings.[2] The proof of 12.3 we give here applies only to affine buildings.

We continue to assume that Π is one of the affine Coxeter diagrams X_ℓ in Figure 1.1, that Δ is a thick building of type Π and that $\mathcal{A}$ is a system of apartments of Δ. Let I be the vertex set of Π.

We now assume as well that Δ' is a second thick building of type Π and that $\mathcal{A}'$ is a system of apartments of Δ'. Moreover, the following conventions, along with those in 11.1, will apply from now on:

Conventions 12.1. For each wall m of the building at infinity $\Delta_{\mathcal{A}}^\infty$, we identify the set of roots of $\Delta_{\mathcal{A}}^\infty$ having a wall m with the set of $\mathcal{A}_m$-ends of the tree T_m via the map ξ_m defined in 10.38. We identify each panel F of $\Delta_{\mathcal{A}}^\infty$ with the set of $\mathcal{A}_F$-ends of T_F via the map ζ_F defined in 11.22.

For each wall m' and each panel F' of the building at infinity $(\Delta')_{\mathcal{A}'}^\infty$, let $(T'_{m'}, \mathcal{A}'_{m'})$ and $(T'_{F'}, \mathcal{A}'_{F'})$ denote the corresponding wall tree and panel tree defined with respect to the pair $(\Delta', \mathcal{A}')$ and let the analogous identifications be made between the set of roots of $(\Delta')_{\mathcal{A}'}^\infty$ having wall m' and the set of $\mathcal{A}'_{m'}$-ends of $T'_{m'}$ and between the panel F' of $(\Delta')_{\mathcal{A}'}^\infty$ and the set of $\mathcal{A}'_{F'}$-ends of $T'_{F'}$.

[1] Our use here of the term "extend" is not completely correct. What we really mean is stated more precisely in 12.3.

[2] In 14.54, we will conclude that Bruhat-Tits buildings are classified by root data with valuation (up to equipollence). In [6] and [35], Bruhat and Tits introduce a more general geometrical structure usually called a *non-discrete building* and show that non-discrete buildings with a Moufang building at infinity are classified by root data with non-discrete valuation (up to equipollence). A *non-discrete valuation* of a root datum is the notion obtained by weakening the assumption in 3.21 that the range of each map ϕ_a is $\mathbb{Z}$ to the assumption that the range of each map ϕ_a is a subset of $\mathbb{R}$ of cardinality at least 3 and replacing $\mathbb{Z}$ by $\mathbb{R}$ in conditions (V1) and (V3). See footnote 1 in Chapter 14.

Definition 12.2. Let τ be an isomorphism from $\Delta_{\mathcal{A}}^{\infty}$ to $(\Delta')_{\mathcal{A}'}^{\infty}$. We will say that τ is *tree-preserving*[3] if

$$\omega_{\tau(m)} \circ \tau = \omega_m$$

for each wall m of $\Delta_{\mathcal{A}}^{\infty}$ and

$$\omega_{\tau(F)} \circ \tau = \omega_F$$

for each panel F of $\Delta_{\mathcal{A}}^{\infty}$, where ω_m, ω_F, etc. are as in 11.25. (This makes sense due to the conventions set out in 12.1.)

Here is the main result of this chapter:

Theorem 12.3. *Let τ be a tree-preserving isomorphism from $\Delta_{\mathcal{A}}^{\infty}$ to $(\Delta')_{\mathcal{A}'}^{\infty}$. Then there exists a unique isomorphism ρ from Δ to Δ' mapping $\mathcal{A}$ to $\mathcal{A}'$ such that*

$$\rho(S)^{\infty} = \tau(S^{\infty}) \tag{12.4}$$

for all $\mathcal{A}$-sectors S of Δ.[4]

Proof. By 9.14 and 10.31 (and 12.1), for each wall m of $\Delta_{\mathcal{A}}^{\infty}$, there exists a unique isomorphism from T_m to $T'_{\tau(m)}$ whose action on the $\mathcal{A}_m$-ends of T_m (i.e. on the set of roots of $\Delta_{\mathcal{A}}^{\infty}$ whose wall is m) is the same as the action of τ on this set; we will call this isomorphism τ_m. By 9.14 and 11.18 (and 12.1), for each face F of $\Delta_{\mathcal{A}}^{\infty}$, there exists a unique isomorphism from T_F to $T'_{\tau(F)}$ whose action on the $\mathcal{A}_F$-ends of T_F (i.e. on F) is the same as the action of τ on F; we will call this isomorphism τ_F.

Notation 12.5. For each $\mathcal{A}$-wall M and each $\mathcal{A}$-face f of Δ, we set $\tau(M) = \tau_m(M)$ for $m = M^{\infty}$ (so m is the parallel class of walls containing M as well as a wall of $\Delta_{\mathcal{A}}^{\infty}$ and M is a vertex of T_m) and $\tau(f^{\circ}) = \tau_F(f^{\circ})$ for $F = f^{\infty}$ (so F is the parallel class of faces containing f as well as a panel of $\Delta_{\mathcal{A}}^{\infty}$ and the $\mathcal{A}$-asymptote class f° is a vertex of T_F).

Notation 12.6. Let $A \in \mathcal{A}$. Then A^{∞} is an apartment of $\Delta_{\mathcal{A}}^{\infty}$ and hence $\tau(A^{\infty})$ is an apartment of $(\Delta')_{\mathcal{A}'}^{\infty}$. Thus by 8.27, there is a unique $A' \in \mathcal{A}'$ such that $(A')^{\infty} = \tau(A^{\infty})$. We set $A' = \tau(A)$.

We now proceed with the proof of 12.3 in a series of steps.

Proposition 12.7. *The map $A \mapsto \tau(A)$ from $\mathcal{A}$ to $\mathcal{A}'$ is a bijection.*

Proof. The map τ is an isomorphism from one spherical building to another. By (i) and (ii) of 29.14, it follows that τ maps the set of apartments of $\Delta_{\mathcal{A}}^{\infty}$ bijectively to the set of apartments of $(\Delta')_{\mathcal{A}'}^{\infty}$. The claim holds, therefore, by 8.27 and 12.6. □

Proposition 12.8. *Let M be an $\mathcal{A}$-wall and let e be an $\mathcal{A}$-asymptote class contained in M. Then $\tau(e)$ is contained in $\tau(M)$.*

---

[3] Perhaps "ecological" would be a better word.

[4] See 12.25 below.

Proof. Let $m = M^\infty$ and let $F = e^\infty$. By 11.1, m is a wall of $\Delta_{\mathcal{A}}^\infty$ and F is panel in m. Let $\phi_{m,F}$ and $\phi_{\tau(m),\tau(F)}$ be as in 11.15 and let $\iota_{m,F}$ and $\iota_{\tau(m),\tau(F)}$ be as in 11.23. Thus

$$\tau \circ \iota_{m,F} \circ \tau^{-1} = \iota_{\tau(m),\tau(F)}.$$

By 11.24 and 12.1, it follows that

$$\tau \circ \phi_{m,F} \circ \tau^{-1} = \phi_{\tau(m),\tau(F)}.$$

Since $M = \phi_{m,F}(e)$, we therefore have $\phi_{\tau(m),\tau(F)}(\tau(e)) = \tau(M)$. □

Proposition 12.9. *Let $A \in \mathcal{A}$, let $A' = \tau(A)$, let M be a wall contained in A and let $m = M^\infty$. Then the following hold:*

(i) *$\tau(M)$ is contained in $\tau(A)$ and if α is one of the two roots of A whose wall is M, then there is a unique root α' of A' such that $\tau(M) = \mu(\alpha')$ and $\tau(\alpha^\infty) = (\alpha')^\infty$.*

(ii) *$A' \in \tau(m)_{A'}$ (as defined in 10.19) and τ induces an isomorphism from the $\mathcal{A}_m$-apartment A_m of the tree T_m to the $\mathcal{A}'_{\tau(m)}$-apartment $A'_{\tau(m)}$ of the tree $T'_{\tau(m)}$.*

(iii) *If α and α' are as in (i), then τ maps the $\mathcal{A}_m$-ray α_m to the $\mathcal{A}'_{\tau(m)}$-ray $\alpha'_{\tau(m)}$ (where the rays α_m and $\alpha'_{\tau(m)}$ are as defined in 10.19).*

Proof. Let α be one of the two roots of A whose wall is M. Let $\alpha_1 = \alpha$ and let α_2 be the root of A opposite α_1. By 10.35, we can choose a third $\mathcal{A}$-root α_3 whose wall is M such that $\alpha_i \cup \alpha_j$ is an $\mathcal{A}$-apartment for all pairs i, j of distinct elements of the interval $[1, 3]$.

Thus the $\mathcal{A}$-wall M is the junction in T_m (as defined in 9.9) of the $\mathcal{A}_m$-ends α_1^∞, α_2^∞ and α_3^∞ (which are also roots of $\Delta_{\mathcal{A}}^\infty$ having m as a wall by 12.1). By 12.5, therefore, $\tau(M)$ is the junction in $T'_{\tau(m)}$ of the $\mathcal{A}'_{\tau(m)}$-ends $\tau(\alpha_1^\infty)$, $\tau(\alpha_2^\infty)$ and $\tau(\alpha_3^\infty)$. Since $\mathcal{A}'_{\tau(m)}$ satisfies 9.5.ii, there thus exist unique $\mathcal{A}'$-roots α'_1, α'_2 and α'_3 in Δ' such that $\tau(M) = \mu(\alpha_i)$ and $(\alpha'_i)^\infty = \tau(\alpha_i^\infty)$ for each $i \in [1, 3]$ and $\alpha'_i \cup \alpha'_j \in \mathcal{A}'$ for all pairs i, j of distinct elements of the index set $[1, 3]$.

The apartment A^∞ is the disjoint union of the roots α_1^∞ and α_2^∞. Thus $\tau(A^\infty)$ is the disjoint union of the roots $(\alpha'_1)^\infty$ and $(\alpha'_2)^\infty$. By 12.6, it follows that $A' = \alpha'_1 \cup \alpha'_2$. Thus (i) holds with $\alpha' = \alpha'_1$. In particular, $\tau(A_m) \subset A'_{\tau(m)}$. Hence (ii) holds since τ_m is an automorphism of T_m.

By 12.5, the image of the ray α_m under τ is a ray of $T'_{\tau(m)}$ contained in the $\mathcal{A}'$-end $\tau(\alpha^\infty) = (\alpha')^\infty$. Since the ray α_m starts at the vertex M, its image under τ starts at the vertex $\tau(M)$. Since $\alpha'_{\tau(m)}$ (where $\alpha' = \alpha'_1$) is also a ray contained in the $\mathcal{A}'$-end $\tau(\alpha^\infty) = (\alpha')^\infty$ that starts at $\tau(M)$, we conclude that these two rays are equal. Thus (iii) holds. □

Proposition 12.10. *Let $A \in \mathcal{A}$ and let $A' = \tau(A)$ (as defined in 12.6). There exists a unique bijection ρ_A from the set of roots of A to the set of roots of A' such that for each root α of A,*

$$\mu(\rho_A(\alpha)) = \tau(\mu(\alpha)) \tag{12.11}$$

and

$$\rho_A(\alpha)^\infty = \tau(\alpha^\infty). \tag{12.12}$$

Furthermore, the following hold:

(i) *If α and β are opposite roots of A, then $\rho_A(\alpha)$ and $\rho_A(\beta)$ are opposite roots of A'.*
(ii) *If α and β are roots of A such that $\alpha \subset \beta$, then $\rho_A(\alpha) \subset \rho_A(\beta)$.*

Proof. By 12.9.i, there exists a unique map ρ_A satisfying 12.11 and 12.12. These two properties imply that (i) holds, ρ_A is a bijection by 12.9.ii and by 12.9.iii, (ii) holds. □

Proposition 12.13. *Let $A \in \mathcal{A}$, let $A' = \tau(A)$, let S be a sector in A, let f be a face of S, and let S' be a sector in A' contained in $\tau(S^\infty)$ (which exists by 12.6). Then S' has a face f' parallel to the $\mathcal{A}'$-asymptote class $\tau(f^\circ)$ (as defined in 11.7).*

Proof. Let $e = f^\circ$, let $F = e^\infty$ and let $F' = \tau(F)$. By 12.5, $F' = \tau(e)^\infty$. Since S^∞ is a chamber in F and S' is contained in the parallel class $\tau(S^\infty)$, it follows that S' has a face parallel to $\tau(e)$. □

Proposition 12.14. *Let $A \in \mathcal{A}$, let S be a sector of A, let*

$$X = \{\tau(f^\circ) \mid f \text{ is a face of } S\}$$

and let $A' = \tau(A)$. Then there is a unique sector S' contained in the apartment A' such that

$$(S')^\infty = \tau(S^\infty)$$

and the set of $\mathcal{A}$-asymptote classes containing faces of S' equals X.

Proof. By 12.6, we can choose a sector S_1' in A' contained in $\tau(S^\infty)$. By 12.13, each $\mathcal{A}'$-asymptote class in X is parallel to an $\mathcal{A}'$-asymptote class containing a face of S_1'. By 12.8 and 12.9.i, each $\mathcal{A}'$-asymptote class in X is contained in A'. By 5.14 and 5.22, it follows that there is a unique sector S' in A' parallel to S_1' such that the set of asymptote classes containing faces of S' is precisely X. □

Definition 12.15. For each $A \in \mathcal{A}$, let A_σ denote the set of sectors in A. For each $A \in \mathcal{A}$ and each $S \in A_\sigma$, we set

$$\tau_A(S) = S',$$

where S' is the unique sector of $\tau(A)$ described in 12.14.

Proposition 12.16. *For each $A \in \mathcal{A}$, the map τ_A defined in 12.15 is invertible.*

Proof. Since the isomorphism τ has an inverse, so does each τ_A. □

Proposition 12.17. *Let $A \in \mathcal{A}$, let $A' = \tau(A)$, let S and S_1 be sectors of A having the same terminus R and suppose that the apex of S and the apex of S_1 are opposite chambers of R. Let $S' = \tau_A(S)$ and $S_1' = \tau_A(S_1)$, where τ_A is as in 12.15, let d' be the apex of S' and let d_1' be the apex of S_1'. Then S' and S_1' have the same terminus R' and d' and d_1' are opposite chambers of R'.*

Proof. Let d be the apex of S, let d_1 be the apex of S_1, let R' be the terminus of S' and let R_1' be the terminus of S_1'. By 29.9.v, each panel of R containing d is opposite a unique panel in R containing d_1 (as defined in 9.8 of [37]). By 29.44, it follows that each panel of R containing d is comural with a unique panel of R containing d_1. Therefore a wall of A contains a face of S if and only if it contains a face of S_1. By 12.8 and 12.14, it follows that a wall of A' contains a face of S' if and only if it contains a face of S_1'. In particular, a root α' of A' lies in the set $[R', d']_{A'}$ (as defined in 7.5) if and only if either α' or its opposite lies in the set $[R_1', d_1']_{A'}$. By 1.12, therefore, $R' = R_1'$. By 10.24, S^∞ and S_1^∞ are opposite. Hence $(S')^\infty = \tau(S^\infty)$ and $(S_1')^\infty = \tau(S_1^\infty)$ are opposite chambers of $(\Delta')^\infty_{A'}$.

Let $(x_0, x_1, \ldots, x_n)$ be a minimal gallery in R' from $x_0 := d'$ to $x_n := d_1'$ and let $\hat{S}_i = \sigma_{A'}(R', x_i)$ for each $i \in [1, n]$. Then

$$(\hat{S}_0^\infty, \hat{S}_1^\infty, \ldots, \hat{S}_n^\infty)$$

is a gallery in $(\Delta')^\infty_{A'}$ from $\hat{S}_0^\infty = (S')^\infty$ to $\hat{S}_n^\infty = (S_1')^\infty$. Thus n is at least as large as $\text{diam}((\Delta')^\infty_{A'})$. Since $(\Delta')^\infty_{A'}$ and R' are buildings of the same type (by 1.7 and 8.24), they have the same diameter (by 29.14.i). It follows that d' and d_1' are opposite in R'. □

Proposition 12.18. *Let $A \in \mathcal{A}$, let $A' = \tau(A)$, let R be a gem of Δ cut by A and let P and Q be panels of R such that $P \cap A$ and $Q \cap A$ are opposite panels of $R \cap A$. Furthermore, let $f = \mu_A(R, P)$ and $f_1 = \mu_A(R, Q)$ (as defined in 8.1), let $e' = \tau(f^\circ)$ and $e_1' = \tau(f_1^\circ)$ and let M be the wall contained in A that contains both P and Q (which exists by 29.44). Then the following hold:*

(i) *For every gem R' of Δ' cut by $\tau(M)$, there exist unique panels P' and Q' of R' such that $\mu_{A'}(R', P') \in e'$ and $\mu_{A'}(R', Q') \in e_1'$.*
(ii) *If R', P' and Q' are as in (i), then $P' \cap A'$ and $Q' \cap A'$ are opposite panels of $R' \cap A'$.*

Proof. Choose $d \in R \cap P$ and let d_1 be the chamber of Q opposite d in $R \cap A$. Let $S = \sigma_A(R, d)$, let $S_1 = \sigma_A(R, d_1)$ and let S', S_1' and R' be as in 12.17. By 12.15, there exist panels P' and Q' of R' cut by A' such that the face $\mu_{A'}(R', P')$ of S' is contained in e' and the face $\mu_{A'}(R', Q')$ of S_1' is contained in e_1'. By 5.9, the panels P' and Q' are uniquely determined. By 12.8, e' and e_1' are both contained in $\tau(M)$. By 5.23, it follows that $P' \cap A'$ and $Q' \cap A'$ are opposite panels of $R' \cap A'$.

Suppose now that R_1' is an arbitrary gem of Δ' cut by $\tau(M)$. By 1.9 and 1.13, there is a translation g of A' mapping the chamber set $M \cap A$ to itself such that $R'_1 \cap A' = (R' \cap A')^g$. Let P'_1 and Q'_1 be the unique panels of R'_1 such that $P'_1 \cap A' = (P' \cap A')^g$ and $Q'_1 \cap A' = (Q' \cap A')^g$. By 1.45 and 5.12, each face contained in M is asymptotic to its translate by g. Thus

$$\mu_{A'}(R'_1, P'_1) \in e' \text{ and } \mu_{A'}(R'_1, Q'_1) \in e'_1. \tag{12.19}$$

By 5.9, P'_1 and Q'_1 are the unique panels of R_1 satisfying 12.19. Thus (i) holds. Since $P' \cap A'$ and $Q' \cap A'$ are opposite panels of $R' \cap A'$, $P'_1 \cap A'$ and $Q'_1 \cap A'$ are opposite panels of $R'_1 \cap A'$. Thus (ii) holds. □

Proposition 12.20. *Let $A \in \mathcal{A}$, let $A' = \tau(A)$, let S and S_1 be distinct sectors of A having the same terminus and sharing a face, let $S' = \tau_A(S)$ and let $S'_1 = \tau_A(S_1)$. Then S' and S'_1 are distinct sectors of A' having the same terminus and sharing a face.*

Proof. Let f be the face shared by S and S_1, let R be the terminus of S and S_1, let M be the unique wall contained in A that contains f, let α_1 and α_2 be the two roots of A whose wall is M and let r denote the reflection of A that interchanges α_1 and α_2. The reflection r also maps $R \cap A$ to itself and interchanges the apex of S with the apex of S_1. Thus $S_1 = S^r$.

By 10.35, we can choose a third $\mathcal{A}$-root α_3 whose wall is M such that $A_i := \alpha_i \cup \alpha_3$ is an apartment in $\mathcal{A}$ for $i = 1$ and 2. Let $A'_i = \tau(A_i)$ for $i = 1$ and 2 and let $M' = \tau(M)$. By 12.9.i, the wall M' is contained in A. By 12.9.iii, there exist distinct $\mathcal{A}'$-roots α'_1, α'_2 and α'_3 whose wall is M' such that $(\alpha'_i)^\infty = \tau(\alpha_i^\infty)$ for all $i \in [1, 3]$. Since $\mathcal{A}_{\tau(m)}$ satisfies 9.5.ii, the union $\alpha'_i \cup \alpha'_j$ is an $\mathcal{A}'$-apartment for all pairs of distinct indices $i, j \in [1, 3]$. By 12.6, we have $A' = \alpha'_1 \cup \alpha'_2$ (since $A^\infty = \alpha_1^\infty \cup \alpha_2^\infty$). Similarly, $A'_i = \alpha'_i \cup \alpha'_3$ for $i = 1$ and 2.

Let $e = f^\circ$, let R' be the terminus of S', let f' be the unique face of S' contained in $e' := \tau(e)$ (which exists by 12.15), let r' be the reflection of A' interchanging the roots α'_1 and α'_2 and let $S'_1 = (S')^{r'}$. The reflection r' maps every panel in M' to itself. By 12.8, e' is contained in M'. Thus S' and S'_1 are distinct sectors of A' having the same terminus R' and sharing the face f'. Our goal is to show that $\tau_A(S_1) = S'_1$.

Let f_1 be a face of S distinct from f and let $f_2 = f_1^r$ (so f_2 is a face of $S_1 = S^r$). By 12.15, there exists a unique face f'_1 of S' such that

$$f'_1 \in \tau(f_1^\circ). \tag{12.21}$$

Let $f'_2 = (f'_1)^{r'}$. For $i = 1$ and 2, let M_i be the unique wall contained in A_i containing f_i and let M'_i be the unique wall contained in A'_i containing f'_i. (We really mean A_i and A'_i, not A or A'.)

For $i = 1$ and 2, let P_i be the panel of R such that $f_i = \mu_A(R, P_i)$. Thus $P_2 \cap A = (P_1 \cap A)^r$. By 29.54, there exists a unique panel Q of R such that $Q \cap A_i$ and $P_i \cap A_i$ are opposite panels of $R \cap A_i$ for both $i = 1$ and 2. Since $P_1 \cap A_1 \subset \alpha_1$, we have $Q \cap A_1 \subset \alpha_3$. By 4.11, therefore, the face $\mu_{A_1}(R, Q)$ is contained in α_3. By 29.34, there is an isomorphism from A_1 to A_2 that acts trivially on α_3. It follows that $\mu_{A_1}(R, Q) = \mu_{A_2}(R, Q)$. We denote this face by f_3. Let $e_i = f_i^\circ$ for all $i \in [1, 3]$.

For $i = 1$ and 2, let P_i' be the panel of R' such that $f_i' = \mu_A(R', P_i')$. Thus $P_2' \cap A' = (P_1' \cap A')^{r'}$. By 29.54 again, there exists a unique panel Q' of R' such that $Q' \cap A_i'$ and $P_i' \cap A_i'$ are opposite panels of $R' \cap A_i'$ for both $i = 1$ and 2. Just as in the previous paragraph, we have $\mu_{A_1'}(R', Q') = \mu_{A_2'}(R', Q')$. Let f_3' denote this face and let e_i' denote the $\mathcal{A}'$-asymptote class of f_i' for all $i \in [1, 3]$.

Since e_1 is contained in M_1, $\tau(e_1)$ is contained in $\tau(M_1)$ by 12.8. By 12.21, $\tau(e_1) = e_1'$. Thus e_1' is contained in both $\tau(M_1)$ and M_1'. Since these are both walls contained in A_1, it follows that $\tau(M_1) = M_1'$. Thus R' is cut by $\tau(M_1)$. By 12.18, it follows that $\tau(e_3) = e_3'$. By 12.8 again, e_3' is contained in $\tau(M_2)$ (since e_3 is contained in M_2). By 29.44, e_3' is also contained in M_2'. Since $\tau(M_2)$ and M_2' are both walls contained in A_2, it follows that $\tau(M_2) = M_2'$. Thus R' is cut by $\tau(M_2)$. By 12.18 again, it follows that $\tau(e_2) = e_2'$.

We conclude that for every face of S_1, τ maps the $\mathcal{A}$-asymptote class that contains it to an $\mathcal{A}'$-asymptote class containing a face of S_1'. By 12.15, we conclude that $\tau_A(S_1) = S_1'$. □

Proposition 12.22. *For each $A \in \mathcal{A}$, there exists a unique isomorphism ν_A from A to $\tau(A)$ such that*

(i) *$\nu_A(S) \in \tau(S^\infty)$ for each sector S contained in A and*
(ii) *$\nu_A(f) \in \tau(f^\circ)$ for each face f contained in A.*[5]

Furthermore,

(iii) *$\nu_A(\alpha) = \rho_A(\alpha)$ for all roots α in A, where ρ_A is as in 12.10, and*
(iv) *$\nu_A(M) = \tau(M)$ for each wall M contained in A.*

Proof. Choose $A \in \mathcal{A}$, let $A' = \tau(A)$ and let τ_A be as in 12.15. Let X be the set of all gems of Δ cut by A and let X' be the set of all gems of Δ' cut by A'. For each $R \in X$, let τ_A^R be the map from $R \cap A$ to A' that sends a chamber d of $R \cap A$ to the apex of $\tau_A(\sigma_A(R, d))$. Choose $R \in X$. If d and e are adjacent chambers of $R \cap A$, then by 12.20, the sectors $\tau_A(\sigma_A(R, d))$ and $\tau_A(\sigma_A(R, e))$ have the same terminus and their apices $\tau_A^R(d)$ and $\tau_A^R(e)$ are adjacent. Since $R \cap A$ is connected, it follows that there exists a unique R' in X' such that

$$\tau_A^R(R \cap A) \subset R' \cap A'$$

and τ_A^R sends panels of $R \cap A$ to panels of $R' \cap A'$. As observed in 12.16, the map τ_A has an inverse. Moreover, 12.20 applies also to this inverse. It follows that there exists a map $\tau_{A'}^{R'}$ from $R' \cap A'$ to $R \cap A$ that is the inverse of τ_A^R and also maps panels to panels. We conclude that for each $R \in X$, there exists $R' \in X'$ such that τ_A^R is an isomorphism from $R \cap A$ to $R' \cap A'$.

Let κ_A be the map from X to X' such that

$$\kappa_A(R) \cap A' = \tau_A^R(R \cap A)$$

[5]By $\nu_A(f)$ we mean the set of panels P' in Δ' containing $\nu_A(P \cap A)$ for some panel $P \in f$. In (iv), $\nu_A(M)$ is to be interpreted analogously as the set of panels P' in Δ' containing $\nu_A(P \cap A)$ from some panel $P \in M$.

for all $R \in X$. Thus

$$\tau_A(\sigma_A(R, d)) = \sigma_{A'}(\kappa(R), \tau_A^R(d)) \tag{12.23}$$

for all $R \in X$ and all chambers $d \in R \cap A$. By 12.16, also κ_A is a bijection.

Let ρ_A be the bijection from the set of roots of A to the set of roots of A' defined in 12.10. Let $R \in X$, let $d \in R \cap A$, let $S = \sigma_A(S, d)$, let $\alpha \in [R, d]_A$ (as defined in 7.5), let $M = \mu(\alpha)$, let f be the face of S contained in M and let $e = f^\circ$. Then the sector $\tau_A(S) = \sigma_{A'}(\kappa(R), \tau_A^R(d))$ has a face contained in the $\mathcal{A}'$-asymptote class $\tau(e)$ (by 12.15) and also contained in the $\mathcal{A}'$-wall $\tau(M)$ (by 12.8). By 12.11, $\rho_A(\alpha)$ is a root of A' whose wall is $\tau(M)$. Thus either $\rho_A(\alpha)$ or its opposite in A' is contained in the set $[\kappa_A(R), \tau_A^R(d)]_{A'}$. By 12.12, $\tau(S^\infty) \in \rho_A(\alpha)^\infty$ (since $S \subset \alpha$). By 4.12.ii, it follows that the root $\rho_A(\alpha)$ contains the apex of the sector $\tau_A(S)$. Therefore

$$\alpha \in [R, d]_A \text{ implies } \rho_R(\alpha) \in [\kappa_A(R), \tau_A^R(d)]_{A'}$$

for $R \in X$ and all $d \in R \cap A$. By 6.6 and 12.10, we conclude that there exists an isomorphism ν_A from A to A' whose restriction to $R \cap A$ equals τ_A^R for all $R \in X$. By 6.6 and 12.10, we also know that ν_A satisfies (iii).

Let R, d, S, α, M, f and e be as in the previous paragraph. Let $R' = \kappa(R)$, let $d' = \nu_A(d)$ and let $S' = \nu_A(S)$. Thus R' is the unique gem of Δ' such that $\nu_A(R \cap A) = R' \cap A$, $d' = \tau_A^R(d)$ and

$$S' = \nu_A(\sigma_A(R, d)) = \sigma_{A'}(R', d')$$

(since ν_A is an isomorphism). Hence $S' = \tau_A(S)$ by 12.23. By 12.15, therefore, $S' \in \tau(S^\infty)$. Thus ν_A satisfies (i). Let $M' = \nu_A(M)$. We have $\nu_A(\alpha) = \rho_A(\alpha)$ by (iii) and hence $M' = \nu_A(M)$ by 12.11. Thus ν_A satisfies (iv). Let P be the panel in R containing d such that $f = \mu_A(R, P)$, let $P' = \nu_A(P)$ and let $f' = \nu_A(f)$. Thus $d' \in \tau_A^R(P) = P'$ and $f' = \mu_{A'}(R', P')$. In particular, f' is a face of S'. Since the panel P is in M, the panel P' is in $\nu_A(M) = M'$. Hence also f' is contained in M'. By 12.8, $\tau(e)$ is also contained in M', and by 12.15, $\tau(e)$ contains a face of $S' = \tau_A(S)$. Since only one face of S' can be contained in M', we conclude that $f' \in \tau(e)$. Thus ν_A satisfies (ii).

Suppose that $\hat{\nu}_A$ is another isomorphism from A to A' satisfying (i) and (ii). By 12.15, $\hat{\nu}_A(S) = \tau_A(S)$ for all sectors S in A. This means that the restriction of $\hat{\nu}_A$ to $R \cap A$ must equal τ_A^R for each gem R of Δ cut by A. Hence $\hat{\nu}_A = \nu_A$. □

Proposition 12.24. *Let u be a chamber of Δ and let A_1 and A_2 be two apartments in $\mathcal{A}$ both containing u. Then $\nu_{A_1}(u) = \nu_{A_2}(u)$ (where ν_{A_1} and ν_{A_2} are as in 12.22).*

Proof. We proceed in steps:

Step 1: Assume first that $A_1 \cap A_2$ is a root α. Let $M = \mu(\alpha)$, let $m = M^\infty$ and let $A_i' = \tau(A_i)$ and $\alpha_i' = \nu_{A_i}(\alpha)$ for $i = 1$ and 2. By 12.10 and 12.22.iii, $\mu(\alpha_1') = \mu(\alpha_2') = \tau(M)$, and by 12.6, $(\alpha_1')^\infty = (\alpha_2')^\infty = \tau(\alpha^\infty)$. By 12.1, this means that α_1' and α_2' are two $\mathcal{A}_m$-rays of T_m in the same end and having the same initial vertex. Thus $\alpha_1' = \alpha_2'$.

Let R be a residue of Δ cut by A such that $R \cap A \subset \alpha$, let α_1 be the unique root of A parallel to α that cuts R and let d be a chamber of $R \cap \alpha_1$ such that all the chambers of $R \cap A$ adjacent to d are also in α_1. Let X denote the set consisting of d and all the chambers of $R \cap A$ adjacent to d. By 4.14, $\sigma_{A_1}(R, e) \subset \alpha$ for every chamber $e \in X$. By 7.3, these sectors are also sectors of A_2. Let S be one of these sectors and let

$$S_i' = \nu_{A_i}(S)$$

for $i = 1$ and 2. By 12.22.i, we have

$$(S_1')^\infty = \tau(S^\infty) = (S_2')^\infty,$$

and by 12.22.ii, the set of $\mathcal{A}'$-asymptote classes containing faces of S_1' is the same as the set of $\mathcal{A}'$-asymptote classes containing faces of S_2'. By 12.14, it follows that $S_1' = S_2'$. By 4.4, we conclude that ν_{A_1} and ν_{A_2} agree on the set X. This implies that ν_{A_1} and ν_{A_2} are both σ-isomorphisms for the same bijection σ from Π to Π'. Let ϕ_i be the restriction of ν_{A_i} to α for $i = 1$ and 2 and let ϕ denote the the composition $\phi_2^{-1} \circ \phi_1$. By the conclusion of the previous paragraph, ϕ is a bijection from α to itself. The map ϕ fixes the chamber d and for each vertex i of Π, two chambers of α are i-adjacent if and only if their images under ϕ are i-adjacent. The root α is of course connected and for each vertex i of Π, each chamber of α is i-adjacent to at most one chamber of α. It follows that ϕ is the identity map. We conclude that ν_{A_1} and ν_{A_2} agree on all of $A_1 \cap A_2$.

Step 2: Let k_0 denote the number of chambers in a root of Δ^∞; by 29.45, this is a constant. Let k denote the number of chambers of $\Delta_{\mathcal{A}}^\infty$ contained in $A_1^\infty \cap A_2^\infty$. Since apartments are convex, the intersection $A_1 \cap A_2$ is convex. By 29.20, therefore, $A_1 \cap A_2$ is contained in a root α of A_1. Hence $A_1^\infty \cap A_2^\infty \subset \alpha^\infty$ and therefore $k \le k_0$. Suppose that $k = k_0$. Then $A_1^\infty \cap A_2^\infty = \alpha^\infty$. By 4.26, $A_1 \cap A_2$ is a root. Thus by Step 1, ν_{A_1} and ν_{A_2} agree on all of $A_1 \cap A_2$.

Step 3: Suppose that $0 < k < k_0$. By 29.55, we can choose an apartment $A_3 \in \mathcal{A}$ such that

(i) $A_1^\infty \cap A_3^\infty$ is a root of A_1^∞;
(ii) $A_1^\infty \cap A_2^\infty \subset A_3^\infty$; and
(iii) $|A_1^\infty \cap A_2^\infty| < |A_2^\infty \cap A_3^\infty|$.

By (i) and (iii),

$$|A_1^\infty \cap A_2^\infty| < |A_i^\infty \cap A_3^\infty|$$

for both $i = 1$ and 2. By induction and Step 2, we can assume, therefore, that ν_{A_i} and ν_{A_3} agree on $A_i \cap A_3$ for both $i = 1$ and $i = 2$. Therefore ν_{A_1} and ν_{A_2} agree on $A_1 \cap A_2 \cap A_3$.

By (i) and 4.26, $A_1 \cap A_3$ is a root of A_1 which we will call α. Let β_i be the root opposite α in A_i for $i = 1$ and 3. By (ii), $A_1 \cap A_2 \cap A_3$ is not empty; let

x be a chamber in this set. By (iii), we can choose a chamber y in $A_2 \cap A_3$ not contained in A_1. Thus $x \in \alpha$ and $y \in \beta_3$. Let γ be a minimal gallery from x to y. By 29.6, γ contains a panel $\{v, w\}$ such that $v \in \alpha$ and $w \in \beta_3$ (and hence $w \notin A_1$ since $A_1 \cap A_3 = \alpha$). Since x and y both lie in A_2, also v and w both lie in A_2. Let P be the panel of Δ containing $\{u, v\}$. Since $P \in \mu(\alpha)$, there exists a unique chamber p in $P \cap \beta_1$. Since $v \in \alpha$, we have $p \neq v$. Since $p \in A_1$, we have $p \neq w$. Since $A_2 \cap P = \{v, w\}$, it follows that $p \notin A_2$. Now let z be an arbitrary chamber in β_1. By 29.6.i, there exists a minimal gallery

$$\gamma_1 = (z, \ldots, p, v)$$

from z to v whose penultimate chamber is p. Since $v \in A_2$ and $p \notin A_2$, it follows that $z \notin A_2$. We conclude that $A_1 \cap A_2 \subset \alpha \subset A_3$. Thus

$$A_1 \cap A_2 \subset A_1 \cap A_2 \cap A_3.$$

By the conclusion of the previous paragraph, therefore, the maps ν_{A_1} and ν_{A_2} agree on $A_1 \cap A_2$.

Step 4: Suppose that $k = 0$. Choose a gem R containing u and let $S_1 = \sigma_{A_1}(R, u)$. Let v be the unique chamber opposite u in $R \cap A_2$ and let $S_2 = \sigma_{A_2}(R, v)$. By 10.26, there exists an apartment $A_3 \in \mathcal{A}$ containing $S_1 \cup S_2$. By Step 3, therefore, ν_{A_1} and ν_{A_3} agree on $A_1 \cap A_3$ and ν_{A_2} and ν_{A_3} agree on $A_2 \cap A_3$. Since $u \in A_1 \cap A_2 \cap A_3$, it follows that

$$\nu_{A_1}(u) = \nu_{A_3}(u) = \nu_{A_2}(u).$$

The proof of 12.24 is now complete. □

We can now conclude the proof of 12.3. For each chamber u of Δ, we choose an apartment $A_u \in \mathcal{A}$ containing u. Let

$$\rho(u) = \nu_{A_u}(u)$$

for all $u \in \Delta$. By 12.24, ρ is independent of the choice of the A_u. Every chamber of Δ' lies in some element of $\mathcal{A}'$. By 12.7, therefore, ρ is surjective.

Suppose that u and v are two chambers of Δ. By 8.4.i, there exists an apartment $A \in \mathcal{A}$ containing both u and v. In fact, we can assume that $A_u = A_v = A$. Since ν_A is an isomorphism from A to $\tau(A)$, we conclude that $\rho(u)$ and $\rho(v)$ are adjacent if u and v are adjacent and that $\rho(u) \neq \rho(v)$ if $u \neq v$. This implies that ρ is an isomorphism from Δ to Δ'. By 12.22.i, $\rho(S)^\infty = \tau(S^\infty)$ for all $\mathcal{A}$-sectors S of Δ.

It remains only to prove the uniqueness of ρ. For this it suffices to show that an automorphism of Δ mapping $\mathcal{A}$ to itself that acts trivially on $\Delta_{\mathcal{A}}^\infty$ is the identity. Suppose that ω is such an automorphism of Δ. By 8.27, ω acts trivially on $\mathcal{A}$. By 10.36, it follows that ω maps each root contained in an $\mathcal{A}$-apartment to itself. Now let d be an arbitrary chamber of Δ. Then there exists an $\mathcal{A}$-apartment A containing d (by 8.4.i), and for each chamber e of A adjacent to d, there exists a root of A containing d but not e (by 29.6.i). By 29.6.ii, $\{d\}$ is the intersection of these roots. Hence ω fixes d. □

Remark 12.25. Let τ and ρ be as in 12.3. Let o be the special vertex of Π used in the construction 8.9 of $\Delta_{\mathcal{A}}^{\infty}$. By 3.21 of [32], ρ is a σ-isomorphism for some isomorphism σ from Π to Π'. Let $o' = \sigma(o)$. By 7.1, ρ maps o-special sectors of Δ to o'-special sectors of Δ'. By 8.12, we can assume that o' is the special vertex of Π' used in the construction 8.9 of $(\Delta')_{\mathcal{A}'}^{\infty}$. By 8.8, it follows that τ is a σ_o-isomorphism from Π_o to $\Pi'_{o'}$, where Π_o and $\Pi'_{o'}$ are as in 7.1 and σ_o is the restriction of σ to Π_o.

The following observation makes it practical to check whether a given isomorphism τ is tree-preserving.

Proposition 12.26. *Let τ be an isomorphism from $\Delta_{\mathcal{A}}^{\infty}$ to $(\Delta')_{\mathcal{A}'}^{\infty}$. Let X be a set containing an arbitrary panel of $\Delta_{\mathcal{A}}^{\infty}$ if Π has only single bonds and let $X = \{F_1, F_2\}$, where F_1 and F_2 are two panels of $\Delta_{\mathcal{A}}^{\infty}$ contained in the two components of the graph Ω described in 29.51, if Π has multiple bonds. Let W be a set of walls of $\Delta_{\mathcal{A}}^{\infty}$ such that every $F \in X$ is contained in m for some $m \in W$. Suppose that*

$$\omega_{\tau(F)} \circ \tau = \omega_F$$

for each panel F contained in X or that

$$\omega_{\tau(m)} \circ \tau = \omega_m$$

for each wall in W. Then τ is tree-preserving.

Proof. Suppose m is a wall of $\Delta_{\mathcal{A}}^{\infty}$ and that F is a panel in m. Thus $\tau(F)$ is a panel in the wall $\tau(m)$ of $(\Delta')_{\mathcal{A}'}^{\infty}$. Let $\iota_{m,F}$ and $\iota_{\tau(m),\tau(F)}$ be as in 11.23. Thus

$$\tau \circ \iota_{m,F} \circ \tau^{-1} = \iota_{\tau(m),\tau(F)}. \tag{12.27}$$

By 11.24 and 12.1, we also have

$$\omega_m \circ \iota_{m,F} = \omega_F \tag{12.28}$$

and

$$\omega_{\tau(m)} \circ \iota_{\tau(m),\tau(F)} = \omega_{\tau(F)}. \tag{12.29}$$

Hence

$$\begin{aligned} \omega_{\tau(m)} \circ \tau \circ \iota_{m,F} &= \omega_{\tau(m)} \circ \iota_{\tau(m),\tau(F)} \circ \tau \\ &= \omega_{\tau(F)} \circ \tau \end{aligned}$$

by 12.27 and 12.29. By 12.28, therefore, $\omega_{\tau(m)} \circ \tau = \omega_m$ if and only if $\omega_{\tau(F)} \circ \tau = \omega_F$. By 29.51 and the hypothesis, it follows that τ is tree-preserving. □

The following is a convenient reformulation of 12.3.

Theorem 12.30. *Let*

$$\mathrm{Aut}(\Delta, \mathcal{A})$$

denote the subgroup of $\mathrm{Aut}(\Delta)$ mapping $\mathcal{A}$ to itself and for each element ρ of this group, let ρ^* be the automorphism of $\Delta_{\mathcal{A}}^{\infty}$ induced by ρ. Then the map $\rho \mapsto \rho^*$ is an isomorphism from $\mathrm{Aut}(\Delta, \mathcal{A})$ to the group of tree-preserving automorphisms of $\Delta_{\mathcal{A}}^{\infty}$.

Proof. Since an element ρ of $\mathrm{Aut}(\Delta, \mathcal{A})$ preserves the tree structure of Δ, its image ρ^* is tree-preserving. The claim holds, therefore, by 12.3. □

The following result will play a central role from now on.

Theorem 12.31. *Suppose that the rank ℓ of $\Delta^\infty_\mathcal{A}$ is at least 2 and let $G^\dagger$ be the subgroup of* $\mathrm{Aut}(\Delta^\infty_\mathcal{A})$ *generated by all the root groups of $\Delta^\infty_\mathcal{A}$ (as defined in 29.15). Then every element of $G^\dagger$ is tree-preserving and thus (by 12.30) every element of $G^\dagger$ is induced by a unique element of* $\mathrm{Aut}(\Delta, \mathcal{A})$.

Proof. Let $A \in \mathcal{A}$, let a be a root of A^∞, let ∂a be its border (as defined in 29.40) and let x be a chamber in $a \backslash \partial a$. Let X be a set of panels containing x, one from each connected component of the graph Ω given by 29.51 applied to $\Delta^\infty_\mathcal{A}$. The group U_a acts trivially on every panel containing a chamber in $a \backslash \partial a$. By 12.26, it follows that every element of U_a is tree-preserving. □

Chapter Thirteen

The Moufang Property at Infinity

In the last chapter we saw (in 12.3) that a pair $(\Delta, \mathcal{A})$ is uniquely determined by its building at infinity $\Delta_{\mathcal{A}}^{\infty}$ and its "tree structure." In this chapter, we introduce the assumption that the building at infinity $\Delta_{\mathcal{A}}^{\infty}$ has the Moufang property (defined in 29.15) and show how the tree structure of $(\Delta, \mathcal{A})$ can be translated into a valuation of the root datum of $\Delta_{\mathcal{A}}^{\infty}$ as defined in 3.21. Our main results are 13.30 and 13.31.

Definition 13.1. A *Bruhat-Tits pair* is a pair $(\Delta, \mathcal{A})$ such that Δ is a building whose Coxeter diagram (or type) is one of the affine Coxeter diagrams $\tilde{\mathsf{X}}_\ell$ in Figure 1.1 (which we call the *type* of the Bruhat-Tits pair), $\mathcal{A}$ is a system of apartments of Δ and the building at infinity $\Delta_{\mathcal{A}}^{\infty}$ has the Moufang property as defined in 29.15 (which requires, in particular, that the rank ℓ of $\Delta_{\mathcal{A}}^{\infty}$ be at least 2). We call a building Δ a *Bruhat-Tits building* if it has a system of apartments $\mathcal{A}$ such that $(\Delta, \mathcal{A})$ is a Bruhat-Tits pair. An *isomorphism* from one Bruhat-Tits pair $(\Delta, \mathcal{A})$ to another Bruhat-Tits pair $(\Delta', \mathcal{A}')$ is an isomorphism from Δ to Δ' that maps $\mathcal{A}$ to $\mathcal{A}'$.

If Δ is an arbitrary affine building of type $\tilde{\mathsf{X}}_\ell$ and $\mathcal{A}$ is a system of apartments of Δ, then $\Delta_{\mathcal{A}}^{\infty}$ is a building of type X_ℓ (by 8.26). By 11.3 and 29.15.i, therefore, $(\Delta, \mathcal{A})$ is automatically a Bruhat-Tits pair if $\ell \geq 3$.

For the rest of this chapter, we assume that $(\Delta, \mathcal{A})$ is a Bruhat-Tits pair of type $\Pi = \tilde{\mathsf{X}}_\ell$ and that A is an apartment in $\mathcal{A}$. Let $\Sigma = A^\infty$, let Ξ be the set of roots of Σ, let $(G^\dagger, \xi)$ be the root datum of $\Delta_{\mathcal{A}}^{\infty}$ based at Σ as defined in 3.4 and let I be the vertex set of the Coxeter diagram Π.

By 12.31, we can identify $G^\dagger$ with its pre-image in $\mathrm{Aut}(\Delta, \mathcal{A})$ under the map $\rho \mapsto \rho^*$ described in 12.30. We are thus considering, in particular, each root group U_a of $\Delta_{\mathcal{A}}^{\infty}$ to be a subgroup of $\mathrm{Aut}(\Delta, \mathcal{A})$.

Proposition 13.2. *Let $a \in \Xi$ and let $a_u = A \cap A^u$ for each $u \in U_a^*$. Then for each $u \in U_a^*$, a_u is a root of A such that $a_u^\infty = a$ on which u acts trivially. In particular, each element of U_a is a special automorphism of Δ.*

Proof. Let $u \in U_a^*$. We have $(A^u)^\infty = (A^\infty)^u$ and thus $A^\infty \cap (A^u)^\infty = a$ by 29.56. By 8.30, $a = \alpha^\infty$ for some root α of A. This means (by 4.8) that each sector contained in α contains subsectors contained in the intersection a_u of A and A^u. By 29.13, a_u is convex. Hence by 4.26, a_u is a root of A parallel to α. Thus $a_u^\infty = \alpha^\infty$.

It remains only to show that u acts trivially on a_u. Let R be a gem cut by A such that $R \cap A \subset a_u$, let β be the unique root of A parallel to a_u that cuts $R \cap A$ and let $\partial\beta$ be the border of β as defined in 29.40. By 29.7.iv, $R\cap\beta$ is a root of $R\cap A$. By 6.13.i of [37], therefore, we can choose a chamber $x \in \beta\cap R$ such that x and every chamber of $R\cap A$ adjacent to x are in $\beta\backslash\partial\beta$. Let d be a chamber of $R\cap A$ equal to or adjacent to x and let e be a chamber of $R \cap A$ adjacent to d. Then d and e are both in β. Let $S = \sigma_A(R, d)$, let $S_1 = \sigma_A(R, e)$, let P be the panel containing d and e, let $f = \mu_A(R, P)$ and let $F = f^\infty$. By 11.1, F is the panel of $\Delta_{\mathcal{A}}^\infty$ containing S^∞ and S_1^∞. By 4.11, S and S_1 are both contained in β. Therefore S^∞ and S_1^∞ are both contained in $a = \beta^\infty$. Since $u \in U_a$, it follows that u acts trivially on F (by the definition of a root group). Thus u acts trivially on the set of $\mathcal{A}_F$-ends of the panel tree $(T_F, \mathcal{A}_F)$ (by 12.1). Hence u acts trivially on T_F itself (by 9.14). In particular, f^u is a face asymptotic to f (as defined in 11.7). By 5.22, we conclude (since e is arbitrary) that $S^u = S$. Hence u fixes the apex d of S. We conclude that u fixes x and all the chambers of $R\cap A$ adjacent to x. It follows, in particular, that u is a special automorphism of Δ. Therefore u is a special isomorphism from A to A^u that fixes the chamber x. By 29.34, therefore, u acts trivially on a_u. □

Definition 13.3. Let $a \in \Xi$. For each residue E cut by A, let

$$U_{a,E} = \{u \in U_a \mid E^u = E\}.$$

Proposition 13.4. *Let $a \in \Xi$ and let E be a proper residue cut by A. Then an element $u \in U_a^*$ lies in $U_{a,E}$ if and only if $E \cap a_u \neq \emptyset$ (where a_u and $U_{a,E}$ are as in 13.2 and 13.3).*

Proof. Let $u \in U_a^*$, let b_u be the root of A opposite a_u, let c_u be the root of A^u opposite a_u and let $A' = b_u \cup c_u$. By 13.2, $b_u \cap c_u = \emptyset$. Hence by 10.34, $A' \in \mathcal{A}$.

By 13.2, u is special and acts trivially on a_u. Thus if $E \cap a_u \neq \emptyset$, then u fixes chambers in E and hence $u \in U_{a,E}$. Suppose, conversely, that $E^u = E$. Suppose as well that $E \cap A \subset b_u$. Then

$$E \cap c_u = E^u \cap b_u^u = (E \cap b_u)^u = (E \cap A)^u,$$

so $|E \cap A'| = 2|E \cap A|$. Thus, in particular, $|E \cap A| \neq |E \cap A'|$. This is impossible by 29.18, however, since both $E \cap A$ and $E \cap A'$ are apartments of E (by 29.13.iv). Therefore $E \cap A \not\subset b_u$. Thus $E \cap a_u \neq \emptyset$. □

Proposition 13.5. *Let $a \in \Xi$, let α be a root of A such that $\alpha^\infty = a$, let P be a panel in the wall $\mu(\alpha)$ and let $U_{a,P}$ be as in 13.3. Then $U_{a,P}$ acts transitively on $P\backslash\{x\}$, where x is the unique chamber in $P\cap\alpha$. In particular, there exists $u \in U_a^*$ such that $\alpha = a_u$.*

Proof. Let y be the unique chamber of $P \cap A$ not in α and let z be an arbitrary chamber of $P\backslash\{x\}$. By 10.28, there exists an apartment A_1 in $\mathcal{A}$ containing $\alpha \cup \{z\}$. By 8.27 and 29.15, there is an element u in U_a mapping A to A_1. By 13.2, u is special and acts trivially on $a_u = A \cap A^u$. Thus u fixes x and maps P to itself. Since y is the unique chamber in $A\cap P$ distinct from x and z is the unique chamber in $A_1 \cap P$ distinct from x, we conclude that u maps y to z and, if $y \neq z$, then $y \notin a_u$. Thus if $y \neq z$, then $a_u = \alpha$ since α is the unique root of A containing x but not y (by 29.6.i). □

Definition 13.6. Let α be a root of A and let $i \mapsto \alpha_i$ be the bijection from $\mathbb{Z}$ to the parallel class $[\alpha]$ described in 1.3. Thus

$$\cdots \subset \alpha_{-2} \subset \alpha_{-1} \subset \alpha_0 = \alpha \subset \alpha_1 \subset \alpha_2 \subset \alpha_3 \subset \cdots.$$

We set

$$\text{dist}(\alpha_i, \alpha_j) = j - i$$

for all $i, j \in \mathbb{Z}$.

Proposition 13.7. *Let α and β be parallel roots of A. Then the following hold:*

(i) $\text{dist}(\alpha, \beta) = -\text{dist}(\beta, \alpha)$*;*
(ii) $\text{dist}(\alpha, \beta) = -\text{dist}(-\alpha, -\beta)$ *(where $-\alpha$ and $-\beta$ are the roots of A opposite α and β); and*
(iii) $\text{dist}(\alpha, \beta) = \text{dist}(\alpha, \delta) + \text{dist}(\delta, \beta)$ *for all roots δ of A parallel to α.*

Proof. These identities follow directly from 13.6. □

The following definition is fundamental.

Definition 13.8. Let R be a gem cut by A. For each $a \in \Xi$, let a_R be the unique root of A such that $a_R^\infty = a$ and a_R cuts $R \cap A$, let $u \mapsto a_u$ be the map from U_a^* to the set of roots of A defined in 13.2 and let ϕ_a denote the map from U_a^* to $\mathbb{Z}$ given by

$$\phi_a(u) = \text{dist}(a_R, a_u)$$

for each $u \in U_a^*$, where the map dist is as defined in 13.6. Note that by 13.5, ϕ_a is surjective for each $a \in \Xi$. Let

$$\phi_R = \{\phi_a \mid a \in \Xi\}.$$

Proposition 13.9. *Let R and $\phi_R = \{\phi_a \mid a \in \Xi\}$ be as in 13.8, let $a \in \Xi$, let m be the wall of a, let b be the root opposite a in Σ and let ω_m be the canonical valuation of the wall tree $(T_m, \mathcal{A}_m)$ as defined in 9.9. We identify the set of roots of $\Delta_{\mathcal{A}}^\infty$ whose wall is m (which contains a and b) with the set of $\mathcal{A}_m$-ends of T_m via the map ξ_m given in 10.38. Then*

$$\omega_m(a, b, b^{u_1}, b^{u_2}) = \phi_a(u_2) - \phi_a(u_1)$$

for all distinct elements $u_1, u_2 \in U_a^*$.

Proof. This holds by 9.9 and 13.8. □

Theorem 13.10. *Let $(\Delta', \mathcal{A}')$ be another Bruhat-Tits building and let τ be an isomorphism from $\Delta_{\mathcal{A}}^\infty$ to $(\Delta')_{\mathcal{A}'}^\infty$. Let A' be the unique apartment in $\mathcal{A}'$ such that $\tau(A^\infty) = (A')^\infty$ and let Ξ' denote the set of roots of A'. Let $\phi_R = \{\phi_a \mid a \in \Xi\}$ for some gem R of Δ cut by A and let $\phi_{R'} = \{\phi_{a'} \mid a' \in \Xi'\}$*

for some gem R' of Δ' cut by A' be as in 13.8. Suppose that for each $a \in \Xi$, there is a constant c_a such that

$$\phi_{a'}(\tau u \tau^{-1}) = \phi_a(u) + c_a \tag{13.11}$$

for each $u \in U_a^*$, where $a' = \tau(a)$ (so $\tau U_a \tau^{-1}$ is the root group $U_{a'}$ of $\Delta'_{\mathcal{A}'}$). Then there is a unique isomorphism ρ from Δ to Δ' mapping $\mathcal{A}$ to $\mathcal{A}'$ such that $\rho(S)^\infty = \tau(S^\infty)$ for all $\mathcal{A}$-sectors S of Δ.

Proof. Let α be a root of A, let β be its opposite, let $M = \mu(\alpha)$, let $m = M^\infty$, let $a = \alpha^\infty$ and let $b = \beta^\infty$. Let u_1, u_2 be distinct elements of U_a^* and let $u_i' = \tau u_i \tau^{-1} \in U_{a'}$ for $i = 1$ and 2. By 13.9,

$$\omega_m(a, b, b^{u_1}, b^{u_2}) = \phi_a(u_2) - \phi_a(u_1)$$

and

$$\omega_{m'}\big(a', b', (b')^{u_1'}, (b')^{u_2'}\big) = \phi_{a'}(u_2') - \phi_{a'}(u_1'),$$

where $a' = \tau(a)$ and $b' = \tau(b)$. By 13.11, therefore,

$$\begin{aligned}\omega_m(a, b, b^{u_1}, b^{u_2}) &= \omega_{m'}\big(a', b', (b')^{u_1'}, (b')^{u_2'}\big)\\ &= \omega_{m'}\big(\tau(a), \tau(b), \tau(b^{u_1}), \tau(b^{u_2})\big).\end{aligned}$$

By 12.31, ω_m is U_a-invariant and $\omega_{m'}$ is $U_{a'}$-invariant. Since a is fixed by U_a and a' is fixed by $U_{a'}$, it follows that

$$\omega_m(a, b^{u_1}, b^{u_2}, b^{u_3}) = \omega_{m'}\big(\tau(a), \tau(b^{u_1}), \tau(b^{u_2}), \tau(b^{u_3})\big) \tag{13.12}$$

for all triples u_1, u_2, u_3 of distinct elements of U_a^*. Let $(T_m)_{\mathcal{A}_m}$ denote the set of $\mathcal{A}_m$-ends of the wall tree T_m, i.e. the set of roots of $\Delta_{\mathcal{A}}^\infty$ having wall m. For each $c \in (T_m)_{\mathcal{A}_m}$ distinct from a, there is a unique apartment with chamber set $a \cup c$ (by 29.48). By the Moufang property, it follows that

$$(T_m)_{\mathcal{A}_m} = \{a\} \cup \{b^u \mid u \in U_a\}.$$

Thus ω_m and $\omega_{m'} \circ \tau$ agree on all 4-tuples of distinct elements of $(T_m)_{\mathcal{A}_m}$ whose first coordinate is a. By 9.13, it follows that $\omega_m = \omega_{m'} \circ \tau$.

Let Ω be the graph obtained by applying 29.51 to $\Delta_{\mathcal{A}}^\infty$ and let x be a chamber of A^∞. By 29.51, there exist panels in each connected component of Ω (of which there are at most two) all containing x. Thus the set of walls W described in 12.26 can be chosen so that every wall in W is contained in A^∞. By 12.26, therefore, τ is tree-preserving. Thus the claim holds by 12.3. □

Our next goal (which we reach in 13.30) is to show that for each gem R of Δ cut by A, the collection of maps

$$\phi_R = \{\phi_a \mid a \in \Xi\}$$

defined in 13.8 is a valuation of the root datum $(G^\dagger, \xi)$ (as defined in 3.21). The first step toward this goal is to introduce a root map ι of Σ (as defined in 3.12). To do this, we need to return to the representation of the Coxeter chamber system $\tilde{\Sigma}_\Pi$ in terms of a root system Φ and alcoves that was the subject of Chapter 2:

Notation 13.13. The Coxeter diagram Π of Δ is one of the diagrams $\tilde{\mathsf{X}}_\ell$ in Figure 1.1. Let Φ be the root system X_ℓ, let V be its ambient space and let $\tilde{\Sigma}_\Phi$ be the chamber system introduced in 2.29. This chamber system is isomorphic to Σ_Π and hence to A. We choose an isomorphism ψ from A to $\tilde{\Sigma}_\Phi$, and from now on, we identify A with $\tilde{\Sigma}_\Phi$ via ψ, so the chambers of A are alcoves and the roots of A are identified with the sets $K_{\alpha,k}$ defined in 2.31. Let R be the gem cut by A such that $R \cap A$ corresponds under this identification to the set of all alcoves whose closure contains 0 and let Σ_Φ be the Coxeter chamber system introduced in 2.9. By 2.10 and 2.13, the roots of Σ_Φ are the sets K_α for $\alpha \in \Phi$ defined in 2.11. The chambers of Σ_Φ (i.e. the Weyl chambers of Φ) can be thought of as the sectors of A with terminus R. By 8.23, there is an isomorphism π from Σ to Σ_Φ such that

$$\pi(\sigma_A(R,d)^\infty) = \sigma_A(R,d) \tag{13.14}$$

for each $d \in R$ (where $\sigma_A(R,d)$ is as defined in 7.4 and $\sigma_A(R,d)^\infty$ is as defined in 8.17). We identify the roots of Σ with the parallel classes of roots in A, i.e. the sets

$$\{K_{\alpha,k} \mid k \in \mathbb{Z}\},$$

via the bijection in 8.30.ii and let ι be the bijection from Ξ to Φ such that

$$\iota\big(\{K_{\alpha,k} \mid k \in \mathbb{Z}\}\big) = \alpha$$

for each $\alpha \in \Phi$. Then $\pi(a) = K_{\iota(a)}$ for each $a \in K$. Thus ι is a root map of Σ with target Φ as defined in 3.12. We now identify Ξ with Φ via ι and apply 3.19: For each pair a, b of roots of A^∞ (i.e. for each pair a, b of roots in Φ) such that $a \neq \pm b$ and for each c in the interval (a, b) as defined in 3.1, there exist (by 3.19) unique positive real numbers p_c and q_c such that

$$c = p_c a + q_c b. \tag{13.15}$$

Suppose that Π is not simply laced. By 2.48, the identification of A with $\tilde{\Sigma}_\Phi$ is unique up to an element of $T_\Phi W_\Phi$. Since T_Φ acts trivially on Ξ, it follows that ι is, up to equivalence as defined in 3.12, independent of the choice of the isomorphism ψ from A to $\tilde{\Sigma}_\Phi$. By 3.14.ii, this is true also if Π is simply laced. Thus, in particular, the coefficients p_c in 13.15 are independent of the choice of ψ (whether or not Π is simply laced).

We emphasize that it is the identification of Ξ with Φ via ι that allows us to make sense of 13.15.

Remark 13.16. In 13.13 it appears that we imply that for a given root system Φ there is a unique equivalence class of root maps of the apartment Σ with target Φ, whereas in 3.14 we showed that there are, for certain choices of Φ, two. We resolve this "paradox" as follows. The main point in 13.13 is that there is a canonical isomorphism π (given in 13.14) from A^∞ to Σ_Φ. Since Σ *equals* A^∞ in 13.13, we obtain from π a unique root map of Σ satisfying 3.13. In 3.14, on the other hand, there is no affine building. We could nevertheless introduce a Coxeter chamber system A of type $\tilde{\Pi}$ and define the isomorphism π from A^∞ to Σ_Φ as in 13.13, but to obtain from π a root map of the Coxeter chamber system Σ with target Φ, we must first choose an isomorphism from Σ to A^∞ which we then compose with π to obtain an isomorphism from Σ to Σ_Φ. It is in the choice of this isomorphism from Σ to A^∞ that the possibility of more than one equivalence class of root maps with target Φ arises.

Notation 13.17. Let $a \in \Phi$ and let $\phi = \phi_R$ be as defined in 13.8, where R is the gem chosen in 13.13. For $a \in \Xi$ (i.e. for each $a \in \Phi$), we extend ϕ_a to U_a by setting $\phi_a(1) = \infty$ and set

$$U_{a,k} = \{u \in U_a \mid \phi_a(u) \geq k\}$$

for each $k \in \mathbb{R}$. Let $k \in \mathbb{Z}$ and let β be the unique root of A such that $\beta^\infty = a$ and $\text{dist}(a_R, \beta) = k$. By 13.2, $U_{a,k}$ is the pointwise stabilizer in U_a of β. Thus, in particular, $U_{a,k}$ is a subgroup of U_a. In other words, ϕ satisfies condition (V1) in 3.21.

Proposition 13.18. *Let $a \in \Phi$, let R and $\phi = \phi_R$ be as in 13.17, let $u \in U_a^*$ and let a_u be as in 13.2. For each $k \in \mathbb{Z}$, let $K_{a,k}$ be as in 13.13, so $K_{a,0} = a_R$ (where a_R is as defined in 13.8), and let $U_{a,k}$ be as in 13.17. Then the following hold:*

(i) $\text{dist}(K_{a,0}, K_{a,-k}) = k$ *for all* $k \in \mathbb{Z}$.
(ii) $a_u = K_{a,-\phi_a(u)}$.
(iii) $u \in U_{a,k}^*$ *if and only if* $K_{a,-k} \subset a_u$.

Proof. Let $k \in \mathbb{Z}$. Then

$$K_{a,0} \subset K_{a,-1} \subset \cdots \subset K_{a,-k}$$

if $k \geq 0$ and

$$K_{a,-k} \subset K_{a,-k-1} \subset \cdots \subset K_{a,0}$$

if $k < 0$. By 13.6, therefore, (i) holds. By 13.8,

$$\phi_a(u) = \text{dist}(K_{a,0}, a_u).$$

Therefore (ii) follows from (i). By (ii) and 13.17, (iii) holds. □

We show next that ϕ satisfies (V2) in 3.21. The proof we give is borrowed from the Diplom-Arbeit of A. von Heydebreck [15].

Proposition 13.19. *Suppose a and b are two roots of A^∞ such that $a \neq \pm b$. For each $c \in (a, b)$, let p_c and q_c be as in 13.15. Then*

$$[U_{a,k}, U_{b,l}] \subset \prod_{c \in (a,b)} U_{c,p_c k + q_c l}$$

for all $k, l \in \mathbb{Z}$ (with the notation in 13.17).

Proof. Choose $k, l \in \mathbb{Z}$. Let $P = \langle a, b \rangle$. Since $a \neq \pm b$, the subspace P of the ambient space V of Φ is two-dimensional. We denote the closure of a subset L of P by $\bar{L}$. Let

$$X = \{v \in P \mid v \cdot a > -k \text{ and } v \cdot b > -l\}. \tag{13.20}$$

Let

$$[a, b] = (a_1, a_2, \ldots, a_s)$$

(so $a_1 = a$ and $a_s = b$). By 3.2.iii, we can assume that $s > 2$.

Let $w \in [U_{a,k}, U_{b,l}]$. By 13.2 and 13.18.iii, $U_{a,k}$ and $U_{b,l}$ act trivially on the set

$$K_{a,-k} \cap K_{b,-l}.$$

Hence w acts trivially on this set. By 3.2.iii again, there exists for each $i \in (1, s)$ an element $g_i \in U_{a_i}$ such that

$$w = g_2 \cdots g_{s-1}. \tag{13.21}$$

Let

$$m_i = \phi_{a_i}(g_i), \tag{13.22}$$

so $K_{a_i,-m_i}$ is the set of chambers in A fixed by g_i (by 13.2 and 13.18.ii), and let

$$D_i = \{v \in P \mid v \cdot a_i > -m_i\} \tag{13.23}$$

for each $i \in (1, s)$. By 13.15, there exist integers n_i such that

$$X \subset \{v \in P \mid v \cdot a_i > -n_i\} \tag{13.24}$$

for all $i \in (1, s)$, where X is as in 13.20.

We claim that

$$X \subset D_2 \cap \cdots \cap D_{s-1}. \tag{13.25}$$

Suppose that 13.25 is false. Let $\tilde{\Phi} = \Phi \times \mathbb{Z}$ (as in 2.16), let

$$\tilde{\Phi}_0 = \{(c, p) \in \tilde{\Phi} \mid P \subset H_{c,p}\},$$

where $H_{c,p}$ is as defined in 2.17, and let $\tilde{\Phi}_1 = \tilde{\Phi} \backslash \tilde{\Phi}_0$. By 13.24 and some plane geometry, there exists a point

$$u \in X \cap \bar{D}_2 \cap \cdots \cap \bar{D}_{s-1}$$

that is contained in $H_{a_j,-m_j}$ for some $j \in (1, s)$ but not in $H_{c,p}$ for any $(c, p) \in \tilde{\Phi}_1$ distinct from $(a_j, -m_j)$. Let w be a point in V orthogonal to the plane P such that $w \notin H_{c,k}$ for any $(c, p) \in \tilde{\Phi}_0$. Then there exists an $\epsilon > 0$ such that the intersection of the ball in V centered at $u + w$ of radius ϵ with the set $\{v \in V \mid v \cdot a_j < -m_j\}$ is contained in a single alcove d. This alcove is not contained in the root $K_{a_j,-m_j}$ but it is contained in $K_{a,-k} \cap K_{b,-l}$ and in $K_{a_i,-m_i}$ for all $i \in (1, s)$ different from j. Therefore d is fixed by w and by g_i for all $i \in (1, s)$ different from j but not by g_j. With this contradiction (to 13.21), we conclude that 13.25 holds. Thus, in particular,

$$x \in \bar{D}_2 \cap \cdots \cap \bar{D}_{s-1}, \tag{13.26}$$

where x is the unique point in

$$P \cap H_{a,k} \cap H_{b,l}. \tag{13.27}$$

Let $c = a_i$ for some $i \in (1, s)$ and let

$$M = p_c k + q_c l,$$

where p_c and q_c are as in 13.15. By 13.23 and 13.26, we have $x \cdot c \geq -m_i$. Therefore

$$\begin{aligned} \phi_c(g_i) &= m_i && \text{by } 13.22 \\ &\geq -x \cdot (p_c a + q_c b) && \text{by } 13.15 \\ &= -p_c(x \cdot a) - q_c(x \cdot b) \\ &= p_c k + q_c l = M && \text{by } 13.27, \end{aligned}$$

and hence $g_i \in U_{c,M}$ by 13.17. □

Proposition 13.28. *Let a be a root of Σ (i.e. $a \in \Xi$), let b denote the root opposite a in Σ, let $u \in U_a^*$ and let v, w be the unique elements of U_b^* such that $m_\Sigma(u) = vuw$, where m_Σ is as defined in 3.8. Then*

$$b_v = b_w = -(a_u),$$

where b_v, b_w and a_u are as defined in 13.2, $-(a_u)$ is the root of A opposite a_u and $m_\Sigma(u)$ interchanges a_u and $-(a_u)$.

Proof. By 3.8, $m_\Sigma(u)$ interchanges a and b. In particular, $m_\Sigma(u)$ maps Σ to itself. By 8.27, therefore, $m_\Sigma(u)$ maps A to itself. Let α be the image of the root $-(a_u) \cap b_v$ under the element u. By 13.2, the element v acts trivially on the root b_v and thus $m_\Sigma(u)$ maps $-(a_u) \cap b_v$ to α^w. Therefore α^w is a root of A parallel to a_u. Let β be the image of the root $-(a_u)$ under the element u. By 10.34 and 13.2, $-(a_u) \cup \beta$ is an $\mathcal{A}$-apartment; we denote it by A_1. By 29.13.v, β and $-(a_u)$ are opposite roots of A_1, $-(a_u) \cap b_w$ is a root of A_1 parallel to $-(a_u)$ and α is a root of A_1 parallel to β. Thus by 8.27, A is the unique $\mathcal{A}$-apartment containing $-(a_u) \cap b_w$ and α^w and A_1 is the unique $\mathcal{A}$-apartment containing $-(a_u) \cap b_w$ and α. Since w acts trivially on $-(a_u) \cap b_w$, it follows that $A_1^w = A$. Thus

$$A \cap A^{w^{-1}} = A \cap A_1 = -(a_u).$$

Hence $b_w = b_{w^{-1}} = -(a_u)$. Replacing u by u^{-1} in this argument, we conclude (by 3.9) that also $b_v = -(a_u)$. □

Proposition 13.29. *Let $\phi = \phi_R$ be as in 13.8 for some gem R cut by A. Let $a, b \in \Xi$, let $u \in U_a^*$, let v be an element of U_b^* such that $\phi_b(v) = 0$, let s_a denote the reflection of Σ that interchanges a and its opposite (as in 3.7) and let*

$$t = \phi_{s_a(b)}(v^{m_u}),$$

where $m_u := m_\Sigma(u)$ (as defined in 3.8). Then the following hold:

(i) $\phi_{s_a(b)}(x^{m_u}) = \phi_b(x) + t$ *for all $x \in U_b^*$. In particular, t is independent of the choice of v.*

(ii) *If $a = b$, then $t = -2\phi_a(u)$ and thus*

$$\phi_{-a}(x^{m_u}) = \phi_a(x) - 2\phi_a(u)$$

for all $x \in U_a^*$.

Proof. Let $x \in U_b^*$, let $y = x^{m_u}$, let $z = v^{m_u}$ and let $c = s_a(b)$. Thus $b_x^{m_u} = c_y$ and $b_v^{m_u} = c_z$ (where b_x, c_y, etc. are as defined in 13.2). By 13.8 and the choice of v, we have $b_v = b_R$. Since m_u preserves containment among roots, it follows that

$$\text{dist}(c_z, c_y) = \text{dist}(b_v, b_x) = \text{dist}(b_R, b_x).$$

Therefore

$$\begin{aligned} \phi_c(y) &= \text{dist}(c_R, c_y) && \text{by 13.8} \\ &= \text{dist}(c_z, c_y) + \text{dist}(c_R, c_z) && \text{by 13.7.iii} \\ &= \text{dist}(b_R, b_x) + \text{dist}(c_R, c_z) \\ &= \phi_a(x) + \phi_c(z) \\ &= \phi_a(x) + t. \end{aligned}$$

Thus (i) holds.

Suppose now that $b = a$. Then $c = s_a(b) = -a$, $b_v = a_R$ and c_R is the opposite of a_R in A, i.e. $c_R = -(a_R)$. Thus

$$\begin{aligned} \text{dist}(-(a_u), c_z) &= \text{dist}(a_u^{m_u}, b_v^{m_u}) && \text{by 13.28} \\ &= \text{dist}(a_u, b_v) \\ &= -\text{dist}(a_R, a_u) && \text{by 13.7.i} \\ &= -\phi_a(u) \end{aligned}$$

and

$$\text{dist}(c_R, -(a_u)) = -\text{dist}(a_R, a_u) = -\phi_a(u)$$

by 13.7.ii. Hence

$$\begin{aligned} t &= \text{dist}(c_R, c_z) \\ &= \text{dist}(c_R, -(a_u)) + \text{dist}(-(a_u), c_z) \\ &= -2\phi_a(u) \end{aligned}$$

by 13.7.iii. Thus (ii) holds. □

Theorem 13.30. *Let $(G^\dagger, \xi)$ be the root datum of Δ_A^∞ based at Σ (as defined in 3.4), let R, ι and V be as in 13.13 and let $\phi = \phi_R$ be as defined in 13.8. Then the following hold:*

(i) *ϕ is a valuation of $(G^\dagger, \xi)$ based at Σ with respect to the root map ι of Σ.*
(ii) *If R_1 is another gem cut by A and v is the center of $R_1 \cap A$ as defined in 2.41 (modulo the identification of A with V), then the valuation ϕ_{R_1} (as defined in 13.8) equals $\phi + v$ (as defined in 3.23).*

(iii) *Every valuation of* $(G^\dagger, \xi)$ *equipollent to* ϕ *is of the form* ϕ_{R_1} *for some gem* R_1 *cut by* A.

Proof. It was observed in 13.8 that the map ϕ_a is surjective for each root a of Σ and in 13.17 that the set ϕ satisfies condition (V1). The set ϕ satisfies condition (V2) by 13.19 and conditions (V3) and (V4) by 13.29. Thus (i) holds.

Let R_1 be an arbitrary gem cut by A, let v be the center of $R_1 \cap A$, let $a \in \Xi$ and let a_R and a_{R_1} be as defined in 13.8. Then $a_R = K_{a,0}$ and $a_{R_1} = K_{a,v\cdot a}$ (since $v \in H_{a,v\cdot a}$). Thus $\text{dist}(a_R, a_{R_1}) = -v \cdot a$ by 13.18.i and hence

$$\begin{aligned}\text{dist}(a_{R_1}, a_u) &= \text{dist}(a_R, a_u) + \text{dist}(a_{R_1}, a_R) && \text{by 13.7.iii}\\ &= \text{dist}(a_R, a_u) + v \cdot a && \text{by 13.7.i}\end{aligned}$$

for all $u \in U_a^*$. Therefore $\phi_{R_1} = \phi_R + v$ by 13.10. Thus (ii) holds.

Suppose, conversely, that $\psi := \{\psi_a \mid a \in \Xi\}$ is an arbitrary valuation of $(G^\dagger, \xi)$ that is equipollent to ϕ. By 3.23, there exists a special point $v \in V$ such that $\psi = \phi + v$. By 2.42, there exists a gem R_1 cut by A such that v is the center of $R_1 \cap A$. By (ii), it follows that $\psi = \phi_{R_1}$. Thus (iii) holds. □

In light of 13.30, we can rephrase 13.10 as follows:

Theorem 13.31. *Let* $(\Delta', \mathcal{A}')$ *be another Bruhat-Tits pair and let* τ *be an isomorphism from* $\Delta_{\mathcal{A}}^\infty$ *to* $(\Delta')_{\mathcal{A}'}^\infty$. *Let* A' *be the unique apartment in* $\mathcal{A}'$ *such that* $\tau(A^\infty) = (A')^\infty$ *and let* Ξ' *be the set of roots of* A'. *Let*

$$\phi_R = \{\phi_a \mid a \in \Xi\}$$

for some gem R *of* Δ *cut by* A *and let*

$$\phi_{R'} = \{\phi_{a'} \mid a' \in \Xi'\}$$

for some gem R' *of* Δ' *cut by* A' *be as in 13.8. For each* $a \in \Xi$, *let*

$$\psi_a(u) = \phi_{\tau(a)}(\tau u \tau^{-1}) \tag{13.32}$$

for each $u \in U_a^*$ *and let*

$$\psi := \{\psi_a \mid a \in \Xi\}. \tag{13.33}$$

Then the following hold:

(i) ψ *is a valuation of* $(G^\dagger, \xi)$ *with respect to the root map* ι *defined in 13.13.*

(ii) *The valuations* ψ *and* ϕ_R *are equipollent to each other if and only if there is an isomorphism* ρ *from* Δ *to* Δ' *mapping* $\mathcal{A}$ *to* $\mathcal{A}'$ *such that* $\rho(S)^\infty = \tau(S^\infty)$ *for all* $\mathcal{A}$*-sectors* S *of* Δ.

Proof. Let ι be as in 13.13. By 13.30.i, ϕ_R is a valuation of the root datum $(G^\dagger, \xi)$ with respect to ι. By 11.23 of [37] and 3.21, it follows that ψ is also a valuation of $(G^\dagger, \xi)$ with respect to ι. Thus (i) holds. Suppose that ρ is an isomorphism from Δ to Δ' mapping $\mathcal{A}$ to $\mathcal{A}'$ such that $\rho(S)^\infty = \tau(S^\infty)$ for all $\mathcal{A}$-sectors S of Δ and let $R_1 = \rho^{-1}(R')$. Then ψ equals the valuation ϕ_{R_1} of $(G^\dagger, \xi)$ (as defined in 13.8). By 13.30.ii, it follows that ψ is equipollent to ϕ_R. The converse holds by 13.10. Thus (ii) holds. □

Here is a slightly less formal reformulation of 13.31:

Theorem 13.34. *Let $(\Delta, \mathcal{A})$ be a Bruhat-Tits pair, let $A \in \mathcal{A}$, let R be a gem cut by A and let $\Sigma = A^\infty$. Then the pair $(\Delta, \mathcal{A})$ is uniquely determined by the root datum $(G^\dagger, \xi)$ of $\Delta_{\mathcal{A}}^\infty$ based at Σ and the equipollence class of the valuation $\phi = \phi_R$ defined in 13.8.*

Proof. By 3.6, the pair $(\Delta_{\mathcal{A}}^\infty, \Sigma)$ is uniquely determined by the root datum $(G^\dagger, \xi)$. The claim holds, therefore, by 13.31. □

Chapter Fourteen

Existence

In this chapter we are concerned with the existence part of the classification of Bruhat-Tits buildings. Our main result is 14.47. Combining this existence result with the uniqueness result 13.31, we obtain as a corollary (in 14.54) one of our major conclusions: that Bruhat-Tits buildings are classified by root data (of Moufang spherical buildings of rank $\ell \geq 2$) with valuation (up to equipollence).

This chapter is essentially a continuation of Chapter 3. We have included it at this point so that we can make the link to 13.8 in 14.47.iii and to 13.31 in 14.54.

The results in this chapter were first proved in Chapters 5 and 6 of [6].[1]

Our plan is to prove 14.38 and then obtain 14.47 as a consequence. We accomplish this in a series of steps.

Notation 14.1. Let Δ be a spherical building whose Coxeter diagram Π is one of the Coxeter diagrams X_ℓ in Figure 1.3 with $\ell > 1$, let Σ be an apartment of Δ and let Ξ denote the set of roots of Σ. We assume that Δ satisfies the Moufang property and let $(G^\dagger, \xi)$ be the root datum of Δ based at Σ as defined in 3.4. Let ι be a root map of Σ with target Φ (as defined in 3.12). This means that there is an isomorphism π from Σ to Σ_Φ such that 3.13 holds. We identify Σ with Σ_Φ via π and Ξ with Φ via ι. Finally, let

$$\psi := \{\psi_a \mid a \in \Phi\}$$

be a valuation of the root datum $(G^\dagger, \xi)$ with respect to ι as defined in 3.21.

Proposition 14.2. *Let $a \in \Phi$, let $u \in U_a^*$, let $m = m_\Sigma(u)$ be as defined in 3.8 and let $k = -\psi_a(u)$. Then for each special point v of the ambient space V of Φ (as defined in 2.26),*

$$(\psi + v)^m = \psi + s_{a,k}(v),$$

where the expression on the left-hand side is as defined in 3.30 and $s_{a,k}$ is the affine reflection of V defined in 2.17.

Proof. Let v be a special point of V. We have $v \cdot s_a(b) = s_a(s_a(v) \cdot b) = s_a(v) \cdot b$ for all $b \in \Phi$, where s_a is the reflection of V interchanging a and $-a$.

[1] The result analogous to 14.47 for root data with a non-discrete valuation (as defined in footnote 2 in Chapter 12) is proved in Chapter 7 of [6]. The proof of this analogous result requires completely different methods.

Therefore (by 3.23)

$$\begin{aligned}((\psi+v)^m)_b(x) &= (\psi+v)_{s_a(b)}(x^m)\\ &= \psi_{s_a(b)}(x^m) + v\cdot s_a(b)\\ &= (\psi^m)_b(x) + s_a(v)\cdot b\end{aligned}$$

for all $x \in U_b^*$ and hence

$$(\psi+v)^m = \psi^m + s_a(v).$$

By 3.31, it follows that

$$\begin{aligned}(\psi+v)^m &= \psi + s_a(v) - 2\psi_a(u)a/(a\cdot a)\\ &= \psi + s_{a,k}(v).\end{aligned}$$

□

Notation 14.3. Let

$$N_a := \langle m_\Sigma(u) \mid u \in U_a^*\rangle$$

and

$$M_a := \langle m_\Sigma(u)m_\Sigma(v) \mid u,v \in U_a^*\rangle$$

for each $a \in \Phi$. Let $N = \langle N_a \mid a \in \Phi\rangle$ and $M = \langle M_a \mid a \in \Phi\rangle$.

Since the group M defined in 14.3 acts trivially on Σ, it normalizes U_a for each $a \in \Phi$. (By 33.9 in [36], M is, in fact, the pointwise stabilizer of Σ in $G^\dagger$.)

Proposition 14.4. *There is a surjective homomorphism π from N to the affine Coxeter group $\tilde{W}_\Phi$ defined in 2.18 such that*

$$\pi\big(m_\Sigma(u)\big) = s_{a,k}$$

for each $a \in \Phi$ and for each $u \in U_a^*$, where $k = -\psi_a(u)$.

Proof. Let V_0 denote the set of special points of V (as defined in 2.26) and let

$$X = \{\psi + v \mid v \in V_0\}.$$

The group $\tilde{W}_\Phi$ acts faithfully on V_0. By 14.2, therefore, there is a surjective homomorphism $g \mapsto \tau_g$ from N to $\tilde{W}_\Phi$ such that

$$(\psi+v)^g = \psi + v^{\tau_g}$$

for each $v \in V_0$ and each $g \in N$ and $\tau_{m_\Sigma(u)} = s_{a,-\psi_a(u)}$ for each $a \in \Phi$ and each $u \in U_a^*$.[2] □

Notation 14.5. Let H denote the kernel of the homomorphism π described in 14.4.

[2]See footnote 12 in Chapter 3.

Proposition 14.6. *Let $L = \langle U_a, U_{-a}\rangle$ and let $m = m_\Sigma(u)$ for some $a \in \Phi$ and some $u \in U_a^*$. Then*

$$L = U_a M_a m \cup U_a M_a U_{-a},$$

where M_a is as in 14.3.

Proof. Let $Q = U_a M_a m \cup U_a M_a U_{-a}$. By 3.8, $m \in U_{-a}^* U_a^* U_{-a}^* \subset L$ and hence $Q \subset L$. It will thus suffice to show that $QL \subset Q$. Since the element m^{-2} is contained in M_a and normalizes U_{-a}, we have $Qm^{-2} \subset Q$. By 3.8 again, we have $U_a^m = U_{-a}$. Hence $L = \langle U_{-a}, m\rangle$ and $QU_{-a} \subset Q$. Since $Qm^{-2} \subset Q$, it therefore suffices to show that $Qm \subset Q$. We claim that, in fact,

$$U_a M_a U_{-a}^* m \subset U_a M_a U_{-a}^*, \tag{14.7}$$

from which $Qm \subset Q$ then follows.

Choose $v \in U_{-a}^*$. By one more application of 3.8, there exist elements $w, w' \in U_a^*$ such that $m_\Sigma(v) = (w')^{-1} v w^{-1}$. By 11.24 in [37], $m_\Sigma(U_{-a}^*) = m_\Sigma(U_a^*)$ and thus $m_\Sigma(v)m \in M_a$. Hence

$$\begin{aligned} U_a M_a v m &= U_a M_a w' m_\Sigma(v) w \cdot m \\ &= U_a M_a w' \cdot m_\Sigma(v) m \cdot w^m \\ &\subset U_a M_a U_{-a}^* \end{aligned}$$

since M_a normalizes U_a (so $U_a M_a w' \subset U_a M_a$) and $U_a^m = U_{-a}$ (so $w^m \in U_{-a}^*$). Thus 14.7 holds as claimed. □

Proposition 14.8. *Let $a \in \Phi$, let $u \in U_a^*$, let $v \in U_{-a}^*$ and suppose that*

$$\psi_a(u) + \psi_{-a}(v) > 0.$$

Then there exist $u_1 \in U_a^*$, $v_1 \in U_{-a}^*$ and $h \in H$ such that $vu = u_1 h v_1$, $\psi_a(u_1) = \psi_a(u)$ and $\psi_{-a}(v_1) = \psi_{-a}(v)$ (where H is as in 14.5).

Proof. Let $m = m_\Sigma(u)$. Suppose first that $vu \in U_a M_a m$. Then there exists $u' \in U_a$ such that $u'vu \in mM_a$. Thus $u'vu$ interchanges a and $-a$. This implies that $u' \neq 1$ (since otherwise $(-a)^{u'vu} = (-a)^u \neq a$). By 3.8, it follows that $u'vu = m_\Sigma(v)$. By 3.25, however, this implies that $\psi_a(u) + \psi_{-a}(v) = 0$. We conclude that $vu \notin U_a M_a m$. By 14.6, therefore, $vu \in U_a M_a U_{-a}$. Hence there exist elements $u_1 \in U_a$, $v_1 \in U_{-a}$ and $h \in M_a$ such that

$$vu = u_1 h v_1. \tag{14.9}$$

Since u and v are both non-trivial, we have $a^v \neq a$ and $(-a)^u \neq -a$ by 29.15.v. Thus the product vu fixes neither a nor $-a$. Since $h \in M_a$, the element h acts trivially on Σ. It follows that u_1 and v_1 are both non-trivial. Let

$$m_1 = m_\Sigma(v_1). \tag{14.10}$$

By 3.8, there exist $u_2, u_3 \in U_a^*$ such that $m_1 = u_3^{-1} v_1 u_2^{-1}$. By 3.25 and 3.29,

$$\psi_a(u_2) = -\psi_{-a}(v_1). \tag{14.11}$$

We have $v = (u_1hv_1)u^{-1}$ by 14.9, hence

$$\begin{aligned} v &= u_1h(u_3m_1u_2)u^{-1} \\ &= (u_1(u_3)^{h^{-1}})hm_1(u_2u^{-1}) \end{aligned}$$

and therefore

$$hm_1 = ((u_3^{-1})^{h^{-1}}u_1^{-1})v(uu_2^{-1}) \in U_a v U_a.$$

Since no element of $U_a v$ or vU_a interchanges a and $-a$, we conclude that the products $u_1^{-1}((u_3^{-1}))^{h^{-1}}$ and uu_2^{-1} are both non-trivial. Hence

$$hm_1 = m_\Sigma(v) \tag{14.12}$$

(by 3.8) and thus (by 3.25 again)

$$\psi_a(uu_2^{-1}) = -\psi_{-a}(v). \tag{14.13}$$

Therefore

$$\psi_a(u) - \psi_a(uu_2^{-1}) = \psi_a(u) + \psi_{-a}(v) > 0$$

(by hypothesis) and hence

$$\psi_a(u) > \min\{\psi_a(uu_2^{-1}), \psi_a(u_2)\}. \tag{14.14}$$

By condition (V1) (in 3.21) and 3.29, ψ_a satisfies the conditions (a) and (b) in 9.18. By 9.18.i and 14.14, we thus have

$$\psi_a(uu_2^{-1}) = \psi_a(u_2). \tag{14.15}$$

Therefore

$$\begin{aligned} \psi_{-a}(v) &= -\psi_a(u_2) \quad \text{by 14.13 and 14.15} \\ &= \psi_{-a}(v_1) \quad \text{by 14.11.} \end{aligned} \tag{14.16}$$

Replacing u by u^{-1} and v by v^{-1} and interchanging a and $-a$ in this argument, we conclude (using 3.29) also that

$$\psi_a(u) = \psi_a(u_1).$$

By 14.10 and 14.12,

$$h = hm_1 \cdot m_1^{-1} = m_\Sigma(v)m_\Sigma(v_1)^{-1},$$

and by 14.5 and 14.16,

$$m_\Sigma(v)m_\Sigma(v_1)^{-1} \in H.$$

Hence $h \in H$. □

Definition 14.17. Let

$$[a, k] = \{v \in V \mid v \cdot a \geq -k\}$$

for each pair $(a, k) \in \tilde{\Phi} := \Phi \times \mathbb{Z}$, let

$$\Omega_a = \{[a, k] \mid k \in \mathbb{Z}\}$$

for each $a \in \Phi$ and let

$$\Omega = \bigcup_{a\in\Phi} \Omega_a.$$

We identify each of the sets $[a,k]$ with the set of alcoves it contains (as defined in 2.19). With this interpretation, the elements of Ω are precisely the roots of the Coxeter chamber system $\tilde{\Sigma}_\Phi$ (by 2.30 and 2.33), $[-a,-k]$ is the root of $\tilde{\Sigma}_\Phi$ opposite $[a,k]$, i.e.

$$-[a,k] = [-a,-k],$$

the affine reflection $s_{a,-k}$ interchanges the roots $[a,k]$ and $[-a,-k]$ for each $(a,k) \in \tilde{\Phi}$ and the subsets Ω_a for $a \in \Phi$ are precisely the parallel classes of roots (by 2.34). For each $\alpha = [a,k] \in \Omega$, we set

$$U_\alpha = U_{a,k},$$

where $U_{a,k}$ is as defined in condition (V1) (in 3.21), i.e.

$$U_{a,k} = \{u \in U_a \mid \psi_a(u) \geq k\}, \tag{14.18}$$

where $\psi_a(1) := \infty$. Note, too, that $[a,k] \subset [a,k+1]$ for all $(a,k) \in \tilde{\Phi}$.

Proposition 14.19. *Let $\alpha = [a,k] \in \Omega$, where Ω is as in 14.17, let $g \in N$ and let*

$$\beta = \alpha^{\pi(g)},$$

where π is as in 14.4. Then $U_\alpha^g = U_\beta$.

Proof. It suffices to assume that $g = m_\Sigma(u)$ for some $b \in \Phi$ and some $u \in U_b^*$. Let $l = \psi_b(u)$. Then

$$\begin{aligned}
\beta &= s_{b,-l}(\alpha) && \text{by 14.4}\\
&= s_{b,-l}\big(\{v \mid v\cdot a \geq -k\}\big) && \text{by 14.17}\\
&= \{v \mid s_{b,-l}(v)\cdot a \geq -k\} && \text{since } s_{b,-l}^2 = 1\\
&= \{v \mid \big(s_b(v) - 2lb/(b\cdot b)\big)\cdot a \geq -k\} && \text{by 2.17}\\
&= \{v \mid s_b(v)\cdot a - 2l(b\cdot a)/(b\cdot b) \geq -k\}\\
&= \{v \mid s_b(v)\cdot s_b(s_b(a)) \geq -\big(k - 2l(b\cdot a)/(b\cdot b)\big)\}\\
&= \{v \mid v\cdot s_b(a) \geq -\big(k - 2l(b\cdot a)/(b\cdot b)\big)\}\\
&= \big[s_b(a), k - 2l(b\cdot a)/(b\cdot b)\big] && \text{by 14.17,}
\end{aligned}$$

and therefore

$$\begin{aligned}
U_\alpha^g &= \{z \in U_{s_b(a)} \mid \psi_a(z^{g^{-1}}) \geq k\} && \text{by 14.18}\\
&= \{z \in U_{s_b(a)} \mid (\psi^{g^{-1}})_{s_b(a)}(z) \geq k\} && \text{by 3.30}\\
&= \{z \in U_{s_b(a)} \mid \psi_{s_b(a)}(z) - 2l(b\cdot s_b(a))/(b\cdot b) \geq k\} && \text{by 3.40}\\
&= \{z \in U_{s_b(a)} \mid \psi_{s_b(a)}(z) - 2l\big(s_b(-b)\cdot s_b(a)\big)/(b\cdot b) \geq k\}\\
&= \{z \in U_{s_b(a)} \mid \psi_{s_b(a)}(z) + 2l(b\cdot a)/(b\cdot b) \geq k\}\\
&= U_{s_b(a),k-2l(b\cdot a)/(b\cdot b)} = U_\beta
\end{aligned}$$

since s_b is an isometry. □

Corollary 14.20. *Let H be as in 14.5 and let Ω be as in 14.17. Then H normalizes U_α for each $\alpha \in \Omega$.*

Proof. This follows directly from 14.19. □

Proposition 14.21. *Let*

$$L_\alpha = \langle U_\alpha, U_{-\alpha}, H\rangle$$

for some $\alpha = [a, k] \in \Omega$, where H is as in 14.5, $-\alpha = [-a, -k]$ and the groups U_α and $U_{-\alpha}$ are as in 14.18. Let $m = m_\Sigma(z)$ for some $z \in U_a^*$ such that $k = \psi_a(z)$, let $\beta = [-a, -k+1]$ and let U_β be as in 14.18. Then

$$L_\alpha = (U_\alpha H m U_\alpha) \cup (U_\alpha H U_\beta).$$

Proof. Let $X = X_1 \cup X_2$, where

$$X_1 = U_\alpha H m U_\alpha \text{ and } X_2 = U_\alpha H U_\beta.$$

By 3.25, we have

$$m \in U_{-\alpha}^* U_\alpha^* U_{-\alpha}^*.$$

Since

$$U_\beta \subset U_{-\alpha}, \tag{14.22}$$

it follows that $X \subset L_\alpha$.

By 14.19,

$$U_{-\alpha}^m = U_\alpha. \tag{14.23}$$

Thus, in particular, $L_\alpha = \langle H, U_\alpha, m\rangle$. To show that $L_\alpha \subset X$, it thus suffices to show that X is closed under right multiplication by H, U_a, m and m^{-1}. By 14.5 and 14.20, H normalizes U_α and U_β and is normalized by $m \in N$. It follows that $XH \subset X$.

Let $u \in U_\alpha^*$ and $v \in U_\beta^*$. Then $\psi_a(u) \geq k$ and $\psi_{-a}(v) \geq -k+1$. By 14.8, therefore, $vu \in X_2$. Hence $X_2 U_a \subset X_2$ and thus $XU_a \subset X$. Since $m^{-2} \in H$ (by 14.5), we also have $Xm^{-2} \subset X$. It therefore remains only to show that $Xm \subset X$.

We have

$$\begin{aligned} X_2 m = U_\alpha H U_\beta m &\subset U_\alpha H U_{-\alpha} m && \text{by 14.22} \\ &\subset U_\alpha H m U_\alpha = X_1 && \text{by 14.23} \end{aligned}$$

and

$$X_1 m = U_\alpha H m U_\alpha m = U_\alpha H U_\alpha^m = U_\alpha H U_{-\alpha}$$

(since $m^2 \in H$). It will thus suffice to show that

$$U_\alpha H U_{-\alpha} \subset X. \tag{14.24}$$

Let $u \in U_{-\alpha}$. If $u \in U_\beta$, then $U_\alpha H u \subset X_2$. Suppose that $u \notin U_\beta$. Thus $\psi_{-a}(u) = -k$. By 3.8 and 3.25, there exist $v, v' \in U_a^*$ such that $\psi_a(v) = \psi_a(v') = k$ (so v and v' are both contained in the subgroup U_α) and $m_\Sigma(u) = vuv'$. By 14.4, $\pi(m_\Sigma(u)) = s_{-a,k} = s_{a,-k} = \pi(m)$. By 14.5, it follows that $vuv'm^{-1} = m_\Sigma(u)m^{-1} \in H$. Hence $u \in v^{-1}Hm \cdot (v')^{-1} \subset X_1$. Therefore $U_\alpha H u \subset X_1$. Thus 14.24 holds. □

Notation 14.25. Let $\{a_1, \ldots, a_\ell\}$ be a basis of Φ (as defined in 2.5), let $\tilde{a}$ be the highest weight with respect to this basis and let

$$x = \{v \in V \mid v \cdot a_i > 0 \text{ for all } i \in [1, \ell] \text{ and } v \cdot \tilde{a} < 1\}.$$

By 2.20.ii, x is an alcove in V. For each $a \in \Phi$, let $[a, k_a]$ be the minimal element of the parallel class Ω_a defined in 14.17 containing x. Thus for each $a \in \Phi$, $[-a, k_{-a}]$ is the root that contains $-[a, k_a] = [-a, -k_a]$ minimally.[3] In other words,

$$k_a + k_{-a} = 1. \tag{14.26}$$

Let $X_a = U_{a,k_a}$ for all $a \in \Phi$ and let

$$U(x) = \langle X_a \mid a \in \Phi \rangle.$$

For each chamber d of the apartment Σ, let Φ_d^+ be the set of roots of Σ (which we have identified with Φ) containing d, let Φ_d^- be the complement of Φ_d^+ in Φ and let

$$U(x)_d^\epsilon = \langle X_a \mid a \in \Phi_d^\epsilon \rangle$$

for $\epsilon = \pm$.

Proposition 14.27. *Let a, b be roots of Φ such that $b \neq \pm a$. Then (with the notation in 14.25)*

$$[X_a, X_b] \subset \prod_{c \in (a,b)} X_c,$$

where (a, b) is as in 3.1.

Proof. We have $X_a = U_{a,k_a}$ and $X_b = U_{b,k_b}$. Choose a point v in the alcove x defined in 14.25 and let $c \in (a, b)$. Then $a \cdot v \geq -k_a$ and $b \cdot v \geq -k_b$ by 14.17 and 14.25. By 3.19, there exist positive real numbers p and q such that $c = pa + qb$. Thus

$$v \cdot c = v \cdot (pa + qb) \geq -(k_a p + k_b q).$$

Since v is arbitrary, it follows that $x \in [c, k_a p + k_b q]$ and hence $k_c \leq k_a p + k_b q$. Thus $U_{c,k_a p + k_b q} \subset U_{c,k_c} = X_c$ by 14.25. The claim holds, therefore, by condition (V2) in 3.21. □

Proposition 14.28. *Let d be a chamber of Σ, let a be a root of Σ containing d and let X_a' denote the group generated by X_b for all $b \in \Phi_d^+$ distinct from a, where X_a and Φ_d^+ are as in 14.25. Then X_a normalizes X_a'.*

Proof. Choose $b \in \Phi_d^+$ distinct from a. Since a and b both contain d, they are not opposite (by 29.9.i). By 3.2.i, all the roots in the interval (a, b) contain d. By 14.27, therefore, $[X_a, X_b] \subset X_a'$. □

---

[3]Here are the details: Let $\alpha = [a, k_a]$ for some $a \in \Phi$ and let β be the root containing $-\alpha$ minimally. Then $-\beta$ is contained properly in α. Hence $x \in \alpha$ but $x \notin -\beta$. Therefore $x \in \beta$ but $x \notin -\alpha$. Thus $\beta = [-a, k_{-a}]$.

Proposition 14.29. *Let a be a root of Σ containing a chamber d but not some chamber d' of Σ adjacent to d. Then $U_a \cap U(x)_d^+ = X_a$.*

Proof. Let $u \in U_a \cap U(x)_d^+$. By 14.27, we have $U(x)_d^+ = X_a X_a'$, where X_a' is as in 14.28. Thus $u = ww'$, where $w \in X_a$ and $w' \in X_a'$. Let P be the unique panel of Δ that contains d and d'. By 29.14.iii and 29.15.v, U_a acts faithfully on P. By 29.6.i, a is the only root of Σ that contains d but not d'. In other words, every root in $\Phi_d^+\backslash\{a\}$ contains both d and d'. By the definition of a root group in 29.15, it follows that U_b acts trivially on P for each $b \in \Phi_d^+\backslash\{a\}$. Thus w' acts trivially on P. Therefore $w^{-1}u$ is an element of U_a acting trivially on P. Hence $u = w \in X_a$. □

Proposition 14.30. *The product $(H \cap U(x))U(x)_d^- U(x)_d^+$ is independent of the choice of the chamber d of Σ (although neither $U(x)_d^+$ nor $U(x)_d^-$ is independent of d).*

Proof. Let $H(x) = H \cap U(x)$ and let d, d' be two chambers of Σ. Since Σ is connected, it will suffice to show that

$$H(x)U(x)_d^- U(x)_d^+ = H(x)U(x)_{d'}^- U(x)_{d'}^+$$

under the assumption that d and d' are adjacent.

By 29.6.i, there is a unique root a in Φ containing d but not d'. Then

$$\Phi_{d'}^+ = \{-a\} \cup \Phi_d^+\backslash\{a\}$$

and

$$\Phi_{d'}^- = \{a\} \cup \Phi_d^-\backslash\{-a\}.$$

Let X_a' be the group generated by $\{X_b \mid b \in \Phi_d^+\backslash\{a\} = \Phi_{d'}^+\backslash\{-a\}\}$ and let X_{-a}' be the group generated by $\{X_b \mid b \in \Phi_d^-\backslash\{-a\} = \Phi_{d'}^-\backslash\{a\}\}$. By 14.28, $U(x)_d^+ = X_a X_a'$ and $U(x)_d^- = X_{-a}' X_{-a}$ as well as $U(x)_{d'}^+ = X_{-a} X_a'$ and $U(x)_{d'}^- = X_{-a}' X_a$. By 14.8 and 14.26,

$$X_{-a} X_a \subset X_a H(x) X_{-a}.$$

Thus by 14.20, we conclude that

$$\begin{aligned} H(x)U(x)_d^- U(x)_d^+ &= H(x)X_{-a}' X_{-a} \cdot X_a X_a' \\ &= H(x)X_{-a}' X_a \cdot X_{-a} X_a' \\ &= H(x)U(x)_{d'}^- U(x)_{d'}^+. \end{aligned}$$

□

Corollary 14.31. *Let d be a chamber of Σ. Then the following hold:*

(i) *There is a unique chamber e of Σ opposite d.*
(ii) *$\Phi_e^\epsilon = \Phi_d^{-\epsilon}$ and hence $U_d^\epsilon = U_e^{-\epsilon}$ for e as in (i) and for $\epsilon = \pm$.*
(iii) *$(H \cap U(x))U(x)_d^- U(x)_d^+ = (H \cap U(x))U(x)_d^+ U(x)_d^-$.*

Proof. By 29.9.i and 29.9.iv, (i) and (ii) hold; (iii) follows from (ii) and 14.30. □

Proposition 14.32. *Let d be a chamber of Σ, let e be as in 14.31.i, let u^+ be an element of $U(x)_d^+$ fixing e and let u^- be an element of $U(x)_d^-$ fixing d. Then $u^+ = u^- = 1$.*

Proof. By 29.14.ii, Σ is the unique apartment of Δ containing both d and e. Thus u^+ maps Σ to itself. Since elements of root groups are special automorphisms of Δ, it follows by 29.4 that, in fact, u^+ acts trivially on Σ. Let P be a panel containing d and let d' be the unique chamber in $P \cap \Sigma$ distinct from d. By 29.6.i, there is a unique root a in Φ_d^+ that does not contain d'. Thus for each $b \in \Phi_d^+ \backslash \{a\}$, the root b contains both d and d' and hence (by the definition of a root group in 29.15) the root group U_b acts trivially on P. By 29.14.iii and 29.15.v, U_a acts sharply transitively on $P \backslash \{d\}$. Since u^+ fixes d', we conclude that u^+ acts trivially on P. Since P is arbitrary, it follows by 29.14.iv that $u^+ = 1$. Replacing d by e in this argument, we conclude (by 14.31.ii) that $u^- = 1$. □

Proposition 14.33. *The following hold (with the notation of 14.25):*

(i) *$U(x) = (H \cap U(x))U(x)_d^- U(x)_d^+$ for each chamber d of Σ.*
(ii) *$N \cap U(x) \subset H$.*
(iii) *$U_a \cap (HU(x)) = X_a$ for all $a \in \Phi$.*

Proof. Let d be a chamber of Σ, let $H(x) = H \cap U(x)$ and let $W = H(x)U(x)_d^- U(x)_d^+$. Then $W \subset U(x)$ and $WU(x)_d^+ \subset W$. By 14.31.iii, also $WU(x)_d^- \subset W$. Hence $WU(x) \subset W$. Thus (i) holds.

Let $g \in N \cap U(x)$. Then $g \in W$, so $g = hu^-u^+$, where $u^\epsilon \in U(x)_d^\epsilon$ for $\epsilon = \pm$ and $h \in H(x)$. Let e be as in 14.31.i. By 14.31.ii, hu^- fixes e. Since u^+ fixes d, it maps e to a chamber opposite d in Δ (by 29.14.i). Since $g \in N$, we have $e^g \subset \Sigma^g = \Sigma$. Thus e^g is a chamber of Σ opposite d. By the uniqueness of e, we conclude that $e^{u^+} = e^g = e$. By 14.32, therefore, $u^+ = 1$. Therefore $u^- = h^{-1}g \in N$ is an element of $U(x)_d^-$ mapping Σ to itself. Since $U(x)_d^-$ fixes e and contains only special automorphisms of Δ, the element u^- must act trivially on Σ. Hence also $u^- = 1$ by 14.32. We conclude that $g = h \in H(x)$. Thus (ii) holds.

Now let $g \in U_a \cap (HU(x))$ for some $a \in \Phi$. Choose adjacent chambers $d, d' \in \Phi$ so that d lies in a but d' does not. By (i), we have $g = hu^-u^+$, where $u^\epsilon \in U(x)_d^\epsilon$ for $\epsilon = \pm$ and $h \in H$. Since the elements g, h and u^+ all fix d, so does the element u^-. Thus $u^- = 1$ by 14.32. Hence $h \in U(x)_d^+$. By 14.32 again, it follows that $h = 1$ (since h acts trivially on Σ). Thus $g \in U_a \cap U(x)_d^+$. By 14.29, we conclude that $g \in X_a$. Thus (iii) holds. □

Notation 14.34. Let $\{a_1, \ldots, a_\ell\}$ and $\tilde{a}$ be as in 14.25. For each $a \in \{a_1, \ldots, a_\ell, -\tilde{a}\}$, choose $u \in U_a^*$ such that $\psi_a(u) = k_a$, where k_a is as in 14.25, and let $m_a = m_\Sigma(u)$. Let

$$S = \{m_{a_i} \mid i \in [1, \ell]\}$$

and

$$\tilde{S} = S \cup \{m_{-\tilde{a}}\}.$$

Notation 14.35. Let R be the residue of $\tilde{\Sigma}_\Phi$ whose center is the origin and let x be the chamber of R defined in 14.25. Let d denote the sector $\sigma(R,x)$ interpreted as a Weyl chamber (i.e. as a chamber of Σ_Φ). Since we have identified the apartment Σ of Δ with the Coxeter chamber system Σ_Φ (in 14.1), we can also think of d as a chamber of Σ. Thus d is contained in a root a of Σ if and only if the Weyl chamber $\sigma(R,x)$ is contained in the root $[a,0]$ of Σ_Φ. By 4.11 and 4.12.ii, $\sigma(R,x) \subset [a,0]$ if and only if $x \in [a,0]$.

Proposition 14.36. *Let π be as in 14.4, let x be as in 14.25, let S and $\tilde{S}$ be as in 14.34 and let d be as in 14.35. Then the following hold:*

(i) *The map $m_a \mapsto d^{\pi(m_a)}$ is a bijection from S to the set of Weyl chambers adjacent to d in Σ_Φ.*
(ii) *The map $m_a \mapsto x^{\pi(m_a)}$ is a bijection from $\tilde{S}$ to the set of alcoves adjacent to x in $\tilde{\Sigma}_\Phi$.*

Proof. We have $k_{a_i} = 0$ and $k_{-\tilde{a}} = 1$. By 14.4, therefore, $\pi(m_{a_i}) = s_{a_i}$ for each $i \in [1,\ell]$ and $\pi(m_{-\tilde{a}}) = s_{-a,-1} = s_{a,1}$. The claims holds, therefore, by 2.8 and 2.28. □

The following fundamental definition is taken from Chapter IV, Section 2.1, of [3].[4]

Definition 14.37. A *Tits system* (or *BN-pair*) is a quadruple (G,B,N,S), where G is a group, B and N are subgroups of G and S is a subset of N, satisfying the following conditions:

(T1) $G = \langle B,N\rangle$ and $B \cap N$ is a normal subgroup of N.
(T2) $N = \langle S, B\cap N\rangle$ and $s^2 \in B$ for all $s \in S$.
(T3) $BwBsB \subset BwB \cup BwsB$ for all $s \in S$ and all $w \in W$.
(T4) $sBs \not\subset B$ for all $s \in S$.

The group $W := N/N \cap B$ is often called the Weyl group of the Tits system (G,B,N,S). By Theorem 2 in Chapter IV, Section 2.4, of [3], (W,S) is a Coxeter system as defined in 29.4. The corresponding Coxeter diagram Π is called the *type* of the Tits system (G,B,N,S). Thus a Tits system (or BN-pair) is *spherical* if its type is spherical and *affine* if its type is affine.

Here is the result we have been aiming at:

Theorem 14.38. *Let $G^\dagger := \langle U_a \mid a \in \Phi\rangle$, let $\tilde{B} = HU(x)$, where H is as defined in 14.5 and $U(x)$ is as defined in 14.25. Let N be as in 14.3 and let $\tilde{S}$ be as in 14.34. Then $(G^\dagger, \tilde{B}, N, \tilde{S})$ is a Tits system as defined in 14.37, $H = \tilde{B} \cap N$ and $N/H \cong \tilde{W}_\Phi$.*

Proof. By 14.5, H is a normal subgroup of N and $N/H \cong \tilde{W}_\Phi$. By 14.33.ii, $\tilde{B} \cap N = H$. Let k_a for each $a \in \Phi$ be as in 14.25. Choose $a \in \Phi$ and

[4] In fact, we have written $BwBsB$ and $BwsB$ in place of $BsBwB$ and $BswB$ in (T3) since we are composing permutations from left to right in this book. Note that the map $x \mapsto x^{-1}$ interchanges these two versions of (T3) since $s^2 \in B$ by (T2).

$u \in U_a^*$, let $k = \psi_a(u)$ and let $[a, k]$ be as in 14.17. Then either $x \in [a, k]$ or $x \in -[a, k] = [-a, -k]$; i.e. $k \geq k_a$ or $k < k_a$. In the first case,

$$u \in U_{a,k} \subset U_{a,k_a} = X_a \subset U(x) \subset \tilde{B},$$

and in the second, $-k \geq -k_a + 1 = k_{-a}$ by 14.26 and hence

$$u^{m_\Sigma(u)} \in U_{-a,-k} \subset U_{-a,k_{-a}} = X_{-a} \subset U(x) \subset \tilde{B}$$

by 14.19. Since $m_\Sigma(u) \in N$, it follows that $U_a \in \langle \tilde{B}, N \rangle$. Since a is arbitrary, we conclude that $G^\dagger = \langle \tilde{B}, N \rangle$. Thus (T1) holds.

By 14.36.ii, $N = \langle \tilde{S}, H \rangle$. Since $\tilde{B} \cap N = H$, in fact $N = \langle \tilde{S}, \tilde{B} \cap N \rangle$. For each $a \in \Phi$ and each $u \in U_a^*$, the element $m_\Sigma(u)^2$ lies in H (by 14.5). Thus, in particular, $s^2 \in H \subset \tilde{B}$ for all $s \in \tilde{S}$. Thus (T2) holds.

We turn now to (T3). Let $s \in \tilde{S}$. Then $s = m_a$ for some

$$a \in \{a_1, \ldots, a_\ell, -\tilde{a}\},$$

where $m_a, a_1, \ldots, a_\ell$ and $\tilde{a}$ are as in 14.34. Let d be a chamber of Σ contained in a, let

$$X_a' = \langle X_b \mid b \in \Phi_d^+ \backslash \{a\} \rangle$$

as in 14.28 and let $\alpha = [a, k_a]$, so $X_a = U_\alpha$. By 14.19, H normalizes U_b for all $b \in \Phi$. Thus

$$\begin{aligned} \tilde{B} = HU(x) &= HU(x)_d^- U(x)_d^+ && \text{by 14.33.i} \\ &= U(x)_d^+ U(x)_d^- H && \text{by 14.31.iii} \\ &= X_a X_a' U(x)_d^- H && \text{by 14.28.} \end{aligned} \tag{14.39}$$

By 14.36.ii, $y := x^{\pi(s)}$ is adjacent to x. By 14.17 and 14.34, $\pi(s)$ interchanges α and $-\alpha$. Thus α contains x but not y. By 29.6.i, therefore, α is the unique root of $\tilde{\Sigma}_\Phi$ that contains x but not y. For each $b \in \Phi \backslash \{a\}$, the root $[b, k_b]$ is distinct from $\alpha = [a, k_a]$ and contains x. Therefore, $y \in [b, k_b]$ and hence $x \in [b, k_b]^{\pi(s)}$ for all $b \in \Phi \backslash \{a\}$. Thus $X_b^s \subset U(x)$ for each $b \in \Phi \backslash \{a\}$ by 14.19. Since s normalizes H, we conclude that

$$\left(X_a' U(x)_d^- H\right)^s \subset \tilde{B}.$$

By 14.39, we therefore have

$$\tilde{B}s \subset X_a X_a' U(x)_d^- Hs \subset X_a s\tilde{B} = U_\alpha s\tilde{B}. \tag{14.40}$$

Let $L_\alpha = \langle U_\alpha, U_{-\alpha}, H \rangle$. Then $s \in U_{-\alpha}^* U_\alpha^* U_{-\alpha}^* \subset L_\alpha$ (by 3.8) and therefore

$$s^i U_\alpha s\tilde{B} \subset L_\alpha \tilde{B} \tag{14.41}$$

for both $i = 0$ and 1.

Now choose $w \in N$. We have $x^{\pi(w)} \in \alpha$ or $-\alpha$. Replacing w by ws if necessary, we can assume that $x^{\pi(w)} \in \alpha$ as long as we remember at the end to verify that (T3) holds for both s and w as well as s and ws. By 14.19 and 14.25, it follows that $U_\alpha^{w^{-1}} \subset U(x) \subset \tilde{B}$ and hence

$$wU_\alpha \subset \tilde{B}w. \tag{14.42}$$

Now let $\beta = [-a, -k_a + 1]$. Then

$$\tilde{B}wL_\alpha = \tilde{B}wU_\alpha HsU_\alpha \cup \tilde{B}wU_\alpha HU_\beta \tag{14.43}$$

by 14.21. We also have

$$U_\beta \subset U(x) \subset \tilde{B} \tag{14.44}$$

(since $\beta = [-a, k_{-a}]$ by 14.26) as well as $sH = Hs$, $HU_\alpha \subset \tilde{B}$ and $HU_\beta \subset \tilde{B}$. By 14.42 and 14.44, it follows that

$$\begin{aligned}\tilde{B}wU_\alpha HsU_\alpha &= \tilde{B}wU_\alpha sHU_\alpha\\ &\subset \tilde{B}wsHU_\alpha\\ &\subset \tilde{B}ws\tilde{B}\end{aligned}$$

and

$$\tilde{B}wU_\alpha HU_\beta \subset \tilde{B}w\tilde{B}.$$

Thus

$$\tilde{B}wL_\alpha \subset \tilde{B}ws\tilde{B} \cup \tilde{B}w\tilde{B} \tag{14.45}$$

by 14.43. Hence

$$\begin{aligned}\tilde{B}ws^i\tilde{B}s\tilde{B} &\subset \tilde{B}ws^iU_\alpha s\tilde{B} && \text{by 14.40}\\ &\subset \tilde{B}wL_\alpha\tilde{B} && \text{by 14.41}\\ &\subset \tilde{B}ws\tilde{B} \cup \tilde{B}w\tilde{B} && \text{by 14.45}\end{aligned}$$

for $i = 0$ and 1. Thus (T3) holds.

By 14.19, $U_\alpha^s = U_{-\alpha}$, where α and s are as in the previous paragraph. Hence $sU_\alpha s = s^2U_{-\alpha}$. By 14.33.iii,

$$U_{-a} \cap \tilde{B} = X_{-a}.$$

By 14.26, $U_{-\alpha} \not\subset X_{-a}$. Therefore $U_{-\alpha} \not\subset \tilde{B}$. Thus $sU_\alpha s \not\subset \tilde{B}$ since $s^2 \in H \subset \tilde{B}$. Hence (T4) holds. □

Before turning to our main theorem, we recall that in addition to the affine Tits system

$$(G^\dagger, \tilde{B}, N, \tilde{S}),$$

we also have the usual spherical Tits system associated with the group $G^\dagger$ and the spherical building Δ:

Proposition 14.46. *Let d be as in 14.35, let Φ_d^+ be as in 14.25, let*

$$U_+ = \langle U_a \mid a \in \Phi_d^+ \rangle,$$

let M and N be as in 14.3, let S be as in 14.34 and let $B = MU_+$. We interpret d as a chamber of Σ and hence a chamber of Δ. Then the following hold:

(i) *The subgroup B is the stabilizer of the chamber d in $G^\dagger$.*

(ii) *$(G^\dagger, B, N, S)$ is a Tits system with $M = B \cap N$ and $N/M \cong W_\Phi$.*[5]

Proof. By 11.35 in [37], B is the stabilizer of d in $G^\dagger$. Thus (i) holds. By 29.15.iii, therefore, (ii) holds. □

We come finally to the main result of this chapter.

Theorem 14.47. *Let $\tilde{\mathsf{X}}_\ell$ be one of the Coxeter diagrams in Figure 1.1 with $\ell \geq 2$, let Φ be the root system X_ℓ, let Δ be a building of type X_ℓ satisfying the Moufang condition,*[6] *let $G^\dagger$ be the subgroup of* Aut(Δ) *generated by all the root groups of Δ, let Σ be an apartment of Δ, let ι be a root map of Σ with target Φ as defined in 3.12 and let*

$$\psi := \{\psi_a \mid a \in \Phi\}$$

be a valuation of the root datum $(G^\dagger, \xi)$ with respect to ι as defined in 3.21. Then there exists a Bruhat-Tits pair $(\Delta^{\rm aff}, \mathcal{A})$ of type $\tilde{\mathsf{X}}_\ell$ such that the following hold:

(i) *The group $G^\dagger$ is a strongly transitive*[7] *subgroup of* Aut°$(\Delta^{\rm aff})$,[8] *the group N defined in 14.3 is the setwise stabilizer in $G^\dagger$ of an apartment A in $\mathcal{A}$ and the group B defined in 14.38 is the stabilizer in $G^\dagger$ of a chamber of A.*
(ii) *$A^\infty = \Sigma$ and $\Delta^{\rm aff}_{\mathcal{A}} \cong \Delta$.*
(iii) *$\psi = \phi_R$ for some gem R cut by A, where ϕ_R is as defined in 13.8, and ι is as described in 13.13.*

Proof. Let $(G^\dagger, \tilde{B}, N, \tilde{S})$ be as in 14.38. By 5.3 in [25] and 14.38, there is a building $\Delta^{\rm aff}$ of type $\tilde{\mathsf{X}}_\ell$ such that $G^\dagger$ is a subgroup of Aut°(Δ), $\tilde{B}$ is the stabilizer of a chamber x in $G^\dagger$, N is the setwise stabilizer of an apartment A containing x, $s \mapsto x^s$ is a bijection from $\tilde{S}$ to the set of chambers of A adjacent to x, $H = \tilde{B} \cap N$ is the stabilizer of x in N and $G^\dagger$ acts strongly transitively on $\Delta^{\rm aff}$. Thus, in particular, (i) holds.

By 2.29, we have $A \cong \tilde{\Sigma}_\Phi$. Since $s^2 \in \tilde{B}$ for each $s \in \tilde{S}$, the elements of $\tilde{S}$ are precisely the reflections of A interchanging x with a chamber adjacent to x. This means that we can identify A with the Coxeter chamber system $\tilde{\Sigma}_\Phi$ so that x is as in 14.25 and for each $g \in N$, the action of g on A is the same as the action of $\pi(g)$ on $\tilde{\Sigma}_\Phi$. The sets $[a, k]$ for $(a, k) \in \tilde{\Phi}$ (as defined in 14.17) are the roots of A and the sets $[a, 0]$ for $a \in \Phi$ are the roots of Σ_Φ. In 14.1, we identified Σ with Σ_Φ and Ξ with Φ. If we undo this second identification, then ι is the map that sends the root $[a, 0]$ of Σ_Φ to a for each $a \in \Phi$. In other words, ι is precisely the root map described in 13.13.

---

[5]In other words, the affine Tits system $(G^\dagger, \tilde{B}, N, \tilde{S})$ is a *double Tits system* as defined in 5.1.1 of [6].

[6]Thus X_ℓ denotes both a root system and a Coxeter diagram in the statement of 14.47.

[7]This means (1) that for each w in the Coxeter group $W_{\tilde{\mathsf{X}}_\ell}$, the group $G^\dagger$ acts transitively on the set of ordered pairs of chambers (x, y) of $\Delta^{\rm aff}$ such that $\delta(x, y) = w$, where δ is as in 29.11, and (2) that the stabilizer in $G^\dagger$ of some apartment acts transitively on that apartment.

[8]This is the group of special automorphisms of $\Delta^{\rm aff}$ as defined in 29.2.

Let $\alpha = [a, k]$ be a root of A and let $\beta = \alpha^{g^{-1}}$ for some $g \in N$. By 14.19, $U_\alpha \subset \tilde{B}^g$ if and only if $U_\beta \subset \tilde{B}$. By 14.33.iii, $U_\beta \subset \tilde{B}$ if and only if $U_\beta \subset X_b$, where $b = a^{g^{-1}}$ and X_b is as in 14.25. By 14.25, $U_\beta \subset X_b$ if and only if $x \in \beta$. Since $\tilde{B}^g$ is the stabilizer of x^g in $G^\dagger$, we conclude that U_α fixes the chamber x^g if and only if $x \in \beta$, i.e. if and only if $x^g \in \alpha$. It follows that the chambers of A fixed by U_α are precisely the chambers contained in α. If $u \in U_\alpha^*$, then $A \cap A^u$ is therefore a convex subset of A (by 29.13.iii) that contains α. Thus for each $u \in U_\alpha^*$,

(14.48) $$A \cap A^u = [a, l]$$

for some $l \geq k$ by 29.20 and

(14.49) $$u \text{ acts trivially on } [a, k].$$

Suppose that P is a panel cutting A such that $P \cap A \subset [a, l]$ and u fixes one of the two chambers in $P \cap A$. Then u maps P to itself (since it is a special automorphism). By 14.48, it follows that u fixes both chambers in $P \cap A$. Since $[a, l]$ is connected, it follows (by 14.49) that, in fact,

(14.50) $$u \text{ acts trivially on } [a, l].$$

Let R be the gem of Δ^{aff} cutting A such that the center of $R \cap A$ is the origin and let $S_x = \sigma_A(R, x)$. Thus S_x is a sector of A. We can also interpret S_x, however, as a Weyl chamber of Σ_Φ, in which case it is the Weyl chamber we called d in 14.35. Since we have identified Σ with Σ_Φ, we can also think of d as a chamber of Σ. Let Φ_d^+ be as in 14.25 (so $a \in \Phi_d^+$ if and only if $S_x \subset [a, 0]$) and let U_+ be as in 14.46. As we observed in 14.35, $a \in \Phi_d^+$ if and only if $x \in [a, 0]$. Thus if $a \in \Phi_d^+$ and $u \in U_a$, then u fixes a subsector of S_x by 4.12.i and 14.49. By 4.8, therefore, every element of U_+ fixes a subsector of S_x.

Let $u \in U_a^*$ for some $a \in \Phi$ and suppose that S_x^u is parallel to S_x. By 8.2, it follows that $S_x \cap S_x^u$ contains a subsector S_x' of S_x. By 14.48, $A \cap A^u = [a, l]$ for some $l \in \mathbb{Z}$. Hence the subsector S_x' is contained in $[a, l]$. By 4.12.ii, it follows that $x \in [a, 0]$, i.e. $a \in \Phi_d^+$.

Let

$$\mathcal{A} = \{A^g \mid g \in G^\dagger\}$$

and let $G_0^\dagger$ be the stabilizer of S_x^∞ in $G^\dagger$, where S_x^∞ denotes the set of $\mathcal{A}$-sectors parallel to S_x. (We will show below that $\mathcal{A}$ is a system of apartments; note, however, that we do not yet know this.) By the conclusions of the previous two paragraphs,

(14.51) $$U_a \subset G_0^\dagger \text{ if and only if } a \in \Phi_d^+.$$

In particular, $U_+ \subset G_0^\dagger$. Let M be as in 14.3. Then $M \subset N$, so M maps A to itself. By 3.10, M normalizes U_a for each $a \in \Phi$. Choose $a \in \Phi$. By condition (V3) in 3.21, for each $h \in M$, there exists a constant t such that $\phi_a(x^h) = \phi_a(x) + t$ for all $x \in U_a^*$. It follows that M maps every root of A

to a parallel root, i.e. that M induces only translations on A. By 4.6, we conclude that

$$MU_+ \subset G_0^\dagger. \tag{14.52}$$

By 14.46.i, MU_+ is the stabilizer in $G^\dagger$ of a chamber d of Δ. By 11.16 in [37] and 14.52, it follows that $G_0^\dagger$ is the stabilizer in $G^\dagger$ of a residue T of Δ containing d. If $T \neq \{d\}$, then $G_0^\dagger$ would contain U_a for each root a cutting T. By 14.51, we conclude that $T = \{d\}$. Thus $G_0^\dagger = MU_+$; i.e. MU_+ *equals* the stabilizer of S_x^∞ in $G^\dagger$. There is thus a unique well-defined injection ρ from the chamber set of Δ to the set of parallel classes of $\mathcal{A}$-sectors such that

$$\rho(d^g) = (S_x^\infty)^g \tag{14.53}$$

for all $g \in G^\dagger$. Since $G^\dagger$ acts transitively on $\mathcal{A}$, the map ρ is, in fact, a bijection.

Let e be a chamber of Δ. Then $\rho(e^g) = \rho(e)^g$ for all $g \in G^\dagger$ and $e \in \Sigma$ if and only if $e = d^g$ for some $g \in N$. If S_0 is an $\mathcal{A}$-sector, then $S_0^\infty \in A^\infty$ if and only if $(S_0)^\infty = (S_x)^g$ for some $g \in N$. It follows from these observations that the image of Σ under ρ is A^∞.

Since $G^\dagger$ acts strongly transitively on Δ, $\mathcal{A}$ satisfies 8.4.i. Let S_1 and S_2 be two $\mathcal{A}$-sectors of $\Delta^{\rm aff}$ and let $e_i = \rho^{-1}(S_i^\infty)$ for $i = 1$ and 2. By 29.13.ii, e_1 and e_2 are contained in an apartment of Δ. By 29.15.iii, therefore, there exists $g \in G^\dagger$ such that $e_i^g \in \Sigma$ for $i = 1$ and 2. Hence

$$(S_i^\infty)^g = \rho(e_i)^g = \rho(e_i^g) \in \rho(\Sigma) = A^\infty$$

for $i = 1$ and 2. Therefore $\mathcal{A}$ satisfies 8.4.ii. Thus $\mathcal{A}$ is a system of apartments of Δ.

Let S be as in 14.34. By 14.36.i, the map $s \mapsto d^s$ is a bijection from the set S to the set of chambers of Σ adjacent to d and $s \mapsto (S_x^\infty)^s$ is a bijection from the set S to the set of chambers of A^∞ adjacent to S_x^∞ in $\Delta_{\mathcal{A}}^\infty$. It follows that the restriction of ρ to Σ is an isomorphism from Σ to A^∞. Hence the restriction of ρ to Σ^g is an isomorphism from Σ^g to $(A^\infty)^g$ for all $g \in G^\dagger$. Since every two chambers of Δ are contained in Σ^g for some g (by 29.13.ii and 29.15.iii), it follows that ρ is an isomorphism from Δ to $\Delta_{\mathcal{A}}^\infty$. Thus (ii) holds.

Now let $u \in U_a^*$ for some $a \in \Phi$, let $k = \psi_a(u)$, let $\alpha = [a, k]$ and let v, v' be the elements of U_{-a}^* such that $m_\Sigma(u) = vuv'$. By 3.25, $\phi_{-a}(v) = \phi_{-a}(v') = -k$. Thus $u \in U_\alpha^*$ and $v, v \in U_{-\alpha}^*$. By 14.19, $m_\Sigma(u)$ maps α to $-\alpha$. Since $m_\Sigma(u)$ maps A to itself, it follows that $m_\Sigma(u)$ also maps $-\alpha$ to α. In particular, $m_\Sigma(u)$ does not fix any chambers in A. By 14.49 applied to $U_{-\alpha}$, both v and v' act trivially on $-\alpha$. By 14.48 and 14.50, $A \cap A^u = [a, l]$ for some $l \geq k$ and u acts trivially on $[a, l]$. Thus any chamber contained in $[a, l] \backslash [a, k]$ would be fixed by $m_\Sigma(u)$ since it would be fixed by v, u and v'. Since we know that $m_\Sigma(u)$ does not fix any chambers in A, it follows that $l = k$ and hence $A \cap A^u = [a, k]$. Thus ψ and ϕ_R (as defined in 13.8) agree at a. Since a is arbitrary, it follows that $\psi = \phi_R$. Thus (iii) holds. □

We can now summarize our results as follows:

Theorem 14.54. *Let* $\hat{\mathsf{X}}_\ell$ *be one of the Coxeter diagrams in Figure 1.1 and let* Φ *be the root system* X_ℓ. *Then Bruhat-Tits pairs* $(\Delta, \mathcal{A})$ *of type* $\tilde{\mathsf{X}}_\ell$ *(as defined in 13.1) are classified by pairs* $((G^\dagger, \xi), [\phi])$, *where* $(G^\dagger, \xi)$ *is the root datum belonging to a Moufang building of type* X_ℓ, ϕ *is a valuation of this root datum with respect to a root map* ι *with target* Φ *and* $[\phi]$ *is its equipollence class.*

Proof. This holds by 13.31, where the correspondence

$$(\Delta, \mathcal{A}) \rightsquigarrow ((G^\dagger, \xi), [\phi])$$

is described explicitly, and 14.47. □

Chapter Fifteen

Partial Valuations

In 14.54, we concluded that Bruhat-Tits buildings are classified by root data with valuation. Our principal goal for the rest of this book is to determine when the root datum of a spherical building has a valuation. In this chapter, we show (in 15.21) that it suffices to consider this question "locally" in the sense spelled out in 15.1 and 15.4.

We adopt all the notation in 3.4, 3.5 and 3.12. In particular, Δ is a building satisfying the Moufang property whose Coxeter diagram Π is one of the spherical diagrams X_ℓ in Figure 1.3 with $\ell \geq 2$, Σ is an apartment of Δ, Ξ is the set of roots of Σ, $(G^\dagger, \xi)$ is the root datum of Δ based at Σ, d is a chamber of Σ, Ξ_d° is the set of roots of Σ containing d but not some panel of Σ containing d, Ξ_d is the set of roots of Σ containing d but not some irreducible residue of rank 2 of Σ containing d and ι is a root map of Σ with target Φ. Let I be the vertex set of Π.

Definition 15.1. A *partial valuation* of the root datum $(G^\dagger, \xi)$ (with respect to ι) based at d is a collection ϕ of maps ϕ_a from U_a^* onto $\mathbb{Z}$, one for each $a \in \Xi_d$, such that the following hold:

(i) For each $a \in \Xi_d^\circ$,

$$U_{a,k} := \{u \in U_a \mid \phi_a(u) \geq k\}$$

is a subgroup of U_a for each $k \in \mathbb{R}$, where we assign to $\phi_a(1)$ the value ∞, so that $1 \in U_{a,k}$ for all k.

(ii) For all pairs a, b of distinct roots in Ξ_d° and all pairs a_1, b_1 of distinct roots in the interval $[a, b]$,

$$[U_{a_1,k}, U_{b_1,l}] \subset \prod_{c \in (a_1,b_1)} U_{c,p_c k + q_c l}$$

for all $k, l \in \mathbb{R}$, where for each $c \in (a_1, b_1)$, p_c and q_c are the unique real numbers (which are positive by 3.19) such that $c = p_c a_1 + q_c b_1$.

Note that the expression $p_c a_1 + q_c b_1$ in 15.1.ii only makes sense because we have fixed a root map ι.

Proposition 15.2. *If ϕ is a partial valuation of $(G^\dagger, \xi)$ based at d and $a \in \Xi_d^\circ$, then*

$$\phi_a(u) = \phi_a(u^{-1})$$

for all $u \in U_a^*$.

Proof. The proof of 3.29 applies verbatim. □

Notation 15.3. Suppose that $\phi = \{\phi_a \mid a \in \Xi_d\}$ is a partial valuation of $(G^\dagger, \xi)$ based at a chamber d of Σ. Let

$$U_a^\circ = \{u \in U_a^* \mid \phi_a(u) = 0\}$$

and

$$H_a = \langle m_\Sigma(u) m_\Sigma(v) \mid u, v \in U_a^\circ \rangle$$

for each $a \in \Xi_d$. Let

$$N_0 = \langle m_\Sigma(U_a^\circ) \mid a \in \Xi_d^\circ \rangle$$

and let

$$H_0 = \langle H_a \mid a \in \Xi_d^\circ \rangle.$$

As in 14.3, we set

$$M_a = \langle m_\Sigma(u) m_\Sigma(v) \mid u, v \in U_a^* \rangle$$

for all $a \in \Xi$ and $M = \langle M_a \mid a \in \Xi \rangle$.

Definition 15.4. Let ϕ be a partial valuation of the root datum $(G^\dagger, \xi)$ based at d as defined in 15.1 and let U_a° be as in 15.3 for all $a \in \Xi_d$. We will say that ϕ is *viable* if the following hold:

(i) For each ordered pair (a, b) of roots in Ξ_d° (distinct or not), there exists an element $m_b \in m_\Sigma(U_b^\circ)$ such that for each $u \in U_b^*$, the quantity

$$\phi_a(x^{m_b m_\Sigma(u)}) - \phi_a(x)$$

is independent of the choice of $x \in U_a^*$; moreover, this quantity equals $2\phi_a(u)$ if $a = b$ and it equals zero if $\phi_b(u) = 0$ (whether or not $a = b$).

(ii) For each ordered pair (a, b) of roots in Ξ_d° such that $|[a, b]| \geq 3$, there exists an element $g \in m_\Sigma(U_a^\circ)$ such that

$$\phi_{c^g}(x^g) = \phi_c(x)$$

for all $c \in (a, b]$ and all $x \in U_c^*$.[1]

Proposition 15.5. *If $\phi = \{\phi_a \mid a \in \Xi\}$ is a valuation of $(G^\dagger, \xi)$ based at d (as defined in 3.21), then the restriction*

$$\phi|_{\Xi_d} := \{\phi_a \mid a \in \Xi_d\}$$

of ϕ to the set Ξ_d is a viable partial valuation.

Proof. By conditions (V1) and (V2) in 3.21, $\phi|_{\Xi_d}$ is a partial valuation as defined in 15.1. Let (a, b) be an ordered pair of roots in Ξ_d° (distinct or not), let $u \in U_b^*$ and let $m_b \in m_\Sigma(U_b^\circ)$. By 3.44, ϕ is invariant under the action of N_0. Thus 15.4.ii holds. By two applications of condition (V3), the quantities

$$\phi_a\left(x^{m_b m_\Sigma(u)}\right) - \phi_{s_b(a)}(x^{m_b})$$

[1]Note that by 3.11, g maps the half-open interval $(a, b]$ to itself.

and

$$\phi_{s_b(a)}(x^{m_b}) - \phi_a(x)$$

are independent of x. Hence the quantity

$$\phi_a\left(x^{m_b m_\Sigma(u)}\right) - \phi_a(x)$$

is also independent of x. By 3.44, this quantity equals zero if $\phi_b(u) = 0$. Suppose that $a = b$, let $x \in U_a^*$ and let v, v' be the unique elements of U_{-a}^* such that $m_\Sigma(u) = vuv'$. Then

$$\begin{aligned}
\phi_a(x^{m_a m_\Sigma(u)}) &= \phi_a(x^{m_a m_\Sigma(v)}) && \text{by 3.26} \\
&= \phi_{-a}(x^{m_a}) - 2\phi_{-a}(v) && \text{by (V4)} \\
&= \phi_{-a}(x^{m_a}) + 2\phi_a(u) && \text{by 3.25} \\
&= \phi_a(x) + 2\phi_a(u) && \text{since } m_a \in N_0.
\end{aligned}$$

Thus 15.4.i holds. □

In 15.21, we will show, conversely, that every viable partial valuation of $(G^\dagger, \xi)$ has a unique extension to a valuation of $(G^\dagger, \xi)$. We begin the proof of this result with a series of preliminary observations.

Proposition 15.6. *Let ϕ be a partial valuation of the root datum $(G^\dagger, \xi)$ based at d and suppose that ϕ is viable as defined in 15.4. Let (a, b) be an ordered pair of roots in Ξ_d° such that $|[a, b]| = 4$ and let*

$$[a, b] = (a_1, a_2, a_3, a_4).$$

Let $U_i = U_{a_i}$ and $\phi_i = \phi_{a_i}$ for all $i \in [1, 4]$. Then there exists $g \in m_\Sigma(U_3^\circ)$ such that $\phi_1(x^g) = \phi_1(x)$ for all $x \in U_1$.

Proof. By 15.4.ii, there exist elements $u \in m_\Sigma(U_1^\circ)$ and $v \in m_\Sigma(U_4^\circ)$ such that

$$\phi_3(x^u) = \phi_3(x) \tag{15.7}$$

and

$$\phi_1(x^v) = \phi_3(x) \tag{15.8}$$

for all $x \in U_3^*$ as well as

$$\phi_3(x^v) = \phi_1(x)$$

for all $x \in U_1^*$. Let $g = u^v$. Then

$$\phi_3(g) = \phi_3(u^v) = \phi_1(u) = 0,$$

so $g \in m_\Sigma(U_3^\circ)$ (by 11.23 in [37]), and

$$\begin{aligned}
\phi_1(x^g) = \phi_1(x^{v^{-1}uv}) &= \phi_3(x^{v^{-1}u}) && \text{by 15.8} \\
&= \phi_3(x^{v^{-1}}) && \text{by 15.7} \\
&= \phi_1(x) && \text{by 15.8}
\end{aligned}$$

for all $x \in U_1^*$. □

Proposition 15.9. *Let ϕ be a viable partial valuation of $(G^\dagger, \xi)$ based at $d \in \Sigma$, let M and H_0 be as in 15.3. Then the following hold:*

(i) $\phi_a(x^h) = \phi_a(x)$ *for all* $a \in \Xi_d^\circ$, *all* $h \in H_0$ *and all* $x \in U_a^*$.
(ii) *For all* $a \in \Xi_d^\circ$ *and all* $h \in M$, *the quantity*

$$\phi_a(x^h) - \phi_a(x)$$

is independent of the choice of $x \in U_a^*$.

Proof. Let (a, b) be an ordered pair of roots in Ξ_d° (distinct or not), let H_b be as in 15.3 and let m_b be as in 15.4.i. By 15.2, we have

$$H_b = \langle m_b m_\Sigma(u) \mid u \in U_b^\circ \rangle.$$

By 15.4.i,

$$\phi_a(x^{m_b m_\Sigma(u)}) = \phi_a(x)$$

for all $u \in U_b^\circ$ and all $x \in U_a^*$. Thus (i) holds.

Let

$$M^\# = \{g \in M \mid \phi_a(x^g) - \phi_a(x) \text{ is independent of the choice of } x \in U_a\}.$$

If $g, h \in M^\#$, then $gh \in M^\#$ since the quantity

$$\phi_a(x^{gh}) - \phi_a(x)$$

is the sum of $\phi_a(x^{gh}) - \phi_a(x^g)$ and $\phi_a(x^g) - \phi_a(x)$, both of which are independent of the choice of $x \in U_a^*$, and $g^{-1} \in M^\#$ since the quantity $\phi_a(x^{g^{-1}}) - \phi_a(x)$ equals

$$\phi_a(x^{g^{-1}}) - \phi_a((x^{g^{-1}})^g),$$

which is also independent of the choice of $x \in U_a^*$. By 11.35 of [37], we have

$$M = \langle M_b \mid b \in \Xi_d^\circ \rangle,$$

where M_b for $b \in \Xi$ is as in 15.3. By 15.4.i, therefore, $M_b \subset M^\#$ for all $b \in \Xi_d^\circ$. Therefore $M = M^\#$. Thus (ii) holds. □

Corollary 15.10. *Let ϕ be a viable partial valuation of $(G^\dagger, \xi)$ based at d. Then the following stronger version of 15.4.i holds: For each ordered pair (a, b) of roots in Ξ_d° (distinct or not), for each $m_b \in m_\Sigma(U_b^\circ)$ and for each $u \in U_b^*$, the quantity*

$$\phi_a(x^{m_b m_\Sigma(u)}) - \phi_a(x)$$

is independent of the choice of $x \in U_a^*$; moreover, this quantity equals $2\phi_a(u)$ if $a = b$ and it equals zero if $\phi_b(u) = 0$ (whether or not $a = b$).

Proof. Let (a, b) be an ordered pair of roots in Ξ_d°, let m_b be as in 15.4.i, let m_b' be an arbitrary element of $m_\Sigma(U_b^\circ)$ and let $h = m_b' m_b^{-1}$. Then $h \in H$. By 15.9.i,

$$\phi_a(x^{m_b' m_\Sigma(u)}) - \phi_a(x) = \phi_a((x^h)^{m_b m_\Sigma(u)}) - \phi_a(x^h)$$

for all $u \in U_b^*$ and all $x \in U_a^*$. By 15.4.i, the quantity

$$\phi_a((x^h)^{m_b m_\Sigma(u)}) - \phi_a(x^h)$$

is independent of the choice of x, equals $2\phi_a(u)$ if $a = b$ and equals zero if $\phi_b(u) = 0$. □

Proposition 15.11. *Let ϕ, H_0, N_0, etc. be as in 15.3. Then N_0 induces the Coxeter group W_Π on Σ and the kernel of this action is H_0.*

Proof. By 3.10, the group N_0 induces the Coxeter group W_Π on Σ and the subgroup H_0 acts trivially on Σ. We thus need to show only that the kernel of the action of N_0 on Σ is contained in H_0.

Choose $a \in \Xi_d^\circ$, $u \in U_a^\circ$ and $h \in H_0$ and let $m = m_\Sigma(u)$. Then h^{-1} normalizes U_a (since H_0 acts trivially on Σ) and $m^{h^{-1}} = m_\Sigma(u^{h^{-1}})$ (by 11.23 of [37]), so $m^{h^{-1}} \in m_\Sigma(U_a^\circ)$ by 15.9.i. We also have $m^{-1} = m_\Sigma(u^{-1})$ by 3.9 and hence $m^{-1}m^{h^{-1}} \in H_a$ by 15.2. Therefore

$$\begin{aligned} h^m &= m^{-1}hmh^{-1} \cdot h \\ &= m^{-1}m^{h^{-1}} \cdot h \\ &\in H_a H_0 \subset H_0. \end{aligned}$$

We conclude that H_0 is a normal subgroup of N_0.

By 15.3, the image of an element in $m_\Sigma(U_a^\circ)$ in N_0/H_0 is of order 2 (for all $a \in \Xi_d^\circ$). By 11.28.ii of [37], it follows that for all pairs of distinct roots a and b in Ξ_d°, the image of

$$(m_a m_b)^{|[a,b]|}$$

in N_0/H_0 is trivial for all $m_a \in m_\Sigma(U_a^*)$ and all $m_b \in m_\Sigma(U_b^*)$. Therefore the quotient group N_0/H_0 is generated by elements satisfying the defining relations of the Coxeter group W_Π. Thus N_0/H_0 is a homomorphic image of W_Π. Since N_0 induces the Coxeter group W_Π on Σ and H_0 acts trivially on Σ, it follows that, in fact, $N_0/H_0 \cong W_\Pi$ and H_0 is the kernel of the action of N_0 on Σ. □

Proposition 15.12. *Let E be a residue of rank 2 containing d, let a and b be the two roots in Ξ_d° that cut E and suppose that $|[a,b]| = 3$ (equivalently, $|E| = 6$). Then there exists an element $g \in N_0$ rotating E and mapping a to b such that $\phi_a(u) = \phi_b(u^g)$ for all $u \in U_a^*$.*

Proof. By 15.4.ii, there exist elements $v \in m_\Sigma(U_a^\circ)$ and $u \in m_\Sigma(U_b^\circ)$ such that the product $g := uv$ has the desired properties. □

Proposition 15.13. *Let ϕ be a viable partial valuation of $(G^\dagger, \xi)$ based at $d \in \Sigma$, let $a, b \in \Xi_d^\circ$ and suppose that a and b are in the same $G^\dagger$-orbit. Then there exists an element $g \in N_0$ mapping a to b (but not necessarily fixing d) such that*

$$\phi_a(u) = \phi_b(u^g)$$

for all $u \in U_a^*$.

Proof. Let y, z be the vertices of Π such that $a = a_y$ and $b = a_z$ (where a_y and a_z are as in 3.5). By 29.52, there is a path from y to z in Π passing only through edges of Π with label 3. The claim holds, therefore, by 15.12. □

Proposition 15.14. *Let ϕ be a viable partial valuation of $(G^\dagger, \xi)$ based at $d \in \Sigma$ and let $a \in \Xi_d^\circ$. Then $\phi_a(u^g) = \phi_a(u)$ for all $u \in U_a^*$ and all $g \in N_0$ that map a to itself.*

Proof. Choose $g \in N_0$ mapping a to itself. Let e be the unique chamber of Σ adjacent to d that is not in the root a, let $x = d^g$ and let $y = e^g$. Since $d \in a$ and $e \notin a$, we have $x \in a$ and $y \notin a$.

Suppose that $d \neq x$. By 29.23, there exist residues $E_1, \dots, E_k$ of rank 2 in Σ for some $k \geq 1$ such that the following hold:

(i) $E_i \cap E_{i+1}$ is a panel $\{u_i, u_i'\}$ with $u_i \in a$ and $u_i' \notin a$ for each $i \in [1, k-1]$.
(ii) $\{u_i, u_i'\}$ is opposite $\{u_{i-1}, u_{i-1}'\}$ in E_i for each $i \in [2, k-1]$.
(iii) Either $k = 1$ and $\{d, e\}$ is the panel of E_1 opposite $\{x, y\}$ or $k \geq 2$, $\{d, e\}$ is the panel of E_1 opposite $\{u_1, u_1'\}$ and $\{u_{k-1}, u_{k-1}'\}$ is the panel of E_k opposite $\{x, y\}$.

We call a sequence $E_1, \dots, E_k$ satisfying (i)–(iii) an *a-path* from d to x and we call k its *length*. We define the *a-distance* from d to x to be the length of a shortest a-path from d to x if $d \neq x$ and 0 if $d = x$. We proceed now by induction with respect to the a-distance from d to x.

Suppose first that the a-distance from d to x is 0. Then $d = x$. Hence $g \in H_0$ by 15.11 (since W_Π acts sharply transitively on Σ) and thus the claim holds by 15.9.i.

Suppose now that the a-distance k from d to x is positive, let $E_1, \dots, E_k$ be an a-path of length k from d to x and let u_i, u_i' be as in (i). Let $u_0 = d$, $u_0' = e$, $u_k = x$ and $u_k' = y$. Let $n_i = |E_i|/2$ for all $i \in [1, k]$ and let j_i be the type of the panel $\{u_i, u_i'\}$ for all $i \in [0, k]$. (Thus $j_0, j_1, \dots, j_k$ are vertices of the Coxeter diagram Π.) The group $N_0 \subset G^\dagger$ is type-preserving. Since $g \in N_0$ maps $\{d, e\}$ to $\{x, y\}$, we have $j_0 = j_k$. For all $i \in [1, k]$, the following hold:

(a) $j_{i-1} = j_i$ if n_i is even.
(b) If $n_i = 3$, then j_{i-1} and j_i are joined by an edge of Π having label 3.
(c) If $i < k$ and $n_i = n_{i+1} = 3$, then $j_{i-1} \neq j_{i+1}$.

By (b) and (c), it is impossible that $n_i = 3$ for all $i \in [1, k]$ (since $j_0 = j_k$ and the Coxeter diagram Π is a tree).

Suppose that $n_i = 2$ for some $i \in [1, k]$. Then $u_{i-1}, u_{i-1}', u_i, u_i'$ are all the chambers in E_i. Let c be the root of Σ containing u_{i-1} but not u_i. By 15.11, there exists $\tilde{g} \in N_0$ mapping u_{i-1} to d. Let $\tilde{c} = c^{\tilde{g}}$. Then $\tilde{c} \in \Xi_d^\circ$. Choose $\hat{g} \in m_\Sigma(U_{\tilde{c}}^\circ)^{\tilde{g}^{-1}}$. Thus $\hat{g} \in N_0$ and, by 11.23 of [37], $\hat{g} \in m_\Sigma(U_c^*)$, so, in particular, $\hat{g}$ interchanges u_{i-1} and u_i. By 11.28.iii of [37], $[m_\Sigma(U_c^*), U_a] = 1$. Hence

$$[\hat{g}, U_a] = 1. \tag{15.15}$$

Since $\hat{g}$ induces the reflection on E_i that interchanges u_{i-1} and u_i, it maps a to itself. Therefore

$$E_1, \dots, E_{i-1}, E_{i+1}^{\hat{g}}, \dots, E_k^{\hat{g}}$$

is an a-path from d to $x^{\hat{g}} = d^{g\hat{g}}$ of length $k-1$. Thus $\phi_a(u^{g\hat{g}}) = \phi_a(u)$ for all $u \in U_a^*$ by induction and $u^{g\hat{g}} = u^g$ for all $u \in U_a$ by 15.15. Hence $\phi_a(u^g) = \phi_a(u)$ for all $u \in U_a^*$. We can thus assume that

$$n_i \neq 2 \text{ for all } i \in [1,k]. \tag{15.16}$$

Suppose next that $n_i = 4$ for some $i \in [1,k]$. Thus $\Pi = \mathsf{B}_\ell$ or F_4. Let γ be the unique non-stuttering gallery from d to $x = u_k$ that goes from d to u_1 in $E_1 \cap a$, from u_1 to u_2 in $E_2 \cap a$, etc. By (a)–(c) and 15.16, the type of γ is a palindrome,

$$(j_0, j_1, \ldots, j_k) = (j_k, j_{k-1}, \ldots, j_1),$$

k is odd, $i = (k+1)/2$ and $n_m = 3$ for all $m \in [1,k]$ distinct from i. By 15.11, there is an element $g_0 \in N_0$ that maps E_i to itself and interchanges $u_{(k-1)/2}$ and $u_{(k+1)/2}$. Since the elements of N_0 are special, the element g_0 must interchange the gallery γ with its inverse γ^{-1}. In particular, it interchanges u_i and u_{k-i} for all $i \in [1, k-1]$. Since the product gg_0^{-1} fixes d, it lies in H_0 (by 15.11 again). By 15.9.i, it suffices to show that $\phi_a(u^{g_0}) = \phi_a(u)$ for all $u \in U_a^\circ$. We can thus assume that $g_0 = g$. In particular,

$$u_1^g = u_{k-1} \tag{15.17}$$

if $k > 1$.

Now suppose that $k > 1$. Then $n_1 = 3$. Let b be the unique root in $\Xi_d^\circ \backslash \{a\}$ that cuts E_1. By 15.12, there exists $h \in N_0$ that maps u_1 to d as well as a and b such that

$$\phi_a(u) = \phi_b(u^h) \tag{15.18}$$

for all $u \in U_a^*$. Let $g_1 = g^h$. Then g_1 maps b to itself, $g_1 \in N_0$,

$$d^{g_1} = d^{h^{-1}gh} = u_1^{gh} = u_{k-1}^h$$

(by 15.17) and

$$E_2^h, \ldots, E_{k-1}^h$$

is a b-path from $d = u_1^h$ to u_{k-1}^h of length $k-2$. Thus $\phi_b(u^{g_1}) = \phi_b(u)$ for all $u \in U_b^*$ by induction. Therefore

$$\phi_a(u^g) = \phi_b(u^{gh}) = \phi_b(u^{hg_1}) = \phi_b(u^h) = \phi_a(u)$$

for all $u \in U_a^*$ by two applications of 15.18.

Next suppose that $k = 1$, so $n_1 = 4$, and let

$$(x_0, x_1, x_2, x_3)$$

be the minimal gallery in E_1 from $x_0 = d$ to $x_3 = x$. Let c be the root in Ξ_d containing x_1 but not x_2. Thus if b is (as above) the unique root in $\Xi_d^\circ \backslash \{a\}$ that cuts E_1 and

$$[a, b] = (a_1, a_2, a_3, a_4),$$

then $a = a_1$ and $c = a_3$. By 15.6, there thus exists $h \in m_\Sigma(U_c^\circ)$ such that

$$\phi_a(u^h) = \phi_a(u) \tag{15.19}$$

for all $u \in U_a^*$. The element h induces the reflection on E_1 that interchanges x_1 and x_2. Thus gh^{-1} fixes d and hence lies in H_0. By 15.9.i and 15.19, it follows that $\phi_a(u^g) = \phi_a(u)$ for all $u \in U_a^*$. We can thus assume that $n_i > 4$ for all $i \in [1, k]$.

Hence $\Pi = \mathsf{G}_2$, $k = 1$ and $n_1 = 6$. Let

$$(x_0, x_1, \ldots, x_5)$$

be the minimal gallery in E_1 from $x_0 = d$ to $x_5 = x$ and let c be the root in Ξ_d containing x_2 but not x_3. Choose $h \in m_\Sigma(U_c^\circ)$. By 16.8 in [36], the root groups corresponding to two orthogonal roots commute elementwise. Thus, in particular, $[U_a, U_c] = [U_a, U_{-c}] = 1$, so

$$[U_a, h] = 1 \tag{15.20}$$

by 3.8. Thus $\phi_a(u^h) = \phi_a(u)$ for all $u \in U_a^*$. Since h induces the reflection on E_1 that interchanges x_2 and x_3, the product gh^{-1} fixes d and hence lies in H_0 (by 15.11). By 15.9.i and 15.20, it follows that $\phi_a(u^g) = \phi_a(u)$ for all $u \in U_a^*$. □

We can now prove the main result of this chapter:

Theorem 15.21. *Every viable partial valuation of $(G^\dagger, \xi)$ (as defined in 15.1 and 15.4) extends to a unique valuation of $(G^\dagger, \xi)$.*

Proof. Let ϕ be a viable partial valuation of $(G^\dagger, \xi)$ based at d. Let b be an arbitrary root of Σ, let ∂b be the border of b as defined in 29.40 and choose $x \in \partial b$. By 15.11, there exists $g \in N_0$ mapping x to d. Let $a = b^g$ (so $a \in \Xi_d^\circ$) and let $\psi_b(u) = \phi_a(u^g)$ for all $u \in U_b^*$. We claim that ψ_b is independent of the choice of x and g and that $\psi_b = \phi_b$ if $b \in \Xi_d$.

Suppose that g_1 is another element of N_0 mapping x to d (so $b^{g_1} = a$ since g_1 is special) and let $h = g^{-1}g_1$. By 15.11, $h \in H_0$. By 15.9.i, therefore, $\phi_a(u^{g_1}) = \phi_a(u^{gh}) = \phi_a(u^g)$ for all $u \in U_b^*$. Thus ψ_b is independent of the choice of g.

We now show that ψ_b is independent of the choice of x. Suppose that y is another chamber in ∂b. Let $z = y^g$. By 15.11, there exists an element $g' \in N_0$ mapping z to d. Let $c = a^{g'}$. Since gg' maps y to d and b to c, we have $c \in \Xi_d^\circ$. Thus a and c are two roots in Ξ_d° such that U_a and U_c are conjugate. By 15.13, there exists an element $\hat{g} \in N_0$ mapping c to a such that $\phi_c(u) = \phi_a(u^{\hat{g}})$ for all $u \in U_c^*$. By 15.14, $\phi_a(u^{g'\hat{g}}) = \phi_a(u)$ for all $u \in U_a^*$. Therefore

$$\phi_a(u^g) = \phi_a(u^{gg'\hat{g}}) = \phi_c(u^{gg'})$$

for all $u \in U_b^*$. Thus ψ_b is independent of the choice of x.

Next suppose that $b \in \Xi_d$. If $b \in \Xi_d^\circ$, then we can choose $x = d$ and $g = 1$ to conclude that $\psi_b = \phi_b$. Suppose that $b \notin \Xi_d^\circ$. By 3.5, there is an irreducible rank 2 residue E containing d and cut by b. By 15.4.ii, we can choose a chamber $x \in \partial b$ and an element $g \in N_0$ that maps E to itself and x to d such that $\phi_b(u) = \phi_a(u^g)$ for all $u \in U_b^*$, where $a = b^g$. Thus $\psi_b = \phi_b$

also in this case. We conclude that $\psi := \{\psi_a \mid a \in \Xi\}$ is an extension of ϕ. Moreover, ψ is N_0-invariant in the sense that $\psi_b(u) = \psi_a(u^g)$ for all $a \in \Xi$, all $g \in N_0$ and all $u \in U_b^*$, where $b = a^g$.

We now show that ψ satisfies the conditions (V1)–(V4). Let a, b be two roots of Σ that are not opposite each other. By 29.25, there exists a residue E of rank 2 cut by both a and b. By 15.11, there exists $g \in N_0$ mapping the unique chamber in $\partial a \cap b \cap E$ to d. Let $a' = a^g$ and $b' = b^g$. Then $a' \in \Xi_d^\circ$, and by 3.5, there exists $c' \in \Xi_d$ such that $b' \in [a', c']$. Since ψ extends ϕ and is N_0-invariant, it follows by 15.1 that ψ satisfies the conditions (V1) and (V2) and that to check that conditions (V3) and (V4) hold, it suffices to verify that they hold when $a \in \Xi_d^\circ$ and $b \in [a, c]$ for some $c \in \Xi_d^\circ$.

Let $a, c \in \Xi_d^\circ$ and let $b \in [a, c]$. Let m_b be as in 15.4.i if $a = b$; otherwise let m_b be an arbitrary element of $m_\Sigma(U_b^\circ)$. Choose $u \in U_b^*$. By 15.9.ii,

$$t := \phi_a(x^{m_b m_\Sigma(u^{-1})}) - \phi_a(x) \tag{15.22}$$

is independent of the choice of $x \in U_a^*$, and by 15.2 and 15.4.i, $t = 2\phi_a(u)$ if $a = b$. By 15.22, also

$$\phi_a(x^{(m_b m_\Sigma(u^{-1}))^{-1}}) - \phi_a(x) = -t$$

for all $x \in U_a^*$. By 3.9 and the N_0-invariance of ψ, we thus have

$$\begin{aligned}\psi_{s_b(a)}(x^{m_\Sigma(u)}) &= \psi_a(x^{m_\Sigma(u)m_b^{-1}})\\ &= \psi_a(x) - t\end{aligned}$$

for all $x \in U_a^*$. Thus conditions (V3) and (V4) hold. We conclude that ψ is a valuation of $(G^\dagger, \xi)$. Uniqueness holds by 3.18 and 3.41.ii. □

We conclude this chapter with one more observation about partial valuations (in 15.25).

Definition 15.23. Let (x, y) be an edge of Π, let

$$[a_x, a_y] = (a_1, a_2, \ldots, a_n)$$

(as defined in 3.1 and 3.5) and let $U_i = U_{a_i}$ for all $i \in [1, n]$. By 3.2.iii and 3.2.iv, every element in $[U_1, U_n]$ can be written uniquely as an element in $U_2 \cdots U_{n-1}$. If

$$[z_1, z_n] = z_2 \cdots z_{n-1}$$

with $z_i \in U_i$ for all $i \in [1, n]$, we call z_i the a_i-component of $[z_1, z_n]$ and denote it by

$$[z_1, z_n]_{a_i}.$$

Thus $[z_1, z_n]_{a_i}$ is the same as $[z_1, z_n]_i$ as defined in 5.10 of [36].

Definition 15.24. Let ϕ be a partial valuation of $(G^\dagger, \xi)$, let $a = a_x$ and $b = a_y$ for some edge $\{x, y\}$ of Π and let $c \in (a, b)$. We will say that ϕ is *exact at* (a,b,c) if there exist positive real numbers e and f depending only on the ordered triple (a,b,c) such that

$$\phi_c([w,u]_c) = e\phi_a(w) + f\phi_b(u)$$

for all $w \in U_a^*$ and all $u \in U_b^*$, where $[w,u]_c$ is as in 15.23. We call the ordered pair (e,f) the *(a,b)-coefficient of ϕ at c.* We call ϕ *exact* if it is exact at (a_x, a_y, c) for all directed edges (x,y) of Π and all $c \in (a_x, a_y)$.

It turns out that all viable partial valuations are, in fact, exact. We give an explanation of this observation in 18.32.

The following result will be used in the verification that a given partial valuation satisfies 15.4.i for $a = b$.[2]

Proposition 15.25. *Let $a = a_x$ and $b = a_y$ for some directed edge (x,y) of Π, let $c \in (a,b)$, let $m_b \in m_\Sigma(U_b^\circ)$ and let ϕ be a partial valuation of $(G^\dagger, \xi)$ based at d. Suppose that ϕ is exact at (a,b,c) and let (e,f) be the (a,b)-coefficient of ϕ at c (as defined in 15.24). Suppose, too, that for all $u \in U_b^*$,*

(i) $\psi_a(u) := \phi_a(w^{m_b m_\Sigma(u)}) - \phi_a(w)$ *is independent of the choice of* $w \in U_a^*$,
(ii) $\psi_c(u) := \phi_c(w^{m_b m_\Sigma(u)}) - \phi_c(w)$ *is independent of the choice of* $w \in U_c^*$ *and*
(iii) $\psi_c(u) = e\psi_a(u) + 2f\phi_b(u)$.

Then

$$\phi_b(v^{m_b m_\Sigma(u)}) - \phi_b(v) = 2\phi_b(u)$$

for all $u, v \in U_b^*$.

Proof. Choose $w \in U_a^*$ and $u, v \in U_b^*$ and let $h = m_b m_\Sigma(u)$ and $y = [w,v]_c$. Since h acts trivially on Σ, it normalizes all the root groups corresponding to roots of Σ. Hence $y^h = [w^h, v^w]_c$. By 15.24, therefore,

$$\phi_c(y) = e\phi_a(w) + f\phi_b(v) \tag{15.26}$$

and

$$\phi_c(y^h) = e\phi_a(w^h) + f\phi_b(v^h).$$

By (i) and (ii), it follows from this last equation that

$$\phi_c(y) + \psi_c(u) = e\big(\phi_a(w) + \psi_a(u)\big) + f\phi_b(v^h).$$

Thus by 15.26,

$$f\phi_b(v) + \psi_c(u) = e\psi_a(u) + f\phi_b(v^h)$$

---

[2] In fact, verification of 15.4.i for $a = b$ can also be accomplished more directly using formulas for the expression

$$x^{m_b m_\Sigma(u)}$$

for $x, u \in U_b^*$ like the ones in 30.36 except when the root datum is the root datum of an exceptional quadrangle of type E_6, E_7 or E_8. In these exceptional cases, there are also formulas for this expression, but they are complicated and do not seem to allow a direct verification of 15.4.i, whereas this task is easily accomplished using 15.25.

and hence

$$f\big(\phi_b(v^h) - \phi_b(v)\big) = \psi_c(u) - e\psi_a(u).$$

By (iii), we conclude that

$$\phi_b(v^h) - \phi_b(v) = 2\phi_b(u)$$

since $f \neq 0$ by 15.24. □

Chapter Sixteen

Bruhat-Tits Theory

We continue to adopt all the notation in 3.4 and 3.12. In particular, Δ is a building whose Coxeter diagram Π is one of the spherical diagrams X_ℓ in Figure 1.3 with $\ell \geq 2$, Δ satisfies the Moufang property, Σ is an apartment of Δ, $(G^\dagger, \xi)$ is the root datum of Δ based at Σ and ι is a root map of Σ with target Φ.

From now on, we rely on the classification of Moufang spherical buildings ([32] and [36]). More precisely, we rely on the description of the root data of Moufang spherical buildings as summarized in 30.14.[1]

In each case of 30.14, the letter K denotes a field or a skew field[2] (or in three exotic cases[3] an octonion division algebra) over which the root datum $(G^\dagger, \xi)$ of Δ is "defined."[4] In 16.4, we show that if the root datum $(G^\dagger, \xi)$ has a valuation ϕ, then for at least one root a of Σ, the root group U_a is isomorphic to the additive group of K (except in a few cases in which this statement must be suitably modified) and ϕ_a is given by a discrete valuation of K (as defined in 9.17). By Bruhat-Tits theory, we mean the complex of results that describe, conversely, the conditions under which a single such ϕ_a determined by a discrete valuation ν of K extends to a valuation of the root datum $(G^\dagger, \xi)$.

In 16.14.i, we show that the answer is "always" if Π is simply laced[5] and in 16.14.ii, we reduce the problem for Π not simply laced to the rank 2 case, i.e. to the case that the root datum $(G^\dagger, \xi)$ is the root datum of a Moufang n-gon for $n = 4$ or 6. In Chapters 19–25 we give necessary and sufficient conditions for the existence of an extension of ϕ_a for each of the seven families of Moufang quadrangles and hexagons.

Conventions 16.1. The building Δ is linked to one of the root group labelings ζ of Π described in 30.14. Without loss of generality, we can assume

---

[1] At this point the reader should read 30.1–30.15 and take a careful look at the notation in 30.8 and 30.15. See also Figure 27.1.

[2] The cases of 30.14 in which K can be a skew field are (i), (iv), (v) and (vii); in cases (iv), (v) and (vii), K *must*, in fact, be non-commutative unless (K, K_0, σ) is a quadratic involutory set of type (ii) or (iii) as defined in 30.21.

[3] The Moufang spherical buildings (of rank $\ell \geq 2$) determined by an octonion division algebra K are $\mathsf{A}_2(K)$, $\mathsf{C}_3^{\mathcal{I}}(K, F, \sigma)$ and $\mathsf{F}_4(K, F)$, where F denotes the center of K and σ its standard involution, i.e. the standard involution of the composition algebra (K, F) as defined in 30.17. These are cases (ii) and (vi) and one part of case (xiii) in 30.14.

[4] See 30.15 and 30.29.

[5] This allows us (in 16.18) to complete the classification of Bruhat-Tits buildings whose Coxeter diagram is simply laced.

from now on that (Δ, Σ, d) is, in fact, the canonical realization of the pair (Π, ζ) as defined in 30.15. Thus, in particular,

$$\Delta = \mathsf{X}_\ell^*(\Lambda)$$

for some parameter system Λ (as defined in 30.15), where $\Pi = \mathsf{X}_\ell$. Let K be the defining field of Δ as defined in 30.15 (so K is, in fact, a field or a skew field or an octonion division algebra.)

Notation 16.2. Let (x, y) be a directed edge of Π with label n and let

$$\Omega_{xy} = (U_+^{(e)}, U_{a_1}, U_{a_2}, \ldots, U_{a_n})$$

be as in 30.1. By 16.1, Ω_{xy} is as in one of the cases (i)–(viii) of 30.8. In the corresponding case 16.1–16.8 of [36] there is a parameter group Λ_i and an isomorphism x_i from Λ_i to $U_i = U_{a_i}$. For each $a \in [a_x, a_y]$, we set $\Lambda_a = \Lambda_i$ and $x_a = x_i$, where i is the index in $[1, n]$ such that $a = a_i$. Since (x, y) is arbitrary, the group Λ_a and the isomorphism x_a are thus defined for all $a \in \Xi_d$; note that by 30.13.ii, Λ_a and x_a are independent of the choice of (x, y) if $a \in \Xi_d^\circ$.[6]

Thus $\Lambda_a = K$ (i.e. the additive group of K) for all $a \in \Xi_d$ if Π is simply laced. Suppose, as an example, that $\Delta = \mathsf{B}_2^{\mathcal{Q}}(\Lambda)$ for some anisotropic quadratic space $\Lambda = (K, L, q)$ (as defined in 30.15). Let (u, v) be the directed edge of Π defined in 30.11 and let $[a_u, a_v] = (a_1, a_2, a_3, a_4)$. Then $\Lambda_{a_i} = K$ for $i = 1$ and 3 and $\Lambda_{a_i} = L$ for $i = 2$ and 4 (by 16.3 in [36]) and $\Lambda_a = K$ for all $a \in \Xi_d$ not in the interval $[a_u, a_v]$.

Notation 16.3. If Π is simply laced, let x be an arbitrary vertex of Π. If Π is not simply laced, let (u, v) be as in 30.11 and then let $x = v$ in case (xiv) of 30.14 and let $x = u$ in every other case. We set $a = a_x$ (as defined in 3.5) and let Λ_a and x_a be as defined in 16.2. Thus either $\Lambda_a = K$ or $\ell = 2$ and one of the following holds:

(i) $\Lambda = (K, L, q)$ is a quadratic space of type E_6, E_7 or E_8 (as defined in 12.31 of [36]), Λ_a is the group S defined in 16.6 of [36] and $\mathsf{X}_\ell^* = \mathsf{B}_2^{\mathcal{E}}$.
(ii) $\Lambda = (K, L, q)$ is a quadratic space of type F_4 (as defined in 14.1 of [36]), Λ_a is the group $W_0 \oplus K$ defined in 16.7 of [36] and $\mathsf{X}_\ell^* = \mathsf{B}_2^{\mathcal{F}}$.
(iii) $\Lambda = (K, K_0, L_0)$ is an indifferent set (as defined in 10.1 of [36]), Λ_a is the group K_0 and $\mathsf{X}_\ell^* = \mathsf{B}_2^{\mathcal{D}}$.

In the next result, we show that every valuation of the root datum $(G^\dagger, \xi)$ is an extension, in some sense, of a valuation of the defining field K.

Theorem 16.4. *Let ψ be a valuation of the root datum $(G^\dagger, \xi)$ and let a, x_a and Λ_a be as in 16.3. Then there exists a valuation ν of K, a valuation ϕ of $(G^\dagger, \xi)$ equipollent to ψ and, in cases (i)–(iii) of 16.3, a positive integer δ such that one of the following holds:*

---

[6]Notice that now x is a vertex of Π, x_a denotes an isomorphism from the group Λ_a to U_a, a_x denotes a root of Σ (defined in 3.5) and a_u for $u \in U_a^*$ denotes a certain root of an affine apartment (defined in 13.2). We hope that this notation is not causing any unnecessary confusion.

(i) $\phi_a(x_a(0,t)) = \delta\nu(t)$ *for all* $t \in K^*$ *in cases 16.3.i and 16.3.ii.*[7]
(ii) $\phi_a(x_a(t)) = \nu(t)/\delta$ *for all* $t \in K_0^*$ *and* $\delta = 1$ *or 2 in case 16.3.iii.*
(iii) $\phi_a(x_a(t)) = \nu(t)$ *for all* $t \in K^*$ *in all other cases.*

Proof. Let ϕ be a valuation equipollent to ψ such that

$$\phi_a(x_a(1)) = 0 \tag{16.5}$$

(where we replace $x_a(1)$ by $x_a(0,1)$ if 16.3.i or 16.3.ii holds). Let

$$h_a(t) := m_\Sigma(x_a(1))m_\Sigma(x_a(t))$$

and

$$\nu^\#(t) = \phi_a(x_a(t))$$

for all $t \in K^*$ (where we replace $x_a(1)$ by $x_a(0,1)$ and $x_a(t)$ by $x_a(0,t)$ if 16.3.i or 16.3.ii holds and we replace "for all $t \in K^*$" by "for all $t \in K_0^*$" if 16.4.iii holds). By condition (V1), $\nu^\#$ satisfies 9.17.ii. By 15.5 and 15.10,

$$\phi_a(w^{h_a(t)}) - \phi_a(w) = 2\nu^\#(t) \tag{16.6}$$

for all $w \in U_a^*$ and all $t \in K^*$ and hence

$$\phi_a(w^{h_a(s)h_a(t)}) - \phi_a(w) = 2\big(\nu^\#(s) + \nu^\#(t)\big) \tag{16.7}$$

for all $w \in U_a^*$ and all $s, t \in K^*$.

Suppose that $\Delta = \mathsf{B}_2^{\mathcal{Q}}(\Lambda)$ for some anisotropic quadratic space

$$\Lambda = (K, L, q).$$

By 30.37.i,

$$h_a(s)h_a(t) = h_a(st)$$

for all $s, t \in K^*$. By 16.6 and 16.7, it follows that

$$\nu^\#(st) = \nu^\#(s) + \nu^\#(t)$$

for all $s, t \in K^*$, i.e. $\nu^\#$ satisfies 9.17.i. Since $\nu^\#(K^*) = \mathbb{Z}$ (by 3.21), we conclude that $\nu^\#$ is a discrete valuation of K. Thus (iii) holds with $\nu = \nu^\#$ in this case.

Now suppose that 16.3.i or 16.3.ii holds. Thus

$$h_a(t) = m_\Sigma(x_a(0,1))m_\Sigma(x_a(0,t))$$

for each $t \in K^*$. By 30.37.iii, we have

$$h_a(s)h_a(t) = h_a(st)$$

for all $s, t \in K^*$. Therefore

$$\nu^\#(st) = \nu^\#(s) + \nu^\#(t)$$

by 16.6 and 16.7 (exactly as in the previous paragraph) and therefore for all $s, t \in K^*$. Hence $\nu^\#(K^*) = \delta\mathbb{Z}$ for some positive integer δ and $\nu^\#/\delta$ is a valuation of K. Thus (i) holds with $\nu = \nu^\#/\delta$.

---

[7] In 21.27 and 22.16 we show that $\delta = 1$ or 2 in these two cases.

Next suppose that 16.4.iii holds. By 10.2 in [36], $K^2 \subset K_0$. We have

$$h_a(t) = m_\Sigma(x_a(1))m_\Sigma(x_a(t))$$

for all $t \in K_0^*$ this time. By 30.37.ii,

$$h_a(s)h_a(t) = h_a(st) \tag{16.8}$$

for all $s, t \in (K^*)^2 \subset K_0^*$ and

$$h_a(s)^2 = h_a(s^2) \tag{16.9}$$

for all $s \in K_0^*$. By 16.6, 16.7 and 16.8, we have

$$\nu^\#(st) = \nu^\#(s) + \nu^\#(t) \tag{16.10}$$

for all $s, t \in (K^*)^2$. As observed above, $\nu^\#$ satisfies 9.17.ii. The restriction of $\nu^\#$ to K^2 thus satisfies both 9.17.i and 9.17.ii. It follows that the restriction of $\nu^\#$ to K^2 equals $\delta_0\nu_0$ for some valuation ν_0 of K^2 and some positive integer δ_0. Let

$$\nu(s) = \nu_0(s^2)$$

for all $s \in K^*$. Then ν is a valuation of K. By 16.6, 16.7 and 16.9, we have

$$\nu^\#(s^2) = 2\nu^\#(s)$$

for all $s \in K_0^*$. Therefore

$$2\nu^\#(s) = \nu^\#(s^2) = \delta_0\nu_0(s^2) = \delta_0\nu(s)$$

for all $s \in K_0^*$. By 3.21, $\nu^\#(p) = 1$ and thus $\nu(p) = 2/\delta_0$ for some $p \in K_0^*$. Hence δ_0 divides 2 and thus (ii) holds with $\delta = 2/\delta_0$.

Now suppose that we are in one of the remaining cases. Let m be the wall of the root a and let X be the set of roots of Δ having wall m. By 14.47, there exists a Bruhat-Tits pair $(\Delta^{\text{aff}}, \mathcal{A})$ whose building at infinity is Δ. Thus m is a parallel class of $\mathcal{A}$-walls of Δ. Let $(T_m, \mathcal{A}_m)$ be the corresponding wall tree and let ω_m be its canonical valuation. We identify the set of $\mathcal{A}_m$-ends of T_m with the set X via the bijection ξ_m defined in 10.38. Let b be the root of Σ opposite a. By 13.9, we have

$$\omega_m(a, b, b^{x_a(1)}, b^{x_a(t)}) = \phi_a(x_a(t)) \tag{16.11}$$

for all $t \in K^*$ distinct from 1. Since U_a acts sharply transitively on $X\backslash\{a\}$ (by 29.15.v and 29.48), we can identify K with $X\backslash\{a\}$ via the map $t \mapsto b^{x_a(t)}$. If we also set $\infty = a$, then

$$\nu^\#(t) = \omega_m(\infty, 0, 1, t)$$

for all $t \in K^*$ distinct from 1 by 16.11. By (i), (ii), (v) and (viii) of 30.36, the stabilizer $G_m^\dagger$ of the wall m in $G^\dagger$ induces a permutation group on X that contains the elements

$$s \mapsto us + t$$

for all $u \in K^*$ and all $t \in K$. By 12.31, ω_m is $G_m^\dagger$-invariant. By 9.24, therefore, we conclude again that $\nu^\#$ is a discrete valuation of K. Thus (iii) holds with $\nu = \nu^\#$. □

Notation 16.12. Suppose that Π is not simply laced (so that Φ has roots of two different lengths by 2.14.iv) and let a be as in 16.3. We will call a root map ι of Σ *long* if $\iota(a)$ is a long root of Φ, respectively, *short* if $\iota(a)$ is a short root of Φ. By 3.14, if ι is long, respectively, short, then so is every root map equivalent to ι.

We come now to our main problem: Suppose that for the root a chosen in 16.3 the map ϕ_a from U_a^* to $\mathbb{Z}$ defined in (i)–(iii) of 16.4 is given. When does there exist a valuation ψ of the root datum (G, ξ) with respect to the root map ι that extends ϕ_a (i.e. such that $\psi_a = \phi_a$)? We will begin to answer this question in 16.14 below.

Notation 16.13. Let d be the chamber of Σ fixed in 16.1, let J be a subset of the vertex set I of Π, let D be the J-residue of Δ containing d, let Π_J be the subdiagram of Π spanned by J and let Φ_J be the intersection of the root system Φ with the subspace of V spanned by the vectors $\{a_x \mid x \in J\}$. Suppose that $|J| \geq 2$ and that Π_J is connected. By 29.10.iii, D is a building of type Π_J. Let Ξ_J be the set of roots in Ξ that cut D and let $G_J^\dagger$ denote the group induced by

$$\langle U_b \mid b \in \Xi_J \rangle$$

on D. By 29.13.iv, $\Sigma \cap D$ is an apartment of D and the restriction ι_D of ι to $\Sigma \cap D$ is a root map of this apartment with target Φ_J. By 29.19, the map $b \mapsto b \cap D$ is a bijection from Ξ_J to the set of roots of the apartment $\Sigma \cap D$. Let $b \in \Xi_J$, let P be a panel in the wall of b that is contained in D and let x be the unique chamber in $P \cap b$. Then the root group U_b acts sharply transitively on $P \backslash \{x\}$ (by 29.14.iii and 29.15.v). In particular, U_b acts faithfully on P. Hence we can identify U_b with its image in $G_J^\dagger$. This image is contained in the root group $U_{b \cap D}$ of D. By 29.14.iii applied to D, therefore, $U_{b \cap D}$ acts transitively on the set of apartments of D containing $b \cap D$. Since b is an arbitrary element of Ξ_J, it follows that D is Moufang. Hence by 29.15.v applied to D, the root group $U_{b \cap D}$ also acts sharply transitively on $P \backslash \{x\}$. We conclude that $U_b = U_{b \cap D}$. Thus if we identify Ξ_J with the set of roots of $\Sigma \cap D$ via the bijection $b \mapsto b \cap D$ and let ξ_J be the restriction of the map ξ (defined in 3.4) to Ξ_J, then $(G_J^\dagger, \xi_J)$ is the root datum of D based at the apartment $\Sigma \cap D$.

Theorem 16.14. *Suppose that either* $\Pi = \mathsf{A}_2$ *or* $\ell \geq 3$. *Let* x, a *and* x_a *be as in 16.3, let* ν *be a valuation of* K *as defined in 9.17 and let* ϕ_a *be the map from* U_a^* *to* $\mathbb{Z}$ *given by*

$$\phi_a(x_a(t)) = \nu(t) \tag{16.15}$$

for all $t \in K^*$. *If* $\Pi \neq \mathsf{F}_4$, *let* J *be the unique maximal subset of the vertex set* I *of* Π *such that the subdiagram* Π_J *is connected and simply laced. If* $\Pi = \mathsf{F}_4$, *let* z *be the unique vertex of* Π *such that* $\{x, z\}$ *is an edge of* Π *with label 3 and let* $J = \{x, z\}$. *Then the following hold:*

(i) *The map ϕ_a extends to a valuation ϕ of the root datum $(G^\dagger_J, \xi_J)$ with respect to the root map ι_J (as defined in 16.13) such that*

$$\phi_b(x_b(t)) = \nu(t) \tag{16.16}$$

for all $b \in \Xi_J \cap \Xi_d$ and for all $t \in K$, where x_b is the isomorphism from the additive group of K to U_b defined in 16.2. If Π is simply laced (so $J = I$ and $\Xi_J = \Xi$), then ϕ is an extension of ϕ_a to a valuation of $(G^\dagger, \xi)$ with respect to ι that is independent of the choice of a in 16.3, and every extension of ϕ_a to a valuation of $(G^\dagger, \xi)$ with respect to ι is equipollent to ϕ.

(ii) *Suppose that Π is not simply laced and let y be the unique vertex of Π such that $E := \{x, y\}$ is the unique edge of Π with label greater than 3. Then the valuation ϕ_a extends to a valuation of the root datum $(G^\dagger, \xi)$ (with respect to ι) if and only if it extends to a valuation of the root datum $(G^\dagger_E, \xi_E)$ with respect to the root map ι_E. If an extension exists, it is unique up to equipollence.*

Proof. Let

$$\phi_b(x_b(t)) = \nu(t)$$

for all $b \in \Xi_J \cap \Xi_d$ and all $t \in K^*$, where x_b is as in (i). (This is, of course, consistent with 16.15.) Let ϕ denote the collection of all such maps ϕ_b.

By 9.17.ii and 9.23.i, ϕ satisfies 15.1.i, and by 9.17.i (as well as 16.1 of [36]), ϕ satisfies 15.1.ii. Thus ϕ is a partial valuation of the root datum $(G^\dagger_J, \xi_J)$ based at d. In order to prove (i), it will suffice, by 15.21, to show that ϕ is viable as defined in 15.4.

Let (u, v) be an arbitrary undirected edge of Π with label 3, let $b = a_u$ and $c = a_v$ and let e be the unique root in the open interval (a, b). By 30.36.i,

$$\phi_b\big(x_b(u)^{m_\Sigma(x_c(1))m_\Sigma(x_c(t))}\big) - \phi_b(x_b(u)) = -\nu(t)$$

and

$$\phi_e\big(x_e(u)^{m_\Sigma(x_c(1))m_\Sigma(x_c(t))}\big) - \phi_e(x_e(u)) = \nu(t)$$

for all $t, u \in K^*$. By 16.1 of [36] and 15.24, ϕ is exact at (b, c, e) and the (b, c)-coefficient of ϕ at e is $(1, 1)$. By 15.25, it follows that

$$\phi_c\big(x_c(u)^{m_\Sigma(x_c(1))m_\Sigma(x_c(t))}\big) - \phi_c(x_c(u)) = 2\phi_c(x_c(t))$$

for all $t \in K^*$. Thus ϕ satisfies 15.4.i. By 30.38, it satisfies 15.4.ii. Hence ϕ is viable. Thus (i) holds.

Suppose now that Π is not simply laced and that ϕ_a extends to a valuation ϕ of $(G^\dagger_E, \xi_E)$, where $E = \{x, y\}$ is as in (ii). Suppose, too, that $\Pi \neq \mathsf{F}_4$. By 15.5, the restriction of this valuation to the set of roots in $\Xi_E \cap \Xi_d = [a_x, a_y]$ is a viable partial valuation of $(G^\dagger_E, \xi_E)$ based at d. By (i), this partial valuation extends to a viable partial valuation of $(G^\dagger, \xi)$ based at d.

Suppose that $\Pi = \mathsf{F}_4$, let $b = a_y$ and let z be the unique vertex of Π such that $J_1 := \{y, z\}$ is the unique edge containing y with label 3. Let (K, F),

σ and ζ be as in 30.14.xiii and let q be the norm of the composition algebra (K, F) as defined in 30.17. Then (F, K, q) is an anisotropic quadratic space and

$$\zeta(x, y) = \mathcal{Q}_\mathcal{Q}(F, K, q),$$

where $\mathcal{Q}_\mathcal{Q}(F, K, q)$ is as in 30.8.iii. Applying 16.4, we conclude that after replacing ϕ by an equipollent valuation of $(G_E^\dagger, \xi_E)$ if necessary, there is a valuation $\nu^\#$ of F such that

$$\phi_b(x_b(t)) = \nu^\#(t)$$

for all $t \in F$.[8] By two applications of (i), there is an extension of ψ_a to a viable partial valuation of $(G_J^\dagger, \xi_J)$ based at d and an extension of ϕ_b to a viable partial valuation of $(G_{J_1}^\dagger, \xi_{J_1})$ based at d. We conclude also in this case that ϕ_a extends to a viable partial valuation of $(G^\dagger, \xi)$ based at d. Thus (ii) holds (by 3.41.iii and 15.21) whether or not $\Pi = \mathsf{F}_4$. □

Notation 16.17. Let Δ be as in (i), (ii), (xi) or (xii) of 30.14, so $\Delta = \mathsf{X}_\ell(K)$ in the notation of 30.15, where $\mathsf{X}_\ell = \mathsf{A}_\ell$, D_ℓ or E_ℓ for $\ell = 6$, 7 or 8, let ν be a valuation of K and let ϕ be the valuation of $(G^\dagger, \xi)$ determined by ν in 16.16. We denote by $\tilde{\mathsf{X}}_\ell(K, \nu)$ the pair $(\Delta^{\text{aff}}, \mathcal{A})$ obtained by applying 14.47 to these data.

We observe explicitly that the pair $\tilde{\mathsf{X}}_\ell(K, \nu)$ defined in 16.17 exists for *every* discrete valuation ν of K.

Theorem 16.18. *Every Bruhat-Tits pair whose Coxeter diagram is* $\tilde{\mathsf{X}}_\ell$ *for* $\mathsf{X} = \mathsf{A}$, D *or* E *(i.e. whose Coxeter diagram is simply laced) is of the form* $\tilde{\mathsf{X}}_\ell(K, \nu)$ *for some field or, but only if* $\mathsf{X} = \mathsf{A}$, *skew field or, but only if* $\mathsf{X}_\ell = \mathsf{A}_2$, *octonion division algebra* K *and some discrete valuation* ν *of* K.

Proof. This holds by 16.4, 16.14 and 16.17. □

* * *

Continuing with all the notation in 16.1 and 16.2, we suppose now that the Coxeter diagram Π of Δ is either B_2 or G_2 and let (x, y) be one of the two directed edges of Π. Thus $\Omega_{xy} = X(\Lambda)$ or $X(\Lambda)^{\text{op}}$, where X is one of the operators

(16.19) $$\mathcal{Q}_\mathcal{Q},\ \mathcal{Q}_\mathcal{D},\ \mathcal{Q}_\mathcal{E},\ \mathcal{Q}_\mathcal{F},\ \mathcal{Q}_\mathcal{I},\ \mathcal{Q}_\mathcal{P},\ \mathcal{H}$$

in 30.8 different from the operator $\mathcal{T}$ and Λ is a parameter system of suitable type. Let a and x_a be as in 16.3, let ν be a valuation of the defining field K of Δ (as defined in 30.15) and let the map ϕ_a from the root group U_a of Δ to $\mathbb{Z} \cup \{\infty\}$ be as follows:

(i) $\phi_a(x_a(0, t)) = \delta\nu(t)$ for some positive integer δ and for all $t \in K^*$ if $X = \mathcal{Q}_\mathcal{E}$ or $\mathcal{Q}_\mathcal{F}$;

[8] We also have $\zeta(x, y) = \mathcal{Q}_\mathcal{I}(K, F, \sigma)$. By 23.3.ii, therefore, $\nu^\#$ is the restriction of ν to F divided by δ, where $\delta = 1$ or 2. We do not, however, need to know this here.

(ii) $\phi_a(x_a(t)) = \nu(t)/\delta$ for $\delta = 1$ or 2 and for all $t \in K_0^*$ if $X = \mathcal{D}$ and $\Lambda = (K, K_0, L_0)$; and

(iii) $\phi_a(x_a(t)) = \nu(t)$ for all $t \in K^*$ in the other four cases.

By 16.4 and 16.14, the main challenge remaining is to determine necessary and sufficient conditions for the map ϕ_a to extend to a valuation of $(G^\dagger, \xi)$ with respect to some root map ι. We will do this for each of the seven operators in 16.19 in seven chapters, from Chapter 19 to Chapter 25.

Chapter Seventeen

Completions

Before we begin to study the problem described at the conclusion of Chapter 16, we would like to consider an important issue. In 14.54, we saw that Bruhat-Tits pairs $(\Delta, \mathcal{A})$ are classified by root data with valuation (up to equipollence). Suppose, however, that we are interested in classifying Bruhat-Tits *buildings* (as defined in 13.1) rather than Bruhat-Tits *pairs.* To investigate the question when two root data with valuation correspond to the same Bruhat-Tits building Δ, but not necessarily the same Bruhat-Tits pair $(\Delta, \mathcal{A})$, we need to introduce completions.

Definition 17.1. Let $(\Delta, \mathcal{A})$ be a Bruhat-Tits pair as defined in 13.1 and let $\hat{\mathcal{A}}$ be the complete system of apartments of Δ (as defined in 8.5). We call the pair $(\Delta, \hat{\mathcal{A}})$ the *completion* of $(\Delta, \mathcal{A})$ (or of Δ) and we will say that $(\Delta, \mathcal{A})$ (or Δ) is *completely Bruhat-Tits* if the pair $(\Delta, \hat{\mathcal{A}})$ is a Bruhat-Tits pair, i.e. if the building at infinity $\Delta^\infty_{\hat{\mathcal{A}}}$ is Moufang. We say that $(\Delta, \mathcal{A})$ is *complete* if $\mathcal{A} = \hat{\mathcal{A}}$.

By 8.26 and 29.15.i, every Bruhat-Tits building is completely Bruhat-Tits if the rank ℓ of $\Delta^\infty_{\hat{\mathcal{A}}}$ is greater than 2. In 17.11, 19.27, 20.8, 21.34, 22.33, 23.13, 24.46 and 25.27, we show that this is also true when $\ell = 2$.

Definition 17.2. Let $\big((G^\dagger, \xi), \phi\big)$ be a root datum with valuation, let $(\Delta, \mathcal{A})$ be the corresponding Bruhat-Tits pair and suppose that $(\Delta, \mathcal{A})$ is completely Bruhat-Tits as defined in 17.1. The *complete form* of $((G^\dagger, \xi), \phi)$ is the root datum with valuation (up to equipollence) that corresponds to the completion $(\Delta, \hat{\mathcal{A}})$. We say that $\big((G^\dagger, \xi), \phi\big)$ is *complete* if it is its own complete form.

Since every Bruhat-Tits building is completely Bruhat-Tits, the root datum with valuation corresponding to an arbitrary Bruhat-Tits pair always has a complete form. Thus two root data with valuation correspond to the same affine building if and only if their complete forms are the same (up to equipollence). To answer the question when two root data with valuation correspond to the same Bruhat-Tits building (rather than Bruhat-Tits pair), it will therefore suffice to show how to identify the complete form of a given root datum with valuation.[1]

[1]Our proof that every Bruhat-Tits pair is completely Bruhat-Tits follows from these case-by-case calculations. See 17.10 and 27.4.

We do this in the case that the Coxeter diagram Π of the building Δ is simply laced in 17.11. The other cases will be treated in Chapters 19–25.

Proposition 17.3. *Let Δ be a Moufang spherical building, let Σ be an apartment of Δ, let $(G^\dagger, \xi)$ be the root datum of Δ based at Σ, let ϕ be a valuation of $(G^\dagger, \xi)$ and let a be a root of Σ. For all u, v in the root group U_a, let*

$$\partial_a(u, v) = \begin{cases} 2^{-\phi_a(u-v)} & \text{if } u \neq v \text{ and} \\ 0 & \text{if } u = v, \end{cases}$$

where we are using additive notation for the root group U_a even though it might not be abelian. Then ∂_a is a metric on U_a.

Proof. By condition (V1) in 3.21 (and 3.29), conditions (a) and (b) in 9.18 hold with ϕ_a in place of ν. The claim holds, therefore, by 9.18.ii. □

The following result says roughly that if one root datum with valuation is "dense" in another, then the corresponding Bruhat-Tits buildings (but not necessarily the corresponding Bruhat-Tits pairs) are the same.

Theorem 17.4. *Let $\hat{\Delta}$ be a spherical building satisfying the Moufang condition, let Δ be a subbuilding whose rank equals the rank of $\hat{\Delta}$, let Σ be an apartment of Δ (which is then also an apartment of $\hat{\Delta}$) and let Ξ be the set of roots of Σ. Then the following hold:*

(i) *Δ is Moufang.*
(ii) *For each root $a \in \Xi$, the root group U_a of Δ can be identified canonically with a subgroup of the root group $\hat{U}_a$ of $\hat{\Delta}$.*

Let $(G^\dagger, \xi)$ and $(\hat{G}^\dagger, \hat{\xi})$ be the root data of Δ and $\hat{\Delta}$ based at Σ and suppose that ϕ and $\hat{\phi}$ are valuations of $(G^\dagger, \xi)$ and $(\hat{G}^\dagger, \hat{\xi})$ with respect to the same root map ι of Σ such that for each root $a \in \Xi$, the following hold:

(a) *ϕ_a is the restriction of $\hat{\phi}_a$ to the subgroup U_a and*
(b) *U_a is dense in $\hat{U}_a$ with respect to the metric $\hat{\partial}_a$ on $\hat{U}_a$ given by the formula in 17.3 (with $\hat{\phi}_a$ in place of ϕ_a).*

Let

$$(\Delta^{\text{aff}}, \mathcal{A}) \rightsquigarrow ((G^\dagger, \xi), [\phi])$$

and

$$(\hat{\Delta}^{\text{aff}}, \hat{\mathcal{A}}) \rightsquigarrow ((\hat{G}^\dagger, \hat{\xi}), [\hat{\phi}])$$

be as in 14.54. Then, up to isomorphism,

(iii) *$\Delta^{\text{aff}} = \hat{\Delta}^{\text{aff}}$ and $\mathcal{A} \subset \hat{\mathcal{A}}$.*

Proof. Assertions (i) and (ii) hold by 29.61. By 14.54, we can identify the building at infinity of the pair $(\hat{\Delta}^{\rm aff}, \hat{\mathcal{A}})$ with the building $\hat{\Delta}$. Since Σ is an apartment of $\hat{\Delta}$, there is a unique apartment A in $\hat{\mathcal{A}}$ such that $A^\infty = \Sigma$. In particular, S^∞ is a chamber of $\Sigma \subset \Delta$ for each sector S contained in A (since Σ is also an apartment of the subbuilding Δ). Let

$$\mathcal{A}_0 = \{A^g \mid g \in G^\dagger\} \subset \hat{\mathcal{A}} \tag{17.5}$$

and let X denote the set of chambers of $\hat{\Delta}$ contained in A_0^∞ for some $A_0 \in \mathcal{A}_0$, i.e.

$$X = \bigcup_{A_0 \in \mathcal{A}_0} A_0^\infty.$$

By 29.15.iii, X is the chamber set of Δ and

$$\mathcal{A}_0^\infty := \{A_0^\infty \mid A_0 \in \mathcal{A}_0\}$$

is the set of apartments of Δ.

Suppose that A_1 and A_2 are two apartments in $\mathcal{A}_0$. Let S_1 be a sector in A_1 and let S_2 be a sector in A_2. Then S_1^∞ and S_2^∞ are chambers of Δ. By 29.13.ii, there exists $A_0 \in \mathcal{A}_0$ such that A_0^∞ contains both S_1^∞ and S_2^∞. Thus A_0 contains subsectors of both S_1 and S_2. We conclude that the set $\mathcal{A}_0$ satisfies 8.4.ii.

Now suppose that $\mathcal{A}_0$ also satisfies 8.4.i. This means that $\mathcal{A}_0$ is a system of apartments. Thus the pair $(\hat{\Delta}^{\rm aff}, \mathcal{A}_0)$ has a building at infinity $\hat{\Delta}^{\rm aff}_{\mathcal{A}_0}$. This building at infinity is a subbuilding of $\hat{\Delta}$ and its chamber set is also X. Therefore $\hat{\Delta}^{\rm aff}_{\mathcal{A}_0} = \Delta$. Hence $(G^\dagger, \xi)$ is the root datum of $\hat{\Delta}^{\rm aff}_{\mathcal{A}_0}$ based at Σ. By 14.47 (and 14.54), $\hat{\phi} = \phi_{\hat{R}}$ for some gem $\hat{R}$ of $\hat{\Delta}$ cutting A (as defined in 13.8). Since ϕ_a is the restriction of $\hat{\phi}_a$ to U_a for each $a \in \Xi$, it follows that $\phi = \phi_R$, where $R = \hat{R} \cap \Delta$. In other words,

$$(\hat{\Delta}^{\rm aff}, \mathcal{A}_0) \rightsquigarrow ((G^\dagger, \xi), [\phi]).$$

By 13.34, it follows that there is an isomorphism from $\Delta^{\rm aff}$ to $\hat{\Delta}^{\rm aff}$ mapping $\mathcal{A}$ to $\mathcal{A}_0$. To prove (iii), it thus suffices to show that $\mathcal{A}_0$ satisfies 8.4.i.

Let P be a panel of $\hat{\Delta}^{\rm aff}$ containing a chamber u in A and let α be the unique root of A containing u but not the other chamber in $P \cap A$. Let v be this other chamber and let $a = \alpha^\infty$. There exists a $k \in \mathbb{Z}$ such that $\hat{U}_{a,l}$ acts trivially on $P \cap A$ for every $l \geq k$ (by 1.35 and 13.17). Let w be an arbitrary chamber of P distinct from u. By 13.5, there exists an element $\hat{g} \in \hat{U}_a$ that fixes u and maps v to w. Since U_a is dense in $\hat{U}_a$, there exists an element $g \in U_a$ such that $\hat{g}g^{-1} \in \hat{U}_{a,k}$. Therefore $\hat{g}g^{-1}$ acts trivially on P. Hence also g fixes u and maps v to w. We conclude that

(17.6) the stabilizer of u in the root group U_a acts transitively on $P \backslash \{u\}$.

By 13.2, the elements of $G^\dagger$ are special automorphisms of $\hat{\Delta}^{\rm aff}$. It follows from 17.5 and 17.6, therefore, that for each panel P containing a chamber u in an apartment in $\mathcal{A}_0$, the stabilizer of u in $G^\dagger$ acts transitively on $P \backslash \{u\}$. Thus if a panel P contains one chamber that is contained in an apartment in $\mathcal{A}_0$, then every chamber in P is contained in some apartment in $\mathcal{A}_0$ and the stabilizer of P in $G^\dagger$ acts transitively (in fact, 2-transitively) on P. Since $\hat{\Delta}^{\rm aff}$ is connected, it follows that $G^\dagger$ acts transitively on the set of chambers of $\hat{\Delta}^{\rm aff}$.

We claim that for every minimal gallery γ of $\hat{\Delta}^{\rm aff}$, there exists an element of $G^\dagger$ mapping γ to a gallery contained in A. Let

$$\gamma = (x_0, x_1, \ldots, x_k)$$

be a minimal gallery of $\hat{\Delta}^{\rm aff}$. We proceed by induction with respect to k. By the conclusion of the previous paragraph, the claim holds for $k = 0$. We can thus assume that $x_0 \in A$ and $k \geq 1$. Let

$$\gamma' = (x_0, x_1', \ldots, x_k')$$

be the unique gallery of A starting at x_0 that has the same type as γ. By 17.6, there exists an element in $G^\dagger$ mapping γ to γ' if $k = 1$. Suppose that $k > 1$. By induction, we can suppose that there is an element g_1 in $G^\dagger$ mapping $(x_0, x_1, \ldots, x_{k-1})$ to $(x_0, x_1', \ldots, x_{k-1}')$. Let α be the root of A containing x_{k-1}' but not x_k' and let $a = \alpha^\infty$. By 17.6 again, the stabilizer of x_{k-1}' in U_a contains an element g mapping $g_1(x_k)$ to x_k'. By 13.2, the element g acts trivially on the root α. By 29.10.i, γ' is minimal. By 29.6.i, α contains every chamber of A that is nearer to x_{k-1}' than to x_k'. Thus $x_i' \in \alpha$ for all $i \in [1, k-1]$. Hence g maps $g_1(\gamma)$ to γ'. This proves the claim. It follows that $\mathcal{A}_0$ satisfies 8.4.i. Thus (iii) holds. □

Proposition 17.7. *Let $(\Delta, \mathcal{A})$ be a Bruhat-Tits pair. Then $(\Delta, \mathcal{A})$ is complete (as defined in 17.1) if and only if for each wall m of $\Delta_{\mathcal{A}}^\infty$, the system of apartments $\mathcal{A}_m$ of the tree T_m (as defined in 10.19) is complete.*

Proof. Suppose that $\mathcal{A}$ is complete, let m be a wall of $\Delta_{\mathcal{A}}^\infty$ and let D be an arbitrary apartment in the tree T_m. Thus m is a parallel class of $\mathcal{A}$-walls of Δ and these $\mathcal{A}$-walls are the vertices of T_m. Let M_0 be a vertex of D and let

$$(M_0, M_1, M_2, \ldots)$$

and

$$(M_0', M_1', M_2', \ldots)$$

be the two rays in D that start at $M_0' := M_0$ and go off in different directions.

By 10.11 and 10.30, for each $i \geq 0$, there exists an apartment $A_i \in \mathcal{A}$ containing roots β_i, β_i' such that $M_i = \mu(\beta_i)$, $M_i' = \mu(\beta_i')$, β_i is a translate in A_i of $-\beta_i'$, $\beta_i \cap \beta_i' \neq \emptyset$ and a wall is in

$$\{M_j \mid j \in [0, i]\} \cup \{M_j' \mid j \in [0, i]\}$$

if and only it is the wall of a root α of A_i such that

$$-\beta_i' \subset \alpha \subset \beta_i.$$

By 10.6.iii,

$$\beta_{i-1} \cap \beta_{i-1}' \subset \beta_i \cap \beta_i' \tag{17.8}$$

for all $i > 0$.

Let $i \mapsto \alpha_i$ be the bijection from $\mathbb{Z}$ to the set of roots of the apartment A_1 parallel to β_1 such that $\alpha_1 = \beta_1$ and $\alpha_i \subset \alpha_{i+1}$ for all i. Similarly, let $i \mapsto \alpha'_i$ be the bijection from $\mathbb{Z}$ to the set of roots of A_1 parallel to β'_1 such that $\alpha'_1 = \beta'_1$ and $\alpha'_i \subset \alpha'_{i+1}$ for all i. For each $i \geq 1$, let κ_i be the unique isomorphism from A_1 to the apartment A_i whose restriction to $\alpha_1 \cap \alpha'_1$ is the identity. By 17.8, the maps κ_{i-1} and κ_i agree on the set $\alpha_{i-1} \cap \alpha'_{i-1}$ for each $i \geq 2$. By 1.41, there thus exists a unique special isomorphism κ from A_1 to Δ (as defined in 29.2) that coincides with κ_i on $\alpha_i \cap \alpha'_i$ for each $i \geq 1$. The image of κ is an apartment A of Δ containing M_i and M'_i for all $i \geq 0$ (by 29.13). Since $\mathcal{A}$ is complete, it contain A. Since D is the unique apartment of T_m containing M_i and M'_i for all $i \geq 0$, we conclude that $D = A_m$ (where A_m is as defined in 10.19). Hence $D \in \mathcal{A}_m$. Thus $\mathcal{A}_m$ is the complete system of apartments of T_m.

Suppose, conversely, that $\mathcal{A}_m$ is the complete system of apartments of the tree T_m for each wall m of $\Delta^\infty_{\mathcal{A}}$ and let $\hat{\mathcal{A}}$ denote the complete set of apartments of Δ. Let F be a panel of $\Delta^\infty_{\mathcal{A}}$ and let m be a wall of $\Delta^\infty_{\mathcal{A}}$ containing F. By 11.1, there is an $\mathcal{A}$-face f of Δ and an $\mathcal{A}$-wall M of Δ containing f such that $m = M^\infty$ and $F = f^\infty$. Let $\hat{m}$ be the wall of $\Delta^\infty_{\hat{\mathcal{A}}}$ containing m (i.e. such that each panel in m is contained in a panel in $\hat{m}$), let $\hat{F}$ be the unique panel of $\Delta^\infty_{\hat{\mathcal{A}}}$ containing F and let $(\hat{T}_{\hat{m}}, \hat{\mathcal{A}}_{\hat{m}})$ and $(\hat{T}_{\hat{F}}, \hat{\mathcal{A}}_{\hat{F}})$ be the corresponding wall tree and panel tree of the pair $(\Delta, \hat{\mathcal{A}})$. If we interpret $\hat{m}$ as a parallel class of $\hat{\mathcal{A}}$-walls as in 11.1, then by 10.32, $\hat{m}$ is the same as m. In other words, $T_m = \hat{T}_{\hat{m}}$. Since the pair $(T_m, \mathcal{A}_m)$ is complete, we have, in fact,

$$(T_m, \mathcal{A}_m) = (\hat{T}_{\hat{m}}, \hat{\mathcal{A}}_{\hat{m}}).$$

By 11.16, it follows that the pair $(T_F, \mathcal{A}_F)$ is also complete and

$$(T_F, \mathcal{A}_F) = (\hat{T}_{\hat{F}}, \hat{\mathcal{A}}_{\hat{F}}).$$

In particular, F and $\hat{F}$ are the same if we interpret them as parallel classes of faces in Δ. We claim that they are also the same as subsets of the chamber set of $\Delta^\infty_{\hat{\mathcal{A}}}$.[2] Let $x \in \hat{F}$. Then $x = S^\infty$, where S is an $\hat{\mathcal{A}}$-sector having a face parallel to f. Let $S_{\hat{F}} = S_F$ be the corresponding $\hat{F}$-ray (equivalently, F-rays) of the tree $\hat{T}_{\hat{F}} = T_F$ as defined in 11.20. Since $(T_F, \mathcal{A}_F)$ is complete, there

[2]There is a subtle point here: Suppose that Δ is an arbitrary thick irreducible affine building and that $\mathcal{A}$ is a system of apartments of Δ. Let $\hat{\mathcal{A}}$ be the complete system of apartments of Δ, let f be an arbitrary $\mathcal{A}$-face of Δ and let M be an $\mathcal{A}$-wall of Δ containing f. Let F be the panel of $\Delta^\infty_{\mathcal{A}}$ that corresponds to the parallel class of $\mathcal{A}$-faces containing f and let $\hat{F}$ be the panel of $\Delta^\infty_{\hat{\mathcal{A}}}$ that corresponds to the parallel class of $\hat{\mathcal{A}}$-faces containing f (in both cases, via the bijection described in 8.36.i). By 10.32, every $\hat{\mathcal{A}}$-wall parallel to M is, in fact, an $\mathcal{A}$-wall. By 11.16, therefore, $T_F = \hat{T}_{\hat{F}}$, i.e. every $\hat{\mathcal{A}}$-face of Δ parallel to f is, in fact, an $\mathcal{A}$-face. Thus both F and $\hat{F}$ correspond to the same parallel class of faces of Δ via the map in 8.36.i even though F is, in general, a proper subset of $\hat{F}$ as sets of chambers of $\Delta_{\hat{\mathcal{A}}}$. This is one place where one might doubt whether it is a good idea to identify panels at infinity with parallel classes of faces as we have done in 11.1.

exists an $\mathcal{A}$-apartment A' containing a sector S' having a face parallel to f such that S'_F is a subray of S_F. By 11.22, $S^\infty = (S')^\infty$. Hence $x \in F$. We conclude the panels F and $\hat{F}$ are the same as claimed. Since F is arbitrary and $\Delta^\infty_{\hat{\mathcal{A}}}$ is connected, it follows that

$$\Delta^\infty_{\mathcal{A}} = \Delta^\infty_{\hat{\mathcal{A}}}.$$

Therefore $\mathcal{A} = \hat{\mathcal{A}}$ (by 8.27). □

Theorem 17.9. *Let $(\Delta, \mathcal{A})$ be a Bruhat-Tits building, let $(G^\dagger, \xi)$, Σ and ϕ be as in 13.30, let d be a chamber of Σ and let Ξ°_d be as in 3.5. For each root a of Σ, let U_a be the root group of $\Delta^\infty_{\mathcal{A}}$ corresponding to a and let ∂_a be the metric on U_a defined in 17.3. Then the system of apartments $\mathcal{A}$ is complete (as defined in 8.5) if and only if for each $a \in \Xi^\circ_d$, U_a is complete with respect to the metric ∂_a.*

Proof. For each root a of Σ, let m_a be the wall of a and let $(T_{m_a}, \mathcal{A}_{m_a})$ be the corresponding wall tree. Choose a root a of Σ. By 10.38 there is a canonical correspondence between the $\mathcal{A}_{m_a}$-ends of T_{m_a} and the set of roots of $\Delta^\infty_{\mathcal{A}}$ having wall m_a. By 29.15.v, the root group U_a acts sharply transitively on the set of roots of $\Delta^\infty_{\mathcal{A}}$ having wall m_a that are distinct from a. By 9.30.ii and 12.31, it follows that the root group U_a is complete with respect to ∂_a if and only if the pair $(\mathcal{A}_{m_a}, T_{m_a})$ is complete.

By 29.15.iii, every root of $\Delta^\infty_{\mathcal{A}}$ can be mapped by an element of $G^\dagger$ to a root in Ξ°_d. For each wall m of $\Delta^\infty_{\mathcal{A}}$, there thus exists an element $g \in G^\dagger$ and a root $a \in \Xi^\circ_d$ such that g maps the wall m to the wall m_a and hence the pair $(T_m, \mathcal{A}_m)$ to the pair $(T_{m_a}, \mathcal{A}_{m_a})$. Thus the pair $(T_m, \mathcal{A}_m)$ is complete for all walls m of $\Delta^\infty_{\mathcal{A}}$ if and only if the pair $(T_{m_a}, \mathcal{A}_{m_a})$ is complete for all $a \in \Xi^\circ_d$. By 17.7 and the conclusion of the previous paragraph, it follows that U_a is complete with respect to ∂_a for all $a \in \Xi^\circ_d$ if and only if the system of apartments $\mathcal{A}$ of Δ is complete. □

Remark 17.10. Let Δ be an arbitrary building. The length of a minimal gallery between two chambers is a metric on the chamber set of Δ. The automorphism group of Δ can thus be given the topology of uniform convergence on bounded subsets. A sequence $(g_i)_{i \geq 1}$ of automorphisms of Δ converges to an automorphism g in this topology if and only if there exists a chamber x such that for each $m \in \mathbb{N}$, the automorphisms g and g_i agree on the set of chambers at a distance of at most m from x for all i sufficiently large. Now suppose that $(\Delta, \mathcal{A})$ is a Bruhat-Tits pair, let $A \in \mathcal{A}$, let $\phi = \phi_R$ be as in 13.8 for some gem R of Δ cutting A, let $\Sigma = A^\infty$ and let a be a root of Σ. By 12.31, we can think of the root group U_a as a subgroup of $\mathrm{Aut}(\Delta)$. For each $u \in U^*_a$, let a_u denote the root $A \cap A^u$ of A (as in 13.2) and for each chamber $x \in a_u$, let $M_{u,x}$ be the distance from x to the chamber set of the wall $\mu(a_u)$. It can be deduced from Proposition 7.4.33 in [6] (see also Section 2.1 in [33]) that there exists a constant C (depending only on Δ) such that for each $u \in U^*_a$ and each $x \in a_u$, u fixes every chamber of Δ at a distance of at most $CM_{u,x}$ from x. This implies that the topology on U_a

induced by the metric defined in 17.3 is the same as the topology of uniform convergence on bounded sets. From this observation, it ought to be possible to devise a proof that every Bruhat-Tits pair is completely Bruhat-Tits that does not involve looking at each case separately. We have not tried to carry this out, however, since in any event we want to show how to determine the complete form of an arbitrary root datum with valuation. This requires case-by-case analysis and yields the conclusion that every Bruhat-Tits pair is completely Bruhat-Tits as a corollary.

Here is the first application of our results:

Theorem 17.11. *Let $(\Delta, \mathcal{A}) = \tilde{\mathsf{X}}_\ell(K, \nu)$ for $\mathsf{X} = \mathsf{A}$, D or E and let $\hat{K}$ be the completion of K with respect to ν (as defined in 9.20). Then Δ is completely Bruhat-Tits and its completion is $\tilde{\mathsf{X}}_\ell(\hat{K}, \nu)$.*

Proof. By 30.16, $\mathsf{X}_\ell(K)$ is a subbuilding of $\mathsf{X}_\ell(\hat{K})$. The claim holds, therefore, by 17.4 and 17.9. □

We close this chapter with three classical results. These results will be needed when we prove analogs to 17.11 for the various families of Bruhat-Tits pairs whose Coxeter diagram is not simply laced.

Proposition 17.12. *If f is a uniformly continuous function from a subset A of a metric space M into a complete metric space N, then f has a unique extension to a continuous map from the closure $\bar{A}$ of A to N and this extension is, in fact, uniformly continuous.*

Proof. This is Theorem 2-83 in [16]. □

Proposition 17.13. *Let V be a finite-dimensional vector space over a field K, let ν be a valuation of K, let Q be a map from V to K, let $\omega(u) = \nu(Q(u))$ for all $u \in V$ and let ∂_ω be the map from $V \times V$ to K defined in 9.19 with V in place of U and ω in place of ν. Let $B = \{v_1, \ldots, v_n\}$ be a basis of V over K and let*

$$\partial_B\Big(\sum_{i=1}^n t_i v_i, \sum_{i=1}^n s_i v_i\Big) = \begin{cases} \max\{2^{-\nu(t_i - s_i)} \mid i \in [1, n]\} & \text{if } x \neq y \text{ and} \\ 0 & \text{if } x = y \end{cases}$$

for all $t_1, \ldots, t_n, s_1, \ldots, s_n \in K$. Suppose that the following hold:

(a) *Q is anisotropic (i.e. $Q(u) = 0$ if and only if $u = 0$).*
(b) *For some $m \geq 1$, $Q(tu) = t^m Q(u)$ for all $t \in K$ and all $u \in V$.*
(c) *$\omega(u + v)) \geq \min\{\omega(u), \omega(v)\}$ for all $u, v \in V$.*

Then the following hold:

(i) *∂_ω and ∂_B are metrics on V.*
(ii) *If K is complete with respect to ν, then the metrics ∂_ω and ∂_B are equivalent and V is complete with respect to both metrics.*

Proof. By (a), $\omega(V^*) \subset \mathbb{Z}$. By (b), $\omega(-u) = \omega(u)$ for all $u \in V$. By (c) and 9.18.ii, therefore, ∂_ω is a metric on V. By 9.17, ∂_B is a metric on V. Thus (i) holds. The assertion (ii) holds by the Lemma in Chapter II, Section 8, of [11]. □

Proposition 17.14. *Let K be a field, let E/K be a finite separable extension of degree n with norm N, let ν be a discrete valuation of K and suppose that K is complete with respect ν. Then the following hold:*

(i) $\nu(N(E^*)) = f\mathbb{Z}$ *for some positive integer f and the map $\omega\colon E^* \to \mathbb{Z}$ given by*

$$\omega(u) = \nu(N(u))/f$$

for all $u \in E^*$ is the unique valuation of E.

(ii) *E is complete with respect to ω.*

(iii) *f is the degree of the extension $\bar{E}/\bar{K}$, where $\bar{E}$ is the residue field of E with respect to ω.*

Proof. By the Theorem in Chapter II, Section 10, of [11], (i) holds, (ii) follows by 17.13.ii and (iii) holds by Proposition 3 in Chapter I, Section 5, of [11]. □

Chapter Eighteen

Automorphisms and Residues

With this chapter we again postpone attacking the problem described at the conclusion of Chapter 16, this time in order to examine more closely the structure of the residues of a Bruhat-Tits building. In particular, we prove various results about the automorphism group of a Bruhat-Tits pair (see 18.4–18.8), use them to show (in 18.18 and 18.19) that the proper irreducible residues of rank at least 2 of a Bruhat-Tits building satisfy the Moufang condition and produce a general method for determining the structure of the gems of a Bruhat-Tits building in 18.25.

In 18.30 and 18.31, we apply these results to determine the structure of the proper irreducible residues in the case that the Coxeter diagram is simply laced. The remaining cases will be examined in Chapters 19–25.

The main results about automorphisms in this chapter are 18.6 and 18.15. See also 26.39.

Throughout this chapter, we adopt the following hypotheses and notation:

Notation 18.1. Let $(\Delta, \mathcal{A})$ be a Bruhat-Tits pair of type $\tilde{\mathsf{X}}_\ell$ (so $\tilde{\mathsf{X}}_\ell$ is one of the Coxeter diagrams in Figure 1.1 with $\ell > 2$). Let $A \in \mathcal{A}$, let $\Sigma = A^\infty$, let Ξ denote the set of roots of Σ, let Φ be the root system X_ℓ and let A be identified with $\tilde{\Sigma}_\Phi$ as in 13.13. Let $(G^\dagger, \xi)$ be the root datum of $\Delta_{\mathcal{A}}^\infty$ based at Σ. Let R_0 denote the gem cut by A such that the center of $R_0 \cap A$ is the origin and let $\phi = \phi_{R_0}$ be as in 13.8. Thus by 13.30.i, ϕ is a valuation of $(G^\dagger, \xi)$. Let H denote the pointwise stabilizer of Σ in $\mathrm{Aut}(\Delta_{\mathcal{A}}^\infty)$. By 37.8 of [36],

$$\mathrm{Aut}^\circ(\Delta_{\mathcal{A}}^\infty) = H \cdot G^\dagger, \tag{18.2}$$

where $\mathrm{Aut}^\circ(\Delta_{\mathcal{A}}^\infty)$ denotes the group of type-preserving (or "special") automorphisms of $\Delta_{\mathcal{A}}^\infty$ (as defined in 29.2).[1] Let G denote the subgroup $\mathrm{Aut}(\Delta, \mathcal{A})$ of $\mathrm{Aut}(\Delta)$ defined in 12.30 and let κ be the injective homomorphism from G to $\mathrm{Aut}(\Delta_{\mathcal{A}}^\infty)$ that maps each element of G to the automorphism of $\Delta_{\mathcal{A}}^\infty$ it induces. (This is the map $\rho \mapsto \rho^*$ in 12.30.) By 12.31, $G^\dagger \subset \kappa(G)$; we continue, in fact, to identify $G^\dagger$ with its pre-image under κ (as we have been doing since Chapter 8). The group H, on the other hand, is not always contained in $\kappa(G)$. We set

$$G_A = \{g \in G \mid \kappa(g) \in H\}$$

and

$$G^\circ = G_A \cdot G^\dagger.$$

[1] By 11.36 in [37], the subgroup $G^\dagger$ is normal in $\mathrm{Aut}(\Delta_{\mathcal{A}}^\infty)$.

By 8.27, G_A consists of those elements of G that map A to itself and act trivially on $\Sigma = A^\infty$. By 18.2, G° is the pre-image of $\text{Aut}^\circ(\Delta_A^\infty)$ in G. Thus the subgroup G° consists of those elements of G that induce type-preserving automorphisms of Δ_A^∞. The elements of G° are not necessarily type-preserving automorphisms of Δ, however, as we will see, for example, in 18.15.

We begin this chapter with a number of results about the group G_A and its image in H.

Proposition 18.3. *Let T_A be the group of all translations of A and let W_A be the group induced on A by all the affine reflections of A. Then the following hold:*

(i) *The group G_A induces a subgroup of T_A on A.*
(ii) *The stabilizer of A in $G^\dagger$ induces a group containing W_A on A.*

Proof. The elements of G_A act trivially on the set of parallel classes of roots of A (by 8.30.ii). By 1.8, therefore, (i) holds. By 13.5 and 13.28, every affine reflection of A is induced by some element of the form $m_\Sigma(u)$. Thus (ii) holds. □

Proposition 18.4. *The elements of $G^\dagger$ are type-preserving automorphisms of Δ and the group $G^\dagger$ acts transitively on the set*

$$\{(A, d) \mid A \in \mathcal{A} \text{ and } d \in A\}.$$

Proof. By 13.2, the elements of $G^\dagger$ are type-preserving automorphisms of Δ. By 29.15.iii, $G^\dagger$ acts transitively on the set of apartments of Δ_A^∞. Then 8.27 implies that $G^\dagger$ acts transitively on $\mathcal{A}$. By 18.3.ii, the stabilizer of A in $G^\dagger$ acts transitively on the set of chambers in A. □

Corollary 18.5. *The group $G^\dagger$ acts transitively on the set of gems of a given type. In particular, $G^\dagger$ acts transitively on the set of all gems if the Coxeter diagram of Δ has only one special vertex.*

Proof. By 18.4, the group $G^\dagger$ acts transitively on the set of chambers of Δ and preserves types. For each chamber d and each subset J of the vertex set of the Coxeter diagram of Δ, there is a unique J-residue containing d. □

By 18.1, H is the pointwise stabilizer of the apartment Σ. Thus, in particular, the group H maps every root of Σ to itself and hence normalizes the root group U_a for all $a \in \Xi$.

The following result gives a necessary and sufficient condition for an element of H to extend to an element of G_A. We will use this result in 26.29–26.37 (when the classification of Bruhat-Tits pairs has been completed) to show that most elements of H (in a suitable sense) are, in fact, induced by elements of G_A.

Theorem 18.6. *Let $\tau \in H$. Then the following are equivalent:*

(i) *There exists an element in G_A that induces τ on Δ_A^∞.*

(ii) *For some* $a \in \Xi$, *the quantity* $\phi_a(u^\tau) - \phi_a(u)$ *is independent of the choice of the element* $u \in U_a^*$.

(iii) *For all* $a \in \Xi$, *the quantity* $\phi_a(u^\tau) - \phi_a(u)$ *is independent of the choice of the element* $u \in U_a^*$.

Proof. Let $\psi_b(u) = \phi_b(u^\tau)$ for all $b \in \Xi$ and all $u \in U_b^*$ and let

$$\psi = \{\psi_b \mid b \in \Xi\}.$$

By 13.31.i, ψ is a valuation of $(G^\dagger, \xi)$, and by 13.31.ii, ψ is equipollent to ϕ if and only if there exists an element in G_A that induces τ on $\Delta_{\mathcal{A}}^\infty$. If ψ is equipollent to ϕ, then for all $a \in \Xi$, the quantity $\phi_a(u^\tau) - \phi_a(u)$ is independent of the choice of the element $u \in U_a^*$ (by 3.22). By 3.45, on the other hand, ψ is equipollent to ϕ if for *some* $a \in \Xi$, the quantity

$$\phi_a(u^\tau) - \phi_a(u)$$

is independent of the choice of the element $u \in U_a^*$. □

Proposition 18.7. *Let* $\rho \in G_A$ *and let* τ *be the element of* H *induced by* ρ. *Then*

$$\mathrm{dist}(\alpha, \alpha^\rho) = \phi_a(u^\tau) - \phi_a(u)$$

for each root α *of* A *and each* $u \in U_a^*$, *where* $a = \alpha^\infty$.

Proof. Let $\psi_b(u) = \phi_b(u^\tau)$ for all $b \in \Xi$ and all $u \in U_b^*$ and let

$$\psi = \{\psi_b \mid b \in \Xi\}.$$

By 13.31.i, ψ is a valuation of $(G^\dagger, \xi)$. Let α be a root of A, let α_0 be the unique root of A parallel to α that cuts the gem R_0 (where R_0 is as in 18.1) and let $a = \alpha^\infty$. By 13.5, we can choose $u \in U_a^*$ such that $a_u = \alpha$, where a_u is as in 13.2. Let $v = u^\tau$. By 13.2, $a_v = a_u^\rho = \alpha^\rho$. By 13.8,

$$\phi_a(u) = \mathrm{dist}(\alpha_0, a_u)$$

and

$$\phi_a(v) = \mathrm{dist}(\alpha_0, a_v).$$

Thus

$$\begin{aligned} \mathrm{dist}(\alpha, \alpha^\rho) &= \mathrm{dist}(a_u, a_v) \\ &= \mathrm{dist}(\alpha_0, a_v) - \mathrm{dist}(\alpha_0, a_u) \\ &= \phi_a(v) - \phi_a(u) \\ &= \psi_a(u) - \phi_a(u). \end{aligned}$$

By 13.31.ii, ψ and ϕ are equipollent. By 3.23, therefore,

$$\psi_a(u) - \phi_a(u) = \psi_a(w) - \phi_a(w)$$

for all $w \in U_a^*$. Hence

$$\mathrm{dist}(\alpha, \alpha^\rho) = \phi_a(u^\tau) - \phi_a(u)$$

for *all* $u \in U_a^*$. □

Proposition 18.8. *Let d be a chamber of Σ and let Ξ_d° be as in 3.5. Let ϕ and R_0 be as in 18.1, let S be the unique sector of A with terminus R_0 such that $S^\infty = d$ and let $x \in R_0$ be the apex of S, so the map*

$$\alpha \mapsto \alpha^\infty$$

is a bijection from the set $[R_0, x]_A$ (as defined in 7.5) to the set Ξ_d°. Let $a \in \Xi_d^\circ$, let ρ be an element of G_A inducing an element τ of H on $\Delta_{\mathcal{A}}^\infty$, let α be the unique root of A cutting R_0 such that $a = \alpha^\infty$ and let

$$m := \phi_a(u^\tau) - \phi_a(u)$$

for some $u \in U_a^*$. Suppose that

$$\phi_b(u^\tau) = \phi_b(u)$$

for all $b \in \Xi_d^\circ$ different from a and all $u \in U_b^*$. Then ρ induces on the apartment A the m-th power of a generator of the subgroup $T_{R_0,x,\alpha} \cong \mathbb{Z}$ of T_A defined in 1.24.

Proof. This holds by 18.7. □

Remark 18.9. To apply these results, we will want to be able to produce elements in H. Since $\Delta_{\mathcal{A}}^\infty$ is Moufang, it is isomorphic to one of the buildings $\mathsf{X}_\ell(\Lambda)$ defined in 30.15. In other words, we can assume that for some chamber d of Σ, the triple $(\Delta_{\mathcal{A}}^\infty, \Sigma, d)$ is the canonical realization of the corresponding pair (Π, ζ) described in 30.14, where $\Pi = \mathsf{X}_\ell$. Suppose, too, that for each edge $e = \{x, y\}$ of Π, π_e is an automorphism of the root group sequence $\zeta(x, y)$ (as defined in 30.3) such that whenever two edges $e = \{x, y\}$ and $f = \{x, z\}$ have a vertex x in common, in which case the first terms of $\zeta(x, y)$ and $\zeta(x, z)$ are the same (by 30.12.iii), the restrictions of π_e and π_f to this first term agree. Then by 3.6 with $\Delta_{\mathcal{A}}^\infty$ in place of both Δ and $\hat{\Delta}$, Σ in place of $\hat{\Sigma}$ and $\sigma = 1$, there exists a unique element τ in H that induces π_e on $\zeta(x, y)$ for each edge $e = \{x, y\}$ of Π.

Before moving on to consider residues, we apply our results about the subgroups H and G_A in the simply-laced case. For this purpose, we first need a result about octonion division algebras with valuation:

Proposition 18.10. *Let K be an octonion division algebra with center F and norm N, let ν be a valuation of K, let X be the group defined in 37.9 of [36][2] and let $w \in K^*$. Then there exists an F-linear element λ of X such that the following hold:*

(i) *$\lambda(1) = w$ and $N(\lambda(u)) = N(u)N(w)$ for all $u \in K$.*
(ii) *$\nu(\lambda(u)) = \nu(u) + \nu(w)$ for all $u \in K^*$.*

[2]Thus X is the set of additive automorphisms ψ of K that are linear or semi-linear over F such that

$$\psi(uv) = \psi(u)\psi(1)^{-1} \cdot \psi(v)$$

for all $u, v \in K$. It is shown in 37.17 of [36] that this set is, in fact, a group with multiplication given by composition of functions.

Proof. Let ψ be the map called ϕ in 20.24 of [36] (which depends on w). Thus ψ is an automorphism of K fixing the identity 1 which is linear over F such that

$$\psi(uv) = \big(\psi(u) \cdot \psi(v)w\big)w^{-1} \tag{18.11}$$

and

$$N(\psi(u)) = N(u) \tag{18.12}$$

for all $u, v \in K$. By 18.12 and Corollary 4 in [22], $\nu(\psi(u)) = \nu(u)$ for all $u \in K^*$. Let $\lambda(u) = \psi(u)w$ for all $u \in K$. Then (i) and (ii) hold. It remains only to show that $\lambda \in X$.

By 9.1.i in [36], $uw \cdot w^{-1} = u$ and $uw^{-1} \cdot w = u$ for all $u \in K$. By 18.11, therefore,

$$\begin{aligned}\lambda(uv) &= \psi(uv)w \\ &= \psi(u) \cdot \psi(v)w \\ &= \lambda(u)w^{-1} \cdot \lambda(v)\end{aligned}$$

for all $u, v \in K$. Thus $\lambda \in X$. □

Proposition 18.13. *Suppose that for* $\mathsf{X} = \mathsf{A}$, D *or* E, *for some field K and for some discrete valuation ν of K,*

$$(\Delta, \mathcal{A}) \cong \tilde{\mathsf{X}}_\ell(K, \nu)$$

and ϕ are as in 16.17. Let $a \in \Xi_d^\circ$, where Ξ_d° is as in 3.5. Then there exists an element $\tau \in H$ such that

$$\phi_a(x_a(t)^\tau) = \phi_a(x_a(t)) + 1$$

for all $t \in K^*$ and

$$\phi_b(x_b(t)^\tau) = \phi_b(x_b(t))$$

for all $b \in \Xi_d^\circ$ different from a and for all $t \in K$.

Proof. Let $\pi \in K$ be a uniformizer (as defined in 9.17). Suppose first that K is an octonion division algebra, in which case $\mathsf{X}_\ell = \mathsf{A}_2$ by 30.14. Let $\mathcal{T}(K)$ be as in 30.8.i. For each $u \in K^*$ and for each $\lambda \in X$, where X is as in 18.10, there exists (by 37.12 in [36]) an automorphism σ of the root group sequence $\mathcal{T}(K)$ such that

$$x_1(t)^\sigma = x_1(u \cdot \lambda(t)\lambda(1)^{-1})$$

and

$$x_4(t)^\sigma = x_4(\lambda(t)u)$$

for all $t \in K$. Now either let $u = 1$ and let λ be as in 18.10 with π in place of w or let $u = \pi$ and let λ be as in 18.10 with π^{-1} in place of w. By 18.9 and 18.10.ii, we conclude that there exists $\tau \in H$ such that for all $t \in K^*$,

$$\phi_a(x_a(t)^\tau) = \phi_a(x_a(t)) + 1$$

and

$$\phi_b(x_b(t)^\tau) = \phi_b(x_b(t)),$$

where b is the other root in Ξ_d°.

Suppose next that K is associative (i.e. a field or a skew field). If K is commutative, then by 18.9, there is an element τ in H that centralizes U_b for all $b \in \Xi_d^\circ$ distinct from a such that

$$x_a(t)^\tau = x_a(t\pi)$$

for all $t \in K$.[3] Suppose, instead, that K is non-commutative. This means that we are in case 30.14.i. We say that a vertex y of X_ℓ is *to the right* of a vertex x if there is a sequence of positively oriented directed edges (as defined in 30.11) going from x to y. Let λ be the inverse of the bijection $x \mapsto a_x$ from the vertex set of X_ℓ to Ξ_d° described in 3.5. We say that a root $c \in \Xi_d^\circ$ is *to the right* of a root $b \in \Xi_d$ if $\lambda(c)$ is to the right of $\lambda(b)$. Again by 18.9, there is an element τ in H such that for all $t \in K$ and all $b \in \Xi_d^\circ$,

$$x_b(t)^\tau = x_b(t)$$

if b is to the left of a,

$$x_b(t)^\tau = x_b(\pi^{-1}t\pi)$$

if b is to the right of a and

$$x_a(t)^\tau = x_a(t\pi).^4$$

Thus (by 9.17.i) the claim holds whether or not K is commutative. □

Theorem 18.14. *If $\tilde{\mathsf{X}}_\ell$ is simply laced, then the group induced by G_A on A is precisely the group T_A of all translations on A.*

Proof. Let X be the group induced by G_A on A. By 18.3.i, $X \subset T_A$. If τ is as in 18.13, then by 18.6, there is an element $\rho \in G_A$ that induces τ on Δ_A^∞. Thus $T_A \subset X$ by 1.24 and 18.8. □

Corollary 18.15. *If $\tilde{\mathsf{X}}_\ell$ is simply laced, then the subgroup G° of G defined in 18.1 acts transitively on the set of gems of Δ.*

Proof. By 18.4, $G^\dagger$ acts transitively on $\mathcal{A}$. The claim holds, therefore, by 1.9 and 18.14. □

---

[3] Here are the details. Let $\mathcal{T}(K) = (U_+, U_1, U_2, U_3)$ be as in 30.8.i. Then the maps $x_i(t) \mapsto x_1(\pi t)$ for $i = 1$ and 2 and $x_3(t) \mapsto x_3(t)$ extend to an automorphism of U_+ as do the maps $x_1(t) \mapsto x_1(t)$ and $x_i(t) \mapsto x_i(t\pi)$ for $i = 2$ and 3 (by 30.4). Furthermore, $t\pi = \pi t$ for all $t \in K$. The automorphism τ exists, therefore, by 18.9 and the relevant case of 30.14.

[4] Here are the details. Let $\mathcal{T}(K) = (U_+, U_1, U_2, U_3)$ be as in 30.8.i. Then the maps $x_1(t) \mapsto x_1(t)$ and $x_i(t) \mapsto x_i(t\pi)$ for $i = 2$ and 3 extend to an automorphism of U_+ as do the maps $x_i(t) \mapsto x_i(t\pi)$ for $i = 1$ and 2 and $x_3(t) \mapsto x_3(\pi^{-1}t\pi)$ (by 30.4). The automorphism τ exists, therefore, by 18.9 and the relevant case of 30.14.

Remark 18.16. Let $\Pi = \tilde{\mathsf{X}}_\ell$. There are canonical homomorphisms from $\mathrm{Aut}(\Delta)$ to $\mathrm{Aut}(\Pi)$ and from $\mathrm{Aut}(A)$ to $\mathrm{Aut}(\Pi)$ that both send an arbitrary σ-automorphism (as defined in 29.2) to σ. Suppose that Π is simply laced, let M be the image of G_A under the first of these homomorphisms and let N be the image of T_A under the second. By 18.14, $M = N$. By 1.28, therefore, M acts sharply transitively on the set of special vertices of Π, and by 1.29, the group M is, in fact, cyclic except when $\Pi = \tilde{\mathsf{D}}_\ell$ for ℓ even.

We now turn to an examination of the structure of the residues of Δ. By 1.7 and 29.10.iii, we know that every proper residue is a spherical building.

To begin, we focus on the gems of Δ. The gems are (like the building at infinity $\Delta_{\mathcal{A}}^\infty$) buildings of type X_ℓ.

Proposition 18.17. *Let R be a gem cut by A, let α be a root of A cutting R, let $a = \alpha^\infty$ and let $\beta = \alpha \cap R$. Let $U_{a,R}$ be as in 13.3 and let $\bar{U}_{a,R}$ be the subgroup of* $\mathrm{Aut}(R)$ *induced by $U_{a,R}$. Then the following hold:*

- (i) *The intersection $A \cap R$ is an apartment of R and β is a root of this apartment.*
- (ii) *The group $U_{a,R}$ acts transitively on the set of apartments of R containing the root β.*
- (iii) *The group $\bar{U}_{a,R}$ is the root group of β in* $\mathrm{Aut}(R)$.

Proof. The assertions in (i) hold by 29.13.iv. Let U_β be the root group in $\mathrm{Aut}(R)$ corresponding to the root β of R (as defined in 29.15). We claim that $\bar{U}_{a,R} \subset U_\beta$. Let $g \in U_{a,R}$. Let P be a panel contained in R that contains two chambers x and y of β and let z be a third chamber in P. Let α_1 be the unique root of A that contains y but not x. By 10.28, there exists an apartment $A' \in \mathcal{A}$ containing α_1 and z. Let $S_1 = \sigma_A(R, x)$, $S_2 = \sigma_A(R, y)$ and $S_3 = \sigma_{A'}(R, z)$. These three sectors share the face $\mu_A(R, P)$ (as defined in 8.1). Thus S_1^∞, S_2^∞ and S_3^∞ are three chambers of $\Delta_{\mathcal{A}}^\infty$ contained in a single panel F. By 4.14, α contains both S_1 and S_2. Thus F is a panel containing two chambers S_1^∞ and S_2^∞ in $a = \alpha^\infty$. Hence the root group U_a acts trivially on F. Therefore g maps S_3^∞ to itself. Since g maps R to itself, the sector S_3^g has terminus R. By 10.17, however, S_3 is the unique sector of Δ with terminus R that is parallel to S_3. Therefore $S_3^g = S_3$. Hence g fixes the apex of S_3, i.e. g fixes z. Thus g acts trivially on P. We conclude that $U_{a,R}$ acts trivially on every panel of R containing two chambers of β. Hence $\bar{U}_{a,R} \subset U_\beta$ as claimed.

By 13.5 and 29.14.iii, the group $U_{a,R}$ acts transitively on the set of roots of R containing β. Therefore (ii) holds and, by the conclusion of the previous paragraph, R is Moufang. By 29.15.v, therefore, the root group U_β acts sharply transitively on the set of apartments of R containing β. Thus (iii) holds. □

Theorem 18.18. *The gems of Δ satisfy the Moufang condition (as defined in 29.15).*

Proof. This holds by 18.17. □

Corollary 18.19. *The proper irreducible residues of rank at least 2 of Δ all satisfy the Moufang condition.*

Proof. Suppose that $\tilde{\mathsf{X}}_\ell \neq \tilde{\mathsf{G}}_2$. Inspecting Figure 1.1, we see that every proper connected subdiagram of $\tilde{\mathsf{X}}_\ell$ that contains every special vertex is contained in a proper connected subdiagram with at least three vertices. Thus every proper irreducible residue is contained in a gem or in a proper irreducible residue of rank at least 3 or in both. By 18.18 and 29.15.i, gems and proper irreducible residues of rank at least 3 are Moufang. The claim holds, therefore, by 29.15.ii. The assertion will be proved for the case $\tilde{\mathsf{X}}_\ell = \tilde{\mathsf{G}}_2$ in 25.33. □

If $\tilde{\mathsf{X}}_\ell$ contains proper connected subdiagrams that contain every special vertex, then $\tilde{\mathsf{X}}_\ell$ is $\tilde{B}_\ell$ (for $\ell \geq 3$), $\tilde{\mathsf{E}}_k$ for $k = 7$ or 8, $\tilde{\mathsf{F}}_4$ or $\tilde{\mathsf{G}}_2$. We will determine the structure of the maximal proper irreducible residues that are not contained in a gem (but which are nevertheless Moufang by 18.19) in 18.31 when $\tilde{\mathsf{X}}_\ell = \tilde{\mathsf{E}}_7$ or $\tilde{\mathsf{E}}_8$, in 24.67 when $\tilde{\mathsf{X}}_\ell = \tilde{\mathsf{B}}_\ell$ for $\ell \geq 3$, in 25.33 when $\tilde{\mathsf{X}}_\ell = \tilde{\mathsf{G}}_2$ and in 26.17 when $\tilde{\mathsf{X}}_\ell = \tilde{\mathsf{F}}_4$.

Proposition 18.20. *Let $a \in \Sigma$, let α_0 be the unique root of A cutting $R_0 \cap A$ such that $\alpha_0^\infty = a$ (i.e. the root of A called a_{R_0} in 13.8), let $k \in \mathbb{Z}$, let α be the unique root of A parallel to α_0 such that* $\mathrm{dist}(\alpha_0, \alpha) = k$ *and let X be the set of gems cut by α. Then $X \neq \emptyset$ and for each gem R in X, $U_{a,R} = U_{a,k}$ (where $U_{a,k}$ is as in 3.21), the integer k is uniquely determined by this equation, the subgroup $U_{a,k+1}$ is normal in $U_{a,k}$ and the group $\bar{U}_{a,R}$ defined in 18.17 is isomorphic to the quotient $U_{a,k}/U_{a,k+1}$.*

Proof. By 1.16, $X \neq \emptyset$. Let $R \in X$. By 3.21 and 13.8, we have

$$\begin{aligned} U_{a,k}^* &= \{u \in U_a^* \mid \phi_a(u) \geq k\} \\ &= \{u \in U_a^* \mid \mathrm{dist}(\alpha_0, a_u) \geq k\} \\ &= \{u \in U_a^* \mid \alpha \subset a_u\}, \end{aligned}$$

where a_u is as in 13.2. By 1.13, $\beta \cap R = \emptyset$ for all roots β properly contained in α and $R \cap A \subset \beta$ for all roots β that contain α properly. Therefore $U_{a,k} = U_{a,R}$ by 13.4 and $u \in U_{a,k}$ acts trivially on $R \cap A$ if and only if $u \in U_{a,k+1}$ by 13.2. By 18.17.iii and 29.15.v, an element of $U_{a,k}$ acts trivially on $R \cap A$ if and only if it acts trivially on R. Thus $U_{a,k+1}$ is the kernel of the action of $U_{a,k}$ on R. □

Notation 18.21. For each $a \in \Xi$ and each $k \in \mathbb{Z}$, let $\bar{U}_{a,k}$ denote the quotient group $U_{a,k}/U_{a,k+1}$.

Proposition 18.22. *Let R be a gem cut by A, let v be the center of the residue $R \cap A$ of A as defined in 2.41 and let $a \in \Xi$. Then the following hold:*

(i) $U_{a,R} = U_{a,-v\cdot a}$.

(ii) $\bar{U}_{a,R} = \bar{U}_{a,-v\cdot a}$, *where* $\bar{U}_{a,R}$ *is as defined in 18.17 and* $\bar{U}_{a,-v\cdot a}$ *is as defined in 18.21.*

Proof. Let $u \in U_a^*$. By 13.4, $u \in U_{a,R}^*$ if and only if v is contained in the closure of a_u (as a subset of the Euclidean space A). By 13.18.ii, $a_u = K_{a,-\phi_a(u)}$ and thus $u \in U_{a,R}^*$ if and only if $v \cdot a \geq -\phi_a(u)$. Thus (i) holds. By 18.20, therefore, (ii) holds. □

Proposition 18.23. *Let R be a gem cut by A and let $\bar{a} = R \cap a_R$ for each $a \in \Xi$, where a_R is as in 13.8. Then the following hold:*

(i) *The map $a \mapsto \bar{a}$ is a bijection from Ξ to the set of roots of the apartment $A \cap R$ of R that preserves intervals.*
(ii) *If a, b are roots in Ξ such that $a \neq \pm b$, then*
$$[\bar{U}_{a,R}, \bar{U}_{b,R}] \subset \prod_{c\in(a,b)} \bar{U}_{c,R}.$$

Proof. By 18.17.i, $A \cap R$ is an apartment of R and $\bar{a}$ is a root of this apartment for each $a \in \Xi$. Let π be the isomorphism from Σ to $A \cap R$ that maps $\sigma_A(R,d)^\infty$ to d for all $d \in A \cap R$. Then $\pi(a) = \bar{a}$ for all $a \in \Xi$ (by 4.12). Since $X \subset Y \subset \Sigma$ if and only if $\pi(X) \subset \pi(Y) \subset A \cap R$, it follows (by 3.2.i) that (i) holds. In other words,
$$c \in [a,b] \text{ if and only if } \bar{c} \in [\bar{a}, \bar{b}]$$
for all $a, b, c \in \Xi$. By 3.2.iii and 18.17.iii, therefore, (ii) holds. □

Remark 18.24. Let o be a special vertex of the Coxeter diagram $\Pi := \tilde{\mathsf{X}}_\ell$, let R be an o-special gem of Δ cut by A, let Π_o be as in 8.21, let d be a chamber of Σ, let S be the unique sector in A whose terminus is R such that $S^\infty = d$ and let u be the apex of S. By 8.12, we can assume without loss of generality that o is the special vertex used to construct $\Delta_{\mathcal{A}}^\infty$ in 8.9. By 8.25, Π_o is then the Coxeter diagram of $\Delta_{\mathcal{A}}^\infty$.[5] Now let x be a vertex of Π_o and, as usual, let a_x be the unique root of Σ containing d but not the unique chamber of Σ that is x-adjacent to d. Then the map $a \mapsto \bar{a}$ defined in 18.23 maps a_x to the unique root of $A \cap R$ containing u but not the unique chamber of $A \cap R$ that is x-adjacent to u.

Theorem 18.25. *Let o be a special vertex of the Coxeter diagram $\Pi := \tilde{\mathsf{X}}_\ell$, let R be an o-special gem of Δ cut by A, let the subdiagram Π_o (as defined in 8.21) be identified with the Coxeter diagram of $\Delta_{\mathcal{A}}^\infty$ as explained in 18.24, let v be the center of the gem $A \cap R$ of A as defined in 2.41, let d be a chamber of Σ and let Ξ_d and the map $x \mapsto a_x$ be as in 3.5. Then R is uniquely determined by the groups $\bar{U}_{a,-a\cdot v}$ for all $a \in \Xi_d$ and the commutator relations*

$$[\bar{U}_{a,-a\cdot v}, \bar{U}_{b,-b\cdot v}] \subset \prod_{c\in(a,b)} \bar{U}_{c,-c\cdot v} \tag{18.26}$$

[5]In other words, for every special vertex o of Π, the subdiagram Π_o of Π can be identified with the Coxeter diagram of $\Delta_{\mathcal{A}}^\infty$ simply by adjusting the coloring of the panels of $\Delta_{\mathcal{A}}^\infty$.

for all distinct $a, b \in \Xi_d$ *contained in* $[a_x, a_y]$ *for some edge* $\{x, y\}$ *of* Π_o. *The "name" of the isomorphism type of* R *can be obtained by comparing these data with 30.14 and 30.15.*

Proof. Let u be as in 18.24, let (x, y) be a directed edge of Π_o and let Ω_{xy} be the root group sequence defined in 30.1 with R in place of Δ, $A \cap R$ in place of Σ and u in place of d. (This makes sense since Π_o is also the Coxeter diagram of R.) By 30.4, every root group sequence $(U_+, U_1, \dots, U_n)$ is uniquely determined by the ordered set of groups $U_1, \dots, U_n$ and all the "commutator relations" expressing an element $[u, v]$ such that $u \in U_i$ and $v \in U_j$ for some $i, j \in [1, n]$ such that $1 \leq i < j \leq n$ as an element of $U_{[i+1,j-1]}$, where $U_{[i+1,j-1]}$ is the subgroup defined in 5.1 in [36]. By 18.22.ii, 18.23 and 18.24, therefore, Ω_{xy} is uniquely determined by the commutator relations in 18.26. Thus the claim holds by 3.6. □

Corollary 18.27. *Let* R_0 *be the gem introduced in 18.1, let* d *be a chamber of* Σ, *let* Ξ_d *and the map* $x \mapsto a_x$ *be as in 3.5, let* o *be the unique special vertex of* $\Pi = \tilde{\mathsf{X}}_\ell$ *such that* R_0 *is* o*-special and let the subdiagram* Π_o *be identified with the Coxeter diagram of* $\Delta_{\mathcal{A}}^\infty$ *as explained in 18.24. Then the gem* R_0 *is uniquely determined by the groups* $\bar{U}_{a,0}$ *for all* $a \in \Xi_d$ *and the commutator relations*

$$[\bar{U}_{a,0}, \bar{U}_{b,0}] \subset \prod_{c \in (a,b)} \bar{U}_{c,0}$$

for all distinct $a, b \in \Xi_d$ *contained in* $[a_x, a_y]$ *for some edge* $\{x, y\}$ *of* Π_o.

Proof. This is simply a special case of 18.25. □

Proposition 18.28. *Suppose that* $\mathsf{X} = \mathsf{B}$ *or* C. *Let* d *be a chamber of* Σ *and let* Ξ_d° *be as in 3.5. Let* b *be the unique long root in* Ξ_d° *if* $\mathsf{X} = \mathsf{C}$ *or* $\ell = 2$ *(which exists by 2.14.iv and 2.14.v). If* $\mathsf{X} = \mathsf{B}$ *and* $\ell \geq 3$, *let* $b = a_x$, *where* x *is the vertex at the left end of the Coxeter diagram* B_ℓ *as it is drawn in Figure 1.3 and* a_x *is as in 3.5. Then the following hold:*

(i) *There exists a unique gem* R_1 *cut by* A *such that* $U_{b,R_1} = U_{b,1}$ *and* $U_{a,R_1} = U_{a,0}$ *for all* $a \in \Xi_d^\circ$ *different from* b.
(ii) *The type of* R_1 *is different from the type of* R_0.

Proof. Let v be the unique special point in V (as defined in 2.26) such that $v \cdot b = -1$ and $v \cdot a = 0$ for all $a \in \Xi_d^\circ$ other than b and let R_1 be the unique gem of Δ such that v is the center of $R_1 \cap A$. Assertion (i) holds by 18.22.i and assertion (ii) holds by 2.49. □

Remark 18.29. Suppose that $\mathsf{X} = \mathsf{B}$ or C and let R_1 be as in 18.28.ii. Then the Coxeter diagram $\tilde{\mathsf{X}}_\ell$ has exactly two special vertices. By 18.5, therefore, there are at most two G°-orbits of gems in Δ (where G° is as in 18.1). By 18.28.iii, it follows that every gem of Δ is in the same G°-orbit as R_0 or R_1 (or possibly both). It will turn out that there is only one G°-orbit of gems when $\mathsf{X} = \mathsf{B}$ and $\ell \geq 3$; see 26.39.

We now apply our results to Bruhat-Tits buildings that are simply laced.

Theorem 18.30. *Suppose that* $\tilde{\mathsf{X}}_\ell$ *is simply laced, so that*

$$(\Delta, \mathcal{A}) \cong \tilde{\mathsf{X}}_\ell(K, \nu)$$

for some K *and some* ν *as described in 16.18. Let* $\bar{K}$ *be the residue field of* K *with respect to* ν *(as defined in 9.22). Then every gem of* Δ *is isomorphic to the spherical building* $\mathsf{X}_\ell(\bar{K})$ *(in the notation of 30.15).*

Proof. By 18.15, every gem is isomorphic to R_0. By 18.27, we have $R_0 \cong \mathsf{X}_\ell(\bar{K})$. □

Proposition 18.31. *Let*

$$(\Delta, \mathcal{A}) \cong \tilde{\mathsf{E}}_\ell(K, \nu)$$

for $\ell = 7$ *or* 8, *for some field* K *and some valuation* ν *of* K *and let* R *be a maximal proper irreducible residue of* Δ *that is not contained in a gem. Then either* $R \cong \mathsf{A}_\ell(\bar{K})$ *or* $\ell = 8$ *and* $R \cong \mathsf{D}_8(\bar{K})$.

Proof. The only maximal connected proper subdiagrams of $\tilde{\mathsf{E}}_\ell$ that contain all special vertices of $\tilde{\mathsf{E}}_\ell$ are (up to isomorphism) A_ℓ and, when $\ell = 8$, D_8. By 30.14, it follows that $R \cong \mathsf{A}_\ell(F)$ or $\mathsf{D}_8(F)$ for some field or skew field F (and, in fact, if $R \cong \mathsf{D}_8(F)$, then F must be commutative). Therefore every irreducible rank 2 residue of R is isomorphic to $\mathsf{A}_2(F)$. By 18.30, on the other hand, every irreducible rank 2 residue of a gem is isomorphic to $\mathsf{A}_2(\bar{K})$. Since R has irreducible rank 2 residues that are also residues of gems, it follows by 35.6 in [36] that $F \cong \bar{K}$. □

Remark 18.32. Suppose that a and b are distinct elements of Ξ_d° such that $n := |[a, b]| \geq 3$ and let R be a gem cut by A. Then R is Moufang (by 18.18) and thus $\bar{U}_{a,R}$ and $\bar{U}_{b,R}$ are the first and last terms U_1 and U_n of a root group sequence

$$\Omega = (U_+, U_1, \ldots, U_n)$$

as defined in 8.16 of [36]. Now suppose that Ω is an *arbitrary* root group sequence, let u_1 and u_n be non-trivial elements of the first term U_1 and the last term U_n of Ω and let $u_i = [u_1, u_n]_i$ for all $i \in [2, n-1]$ be as defined in 15.23 (or 5.10 of [36]). Then $u_i \neq 1$ for all $i \in [2, n-1]$ by repeated application of 6.4 of [36]. These observations explain why every valuation of a root datum restricts only to partial valuations that are exact as defined in 15.24.

Remark 18.33. The residues of a Bruhat-Tits building do not depend on any particular system of apartments. To study residues, it is sometimes useful to assume that the system of apartments is, in fact, complete. It turns out that if $(\Delta, \mathcal{A})$ is complete, then the defining field K of $\Delta_{\mathcal{A}}^\infty$ is also complete. (So far, we have shown this, in 17.11, only for $\tilde{\mathsf{X}}_\ell$ simply laced.) This will allow us to apply, for example, the following result.[6]

---

[6]The proof of 18.34 is based on the proof of Theorem 13.3 in [24]; see also [30].

Proposition 18.34. *Let K be a field that is complete with respect to a valuation ν, let $\bar{K}$ be the residue field of K and let L be a vector space over K of finite dimension. Suppose that Q is an anisotropic form on L of degree m for some positive integer m, i.e. that*

$$Q(tu) = t^m Q(u) \tag{18.35}$$

for all $u \in L$ and $Q(u) = 0$ if and only if $u = 0$. Suppose as well that

$$\nu(Q(u+v)) \geq \min\{\nu(Q(u)), \nu(Q(v))\} \tag{18.36}$$

for all $u, v \in L$. Let

$$L_e = \{u \in L \mid \nu(Q(u)) \geq e\}$$

for all $e \in \mathbb{Z}$. Then the following hold for all $e \in \mathbb{Z}$:

(i) *L_e is an additive subgroup of L.*
(ii) *The map $\bar{t}\bar{v} = \overline{tv}$ is a well-defined scalar multiplication on the quotient L_e/L_{e+m}, where $t \mapsto \bar{t}$ is the natural homomorphism from the ring of integers $\mathcal{O}_K$ to the quotient field $\bar{K}$ and $v \mapsto \bar{v}$ is the natural homomorphism from L_e to L_e/L_{e+m}.*
(iii) *$\dim_K L = \dim_{\bar{K}} L_e/L_{e+m}$.*

Proof. Let $\omega(u) = \nu(Q(u))$ for all $u \in L^*$. By 9.23.i and 18.35, we have $\nu(Q(-u)) = \nu(Q(u))$ for all $u \in L$. By 9.18 and 18.36, therefore, (i) holds and the map ∂_ω from $L \times L$ to K defined in 9.19 (with L in place of U and ω in place of ν) is a metric on L. Assertion (ii) follows from 18.35.

Choose $e \in \mathbb{Z}$ and let $V = L_e/L_{e+m}$. Let

$$B := \{a_i \mid i \in M\}$$

for some index set M be a set of elements of L_e such that $\{\bar{a}_i \mid i \in M\}$ is a basis of V. To prove (iii), it will suffice to show that B is a basis of L over K. The set B is linearly independent over K since otherwise we could choose scalars t_i not all zero such that

$$\sum_{i \in M} t_i a_i = 0$$

and $\min\{\nu(t_i) \mid i \in [1,n]\} = 0$, i.e. such that the images $\bar{t}_i$ in $\bar{K}$ are not all zero. Since L is finite-dimensional, it follows that $|B|$ is finite. We can thus assume that $M = [1,n]$ for some n, i.e. that $B = \{a_1, \ldots, a_n\}$. It will suffice now to show that the set B spans L.

Choose $u \in L^*$ and let p be an element in K^* such that $\nu(p) = 1$. Thus

$$\omega(u) - k_1 m \in [e, e+m-1] \tag{18.37}$$

for some integer k_1. Let $v = p^{-k_1}u$. By 18.35 and 18.37, $v \in L_e$. There thus exist scalars $t_i^{(1)} \in \mathcal{O}_K$ for all $i \in [1,n]$ such that

$$\bar{v} = \bar{t}_1^{(1)}\bar{a}_1 + \cdots + \bar{t}_n^{(1)}\bar{a}_n.$$

In other words,

$$\omega\Big(v - \sum_{i \in [1,n]} t_i^{(1)} a_i\Big) \geq e + m.$$

Thus if

$$u_1 = u - \sum_{i \in [1,n]} p^{k_1} t_i^{(1)} a_i,$$

then $\omega(u_1) \geq k_1 m + (e + m) > \omega(u)$ (by 18.37). If $u_1 = 0$, then u is in the span of B. We can thus assume that $u_1 \neq 0$.

By induction and the conclusion of the previous paragraph, we can assume that there exist elements $k_1, k_2, k_3, \ldots$ of $\mathbb{Z}$, elements $u_1, u_2, u_3, \ldots$ of L^* and elements

$$t_i^{(1)}, t_i^{(2)}, t_i^{(3)}, \ldots$$

of $\mathcal{O}_K$ for all $i \in [1, n]$ such that

$$u - u_s = \sum_{i \in [1,n]} (p^{k_1} t_i^{(1)} + p^{k_2} t_i^{(2)} + \cdots + p^{k_s} t_i^{(s)}) a_i \tag{18.38}$$

as well as

$$\omega(u) < \omega(u_1) < \omega(u_2) < \cdots < \omega(u_s) \tag{18.39}$$

and

$$\omega(u_{s-1}) - m k_s \in [e, e + m - 1] \tag{18.40}$$

for all $s \geq 1$ (where $u_0 = u$). By 18.39 and 18.40,

$$k_1 \leq k_2 \leq k_2 \leq \cdots \tag{18.41}$$

and no value is repeated more than $m - 1$ times in the sequence 18.41. Hence

$$\lim_{s \to \infty} k_s = \infty. \tag{18.42}$$

By 18.39, we also have

$$\lim_{s \to \infty} (u - u_s) = u \tag{18.43}$$

with respect to the metric ∂_ω. For each $i \in [1, n]$ and each $s \geq 1$, let $r_i^{(s)}$ denote the coefficient of a_i on the right-hand side of 18.38. By 18.42, $(r_i^{(s)})_{s \geq 1}$ is a Cauchy sequence (for each $i \in [1, n]$) and hence has a limit in K (since K is complete) which we denote by f_i. Thus

$$\lim_{s \to \infty} r_i^{(s)} a_i = f_i a_i$$

for each $i \in [1, n]$ by 18.35 and hence

$$\lim_{s \to \infty} \Big(\sum_{i \in [1,n]} (p^{k_1} t_i^{(1)} + p^{k_2} t_i^{(2)} + \cdots + p^{k_s} t_i^{(s)}) a_i \Big) = f_1 a_1 + \cdots + f_n a_n$$

by 18.36. Therefore

$$u = f_1 a_1 + \cdots + f_n a_n$$

by 18.38 and 18.43. □

The result 18.34 will be applied in 19.36 with $m = 2$ and in 25.30 with $m = 3$; see also 24.59.

Chapter Nineteen

Quadrangles of Quadratic Form Type

In this chapter, we consider the case $X = \mathcal{Q}_{\mathcal{Q}}$ of the problem framed at the end of Chapter 16. Our main result is 19.18. We then use this result to complete the classification of Bruhat-Tits pairs whose building at infinity is $\mathsf{B}_\ell^{\mathcal{Q}}(\Lambda)$ for some anisotropic quadratic space Λ in 19.23. See 12.2 and 12.4 in [36] for the definition of an anisotropic quadratic space and 30.15 for the definition of the building $\mathsf{B}_\ell^{\mathcal{Q}}(\Lambda)$.

In this and the next seven chapters, we also investigate the completions and residues of each family of Bruhat-Tits pairs under consideration using 17.4 and 17.9 (for completions) and 18.25 (for residues). The main results about completions and residues for the case considered in this chapter are 19.27, 19.28 and 19.35.[1]

Proposition 19.1. *Let (K, L, q) be a quadratic space (as defined in 12.2 of [36]), let f be the associated bilinear form and let E be a field containing K. Let $L_E = L \otimes_K E$. Let f_E denote the unique bilinear (over E) map from $L_E \times L_E$ to E such that*

$$f_E(u \otimes s, v \otimes t) = f(u, v)st \tag{19.2}$$

for all $u, v \in L$ and all $s, t \in E$. Then there exists a unique map q_E from L_E to E such that

$$q_E(\sum_{i=1}^{d} v_i \otimes t_i) = \sum_{i=1}^{d} t_i^2 q(v_i) + \sum_{i<j} t_i t_j f(v_i, v_j) \tag{19.3}$$

for all $v_1, \ldots, v_d \in L$, for all $t_1, \ldots, t_d \in E$ and for all d. Moreover,

$$\Lambda_E := (E, L_E, q_E)$$

is a quadratic space and f_E is the associated bilinear form.[2]

Proof. Choose a basis B of L over K and define q_E using 19.3 for all finite subsets $\{v_1, \ldots, v_d\}$ of B. Since $f(v, v) = 2q(v)$ for all $v \in L$, it follows (by

---

[1] Although the details vary from case to case, each of the next seven chapters follows the general pattern set out in this one. Furthermore, the case considered in this chapter is in many ways the most typical. We have made the assumption, therefore, that the reader more interested in one of the later chapters will nevertheless first have read this one. Consequently, we have tried to include every step of every calculation in this chapter but allowed ourselves to be a bit more brief in subsequent chapters where similar calculations are required.

[2] We will sometimes call q_E and f_E the *natural extension of q and f to L_E*.

a bit of calculation) that Λ_E is a quadratic space and f_E is the associated bilinear form. The identity 19.3 then holds in general by induction with respect to d. □

The following result is a special case of Theorem 10.1.15 of [6] (as is 24.9 below). Antecedents can be found in the paper [30] of T. A. Springer .

Proposition 19.4. *Let (K, L, q) be an anisotropic*[3] *quadratic space, let f be the associated bilinear form on L, let ν be a valuation of K,*[4] *let $\hat{K}$ be the completion of K with respect to ν, let $L_{\hat{K}} = L \otimes_K \hat{K}$, let $f_{\hat{K}}$ and $q_{\hat{K}}$ be as in 19.1 and let*

$$L_{\hat{K}}^{\perp} = \{\hat{u} \in L_{\hat{K}} \mid f_{\hat{K}}(\hat{u}, L) = 0\}.$$

Then the following assertions are equivalent:

(i) $\nu(f(u,v)) \geq \min\{\nu(q(u)), \nu(q(v))\}$ *for all* $u, v \in L^*$.
(ii) $\nu(q(u+v)) \geq \min\{\nu(q(u)), \nu(q(v))\}$ *for all* $u, v \in L^*$.
(iii) $\nu(f(u,v)) \geq \big(\nu(q(u)) + \nu(q(v))\big)/2$ *for all* $u, v \in L^*$.
(iv) $\{\hat{u} \in L_{\hat{K}} \mid q_{\hat{K}}(\hat{u}) = 0\} \subset L_{\hat{K}}^{\perp}$.

Proof. We have

$$q(u+v) = q(u) + q(v) + f(u,v)$$

for all $u, v \in L$. By 9.17.ii, therefore, (i) and (ii) are equivalent. Since the average of two numbers is at least as large as their minimum, (iii) implies (i).

Suppose now that (iii) does *not* hold. There thus exist $u, v \in L$ such that

$$\eta := \nu(q(u)) + \nu(q(v)) - 2\nu(f(u,v)) > 0. \tag{19.5}$$

In particular, $f(u,v) \neq 0$. Let $v_0 = v$ and let

$$\lambda_1 = -f(u,v)^{-1}q(v). \tag{19.6}$$

We have $f(u,u) = 2q(u)$, hence $\nu(f(u,u)) \geq \nu(q(u))$ and thus

$$\begin{aligned} \nu(f(\lambda_1 u, u)) &\geq \nu(\lambda_1) + \nu(q(u)) \\ &= -\nu(f(u,v)) + \nu(q(v)) + \nu(q(u)) \\ &= \eta + \nu(f(u,v)) \end{aligned}$$

since $\nu(f(u,v)^{-1}) = -\nu(f(u,v))$ by 9.17.i. Hence $\nu(f(\lambda_1 u, u)) > \nu(f(u,v))$ since $\eta > 0$. By 9.23.ii, therefore,

$$\nu(f(u, \lambda_1 u + v)) = \nu(f(u,v)). \tag{19.7}$$

By 19.6, we have $f(\lambda_1 u, v) = -q(v)$. Hence

$$\begin{aligned} \nu(q(\lambda_1 u + v)) &= \nu\big(\lambda_1^2 q(u) + q(v) + f(\lambda_1 u, v)\big) \\ &= \nu(\lambda_1^2 q(u)) \\ &= -2\nu(f(u,v)) + 2\nu(q(v)) + \nu(q(u)) \end{aligned}$$

---

[3]A quadratic space (K, L, q) is *anisotropic* if $q(u) = 0$ for some $u \in L$ if and only if $u = 0$.

[4]We observe that the statement and proof of 19.4 require only that ν be real-valued (rather than discrete). The same remark applies also to 24.9 and 25.5.

and thus

$$\nu(q(\lambda_1 u + v)) = \eta + \nu(q(v)). \tag{19.8}$$

Next we define inductively

$$\lambda_k = -f(u, v_{k-1})^{-1} q(v_{k-1}) \tag{19.9}$$

for all $k \geq 2$ and

$$v_k = \lambda_k u + v_{k-1} \tag{19.10}$$

for all $k \geq 1$. To show that λ_k is well defined, we need to show that $f(u, v_k) \neq 0$ for each $k \geq 1$. We will show by induction on k that, in fact,

$$\nu(f(u, v_k)) = \nu(f(u, v)), \tag{19.11}$$

so indeed $f(u, v_k) \neq 0$, as well as

$$\nu(\lambda_k) = (2^{k-1} - 1)\eta + \nu(q(v)) - \nu(f(u, v)) \tag{19.12}$$

and

$$\nu(q(v_k)) = (2^k - 1)\eta + \nu(q(v)) \tag{19.13}$$

for all $k \geq 1$. These claims hold for $k = 1$ by 19.6–19.8. Let $k > 1$ and assume that all three claims hold for smaller values of k. Then

$$\begin{aligned} \nu(\lambda_k) &= -\nu(f(u, v_{k-1})) + \nu(q(v_{k-1})) \\ &= -\nu(f(u, v)) + (2^{k-1} - 1)\eta + \nu(q(v)), \end{aligned} \tag{19.14}$$

hence

$$\begin{aligned} \nu(f(\lambda_k u, u)) &\geq \nu(\lambda_k) + \nu(q(u)) \\ &= -\nu(f(u, v)) + (2^{k-1} - 1)\eta + \nu(q(v)) + \nu(q(u)) \\ &= \nu(f(u, v)) + 2^{k-1}\eta > \nu(f(u, v)), \end{aligned}$$

and so

$$\begin{aligned} \nu(f(u, v_k)) &= \nu\big(f(\lambda_k u, u) + f(u, v)\big) \\ &= \nu(f(u, v)). \end{aligned}$$

We then have

$$q(v_k) = q(\lambda_k u + v_{k-1}) = \lambda_k^2 q(u)$$

since $q(v_{k-1}) = -f(\lambda_k u, v_{k-1})$ by 19.9 and thus

$$\begin{aligned} \nu(q(v_k)) &= 2\nu(\lambda_k) + \nu(q(u)) \\ &= -2\nu(f(u, v)) + 2\big((2^{k-1} - 1)\eta + \nu(q(v))\big) + \nu(q(u)) \\ &= \eta + 2(2^{k-1} - 1)\eta + \nu(q(v)) = (2^k - 1)\eta + \nu(q(v)) \end{aligned}$$

by 19.14. Thus all three claims 19.11–19.13 hold for all $k \geq 1$.

By 19.11 and 19.13, we can choose N such that $\nu(f(u, v_N)) = \nu(f(u, v))$, $\nu(q(v_N)) > \nu(f(u, v))$ and $\nu(q(v_N)) \geq \nu(q(v))$. Thus

$$\nu(q(u)) + \nu(q(v_N)) - 2\nu(f(u, v_N)) \geq \eta > 0.$$

Replacing u by v_N and v by u in the previous paragraph, we conclude that there exists an element u_M in the subspace of L spanned by u and v such that

$$\nu(f(u_M, v_N)) = \nu(f(u, v_N)) = \nu(f(u, v))$$

and

$$\nu(q(u_M)) > \nu(f(u, v)).$$

Thus $\nu(f(u_M, v_N)) < \min\{\nu(q(u_M)), \nu(q(v_N))\}$. Therefore (ii) implies (iii).

We continue to assume that 19.5 holds and keep the previous notation. By 19.10,

$$v_k = (\lambda_k + \lambda_{k-1} + \cdots + \lambda_1)u + v$$

for all $k \geq 1$. By 19.12, there exists $\xi \in \hat{K}$ such that

$$\lim_{k \to \infty} \xi_k = \xi,$$

where

$$\xi_k = \lambda_k + \lambda_{k-1} + \cdots + \lambda_1$$

for each $k \geq 1$. Let $\hat{z} = \xi u + v$. Then

$$\lim_{k \to \infty} f_{\hat{K}}(u, v_k - \hat{z}) = \lim_{k \to \infty} (\xi_k - \xi) f(u, u) = 0.$$

By 19.11, therefore, $f_{\hat{K}}(u, \hat{z}) \neq 0$ and hence $\hat{z} \notin L_{\hat{K}}^{\perp}$. We also have

$$q_{\hat{K}}(\hat{z}) - q(v_k) = (\xi^2 - \xi_k^2)q(u) + (\xi - \xi_k)f(u, v).$$

Therefore

$$\lim_{k \to \infty} q(v_k) = q_{\hat{K}}(\hat{z})$$

in $\hat{K}$. By 19.13, on the other hand, we have

$$\lim_{k \to \infty} q(v_k) = 0.$$

Thus $q_{\hat{K}}(\hat{z}) = 0$. We conclude that also (iv) implies (iii).

It remains only to show that (iii) implies (iv). Suppose that there exists a vector $\hat{u} \in L_{\hat{K}}^*$ such that $q_{\hat{K}}(\hat{u}) = 0$ but $f_{\hat{K}}(\hat{u}, L_{\hat{K}}) \neq 0$. By 19.2, there exists $w \in L$ such that $f_{\hat{K}}(\hat{u}, w) \neq 0$. Let $B := (v_1, v_2, \ldots, v_d)$ be an ordered finite subset of a basis of L such that both $\hat{u}$ and w lie in the subspace of $L_{\hat{K}}$ spanned by the vectors in B. Let $\hat{t}_1, \ldots, \hat{t}_d$ be the coordinates of $\hat{u}$ with respect to B. Choose sequences $(t_{i,k})_{k \geq 1}$ of elements in K such that

$$\lim_{k \to \infty} t_{i,k} = \hat{t}_i$$

for all $i \in [1, d]$ and let

$$u_k = t_{1,k}v_1 + \cdots + t_{d,k}v_d$$

for all $k \geq 1$. By 19.3, we have

$$\lim_{k \to \infty} q(u_k) = q_{\hat{K}}(\hat{u}) = 0$$

and hence

$$\lim_{k\to\infty} \nu(q(u_k)) = \infty.$$

From 19.2 it follows similarly that

$$\lim_{k\to\infty} f(u_k, w) = f_{\hat{K}}(\hat{u}, w) \neq 0.$$

By 9.18.iii, therefore, $\nu(f(u_k, w))$ is a constant for all k sufficiently large. Therefore

$$\nu(f(u_k, w)) < \big(\nu(q(u_k)) + \nu(q(w))\big)/2$$

for all k sufficiently large. Thus (iii) implies (iv). □

Lemma 19.15. *Let $\Lambda = (K, L, q)$ be an anisotropic quadratic space, let $s \in K^*$ and let $\Lambda_s = (K, L, sq)$. Then the maps $x_1(t) \mapsto x_1(t)$, $x_3(t) \mapsto x_3(st)$ and $x_i(u) \mapsto x_i(u)$ for $i = 2$ and 4 extend to an isomorphism from the root group sequence $\mathcal{Q}_\mathcal{Q}(\Lambda)$ to the root group sequence $\mathcal{Q}_\mathcal{Q}(\Lambda_s)$ (by 30.4).*

Proof. This holds by the definition of the operator $\mathcal{Q}_\mathcal{Q}$ in 16.3 of [36]. □

Definition 19.16. A quadratic space (K, L, q) is *unitary* if $1 \in q(L)$.

By 19.15, we can assume (without loss of generality) that the quadratic space (K, L, q) in 30.14.iii is unitary.

Definition 19.17. Let $\Lambda = (K, L, q)$ be an anisotropic quadratic space that is unitary as defined in 19.16. Then $(K^*)^2 \subset q(L^*)$. Let ν be a discrete valuation of K. Note that if $\nu(q(u))$ is odd for some $u \in L^*$, then every odd number is of the form $\nu(q(tu))$ for some $t \in K$; thus $\nu(q(L^*))$ is either $2\mathbb{Z}$ or $\mathbb{Z}$. Let

$$\delta = |\mathbb{Z}/\nu(q(L^*))|.$$

We will call δ the *ν-index* of (K, L, q). We will say that (K, L, q) is *ramified* (respectively, *unramified*) *at* ν if $\delta = 1$ (respectively, $\delta = 2$).[5] We will say that (K, L, q) is *ν-compatible* if the four equivalent conditions in 19.4 hold.

Theorem 19.18. *Let*

$$\Lambda = (K, L, q)$$

be a unitary anisotropic quadratic space, let ν be a discrete valuation of K, let

$$\mathcal{Q}_\mathcal{Q}(\Lambda) = (U_+, U_1, \ldots, U_4)$$

and $x_1, \ldots, x_4$ be as in 30.8.iii, let (Δ, Σ, d) be the canonical realization of $\mathcal{Q}_\mathcal{Q}(\Lambda)$ as defined in 30.6 and let $(G^\dagger, \xi)$ be the root datum of Δ based at Σ. Let δ be the ν-index of Λ as defined in 19.17 and let

$$\phi_1(x_1(t)) = \nu(t) \tag{19.19}$$

for all $t \in K^*$. Then the following hold:

[5] These notions coincide with the usual notions of a (totally) ramified and unramified extension (as defined, for example, in Chapter 3, Section 5, of [28]) in the special case that L/K is a separable quadratic extension, q is its norm, K is complete with respect to ν and $\bar{L}/\bar{K}$ is separable (where $\bar{L}$ is defined with respect to the valuation ω obtained by setting $L = E$ in 17.14).

(i) *The map* ϕ_1 *extends to a valuation of* $(G^\dagger, \xi)$ *if and only if* (K, L_0, q) *is* ν*-compatible as defined in 19.17.*

(ii) *Suppose that* ϕ_1 *extends to a valuation* ϕ *of* $(G^\dagger, \xi)$ *and let* ι *be the associated root map of* Σ *(with target* $\mathsf{B}_2 = \mathsf{C}_2$*).*[6] *Then* ι *is long (as defined in 16.12) if and only if* $\delta = 2$*; and after replacing* ϕ *by an equipollent valuation if necessary,*

$$\phi_4(x_4(u)) = \nu(q(u))/\delta \tag{19.20}$$

for all $u \in L^*$.

Proof. Let Ξ_d and Ξ_d° be as in 3.5. The four roots $a_1, \ldots, a_4$ in Ξ_d can be ordered so that $U_i = U_{a_i}$ for all $i \in [1, 4]$. We then have $\Xi_d^\circ = \{a_1, a_4\}$ and $[a_1, a_4] = (a_1, a_2, a_3, a_4)$, where $[a_1, a_4]$ is as defined in 3.1.

Suppose first that (K, L, q) is ν-compatible. Let $\phi_3(x_3(t)) = \nu(t)$ for all $t \in K^*$ and let

$$\phi_i(x_i(u)) = \nu(q(u))/\delta$$

for $i = 2$ and 4 and all $u \in L^*$. Then $\phi_i(U_i^*) = \mathbb{Z}$ for all $i \in [1, 4]$. Let

$$\phi = \{\phi_1, \phi_2, \phi_3, \phi_4\}.$$

By 19.4.ii, ϕ satisfies 15.1.i. Let ι be a root map of Σ (with target B_2) that is long if $\delta = 2$ and short if $\delta = 1$ (as defined in 16.12). Thus U_1 is a long root group (i.e. the root a_1 is long) if and only if ι is long. By 16.3 of [36] and 19.4.iii, ϕ satisfies 15.1.ii with respect to ι and is, in fact, exact as defined in 15.24.[7] By 15.25 and 30.36.iii, ϕ satisfies 15.4.i.[8] By 30.38.iii, ϕ

---

[6] See the comments following 3.21.

[7] Here are the details. Choose $t \in K^*$ and $u, v \in L^*$. By 16.3 in [36], $[x_2(u), x_4(v)^{-1}] = x_3(f(u, v))$, where f is the bilinear form associated with q, and $[x_1(t), x_4(v)^{-1}] = x_2(tv)x_3(tq(v))$. Suppose first that ι is long. This means that a_1 is a long root of the root system B_2 and that $\phi_i(x_i(w)) = \nu(q(w))/2$ for $i = 2$ and 4 and for all $w \in L^*$. Thus $a_2 = a_1 + a_4$ and $a_3 = a_1 + 2a_4 = a_2 + a_4$ as well as

$$\phi_3\big(x_3(f(u, v))\big) \geq \phi_2(x_2(u)) + \phi_4(x_4(v))$$

(by 19.4.iii),

$$\phi_2(x_2(tv)) = q(tv)/2 = \phi_1(x_1(t)) + \phi_4(x_4(v))$$

and

$$\phi_3\big(x_3(tq(v))\big) = \phi_1(x_1(t)) + 2\phi_4(x_4(v)).$$

Thus ϕ satisfies 15.1.ii and is exact. Now suppose that ι is short. This means that a_1 is a short root in the root system B_2 and that $\phi_i(x_i(w)) = \nu(q(w))$ for $i = 2$ and 4 and for all $w \in L^*$. Thus $a_2 = 2a_1 + a_4$ and $a_3 = a_1 + a_4 = (a_2 + a_4)/2$ as well as

$$\phi_3\big(x_3(f(u, v))\big) \geq \big(\phi_2(x_2(u)) + \phi_4(x_4(v))\big)/2$$

(by 19.4.iii),

$$\phi_2(x_2(tv)) = q(tv) = 2\phi_1(x_1(t)) + \phi_4(x_4(v))$$

and

$$\phi_3\big(x_3(tq(v))\big) = \phi_1(x_1(t)) + \phi_4(x_4(v)).$$

Thus ϕ satisfies 15.1.ii and is exact also in this case.

[8] Here are the details. Choose $t \in K^*$ and $u \in L^*$. Then

$$\phi_3([x_1(t), x_4(u)^{-1}]_3) = \phi_3\big(x_3(tq(u))\big) = \phi_1\big(x_1(t)\big) + \delta\phi_4\big(x_4(u)\big),$$

satisfies 15.4.ii. Thus ϕ is a viable partial valuation. By 15.21, ϕ extends to a valuation of $(G^\dagger, \xi)$.

Suppose, conversely, that ϕ is a valuation of $(G^\dagger, \xi)$ with respect to a root map ι of Σ (with target B_2) that extends the map ϕ_1 (so $\phi_{a_1} = \phi_1$). Let $\phi_i = \phi_{a_i}$ for all $i \in [2,4]$. By 2.50.i and 3.41.i with $a = a_1$ and $b = a_4$, we have

$$\phi_4(x_4(u)) = -\Big(\phi_1\big(x_3(1)^{m_\Sigma(x_4(u))}\big) - \phi_3(x_3(1))\Big)/\delta_0$$

for all $u \in L^*$, where

$$\delta_0 = \begin{cases} 2 & \text{if } \iota \text{ is long and} \\ 1 & \text{if } \iota \text{ is short.} \end{cases}$$

By 32.7 in [36],

$$x_3(1)^{m_\Sigma(x_4(u))} = x_1(1/q(u))$$

for all $u \in L^*$. Thus

$$\phi_4(x_4(u)) = \big(\nu(q(u)) + \phi_3(x_3(1))\big)/\delta_0$$

for all $u \in L^*$. By 3.21, $\phi_i(U_i^*) = \mathbb{Z}$ for each $i \in [1,4]$. By 19.17, $\nu(q(L^*)) = \delta\mathbb{Z}$. It follows that $\delta_0 = \delta$ and $\phi_3(x_3(1))/\delta \in \mathbb{Z}$. By 3.43, therefore, we can assume that 19.20 holds. Since ϕ satisfies condition (V1), 19.4.ii holds. Thus (K, L, q) is ν-compatible. This completes the proof of (i) and (ii). □

Theorem 19.21. *Let (Δ, Σ, d) be a canonical realization of type 30.14.iii (as defined in 30.15). Thus*

$$\Delta = \mathsf{B}_\ell^{\mathcal{Q}}(\Lambda) = \mathsf{C}_\ell^{\mathcal{Q}}(\Lambda)$$

---

where the subscript 3 on the left-hand side is as in 15.23. Thus the (a_1, a_4)-coefficient of ϕ at a_3 (as defined in 15.24) is $(1, \delta)$ and the (a_4, a_1)-coefficient of ϕ at a_3 is $(\delta, 1)$. By 30.36.iii (where $h_1(s)$ and $h_4(u)$ are defined), we have

$$\phi_1\big(x_1(s)^{h_4(u)}\big) - \phi_1\big(x_1(s)\big) = -\delta\phi_4\big(x_4(u)\big)$$

and

$$\phi_4\big(x_4(u)^{h_1(s)}\big) - \phi_4\big(x_4(u)\big) = -2\phi_1\big(x_1(s)\big)/\delta.$$

Thus 15.4.i holds if $a \neq b$. Now let $a = a_1$, $b = a_4$ and $c = a_3$ and let ψ_a and ψ_c be as in 15.25. Then $\psi_a\big(x_4(u)\big) = -\delta\phi_4\big(x_4(u)\big)$ and (again by 30.36.iii)

$$\psi_c\big(x_4(u)\big) = \phi_3\big(x_3(s)^{h_4(u)}\big) - \phi_3(x_3(s)) = \delta\phi_4\big(x_4(u)\big).$$

Thus

$$\psi_c\big(x_4(u)\big) = \psi_a\big(x_4(u)\big) + 2\delta\phi_4\big(x_4(u)\big)$$

and hence 15.25.iii holds. By 15.25, therefore, 15.4.i holds when $a = b = a_4$. Now let $a = a_4$, $b = a_1$ and $c = a_3$ and again let ψ_a and ψ_c be as in 15.25. Then $\psi_a\big(x_1(s)\big) = -\phi_1\big(x_1(s)\big)/\delta$ and (again by 30.36.iii)

$$\psi_c\big(x_1(s)\big) = \phi_3\big(x_3(s)^{h_1(s)}\big) - \phi_3(x_3(s)) = 0.$$

Thus

$$\psi_c\big(x_1(s)\big) = \delta\psi_a\big(x_1(s)\big) + \phi_1\big(x_1(s)\big)$$

and hence 15.25.iii holds again. By 15.25, therefore, 15.4.i holds when $a = b = a_1$.

for some anisotropic quadratic space $\Lambda = (K, L, q)$ and some $\ell \geq 2$ in the notation of 30.15. Suppose, too, that Λ is unitary as defined in 19.16.[9] *Let $(G^\dagger, \xi)$ be the root datum of Δ based at Σ and let ν be a valuation of K. Let ι_{B} be a long root map of Σ and let ι_{C} be a short root map of Σ as defined in 16.12.*[10] *Let a and x_a be as in 16.2 and 16.3 and let ϕ_a be given by*

$$\phi_a(x_a(t)) = \nu(t)$$

for all $t \in K^*$. Then the following hold:

(i) *If Λ is not ν-compatible, then ϕ_a does not extend to a valuation of $(G^\dagger, \xi)$.*
(ii) *If Λ is ν-compatible and Λ is unramified at ν, then ϕ_a extends to a valuation ϕ of $(G^\dagger, \xi)$ with respect to ι_{B} (but not with respect to ι_{C}).*
(iii) *If Λ is ν-compatible and Λ is ramified at ν, then ϕ_a extends to a valuation ϕ of $(G^\dagger, \xi)$ with respect to ι_{C} (but not with respect to ι_{B}).*

In cases (ii) and (iii), the extension ϕ is unique up to equipollence.

Proof. This holds by 16.14.ii, 19.17 and 19.18. □

Notation 19.22. Let Δ, Λ, ν, etc., be as in 19.21, suppose that Λ is ν-compatible and let ϕ be as in (ii) or (iii) of 19.21. We denote the pair $(\Delta^{\rm aff}, \mathcal{A})$ obtained by applying 14.47 to this data by $\tilde{\mathsf{X}}_\ell^{\mathcal{Q}}(\Lambda, \nu)$, where $\mathsf{X} = \mathsf{B}$ in case (ii) and $\mathsf{X} = \mathsf{C}$ in case (iii).

Thus the pair $\tilde{\mathsf{X}}_\ell^{\mathcal{Q}}(\Lambda, \nu)$ for $\mathsf{X} = \mathsf{B}$ or C defined in 19.22 exists precisely when the quadratic space Λ is ν-compatible.

Theorem 19.23. *Every Bruhat-Tits pair whose building at infinity is*

$$\mathsf{B}_\ell^{\mathcal{Q}}(\Lambda) = \mathsf{C}_\ell^{\mathcal{Q}}(\Lambda)$$

for some unitary anisotropic quadratic space $\Lambda = (K, L, q)$ and some $\ell \geq 2$ is of the form $\tilde{\mathsf{B}}_\ell^{\mathcal{Q}}(\Lambda, \nu)$ (if Λ is unramified at ν) or $\tilde{\mathsf{C}}_\ell^{\mathcal{Q}}(\Lambda, \nu)$ (if Λ is ramified at ν) for some valuation ν of K such that Λ is ν-compatible.

Proof. This holds by 16.4, 19.21 and 19.22. □

We turn our attention next to completions.

Proposition 19.24. *Let $\Lambda = (K, L, q)$ be an anisotropic quadratic space, let ν be a discrete valuation of K and suppose that Λ is ν-compatible (as defined in 19.17). For all $u, v \in L$, let*

$$\partial_\nu(u, v) = \begin{cases} 2^{-\omega(u-v)} & \textit{if } u \neq v \textit{ and} \\ 0 & \textit{if } u = v, \end{cases}$$

where

$$\omega(w) = \nu(q(w))$$

for all $w \in L$. Then the following hold:

[9] By 19.15, this is no real restriction.
[10] Thus by 2.14.v, ι_{X} has target X_ℓ for $\mathsf{X} = \mathsf{B}$ and C.

(i) ∂_ν *is a metric on* L;

(ii) *The completion* $\hat{L}$ *of* L *with respect to* ∂_ν *is a vector space over* $\hat{K}$ *in which*

$$\hat{t}\hat{u} = \lim_{k\to\infty} (t_k u_k) \tag{19.25}$$

for all $\hat{t} \in \hat{K}$, *all* $\hat{u} \in \hat{L}$, *all sequences* $(t_k)_{k\geq 1}$ *of elements of* K *that converge to* $\hat{t}$ *and all sequences* $(u_k)_{k\geq 1}$ *of elements of* L *that converge to* $\hat{u}$ *(with respect to* ∂_ν*).*

(iii) $\hat{\Lambda} := (\hat{K}, \hat{L}, \hat{q})$ *is an anisotropic quadratic space,*[11] *where* $\hat{q}$ *is the extension of* q *to* $\hat{L}$ *by continuity, i.e.*

$$\hat{q}(\hat{u}) := \lim_{k\to\infty} q(u_k) \tag{19.26}$$

whenever $(u_k)_{k\geq 1}$ *is a sequence of elements* u_k *of* L *that converges to an element* $\hat{u}$ *of* $\hat{L}$.

(iv) $\hat{\Lambda}$ *is* ν*-compatible.*[12]

Proof. Let Δ, Σ and $(G^\dagger, \xi)$ be as in 19.18 and let ϕ be the valuation of the root datum $(G^\dagger, \xi)$ obtained by applying 19.18 to ν. By 17.3 and 19.20 applied to ϕ, we conclude that ∂_ν is a metric. Thus (i) holds.

There is a unique extension (by 17.12) of the addition on L by continuity that makes $\hat{L}$ into an abelian group. Let $\hat{t} \in \hat{K}$, let $(t_k)_{k\geq 1}$ be a sequence of elements of K that converges to $\hat{t}$, let $\hat{u} \in \hat{L}$ and let $(u_k)_{k\geq 1}$ be a sequence of elements of L that converges to $\hat{u}$ (with respect to ∂_ν). Then

$$t_k u_k - t_l u_l = t_k(u_k - u_l) + (t_k - t_l)u_l$$

and

$$\omega(t_k(u_k - u_l)) = 2\nu(t_k) + \omega(u_k - u_l)$$

as well as

$$\omega((t_k - t_l)u_l) = 2\nu(t_k - t_l) + \omega(u_l)$$

for all $k, l \geq 0$. Furthermore, the sequences $(\nu(t_k))_{k\geq 1}$ and $(\omega(u_k))_{k\geq 1}$ are bounded below (by 9.18.iii and 19.4.ii). It follows that $(t_k u_k)_{k\geq 1}$ is a Cauchy sequence with respect to ω. Any two sequences converging to $\hat{t}$ can be interleaved to make a "larger" sequence converging to $\hat{t}$. Since the same remark holds for $\hat{u}$, it follows that

$$\lim_{k\to\infty} t_k u_k$$

depends only on $\hat{t}$ and $\hat{u}$ and not on the sequences $(t_k)_{k\geq 1}$ and $(u_k)_{k\geq 1}$. Thus 19.25 yields a well-defined map from $\hat{K} \times \hat{L}$ to $\hat{L}$ that gives $\hat{L}$ the structure of a vector space over $\hat{K}$. Thus (ii) holds.

---

[11] The quadratic space $\hat{\Lambda}$ is not always the same as the quadratic space $(\hat{K}, L_{\hat{K}}, q_{\hat{K}})$ as defined in 19.1; see 19.29–19.32 below.

[12] We are using the usual convention here (see 9.20) that the unique extension of ν to a valuation of the completion $\hat{K}$ of K is denoted by the same letter ν.

Again let $(u_k)_{k\ge1}$ be a sequence of elements of L that converges to an element $\hat{u}$ in $\hat{L}$ and let f be the bilinear form associated with q. By 9.18.iii and 19.4.ii again, the sequence $(\omega(u_k))_{k\ge1}$ is bounded below. Let $N > 0$ be arbitrary. Since

$$\nu(q(u_k - u_l)) = \omega(u_k - u_l) > N$$

for all k, l sufficiently large, it follows that

$$\nu(f(u_k - u_l, u_l)) > N$$

for all k, l sufficiently large by 19.4.iii. Since

$$q(u_k) - q(u_l) = q(u_k - u_l) + f(u_k - u_l, u_l)$$

for all $k, l \ge 1$, we conclude that $\nu\big(q(u_k) - q(u_l)\big) > N$ for all k, l sufficiently large. Thus $(q(u_k))_{k\ge1}$ is a Cauchy sequence in K whose limit $\hat{t} \in \hat{K}$ depends only on $\hat{u}$ and not on the sequence $(u_k)_{k\ge1}$. Hence 19.26 yields a well-defined map $\hat{q}$ from $\hat{L}$ to $\hat{K}$ which (again by 9.18.iii and 19.4.ii) maps non-zero elements of $\hat{L}$ to non-zero elements of $\hat{K}$. Let

$$\hat{f}(\hat{u}, \hat{v}) := \hat{q}(\hat{u} + \hat{v}) - \hat{q}(\hat{u}) - \hat{q}(\hat{v})$$

for all $\hat{u}, \hat{v} \in \hat{L}$. By 19.25 and 19.26, it follows that $\hat{f}$ is bilinear and that

$$\hat{q}(\hat{t}\hat{u}) = \hat{t}^2\hat{q}(\hat{u})$$

for all $\hat{t} \in \hat{K}$ and all $\hat{u} \in \hat{L}$. Hence $(\hat{K}, \hat{L}, \hat{q})$ is an anisotropic quadratic space. Thus (iii) holds. Since $\hat{K}$ is complete, $\hat{\Lambda}$ satisfies 19.4.iv. Hence $\hat{\Lambda}$ is ν-compatible. Thus (iv) holds. □

Theorem 19.27. *Let $\Lambda = (K, L, q)$ be a unitary anisotropic quadratic space, let ν be a discrete valuation of K and suppose that Λ is ν-compatible (as defined in 19.17). Let $\hat{\Lambda}$ be as in 19.24.iii and let $(\Delta, \mathcal{A}) = \tilde{\mathsf{X}}_\ell^{\mathcal{Q}}(\Lambda, \nu)$ be as in 19.22, where $\mathsf{X} = \mathsf{B}$ or C. Then $\hat{\Lambda}$ is ν-compatible, Δ is completely Bruhat-Tits and its completion is $\tilde{\mathsf{X}}_\ell^{\mathcal{Q}}(\hat{\Lambda}, \nu)$.*

Proof. By 30.16, $\mathsf{X}_\ell^{\mathcal{Q}}(\Lambda)$ is a subbuilding of $\mathsf{X}_\ell^{\mathcal{Q}}(\hat{\Lambda})$. By 19.24.iv, $\hat{\Lambda}$ is ν-compatible. By 9.18.iii and 19.17, Λ is ramified at ν if and only if $\hat{\Lambda}$ is. The claim holds, therefore, by 17.4, 17.9 and 19.24. □

Corollary 19.28. *Let $\Lambda = (K, L, q)$ be an anisotropic quadratic space, let ν be a valuation of K and suppose that Λ is ν-compatible. Then the following hold:*

(i) *The Bruhat-Tits pair $\tilde{\mathsf{X}}_\ell^{\mathcal{Q}}(\Lambda, \nu)$ is complete if and only if K is complete with respect to ν and L is complete with respect to the metric ∂_ν defined in 19.24.*

(ii) *If $\dim_K L$ is finite, then $\tilde{\mathsf{X}}_\ell^{\mathcal{Q}}(\Lambda, \nu)$ is complete if and only if K is complete with respect to ν.*

Proof. This holds by 17.13 and 19.27. □

In the following result, we compare the quadratic spaces $\Lambda_{\hat{K}}$ in 19.4 and $\hat{\Lambda}$ in 19.24.iii.

Proposition 19.29. *Let $\Lambda = (K, L, q)$ be an anisotropic quadratic space, let f be the bilinear form belonging to q and let ν be a discrete valuation of K. Suppose that Λ is ν-compatible as defined in 19.17. Let $(\hat{K}, L_{\hat{K}}, q_{\hat{K}})$ be as in 19.4 and let $(\hat{K}, \hat{L}, \hat{q})$ be as in 19.24. Let ρ be the unique $\hat{K}$-linear map from $L_{\hat{K}}$ to $\hat{L}$ that sends $v \otimes \hat{t}$ to $v\hat{t}$ for all elements $v \in L$ and $\hat{t} \in \hat{K}$. Then the following hold:*

(i) *$\hat{q}(\rho(\hat{u})) = q_{\hat{K}}(\hat{u})$ for all $\hat{u} \in L_{\hat{K}}$.*
(ii) *The map ρ is injective if $q_{\hat{K}}$ is anisotropic.*
(iii) *If f is non-degenerate, then $q_{\hat{K}}$ is anisotropic.*
(iv) *The map ρ is surjective if $\dim_K L$ is finite.*
(v) *If $q_{\hat{K}}$ is anisotropic and $\dim_K L$ is finite, then ρ is an isomorphism of quadratic spaces from $(\hat{K}, L_{\hat{K}}, q_{\hat{K}})$ to $(\hat{K}, \hat{L}, \hat{q})$.*

Proof. Let $\hat{u} \in L_{\hat{K}}$. There exists a finite linearly independent set $v_1, \ldots, v_d$ such that $\hat{u} = v_1 \otimes \hat{t}_1 + \cdots v_d \otimes \hat{t}_d$ for some $\hat{t}_1, \ldots, \hat{t}_d$ in $\hat{K}$. Let $(t_{i,k})_{k \geq 1}$ be a sequence of elements of K that converges to t_i for each $i \in [1, d]$ and let

$$u_k = v_1 \otimes t_{1,k} + \cdots + v_d \otimes t_{d,k}$$

for each $k \geq 1$. Thus

$$\hat{q}(\rho(u_k)) = q(u_k) = q_{\hat{K}}(u_k)$$

for all $k \geq 1$, and by 19.3 and 19.4.ii,

$$q_{\hat{K}}(\hat{u}) = \lim_{k \to \infty} q(u_k).$$

By 19.26, it follows that

$$\hat{q}(\rho(\hat{u})) = q_{\hat{K}}(\hat{u}). \tag{19.30}$$

Thus (i) holds. The assertion (ii) follows from (i) since $\hat{q}$ is anisotropic (by 19.24.iii). Assertion (iii) holds by 19.4.iv.

Suppose that $\dim_K L$ is finite and suppose that $v_1, \ldots, v_d$ is a basis of L over K. If $(u_k)_{k \geq 1}$ is a Cauchy sequence in $\rho(L)$, then (by the Lemma in Chapter II, Section 8, of [11]), there exist Cauchy sequences $(t_{i,k})_{k \geq 1}$ for all $i \in [1, d]$ such that

$$u_k = v_1 t_{1,k} + \cdots + v_d t_{d,k}$$

for each k. It follows that $\rho(L_{\hat{K}})$ is a closed subset of $\hat{L}$ containing L. Thus (iv) holds. Assertion (v) is a consequence of (i), (ii) and (iv). □

Example 19.31. Let k be a field of characteristic 2, let t be a transcendental over k, let α be an element in $k((t))$ that is transcendental over $k(t)$, let $K = k(t, \alpha^2)$, let $L = k(t, \alpha)$, let $q(x) = x^2$ for all $x \in L$ and let ν be the restriction to K of the unique discrete valuation $k((t))$. Then $\bar{L} = \bar{K} = k$ but $\dim_{\hat{K}} L \otimes \hat{K} = \dim_K L = 2$, so the map ρ defined in 19.29 is not injective (and the quadratic form $q_{\hat{K}}$ is not anisotropic). Thus the hypothesis in 19.29.ii is necessary.

Example 19.32. Suppose that (K, L, q) is an anisotropic quadratic space of infinite dimension such that the bilinear form f belonging to q is non-degenerate and suppose that ν is a discrete valuation of K such that (K, L, q) is ν-compatible.[13] Let ω be as in 19.24 and let ρ be as in 19.29. Since f is non-degenerate, we can choose a countable subset

$$\{v_k \mid k \geq 1\}$$

of L^* that is orthogonal if $\text{char}(K) \neq 2$ and "symplectically orthogonal" if $\text{char}(K) = 2$, by which we mean that $f(v_k, v_l)$ is non-zero if k is odd and $l = k + 1$ and $f(v_k, v_l) = 0$ for all other pairs k, l with $k < l$. Adjusting by scalars, we can assume that $\omega(v_k) \geq 0$ for all $k \geq 1$. Choose a uniformizer $\pi \in K^*$ (as defined in 9.17) and let

$$u_d = \sum_{k=1}^{d} \pi^k v_k$$

for all $d \geq 1$. Then $(u_d)_{d \geq 1}$ is a Cauchy sequence in L (with respect to ω); let $u^* \in \hat{L}$ be its limit. For each $k \geq 1$, $f(v_k, u_d)$ is a non-zero constant for all $d > k$. Thus $\hat{f}(v_k, u^*) \neq 0$ for all $k \geq 1$. It follows that $u^* \notin \rho(L_{\hat{K}})$. This shows that the hypothesis in 19.29.iv is necessary.

We turn now to gems.

Definition 19.33. Let

$$\Lambda = (K, L, q)$$

be a ν-compatible anisotropic quadratic space, let $\bar{K}$ and $\mathcal{O}_K$ be the residue field and the ring of integers of K with respect to ν (as defined in 9.22), let π be a uniformizer in K (as defined in 9.17) and let $t \mapsto \bar{t}$ be the natural homomorphism from $\mathcal{O}_K$ to $\bar{K}$. For each $i \in \mathbb{Z}$, let

$$L_i = \{v \in L \mid \nu(q(v)) \geq i\}.$$

By 19.4.ii, L_i is an additive subgroup of L for each $i \in \mathbb{Z}$. Let $\bar{L}_i = L_i / L_{i+1}$ for each $i \in \mathbb{Z}$. Choose $i \in \mathbb{Z}$. Let $\bar{a}$ denote the image of a under the natural homomorphism from L_i to $\bar{L}_i$. Let

$$\bar{t} \cdot \bar{a} = \overline{ta}$$

for all $t \in \mathcal{O}_K$ and all $a \in L_i$. This map is well defined and gives $\bar{L}_i$ the structure of a vector space over $\bar{K}$. By 19.4.iii, the maps $\bar{f}_i \colon \bar{L}_i \times \bar{L}_i \to \bar{K}$ and $q_i \colon \bar{L}_i \to \bar{K}$ given by

$$\bar{f}_i(\bar{a}, \bar{b}) = \overline{f(a, b)/\pi^i}$$

and

$$\bar{q}_i(\bar{a}) = \overline{q(a)/\pi^i}$$

[13] For example, we can set $E = \mathbb{C}((t))$ and $K = \mathbb{R}((t))$ and let L be the orthogonal sum of arbitrarily many copies of the quadratic space (K, E, N), where N is the norm of the extension E/K.

for all $a, b \in L_i$ are well defined. Moreover, $\bar{q}_i$ is an anisotropic quadratic form on $\bar{L}_i$ and $\bar{f}_i$ is the corresponding bilinear form. More precisely, either $\bar{L}_i = 0$ or

$$\bar{\Lambda}_i := (\bar{K}, \bar{L}_i, \bar{q}_i)$$

is an anisotropic quadratic space. Note that if Λ is unitary as defined in 19.16, then $\bar{L}_0 \neq 0$. Note, too, that the quadratic space $\bar{\Lambda}_i$ depends on the choice of π if $i \neq 0$, but it is nevertheless unique up to similarity (as defined in 12.8 in [36]) and hence also the building $\mathsf{C}_\ell^{\mathcal{Q}}(\bar{\Lambda}_i)$ is independent of the choice of π (by 3.6, 19.15 and 30.14.iii). We call the quadratic space $\bar{\Lambda}_i$ the *i-th residue* of Λ (with respect to ν).

Remark 19.34. We continue with all the notation in 19.33. Let ρ be the linear automorphism of L given by $\rho(v) = \pi v$ for all $v \in L$. Then ρ induces an isomorphism from $\bar{\Lambda}_i$ to $\bar{\Lambda}_{i+2}$ for all i. Thus the quadratic spaces $\bar{\Lambda}_i$ and $\bar{\Lambda}_j$ are isomorphic whenever i and j have the same parity. Suppose that Λ is unitary. Then $\bar{\Lambda}_0 \neq 0$, and $\delta = 2$ if and only if $\bar{\Lambda}_1 = 0$. If $\delta = 1$, then $\bar{L}_i$ is non-zero for all i, but $\bar{\Lambda}_0$ and $\bar{\Lambda}_1$ need not be isomorphic (or even of the same dimension). For more about this, see the article by T. A. Springer [30], where the quadratic spaces $\bar{\Lambda}_i$ were, we believe, first introduced. See also 19.36.

Theorem 19.35. *Let*

$$(\Delta, \mathcal{A}) = \tilde{\mathsf{X}}_\ell^{\mathcal{Q}}(\Lambda, \nu)$$

for some unitary anisotropic quadratic space $\Lambda = (K, L, q)$ *and for some valuation* ν *of* K *such that* Λ *is* ν*-compatible, where* $\mathsf{X} = \mathsf{B}$ *if* Λ *is unramified at* ν *and* $\mathsf{X} = \mathsf{C}$ *if* Λ *is ramified at* ν*. Let* $\bar{\Lambda}_i$ *for all* i *be as in 19.33, let the apartment* A*, the gem* R_0 *and the group* G° *be as in 18.1 and let the gem* R_1 *be as in 18.28. Then the following hold:*

(i) *If* $\mathsf{X} = \mathsf{B}$*, then* $R_0 \cong \mathsf{B}_\ell^{\mathcal{Q}}(\bar{\Lambda}_0)$ *and all gems are in the same* G°*-orbit as* R_0*.*

(ii) *If* $\mathsf{X} = \mathsf{C}$*, then* $R_0 \cong \mathsf{C}_\ell^{\mathcal{Q}}(\bar{\Lambda}_0)$*,* $R_1 \cong \mathsf{C}_\ell^{\mathcal{Q}}(\bar{\Lambda}_1)$ *and every gem is in the same* G°*-orbit as* R_0 *or* R_1 *(or both).*

(iii) *Suppose that* $\mathsf{X} = \mathsf{C}$*. If there exists a linear similitude* ψ *of* Λ *such that* $\nu(q(\psi(u))) = \nu(q(u)) + 1$ *for all* $u \in L^*$*, then all gems are in the same* G°*-orbit as* R_0*.*[14]

Proof. Let $\Sigma = A^\infty$, let d be a chamber of Σ, let $(G^\dagger, \xi)$ be the root datum of $\Delta_{\mathcal{A}}^\infty$ based at Σ and let $\phi = \phi_{R_0}$ be the valuation of $(G^\dagger, \xi)$ described in 13.8. Let c be the unique root in Ξ_d that is long (with respect to the root map ι_C) if $\mathsf{X} = \mathsf{C}$ and short (with respect to the root map ι_B) if $\mathsf{X} = \mathsf{B}$. By 19.22, we can assume that $(\Delta_{\mathcal{A}}^\infty, \Sigma, d)$ is the canonical realization of one of

---

[14] Suppose that (L, K) is a composition algebra with norm q as defined in 30.17 and that the anisotropic quadratic space (K, L, q) is ramified, so there exists $w \in L^*$ such that $\nu(q(w)) = 1$. Let $\psi(u) = uw$ for all $u \in L$. Then $\nu(q(\psi(u))) = \nu(q(uw)) = \nu(q(u)) + 1$ for all $u \in L^*$ since q is multiplicative.

the root group labelings described in 30.14.iii, that $\phi_a(x_a(t)) = \nu(t)$ for all $a \in \Xi_d^\circ\backslash\{c\}$ and all $t \in K^*$ and that $\phi_c(x_c(u)) = \nu(q(u))/\delta$ for all $u \in L^*$, where δ is the ν-index of Λ as defined in 19.17. By 18.27,

$$R_0 \cong \mathsf{X}_\ell^{\mathcal{Q}}(\bar{\Lambda}_0).$$

Let $b \in \Xi_d^\circ$ be as in 18.28. Since b is long, we have $b = c$ if $\mathsf{X} = \mathsf{C}$. Let H be as in 18.1 and let $\pi \in K$ be a uniformizer.

Suppose that $\mathsf{X} = \mathsf{B}$. By 18.9, there exists an element $\tau \in H$ centralizing U_a for all $a \in \Xi_d^\circ\backslash\{b\}$ such that $x_b(t)^\tau = x_b(\pi t)$ for all $t \in K$.[15] Hence

$$\phi_b(x_b(t)^\tau) = \phi_b(x_b(t)) + 1$$

for all $t \in K$ and ϕ_a is τ-invariant for all $a \in \Xi_d^\circ\backslash\{b\}$. By 18.6, there is an element $\rho \in G_A$ inducing τ on $\Delta_{\mathcal{A}}^\infty$. By 18.28.i, it follows that ρ maps R_0 to R_1. Thus (i) holds by 18.29.

Suppose now that $\mathsf{X} = \mathsf{C}$, so that $\delta = 1$. By 18.28.i and 19.33, we have $R_1 \cong \mathsf{C}_\ell(\bar{\Lambda}_1)$.[16] By 18.29, therefore, (ii) holds.

Suppose, finally, that there exists ψ as in (iii). By 18.9, there is an element $\tau' \in H$ centralizing U_a for all $a \in \Xi_d^\circ\backslash\{b\}$ such that $x_b(u)^{\tau'} = x_b(\psi(u))$ and hence

$$\phi_b(x_b(u)^{\tau'}) = \phi_b(x_b(u)) + 1$$

for all $u \in L$.[17] By 18.6, there is an element $\rho' \in G_A$ inducing τ' on $\Delta_{\mathcal{A}}^\infty$. By 18.28.i, the elements ρ' map R_1 to R_0. By 18.29, therefore, (iii) holds. □

---

[15] Here are the details. If

$$\mathcal{Q}_{\mathcal{Q}}(\Lambda) = (U_+, U_1, U_2, U_3, U_4)$$

is the root group sequence in 30.8.iii, then (by 30.4) there is an automorphism of U_+ extending the maps $x_i(t) \mapsto x_i(\pi t)$ for $i = 1$ and 3, $x_2(u) \mapsto x_2(\pi u)$ and $x_4(u) \mapsto x_4(u)$. Thus the automorphism τ exists by 18.9 if $\ell = 2$ (in which case $U_b = U_1$). Suppose that $\ell \geq 3$ and let

$$\mathcal{T}(K) = (U_+, U_1, U_2, U_3)$$

be the root group sequence in 30.8.i. Then there exists an automorphism of U_+ extending the maps $x_i(t) \mapsto x_i(\pi t)$ for $i = 1$ and 2 and $x_3(t) \mapsto x_3(t)$ (again by 30.4). Thus the automorphism τ exists by 18.9 and 30.14.iii.

[16] Here are the details. Let (u, v) be the directed edge of the Coxeter diagram C_ℓ defined in 30.11. Thus $a = a_u$ and $b = a_v$, where a_u and a_v are as in 3.5 and a is the root in Ξ_d° chosen in 16.3. Furthermore

$$\Omega_{uv} = \mathcal{Q}_{\mathcal{Q}}(\Lambda) = (U_+, U_1, U_2, U_3, U_4),$$

where Ω_{uv} is the root group sequence defined in 30.1 and $\mathcal{Q}_{\mathcal{Q}}(\Lambda)$ is as in 30.8.iii. Thus, in particular, $U_i = U_{a_i}$ for all $i \in [1, 4]$, where $[a, b] = (a_1, a_2, a_3, a_4)$, so $U_1 = U_a$ and $U_4 = U_b$. By 19.22, a_1 is short, so $a_2 = 2a_1 + a_4$ and $a_3 = a_1 + a_4$. By 18.22 and 18.28.i, therefore, $U_{c,R} = U_{c,0}$ for all roots c in $\Xi_d^\circ\backslash\{b\}$ including a and $U_{a_i,R} = U_{a_i,1}$ for $i \in [2, 4]$. By 3.6, 18.23, 19.33 and 30.4, it follows that $R_1 \cong \mathsf{C}_\ell^{\mathcal{Q}}(\bar{\Lambda}_1)$.

[17] Here are the details. Since ψ is a similitude of q, there exists $\gamma \in K^*$ such that $q(\psi(u)) = \gamma q(u)$ for all $u \in L$. If

$$\mathcal{Q}_{\mathcal{Q}}(\Lambda) = (U_+, U_1, U_2, U_3, U_4)$$

is as in 30.8.iii, then there is an automorphism of U_+ extending the maps $x_1(t) \mapsto x_1(t)$, $x_i(u) \mapsto x_i(\psi(u))$ for $i = 2$ and 4 and $x_3(t) \mapsto x_3(\gamma t)$ (by 30.4). Thus τ' exists by 18.9 and 30.14.iii.

In light of 18.33 and 19.27, the following result about quadratic forms is really a result about the gems of the building Δ in 19.35 in the special case that the vector space L is finite-dimensional (which includes the special case that Δ is algebraic as defined in 30.31).

Proposition 19.36. *Let*

$$\Lambda = (K, L, q)$$

be an anisotropic quadratic space of finite dimension, let ν be a discrete valuation of K and suppose that K is complete with respect to ν. Then

$$\dim_K L = \dim_{\bar{K}} \bar{L}_0 + \dim_{\bar{K}} \bar{L}_1,$$

where $\bar{L}_0$ and $\bar{L}_1$ are as in 19.33.

Proof. By 19.4, q satisfies 18.36. The claim holds, therefore, by 18.34 (with $Q = q$ and $m = 2$). □

Remark 19.37. As we observed in 19.34, both $\bar{L}_1$ and $\bar{L}_2$ are non-zero if the quadratic space Λ in 19.36 is ramified at ν (i.e. if its index δ at ν is 1), but if Λ is unramified at ν, then, in fact,

$$\dim_K L = \dim_{\bar{K}} \bar{L}_0$$

by 19.36 since in this case $\bar{L}_1 = 0$.

The result 19.36 (including 19.37) will play an important role in Chapter 28.

Suppose that $\Lambda = (K, L, q)$ is a finite-dimensional anisotropic quadratic space. Suppose, too, that the field K is complete with respect to a discrete valuation ν and that its residue field $\bar{K}$ is perfect if its characteristic is 2. T. A. Springer [30] showed that under these hypotheses, the quadratic space Λ is uniquely determined by the field K and the quadratic spaces $\bar{\Lambda}_0$ and $\bar{\Lambda}_1$ defined in 19.33. Similar results for hexagonal systems (but with a negative conclusion in one case) were proved by H. P. Petersson in Theorems 2, 3 and 6 in [19]. By Theorem 2 in Chapter II, Section 4, of [28] (and the remark that follows it), a field K that is complete with respect to a discrete valuation is uniquely determined by its residue field $\bar{K}$ if it has the same characteristic as $\bar{K}$. These results (combined with 19.35) indicate that there are many interesting questions regarding the extent to which a Bruhat-Tits building is determined by its residues.[18] We will not, however, pursue these issues in this book.

[18] In Theorem 4.1.2 of [32] (restated in 29.15.i), Tits showed that irreducible *spherical* buildings are, in contrast, *always* uniquely determined by their residues of rank 2.

Chapter Twenty

Quadrangles of Indifferent Type

In this chapter, we consider the case $X = \mathcal{Q}_{\mathcal{D}}$ of the problem framed at the end of Chapter 16. Our main results are 20.2 and 20.6 in which we give the classification of Bruhat-Tits pairs whose building at infinity is $\mathsf{B}_2^{\mathcal{D}}(\Lambda)$ for some indifferent set Λ. See 10.1 in [36] for the definition of an indifferent set and 30.15 for the definition of the building $\mathsf{B}_2^{\mathcal{D}}(\Lambda)$.

Definition 20.1. Let (K, K_0, L_0) be an indifferent set and let ν be a discrete valuation of K. By 10.1 and 10.2 in [36], $1 \in L_0 \subset K_0$, $K^2 L_0 \subset L_0$ and $K^2 K_0 \subset K_0$. By 9.17.i, it follows that $\nu(K_0^*) = \mathbb{Z}$ or $2\mathbb{Z}$, $\nu(L_0^*) = \mathbb{Z}$ or $2\mathbb{Z}$ and $\nu(L_0^*) \subset \nu(K_0^*)$. Let $\delta_L = |\mathbb{Z}/\nu(L_0^*)|$ and $\delta_K = |\mathbb{Z}/\nu(K_0^*)|$. Then $\delta_K \le \delta_L \le 2$, so, in particular, $\delta_L/\delta_K = 1$ or 2. We will call the pair (δ_K, δ_L) the *ν-index* of (K, K_0, L_0).

Theorem 20.2. *Let*

$$\Lambda = (K, K_0, L_0)$$

be an indifferent set, let ν be a discrete valuation of K, let

$$\mathcal{Q}_{\mathcal{D}}(\Lambda) = (U_+, U_1, \ldots, U_4)$$

and $x_1, \ldots, x_4$ be as in 30.8.iv, let (Δ, Σ, d) be the canonical realization of $\mathcal{Q}_{\mathcal{D}}(\Lambda)$ as defined in 30.6 and let $(G^\dagger, \xi)$ be the root datum of Δ based at Σ. Let (δ_K, δ_L) be the ν-index of (K, K_0, L_0) as defined in 20.1 and let

$$\phi_1(x_1(t)) = \nu(t)/\delta_K \tag{20.3}$$

for all $t \in K_0^*$. Then the following hold:

(i) *The map ϕ_1 extends to a valuation ϕ of $(G^\dagger, \xi)$.*
(ii) *Let ι be the root map of Σ (with target B_2) associated to ϕ. Then ι is long if and only if $\delta_L/\delta_K = 2$; and, after replacing ϕ by an equipollent valuation if necessary,*

$$\phi_4(x_4(t)) = \nu(t)/\delta_L \tag{20.4}$$

for all $t \in L_0^*$.

Proof. Let Ξ_d and Ξ_d° be as in 3.5. The four roots $a_1, \ldots, a_4$ in Ξ_d can be ordered so that $U_i = U_{a_i}$ for all $i \in [1, 4]$. We then have $\Xi_d^\circ = \{a_1, a_4\}$ and $[a_1, a_4] = (a_1, a_2, a_3, a_4)$, where $[a_1, a_4]$ is as defined in 3.1. Let

$$\phi_3(x_3(t)) = \nu(t)/\delta_K$$

for all $t \in K_0^*$, let

$$\phi_i(x_i(t)) = \nu(t)/\delta_L$$

for $i = 2$ and 4 and for all $t \in L_0^*$ and let

$$\phi = \{\phi_1, \phi_2, \phi_3, \phi_4\}.$$

By 20.1, $\phi_i(U_i^*) = \mathbb{Z}$ for all $i \in [1, 4]$. By 9.17.ii, ϕ satisfies 15.1.i. Let ι be a root map of Σ (with target B_2) that is long (as defined in 16.12) if $\delta_L/\delta_K = 2$ and short if $\delta_L/\delta_K = 1$. Thus U_1 is a long root group (i.e. the root a_1 is long) if and only if ι is long. By 16.4 of [36] and 9.17.i, ϕ satisfies 15.1.ii with respect to ι, ϕ is exact as defined in 15.24, the (a_1, a_4)-coefficient of ϕ at a_2 (as defined in 15.24) is $(2\delta_K/\delta_L, 1)$ and the (a_4, a_1)-coefficient of ϕ at a_3 is $(\delta_L/\delta_K, 1)$. By 15.25 and 30.36.iv, it follows that ϕ satisfies 15.4.i. By 30.38.iv, ϕ satisfies 15.4.ii. Thus ϕ is a viable partial valuation of $(G^\dagger, \xi)$. By 15.21, ϕ extends to a valuation of $(G^\dagger, \xi)$. Hence (i) holds.

Now suppose, conversely, that ϕ is a valuation of $(G^\dagger, \xi)$ with respect to a root map ι of Σ (with target B_2) that extends the map ϕ_1 (so $\phi_{a_1} = \phi_1$). Let $\phi_i = \phi_{a_i}$ for all $i \in [2, 4]$. By 2.50.i and 3.41.i with $a = a_1$ and $b = a_4$, we have

$$\phi_4(x_4(u)) = -\Big(\phi_1\big(x_3(1)^{m_\Sigma(x_4(u))}\big) - \phi_3(x_3(1))\Big)/\delta_0$$

for all $u \in L_0$, where

$$\delta_0 = \begin{cases} 2 & \text{if } \iota \text{ is long and} \\ 1 & \text{if } \iota \text{ is short.} \end{cases}$$

By 32.8 in [36],

$$x_3(1)^{m_\Sigma(x_4(u))} = x_1(1/u)$$

for $u \in L_0^*$. Thus

$$\begin{aligned} \phi_4(x_4(u)) &= -\big(\phi_1(x_1(1/u)) - \phi_3(x_3(1))\big)/\delta_0 \\ &= \big(\nu(u) + \delta_K\phi_3(x_3(1))\big)/\delta_K\delta_0 \end{aligned}$$

for $u \in L_0^*$. By 3.21, $\phi_4(U_4^*) = \mathbb{Z}$, and by 20.1, $\nu(L^*) = \delta_L\mathbb{Z}$. It follows that $\delta_K\delta_0 = \delta_L$ and $\phi_3(x_3(1))/\delta_0 \in \mathbb{Z}$. Therefore we can assume that 20.4 holds (by 3.43) and $\delta_L/\delta_K = 2$ if and only if $\delta_0 = 2$ if and only if ι is long. Thus (ii) holds. □

Notation 20.5. Let Δ, Λ, ν, etc. be as in 20.2 and let ϕ be the unique valuation of $(G^\dagger, \xi)$ satisfying 20.3 and 20.4. We denote by

$$\tilde{\mathsf{B}}_2^{\mathcal{D}}(\Lambda, \nu) \text{ or } \tilde{\mathsf{C}}_2^{\mathcal{D}}(\Lambda, \nu)$$

the pair $(\Delta^{\text{aff}}, \mathcal{A})$ obtained by applying 14.47 to these data.

Thus the pair $\tilde{\mathsf{B}}_2^{\mathcal{D}}(\Lambda, \nu)$ exists for *every* discrete valuation ν of K. Note that we do not distinguish between $\mathsf{B}_2^{\mathcal{D}}(\Lambda, \nu)$ and $\mathsf{C}_2^{\mathcal{D}}(\Lambda, \nu)$; this is in contrast to the analogous situation in 19.22.

Theorem 20.6. *Every Bruhat-Tits pair whose building at infinity is*

$$\mathsf{B}_2^{\mathcal{D}}(\Lambda) = \mathsf{C}_2^{\mathcal{D}}(\Lambda)$$

for some indifferent set $\Lambda = (K, K_0, L_0)$ *is of the form* $\tilde{\mathsf{B}}_2^{\mathcal{D}}(\Lambda, \nu) = \tilde{\mathsf{C}}_2^{\mathcal{D}}(\Lambda, \nu)$ *for some valuation* ν *of* K.

Proof. This holds by 3.41.ii, 16.4, 20.2 and 20.5. □

We now take a quick look at completions and gems.

Proposition 20.7. *Let* (K, K_0, L_0) *be an indifferent set, let* ν *be a discrete valuation of* K, *let* $\hat{K}$ *be the completion of* K *with respect to* ν, *let* $\hat{K}_0$ *be the closure of* K_0 *in* $\hat{K}$ *and let* $\hat{L}_0$ *be the closure of* L_0 *in* $\hat{K}$. *Then* $(\hat{K}, \hat{K}_0, \hat{L}_0)$ *is an indifferent set.*

Proof. This follows directly from the definition of an indifferent set in 10.1 in [36]. □

Theorem 20.8. *Let* $\Lambda = (K, K_0, L_0)$ *be an indifferent set, let* ν *be a discrete valuation of* K, *let* $\hat{\Lambda} = (\hat{K}, \hat{K}_0, \hat{L}_0)$ *be as in 20.7 and let*

$$(\Delta, \mathcal{A}) := \tilde{\mathsf{X}}_\ell^{\mathcal{D}}(\Lambda, \nu)$$

be as in 20.5 for $\mathsf{X} = \mathsf{B}$ *or* C. *Then* Δ *is completely Bruhat-Tits and*

$$\tilde{\mathsf{X}}_\ell^{\mathcal{D}}(\hat{\Lambda}, \nu)$$

is its completion.

Proof. By 30.16, $\mathsf{X}_2^{\mathcal{D}}(\Lambda)$ is a subbuilding of $\mathsf{X}_2^{\mathcal{D}}(\hat{\Lambda})$. The claim holds, therefore, by 17.4 and 17.9. □

Definition 20.9. Let $\Lambda = (K, K_0, L_0)$ be an indifferent set, let ν be a valuation of K and let $\bar{K}$ be the residue field of K with respect to ν. Let $\bar{K}_0$ be the image of $K_0 \cap \mathcal{O}_K$ in $\bar{K}$ (where $\mathcal{O}_K$ is the ring of integers with respect to ν), let $\bar{L}_0$ be the image of $L_0 \cap \mathcal{O}_K$ in $\bar{K}$ and let

$$\bar{\Lambda} = (\bar{K}, \bar{K}_0, \bar{L}_0).$$

Then $\bar{\Lambda}$ is an indifferent set which we call the *residue* of Λ with respect to ν.

By 18.5, we know that the group G° defined in 18.1 has at most two orbits in the set of gems of Δ.

Theorem 20.10. *Let*

$$(\Delta, \mathcal{A}) = \tilde{\mathsf{X}}_2^{\mathcal{D}}(\Lambda, \nu)$$

for some indifferent set $\Lambda = (K, K_0, L_0)$ *and some valuation* ν *of* K *(as defined in 20.5), where* $\mathsf{X} = \mathsf{B}$ *or* C. *Let* (δ_K, δ_L) *be the* ν*-index of* Λ *(as defined in 20.1). Let* p *be an element of* K_0 *such that* $\nu(p) = \delta_K$ *and let* q *be an element of* L_0 *such that* $\nu(q) = \delta_L$. *Let* $\Lambda_K = (K, pK_0, L_0)$ *and* $\Lambda_L = (K, K_0, qL_0)$ *(so* Λ_K *and* Λ_L *are both indifferent sets by 10.6 of [36][1]) let* $\bar{\Lambda}$, $\overline{\Lambda_K}$ *and* $\overline{\Lambda_L}$ *be the residues of* Λ, Λ_K *and* Λ_L *as defined in 20.9. Then the following hold:*

---

[1]The indifferent sets Λ_K and Λ_L are, in fact, *translates* of Λ as defined in 10.6 of [36].

(i) *If* $\delta_K \neq \delta_L$, *then every gem of* Δ *is isomorphic to* $\mathsf{B}_2^{\mathcal{D}}(\bar{\Lambda}) = \mathsf{C}_2^{\mathcal{D}}(\bar{\Lambda})$ *or* $\mathsf{B}_2^{\mathcal{D}}(\overline{\Lambda_K}) = \mathsf{C}_2^{\mathcal{D}}(\overline{\Lambda_K})$.

(ii) *If* $\delta_K = \delta_L$, *then every gem of* Δ *is isomorphic to* $\mathsf{B}_2^{\mathcal{D}}(\bar{\Lambda}) = \mathsf{C}_2^{\mathcal{D}}(\bar{\Lambda})$ *or* $\mathsf{B}_2^{\mathcal{D}}(\overline{\Lambda_L}) = \mathsf{C}_2^{\mathcal{D}}(\overline{\Lambda_L})$.

Proof. Let $A \in \mathcal{A}$, let R_0 be as in 18.1 and let R_1 be the gem of Δ described in 18.28. Let $\Sigma = A^\infty$, let d be a chamber of Σ, let $(G^\dagger, \xi)$ be the root datum of $\Delta_{\mathcal{A}}^\infty$ based at Σ and let $\phi = \phi_{R_0}$ be the valuation of $(G^\dagger, \xi)$ described in 13.8. By 20.5, we can assume that $(\Delta_{\mathcal{A}}^\infty, \Sigma, d)$ is the canonical realization of

$$\mathcal{Q}_{\mathcal{D}}(\Lambda) = (U_+, U_1, U_2, U_3, U_4)$$

as defined in 30.6, that $\phi_1(x_1(t)) = \nu(t)/\delta_K$ for all $t \in K_0^*$ and that $\phi_4(x_4(u)) = \nu(q(u))/\delta_L$ for all $u \in L_0^*$. By 18.27, $R_0 \cong \mathsf{B}_2^{\mathcal{D}}(\bar{\Lambda})$.

Suppose that $\delta_K \neq \delta_L$. By 20.1, we have $\delta_K = 1$ and $\delta_L = 2$. By 20.2, U_1 is long root group. By 10.2 in [36], $p^2 L_0 = L_0$. By 18.9, 18.28.i and 30.4, therefore, the maps $x_i(s) \mapsto x_i(ps)$ for $i = 1$ and 3, $x_2(t) \mapsto x_2(p^2 t)$, $x_4(t) \mapsto x_4(t)$ induce an isomorphism from R_1 to $\mathsf{B}_2^{\mathcal{D}}(\overline{\Lambda_K})$. Thus (ii) holds by 18.29.

Suppose that $\delta_K = \delta_L$. By 20.2, U_4 is a long root group. By 10.2 in [36], $qK_0 = K_0$. Therefore, the maps $x_1(s) \mapsto x_1(s)$, $x_3(s) \mapsto x_3(qs)$ and $x_i(t) \mapsto x_i(qt)$ for $i = 2$ and 4 induce an isomorphism from R_1 to $\mathsf{B}_2^{\mathcal{D}}(\overline{\Lambda_L})$ (again by 18.9, 18.28.i and 30.4). Thus (iii) holds, again by 18.29. □

Chapter Twenty One

Quadrangles of Type E_6, E_7 and E_8

In this chapter, we consider the case $X = \mathcal{Q}_{\mathcal{E}}$ of the problem framed at the end of Chapter 16. Our main results are 21.27 and 21.35 in which we give the classification of Bruhat-Tits pairs whose building at infinity is $\mathsf{B}_2^{\mathcal{E}}(\Lambda)$ (as defined in 30.15) for some quadratic space Λ of type E_6, E_7 or E_8 (as defined in 21.1).

We begin by repeating the definition of a quadratic space of type E_6, E_7 or E_8 (taken from 12.31 of [36]):

Definition 21.1. Let $\Lambda = (K, L, q)$ be a quadratic space. Then Λ is a *quadratic space of type* E_k for $k = 6$, 7 or 8 if q is anisotropic, there is a separable quadratic extension E/K with norm N such that q is, up to isomorphism, an orthogonal sum of the form

$$s_1N + s_2N + \cdots + s_dN$$

for scalars $s_1, s_2, \ldots, s_d \in K^*$, where $d = 3$ if $k = 6$, $d = 4$ if $k = 7$ and $d = 6$ if $k = 8$, with the additional conditions that the orthogonal sum

$$N - s_1s_2s_3s_4N \tag{21.2}$$

is anisotropic if $k = 7$ and the orthogonal sum

$$N + s_1s_2 \cdots s_6N \tag{21.3}$$

is *not* anisotropic if $k = 8$.

Notation 21.4. We will assume throughout this chapter that $\Lambda = (K, L, q)$ is a quadratic space of type E_k for $k = 6$, 7 or 8 and that 1 is an element of L such that $q(1) = 1$. (Thus q is unitary as defined in 19.16.) Let f be the bilinear form belonging to q. Since the extension E/F in 21.1 is separable, the form f is non-degenerate. Let

$$\Upsilon = (K, L, q, 1, X, \cdot, h, \theta) \tag{21.5}$$

be the *quadrangular algebra* (as defined in 1.17 of [38]) that is determined by Λ and the "basepoint" $1 \in L$.[1] Since f is non-degenerate, Υ is proper as defined in 1.27 in [38]. By 4.2 in [38], we can assume that Υ is δ*-standard* for some $\delta \in L$ as defined in 4.1 in [38]. This means that Υ satisfies all the

---

[1] In fact, Υ is only determined up to equivalence as defined in 1.22 of [38]. Equivalent quadrangular algebras give rise, however, to the same Moufang quadrangle (up to isomorphism). By 11.13 in [38], this quadrangle is also independent of the choice of the basepoint $1 \in L$. The existence of Υ is assured by 10.1 in [38].

identities in Chapter 4 of [38].[2] Let π be the map from X to L defined in 1.17(D1) of [38], let g be the map from $X \times X$ to K defined in 1.7(C3) of [38] and let ϕ be the map from $X \times L$ to K defined in 1.7(C4) of [38].[3] We also assume throughout this chapter that ν is a discrete valuation of K.

Convention 21.6. We always identify K with its image in L under the map $t \mapsto t \cdot 1$, where $1 \in L$ is the basepoint of Υ.

Notation 21.7. Let $S = S_\Lambda$ be the group defined in 11.10 in [38].[4] Thus S is the group whose underlying set is the product $X \times K$ and whose multiplication is given by

$$(a,t) \cdot (b,s) = \big(a + b, s + t + g(b,a)\big)$$

for all $(a,t),(b,s) \in S$, where g is as in 21.4. Moreover,

$$(a,t)^{-1} = \big(-a, -t + g(a,a)\big)$$

for all $(a,t) \in S$.

Proposition 21.8. *Let (K, L, q), θ, h, g, etc. be as in 21.4, let F be a field containing K, let (F, L_F, q_F) be as in 19.1 and let $X_F = X \otimes_K F$. Let $\cdot_F$ be the unique bilinear map from $X_F \times L_F$ to X_F such that*

$$(a \otimes s) \cdot_F (v \otimes t) = (a \cdot v) \otimes st$$

for all $(a,v) \in X \times L$ and all $s,t \in F$. Let h_F be the unique bilinear map from $X_F \times X_F$ to L_F such that

$$h_F(a \otimes s, b \otimes t) = h(a,b) \otimes st$$

for all $a, b \in X$ and all $s,t \in F$. Then there exists a unique map θ_F from $X_F \times L_F$ to L_F that is linear in the second variable such that

$$\begin{aligned} \theta_F\Big(\sum_{i=1}^m x_i \otimes s_i, v \otimes t\Big) &= \sum_{i=1}^m \theta(x_i, v) \otimes s_i^2 t \\ &\quad + \sum_{i<i'} h(x_i, x_{i'} \cdot v) \otimes s_i s_{i'} t \\ &\quad - \sum_{i<i'} g(x_i, x_{i'}) v \otimes s_i s_{i'} t \end{aligned} \tag{21.9}$$

for all $x_1, \ldots, x_m \in X$, for all $s_1, \ldots, s_m, t \in F$, for all m and for all $v \in L$. Furthermore,

$$\Upsilon_F := (F, L_F, q_F, 1, X_F, \cdot_F, h_F, \theta_F)$$

satisfies all the conditions in 1.17 of [38] defining a quadrangular algebra except perhaps the last one called (D2).

---

[2]The notation and the use of letters in [38] is the same as in Chapters 12–13 of [36] except that the basepoint $1 \in L$ of Υ is called ϵ in [36] and the map g defined in 13.26 of [36] and the map g defined in 1.17(C3) of [38] are "opposites" as explained in comment (viii) in [38, p. 7]. As in both [36] and [38], we often use the letter a in this chapter to denote a typical element in X; this is, of course, not the same a as in 16.3.

[3]There should be no danger later in this chapter of confusing the map ϕ with a valuation of a root datum (or the the map π with a uniformizer in K).

[4]By comment (viii) on page 7 of [38], this is the same as the group S defined in 16.6 in [36].

Proof. Choose a basis $x_1, \ldots, x_m$ of X and a basis $v_1, \ldots, v_d$ of L over K and define θ_F using these bases and the formula 21.9. By some calculation, Υ_F satisfies conditions (A1)–(D1) in 1.17 of [38]. The identity 21.9 then holds in general by induction with respect to m. □

Proposition 21.10. *The following identities hold:*

(i) $q(\theta(a,u) + tu) = q(\pi(a) + t)q(u)$ *and*
(ii) $q\big(\pi(au) + tq(u) + \phi(a,u)\big) = q(\pi(a) + t)q(u)^2$

for all $a \in X$, $u \in L$ *and* $t \in K$.[5]

Proof. Choose $a \in X$, $u \in L$ and $t \in K$. By 4.9 of [38],

$$f(\theta(a,u), u) = f(\pi(a), 1)q(u), \tag{21.11}$$

and by 4.22 of [38],

$$q\big(\pi(\theta(a,u))\big) = q(\pi(a))q(u). \tag{21.12}$$

Therefore

$$\begin{aligned} q\big(\theta(a,u) + tu\big) &= q(\theta(a,u)) + tf(\theta(a,u), u) + t^2 q(u) \\ &= \big(q(\pi(a)) + tf(\pi(a), 1) + t^2\big)q(u) \\ &= q(\pi(a) + t)q(u). \end{aligned}$$

Thus (i) holds.

Let $v^\sigma = f(v, 1) - v$ for all $v \in L$ (so, in particular, $1^\sigma = 1$). By 1.4 in [38], both q and f are σ-invariant. By 1.17(C4) of [38], we thus have

$$\pi(au) = \pi(a)^\sigma q(u) - f(u,1)\theta(a,u)^\sigma + f(\theta(a,u),1)u^\sigma + \phi(a,u).$$

By 4.5.iii of [38], the map ϕ is identically zero if $\operatorname{char}(K) \neq 2$. It follows (in all characteristics) that

$$\pi(au) + tq(u) + \phi(a,u) = (\pi(a)^\sigma + t)q(u) - v,$$

where

$$v = f(u,1)\theta(a,u)^\sigma - f(\theta(a,u),1)u^\sigma. \tag{21.13}$$

To prove (ii), it thus suffices to show that

$$q(v) = f(\pi(a)^\sigma + t, v)q(u). \tag{21.14}$$

We have

$$f(\theta(a,u), 1) = -f(\pi(a), u) + f(\pi(a), 1)f(u, 1) \tag{21.15}$$

by 4.9.iii of [38] and

$$f(\pi(a), \theta(a,u)) = q(\pi(a))f(u,1) \tag{21.16}$$

by 4.19 of [38]. By 21.13 (and the σ-invariance of f and q), we have

$$f(\pi(a)^\sigma, v) = f(u,1)f(\pi(a), \theta(a,u)) - f(\pi(a), u)f(\theta(a,u), 1)$$

[5] The terms t and $\phi(a,u)$ in these identities are to be interpreted as elements of L as indicated in 21.6.

and

$$q(v) = f(u,1)^2 q(\theta(a,u)) \\ + f(\theta(a,u),1)^2 q(u) - f(u,1) f(\theta(a,u),1) f(\theta(a,u),u).$$

Hence

$$f(\pi(a)^\sigma, v) = f(u,1)^2 q(\pi(a)) - f(\pi(a),u) f(\theta(a,u),1)$$

by 21.16 and

$$q(v) = f(u,1)^2 q(\pi(a)) q(u) \\ + f(\theta(a,u),1)^2 q(u) - f(u,1) f(\theta(a,u),1) f(\pi(a),1) q(u)$$

by 21.11 and 21.12. Since

$$f(\theta(a,u),1)^2 = f(\theta(a,u),1) f(\pi(a),1) f(u,1) - f(\theta(a,u),1) f(\pi(a),u)$$

by 21.15, it follows that

$$f(\pi(a)^\sigma, v) q(u) = q(v). \tag{21.17}$$

By 21.13, we have

$$f(1,v) = f(u,1) f(1,\theta(a,u)) - f(\theta(a,u),1) f(1,u) = 0$$

and hence $f(\pi(a)^\sigma, v) = f(\pi(a)^\sigma + t, v)$. By 21.17, therefore, 21.14 holds. □

Proposition 21.18. *Let ν, π and h be as in 21.4, let S be as in 21.7 and let*

$$\omega(a,t) = \nu\big(q(\pi(a)+t)\big) \tag{21.19}$$

for all $(a,t) \in S^*$. Suppose that Λ is ν-compatible as defined in 19.17 and that

$$\nu(q(h(a,b))) \geq \big(\omega(a,t) + \omega(b,s)\big)/2 \tag{21.20}$$

for all $(a,t),(b,s) \in S^*$. Then

$$\omega\big((a,t)\cdot(b,s)\big) \geq \min\{\omega(a,t), \omega(b,s)\}$$

for all $(a,t),(b,s) \in S^*$.

Proof. Let $(a,t),(b,s) \in S^*$ and let the map g from $X \times X$ to K be as in 21.4. Then

$$\pi(a+b) + t + s + g(b,a) = \pi(a) + \pi(b) + h(b,a) + t + s \tag{21.21}$$

by 1.17(C3) in [38] and

$$(a,t)\cdot(b,s) = \big(a+b, t+s+g(b,a)\big)$$

by 21.7. Therefore

$$\begin{aligned} \omega\big((a,t)\cdot(b,s)\big) &= \nu\Big(q\big(\pi(a+b)+t+s+g(b,a)\big)\Big) && \text{by 21.19} \\ &= \nu\Big(q\big(\pi(a)+\pi(b)+h(b,a)+t+s\big)\Big) && \text{by 21.21.} \end{aligned}$$

Thus

$$\omega\big((a,t)\cdot(b,s)\big) \geq \min\Big\{\nu\big(q(\pi(a)+t)\big), \nu\big(q(\pi(b)+s)\big), \nu\big(q(h(b,a))\big)\Big\}$$

by 19.4.ii. By 3.6 in [38], $q(h(a,b)) = q(h(b,a))$. Hence

$$\omega\big((a,t)\cdot(b,s)\big) \geq \min\big\{\nu\big(q(\pi(a)+t)\big), \nu\big(q(\pi(b)+s)\big)\big\}$$

by 21.20. □

Definition 21.22. We call $\Lambda = (K, L, q)$ *exceptionally ν-compatible* if the following conditions hold:

(i) Λ is ν-compatible as defined in 19.17.
(ii) The inequality 21.20 holds for all $(a, t), (b, s) \in S^*$.

Conjecture 21.23. We conjecture that condition (ii) of 21.22 follows from condition (i), in other words, that ν-compatibility implies exceptional ν-compatibility.

Proposition 21.24. *Let*

$$B = \{\nu\big(q(\pi(a) + t)\big) \mid (a, t) \in S^*\}.$$

Then $B = \mathbb{Z}$ or $2\mathbb{Z}$.

Proof. Since

$$q(\pi(sa) + s^2 t) = s^4 q(\pi(a) + t)$$

for all $(a, t) \in S^*$ and all $s \in K^*$, the set B is a union of congruence classes modulo 4. Since $q(\pi(0) + t) = t^2$ for each $t \in K$, we have $2\mathbb{Z} \subset B$. If $u = \pi(b) + s$ and $\eta = \nu(q(u))$ for some $(b, s) \in S^*$, then

$$\nu\Big(q\big(\pi(bu) + sq(u) + \phi(b, u)\big)\Big) = \eta + 2\nu(q(u)) = 3\eta$$

by 21.10.ii. Thus if $\nu\big(q(\pi(b) + s)\big)$ is odd for some $(b, s) \in S^*$, then $B = \mathbb{Z}$. □

Definition 21.25. Let $\delta_\Upsilon = |\mathbb{Z}/B|$, where B is as in 21.24. Thus $\delta_\Upsilon = 1$ or 2. We call δ_Υ the *ν-index* of the quadrangular algebra Υ.

Notation 21.26. Let δ_Υ be the ν-index of the quadrangular algebra Υ as defined in 21.25 and let δ_Λ be the ν-index of the quadratic space $\Lambda = (K, L, q)$ as defined in 19.17. By 19.17 and 21.24, we have

$$\delta_\Upsilon / \delta_\Lambda = 1 \text{ or } 2.$$

Theorem 21.27. *Let $\Lambda = (K, L, q)$ be a quadratic space of type E_6, E_7 or E_8, let ν be a discrete valuation of K, let*

$$\mathcal{Q}_\mathcal{E}(\Lambda) = (U_+, U_1, \ldots, U_4)$$

and $x_1, \ldots, x_4$ be as in 30.8.vi, let (Δ, Σ, d) be the canonical realization of $\mathcal{Q}_\mathcal{E}(\Lambda)$ as defined in 30.6 and let $(G^\dagger, \xi)$ be the root datum of Δ based at Σ. Let δ_Υ and δ_Λ be as in 21.26 (where Υ is as in 21.4), let δ be a positive integer and let

$$\phi_1(x_1(0, t)) = \delta\nu(t) \tag{21.28}$$

for all $t \in K^*$. Then the following hold:

(i) *The map ϕ_1 extends to a valuation ϕ of $(G^\dagger, \xi)$ if and only if Λ is exceptionally ν-compatible and $\delta = 2/\delta_\Upsilon$.*

(ii) *Suppose that* ϕ_1 *extends to a valuation* ϕ *of* $(G^\dagger, \xi)$ *and let* ι *be the associated root map of* Σ *(with target* $\mathsf{B}_2 = \mathsf{C}_2$*). Then* ι *is short if and only if* $\delta_\Upsilon/\delta_\Lambda = 2$*; and, after replacing* ϕ *by an equipollent valuation if necessary,*

$$\phi_1(x_1(a,t)) = \nu(q(\pi(a)+t))/\delta_\Upsilon \tag{21.29}$$

for all $(a,t) \in S^*$ *and*

$$\phi_4(x_4(u)) = \nu(q(u))/\delta_\Lambda \tag{21.30}$$

for all $u \in L^*$.

Proof. Let Ξ_d and Ξ_d° be as in 3.5. The four roots $a_1, \ldots, a_4$ in Ξ_d can be ordered so that $U_i = U_{a_i}$ for all $i \in [1,4]$. We then have $\Xi_d^\circ = \{a_1, a_4\}$ and $[a_1, a_4] = (a_1, a_2, a_3, a_4)$, where $[a_1, a_4]$ is as defined in 3.1.

Suppose first that Λ is exceptionally ν-compatible and that $\delta = 2/\delta_\Upsilon$. Let

$$\phi_i(x_i(a,t)) = \nu(q(\pi(a)+t))/\delta_\Upsilon$$

for $i = 1$ and 3 and for all $(a,t) \in S^*$ and let

$$\phi_i(x_i(u)) = \nu(q(u))/\delta_\Lambda$$

for $i = 2$ and 4 and all $u \in L^*$. Then $\phi_i(U_i^*) = \mathbb{Z}$ for all $i \in [1,4]$ and 21.28 holds. Let

$$\phi = \{\phi_1, \phi_2, \phi_3, \phi_4\}.$$

By 19.4.ii, 21.18 and 21.22, ϕ satisfies 15.1.i. Let ι be a root map of Σ (with target B_2) that is long (as defined in 16.12) if $\delta_\Upsilon/\delta_\Lambda = 1$ and short if $\delta_\Upsilon/\delta_\Lambda = 2$. Thus U_1 is a long root group (i.e. the root a_1 is long) if and only if ι is long. By 16.6 of [36], 19.4.iii, 21.10 and 21.22.ii, ϕ satisfies 15.1.ii with respect to ι, ϕ is exact as defined in 15.24, the (a_1, a_4)-coefficient of ϕ at a_3 (as defined in 15.24) is $(1, 2\delta_\Lambda/\delta_\Upsilon)$ and the (a_4, a_1)-coefficient of ϕ at a_2 is $(1, \delta_\Upsilon/\delta_\Lambda)$. By 15.25, 21.10 and 30.36.vi, it follows that ϕ satisfies 15.4.i. By 30.38.vi, ϕ satisfies 15.4.ii. Thus ϕ is a viable partial valuation. By 15.21, ϕ extends to a valuation of $(G^\dagger, \xi)$.

Suppose, conversely, that ϕ is a valuation of $(G^\dagger, \xi)$ with respect to a root map ι of Σ (with target B_2) such that 21.28 holds for some positive integer δ (with $\phi_1 = \phi_{a_1}$). Let $\phi_i = \phi_{a_i}$ for all $i \in [2,4]$. By 2.50.i and 3.41.i, we have

$$\phi_4(x_4(u)) = -\Big(\phi_1\big(x_3(0,1)^{m_\Sigma(x_4(u))}\big) - \phi_3(x_3(0,1))\Big)/\delta_0$$

for all $u \in L^*$, where

$$\delta_0 = \begin{cases} 2 & \text{if } \iota \text{ is long and} \\ 1 & \text{if } \iota \text{ is short.} \end{cases}$$

By 32.10 in [36],

$$x_3(0,1)^{m_\Sigma(x_4(u))} = x_1\big(0, 1/q(u)\big)$$

for all $u \in L^*$. Thus

$$\phi_4(x_4(u)) = \big(\delta\nu(q(u)) + \phi_3(x_3(0,1))\big)/\delta_0$$

for all $u \in L^*$ by 21.28. By 3.21, $\phi_i(U_i^*) = \mathbb{Z}$ for all $i \in [1,4]$. By 19.17, $\nu(q(L^*)) = \delta_\Lambda\mathbb{Z}$. It follows that

$$\delta_0/\delta = \delta_\Lambda \tag{21.31}$$

and $\phi_3(x_3(0,1))/\delta_0 \in \mathbb{Z}$. By 3.43, therefore, we can assume that 21.30 holds. Since ϕ satisfies condition (V1), 19.4.ii holds. Thus (K, L, q) is ν-compatible.

By 2.50.i and 3.41.i again,

$$\phi_1(x_1(a,t)) = -\delta_0\Big(\phi_4\big(x_2(1)^{m_\Sigma(x_1(a,t))}\big) - \phi_2(x_2(1))\Big)/2$$

for all $(a,t) \in S^*$. By 32.10 of [36] again, we have

$$x_2(1)^{m_\Sigma(x_1(a,t))} = x_4\big((\pi(a)+t)/q(\pi(a)+t)\big)$$

for all $(a,t) \in S^*$. Thus by 21.30, we conclude that

$$\phi_1(x_1(a,t)) = \delta_0\nu\big(q(\pi(a)+t)\big)/(2\delta_\Lambda) + \delta_0\phi_2(x_2(1))/2$$

for all $(a,t) \in S^*$. By 21.28, $\phi_1(x_1(0,1)) = 0$. Therefore $\phi_2(x_2(1)) = 0$. Since $\phi_1(U_1^*) = \mathbb{Z}$, it follows that 21.29 holds and $\delta_\Upsilon = 2\delta_\Lambda/\delta_0$. Therefore

$$\delta_0 = 2\delta_\Lambda/\delta_\Upsilon. \tag{21.32}$$

Hence $\delta_\Upsilon/\delta_\Lambda = 2$ if and only if $\delta_0 = 1$ if and only if ι is short. Thus (ii) holds.

By 21.31 and 21.32, $\delta = 2/\delta_\Upsilon$. To complete the proof of (i), it thus remains only to show that Λ is exceptionally ν-compatible, i.e. that 21.20 holds. By 16.6 in [36],

$$[x_1(a,t), x_3(b,s)^{-1}] = x_2(h(a,b)) \tag{21.33}$$

for all $(a,t),(b,s) \in S$. By 3.44 and 30.38.vi applied to 21.29 and 21.30, we have

$$\phi_2(x_2(u)) = \nu(u)/\delta_\Lambda$$

for all $u \in L^*$ and

$$\phi_3(x_3(a,t)) = \nu(q(\pi(a)+t))/\delta_\Upsilon$$

for all $(a,t) \in S^*$. By condition (V2) applied to 21.33, we have

$$\phi_2\Big(x_2\big(h(a,b)\big)\Big) \geq \Big(\phi_1\big(x_1(a,t)\big) + \phi_3\big(x_3(b,s)\big)\Big)/2$$

for all $(a,t),(b,s) \in S$ if ι is long and

$$\phi_2\Big(x_2\big(h(a,b)\big)\Big) \geq \phi_1\big(x_1(a,t)\big) + \phi_3\big(x_3(b,s)\big)$$

for all $(a,t),(b,s) \in S$ if ι is short. It follows that

$$\phi_2\Big(x_2\big(h(a,b)\big)\Big) \geq \Big(\phi_1\big(x_1(a,t)\big) + \phi_3\big(x_3(b,s)\big)\Big)\delta_\Upsilon/(2\delta_\Lambda)$$

and hence

$$\nu\Big(q\big(h(a,b)\big)\Big) \geq \Big(\nu\big(q(\pi(a)+t)\big) + \nu\big(q(\pi(b)+s)\big)\Big)/2$$

for all $(a,t),(b,s) \in S$ whether ι is short or long. □

Notation 21.34. Let Λ, Δ, ν, etc. be as in 21.27, suppose that Λ is exceptionally ν-compatible and let ϕ be the unique valuation of $(G^\dagger, \xi)$ satisfying 21.29 and 21.30. We denote the pair $(\Delta^{\rm aff}, \mathcal{A})$ obtained by applying 14.47 to these data by

$$\tilde{\mathsf{B}}_2^{\mathcal{E}}(\Lambda, \nu) \text{ or } \tilde{\mathsf{C}}_2^{\mathcal{E}}(\Lambda, \nu).$$

Thus the pair $\tilde{\mathsf{B}}_2^{\mathcal{E}}(\Lambda, \nu)$ defined in 21.34 exists precisely when the quadratic space Λ is exceptionally ν-compatible. Note that we do not distinguish between $\tilde{\mathsf{B}}_2^{\mathcal{E}}(\Lambda, \nu)$ and $\tilde{\mathsf{C}}_2^{\mathcal{E}}(\Lambda, \nu)$.

Theorem 21.35. *Every Bruhat-Tits pair whose building at infinity is*

$$\mathsf{B}_2^{\mathcal{E}}(\Lambda) = \mathsf{C}_2^{\mathcal{E}}(\Lambda)$$

for some quadratic space $\Lambda = (K, L, q)$ of type E_6, E_7 or E_8 is of the form $\tilde{\mathsf{B}}_2^{\mathcal{E}}(\Lambda, \nu) = \tilde{\mathsf{C}}_2^{\mathcal{E}}(\Lambda, \nu)$ for some valuation ν of K such that Λ is exceptionally ν-compatible.

Proof. This holds by 3.41.ii, 16.4, 21.27 and 21.34. □

We turn now to completions. As usual, we let $\hat{K}$ be the completion of K with respect to ν.

Proposition 21.36. *Let Λ be a quadratic space of type E_k for $k = 6$, 7 or 8 and suppose that Λ is exceptionally ν-compatible as defined in 21.22. Let $\Lambda_{\hat{K}} = (\hat{K}, L_{\hat{K}}, q_{\hat{K}})$ and*

$$\Upsilon_{\hat{K}} = (\hat{K}, L_{\hat{K}}, q_{\hat{K}}, 1, X_{\hat{K}}, \cdot_{\hat{K}}, h_{\hat{K}}, \theta_{\hat{K}})$$

(as defined in 19.1 and 21.8). Then the following hold:

(i) *$\Upsilon_{\hat{K}}$ is a quadrangular algebra.*
(ii) *$\Lambda_{\hat{K}}$ is an exceptionally ν-compatible quadratic space of type E_k.*

Proof. Let $\pi_{\hat{K}}(\hat{a}) = \theta_{\hat{K}}(\hat{a}, 1)$ for all $\hat{a} \in X_{\hat{K}}$ and let $\hat{K}$ be identified with its image in $L_{\hat{K}}$ under the map $t \mapsto t \cdot_{\hat{K}} 1$. To prove (i), it suffices, by 21.8, to show that $\Upsilon_{\hat{K}}$ satisfies condition (D2) in 1.17 of [38], i.e. that $\pi_{\hat{K}}(\hat{a}) \in \hat{K}$ only if $\hat{a} = 0$. Suppose that $\hat{a}$ is a non-zero element of $X_{\hat{K}}$ such that $\pi_{\hat{K}}(\hat{a}) \in \hat{K}$. Let $\hat{t} = \pi_{\hat{K}}(\hat{a})$. By 13.16 in [36], there exists $b \in X$ such that $h_{\hat{K}}(\hat{a}, b) \neq 0$. Let $x_1, \ldots, x_m$ be an ordered basis of X over K and let $\hat{t}_1, \ldots, \hat{t}_m \in \hat{K}$ be the coordinates of $\hat{a}$ with respect to this ordered basis. Choose $t_{i,k}$ and t_k in K such that

$$\lim_{k \to \infty} t_{i,k} = \hat{t}_i$$

for all $i \in [1, m]$ and

$$\lim_{k \to \infty} t_k = -\hat{t}$$

and let

$$a_k = t_{1,k}x_1 + \cdots + t_{m,k}x_m$$

for all $k \ge 1$. By 19.4.ii and 21.9, we have

$$\lim_{k\to\infty} \pi(a_k) = \pi_{\hat{K}}(\hat{a}) = \hat{t}.$$

This means that

$$\lim_{k\to\infty} \nu\big(q(\pi(a_k)+t_k)\big) = \infty.$$

We have

$$\lim_{k\to\infty} h(a_k, b) = h_{\hat{K}}(\hat{a}, b) \neq 0$$

and hence $\nu(h(a_k, b))$ is a constant for all k sufficiently large. Thus $(a_k, t_k) \in S$ for all k, $(b, 0) \in S$ and

$$\nu(h(a_k, b)) < \big(\nu(q(\pi(a_k)+t_k)) + \nu(\pi(b))\big)/2$$

for all k sufficiently large. This is impossible by 21.22.ii. With this contradiction, we conclude that (i) holds.

Since $\dim_{\hat{K}} L_{\hat{K}} > 4$, the quadrangular algebra $\Upsilon_{\hat{K}}$ is not special as defined in 1.28 of [38]. Since the bilinear form f is non-degenerate, $\Upsilon_{\hat{K}}$ is regular as defined in 1.27 of [38]. By 3.2 in [38], it follows that $\Lambda_{\hat{K}}$ is of type E_k. By 19.4, $\Lambda_{\hat{K}}$ satisfies 21.22.i. It satisfies 21.22.ii by 21.9 and continuity. Thus (ii) holds. □

Notation 21.37. Suppose that Λ is exceptionally ν-compatible as defined in 21.22 and let S be as in 21.7. Let $\partial_\omega\colon S \times S \to \mathbb{R}$ be given as follows:

$$\partial_\omega\big((a,t),(b,s)\big) = \begin{cases} 2^{-\omega\left((a,t)(b,s)^{-1}\right)} & \text{if } (a,t) \neq (b,s) \text{ and} \\ 0 & \text{if } (a,t) = (b,s), \end{cases}$$

where $\omega\colon S^* \to \mathbb{Z}$ is as in 21.19. By 17.3 and 21.29, ∂_ω is a metric on the group S.

Lemma 21.38. *Suppose that Λ is exceptionally ν-compatible and that K is complete. Then the following hold:*

(i) *X is complete with respect to ω (i.e. with respect to the restriction of the metric in 21.37 to the image of $a \mapsto (a, 0)$ of X in S).*
(ii) *If $(a_i, t_i)_{i\ge1}$ is a Cauchy sequence in S, then $(a_i)_{i\ge1}$ is a Cauchy sequence in X and $(t_i)_{i\ge1}$ is a Cauchy sequence in K.*
(iii) *If a sequence $(a_i)_{i\ge1}$ of elements in X converges to $a \in X$ with respect to ω and a sequence $(t_i)_{i\ge1}$ of elements in K converges to an element $t \in K$ with respect to ν, then the sequence $(a_i, t_i)_{i\ge1}$ of elements in S converges to (a, t) with respect to ω.*
(iv) *S is complete with respect to ω (i.e. with respect to the metric ∂_ω defined in 21.37).*

Proof. Assertion (i) holds by 17.13 (with ω in place of Q and 4 in place of m). Let g and δ be as in 21.4. Then

$$g(a,b) = f(h(a,b),\delta) \tag{21.39}$$

for all $a, b \in X$ by 4.3 in [38]. Suppose that $(a_i, t_i)_{i\geq 1}$ is a Cauchy sequence of elements of S with respect to ω. By 21.7,

$$(a_j, t_j) \cdot (a_i, t_i)^{-1} = \big(a_j - a_i, t_j - t_i - g(a_i, a_j - a_i)\big) \tag{21.40}$$

for all i, j. By 4.1 in [38], $f(\pi(a), \delta) = 0$ for all $a \in X$ but $f(1, \delta) \neq 0$. By 19.4.i, for each $N > 0$, there exists an M such that

$$\begin{aligned}\nu\big(f(\pi(a_j - a_i) + t_j - t_i - g(a_i, a_j - a_i), \delta)\big)\\ = \nu\big((t_j - t_i - g(a_i, a_j - a_i))f(1, \delta)\big) \geq N\end{aligned}$$

for all $i, j \geq M$. By 21.40, it follows that $(a_i)_{i\geq 1}$ is a Cauchy sequence in X. In particular $\nu(q(\pi(a_i)))$ is bounded from below. It follows from 19.4.iii, 21.22.i and 21.39 that for each $N > 0$, there exists M such that

$$\nu(g(a_i, a_j - a_i)) \geq N$$

for all $i, j \geq M$. It follows that also $(t_i)_{i\geq 1}$ is a Cauchy sequence. Thus (ii) holds.

Let $(a_i)_{i\geq 1}$ be a sequence of elements in X that converges to $a \in X$ with respect to ω and let $(t_i)_{i\geq 1}$ be a sequence of elements in K that converges to an element $t \in K$ with respect to ν. The sequence $(\omega(a_i))_{i\geq 1}$ is bounded from below. It follows by 19.4.iii, 21.22.i and 21.39 (as in the previous paragraph) that for each $N > 0$, there exists M such that

$$\nu(g(a_i, a - a_i)) \geq N$$

for all $i \geq M$. Since

$$(a, t) \cdot (a_i, t_i)^{-1} = \big(a - a_i, t - t_i - g(a_i, a - a_i)\big)$$

for all $i \geq 1$, we conclude that the sequence $(a_i, t)_{i\geq 1}$ converges to (a, t) with respect to ω. Thus (iii) holds.

Now suppose that $(a_i, t_i)_{i\geq 1}$ is a Cauchy sequence of elements in S with respect to ω. By (ii), $(a_i)_{i\geq 1}$ is a Cauchy sequence in X with respect to ω and $(t_i)_{i\geq 1}$ is a Cauchy sequence in K with respect to ν. By (i), therefore, there are elements $a \in X$ and $t \in K$ such that $(a_i)_{i\geq 1}$ converges to a with respect to ω and $(t_i)_{i\geq 1}$ converges to t with respect to ν. By (iii), it follows that $(a_i, t_i)_{i\geq 1}$ converges to (a, t) with respect to ω. Thus (iv) holds. □

Proposition 21.41. *Suppose that Λ is exceptionally ν-compatible. Let $\Lambda_{\hat{K}}$ and $\Upsilon_{\hat{K}}$ be as as defined in 19.1 and 21.8, where $\hat{K}$ is the completion of K with respect to ν. Let $\hat{S}$ be the group, let $\hat{\omega}$ be the map and let $\partial_{\hat{\omega}}$ be the metric on $\hat{S}$ obtained by applying 16.6 of [36], 21.19 and 21.37 to $\Upsilon_{\hat{K}}$. Then $\partial_{\hat{\omega}}$ is an extension of ∂_ω, $\hat{S}$ is complete with respect to $\partial_{\hat{\omega}}$ and S is a dense subgroup of $\hat{S}$.*

Proof. By 21.38.iv, $\hat{S}$ is complete. By 17.13, X is dense in $X_{\hat{K}} = X \otimes_K \hat{K}$ with respect to ω. By 21.38.iii, it follows that S is dense in $\hat{S}$. □

Theorem 21.42. *Suppose that Λ is a quadratic space of type E_k for $k = 6$, 7 or 8 and suppose that Λ is exceptionally ν-compatible (as defined in 21.22).*

Let $\hat{K}$ be the completion of K with respect to ν, let $\Lambda_{\hat{K}}$ be as in 19.1 and let $(\Delta, \mathcal{A}) = \tilde{\mathsf{X}}_2^{\mathcal{E}}(\Lambda, \nu)$ be as in 21.35 for $\mathsf{X} = \mathsf{B}$ or C. Then $\Lambda_{\hat{K}}$ is also an exceptionally ν-compatible quadratic space of type E_k, Δ is completely Bruhat-Tits and its completion is $\tilde{\mathsf{X}}_2^{\mathcal{E}}(\Lambda_{\hat{K}}, \nu)$.

Proof. By 30.16, $\mathsf{X}_2^{\mathcal{E}}(\Lambda)$ is a subbuilding of $\mathsf{X}_2^{\mathcal{E}}(\Lambda_{\hat{K}})$. The claim holds, therefore, by 17.4, 17.9, 21.36 and 21.41. □

Remark 21.43. Many calculations with quadrangular algebras are required to determine the structure of the gems of a building Δ such that $(\Delta, \mathcal{A}) = \tilde{\mathsf{X}}_2^{\mathcal{E}}(\Lambda, \nu)$ for some pair (Λ, ν). Since the results are not needed anywhere else in this book, we have decided to publish them separately at a later date.

Chapter Twenty Two

Quadrangles of Type F_4

In this chapter, we consider the case $X = \mathcal{Q}_{\mathcal{F}}$ of the problem framed at the end of Chapter 16. Our main results are 22.16 and 22.29 in which we give the classification of Bruhat-Tits pairs whose building at infinity is $\mathsf{B}_2^{\mathcal{F}}(\Lambda)$ (as defined in 30.15) for some quadratic space Λ of type F_4. See 14.1 of [36] for the definition of a quadratic space of type F_4 and 30.15 for the definition of the building $\mathsf{B}_2^{\mathcal{F}}(\Lambda)$.

Throughout this chapter, we make the following assumptions:

Notation 22.1. Let

$$\Lambda = (K, L, q)$$

be a quadratic space of type F_4 and let ν be a valuation of K. Thus, in particular, $\text{char}(K) = 2$. We set $\Lambda_K = \Lambda$ and $q_K = q$. Let the vector spaces X_0 and W_0, the subfield F of K and the quadratic forms q_1 and q_2 be as in in 14.3–14.12 of [36]. By 14.4 of [36],

$$K^2 \subset F, \tag{22.2}$$

so F can be considered as a vector space over K with respect to the scalar multiplication $*$ given by $s * t = st^2$ for all $(s,t) \in F \times K$ and we can identify L with $W_0 \oplus F$ so that

$$\Lambda_K = (K, W_0 \oplus F, q_K)$$

and

$$q_K(b, s) = q_1(b) + s \tag{22.3}$$

for all $(b, s) \in W_0 \oplus F$. Since $\hat{q}$ has another meaning here, we will denote by q_F the quadratic form called $\hat{q}$ in 14.12 of [36] and let

$$\Lambda_F = (F, X_0 \oplus K, q_F). \tag{22.4}$$

Thus

$$q_F(a, t) = q_2(a) + t^2 \tag{22.5}$$

for all $(a, t) \in W_0 \times K$.[1] Let Θ, Υ, ψ and ν be the maps defined in 14.15 and 14.16 of [36]. We rename the last of these four maps ω since ν already denotes the valuation of K chosen above.

---

[1] As in [36], we often use the letter a in this chapter to denote a typical element in X_0; this is, of course, not the same a as in 16.3.

Remark 22.6. By 14.13 of [36], Λ_F is also a quadratic form of type F_4; in particular, it is anisotropic. In fact, the quadratic spaces Λ_K and Λ_F are opposites as defined in 28.46 of [36]. The quadratic space Λ_F is obtained from Λ_K by the formulas given in 14.6 and 14.12 of [36]. If we start with Λ_F and apply these same formulas, we obtain a quadratic space that is similar to Λ_K.

Proposition 22.7. *The following identities hold:*

(i) $q_K\big(\Theta(a,b)+tb, q_F(a,t)s+\psi(a,b)\big) = q_K(b,s)q_F(a,t)$;
(ii) $q_F\big(\Upsilon(a,b)+sa, q_K(b,s)t+\omega(a,b)\big) = q_K(b,s)^2 q_F(a,t)$;

for all $(b,s) \in W_0 \oplus F$ *and all* $(a,t) \in X_0 \oplus K$.

Proof. Keeping 22.3 and 22.5 in mind, we have

$$\begin{aligned} &q_K\big(\Theta(a,b)+tb, q_F(a,t)s+\psi(a,b)\big)\\ &\qquad = q_1(\Theta(a,b)+tb) + q_F(a,t)s + \psi(a,b)\\ &\qquad = q_1(\Theta(a,b)) + t^2q_1(b) + q_F(a,t)s + \psi(a,b)\\ &\qquad\qquad \text{by 14.18.i in [36]}\\ &\qquad = q_1(b)q_2(a) + t^2q_1(b) + q_F(a,t)s\\ &\qquad\qquad \text{by 14.18.xi in [36]}\\ &\qquad = q_K(b,s)q_F(a,t) \end{aligned}$$

and

$$\begin{aligned} &q_F\big(\Upsilon(a,b)+sa, q_K(b,s)t+\omega(a,b)\big)\\ &\qquad = q_2(\Upsilon(a,b)+sa) + q_K(b,s)^2t^2 + \omega(a,b)^2\\ &\qquad = q_2(\Upsilon(a,b)) + s^2q_2(a) + q_K(b,s)^2t^2 + \omega(a,b)^2\\ &\qquad\qquad \text{by 14.18.ii in [36]}\\ &\qquad = q_1(b)^2q_2(a) + s^2q_2(a) + q_K(b,s)^2t^2\\ &\qquad\qquad \text{by 14.18.xii in [36]}\\ &\qquad = q_K(b,s)^2q_F(a,t) \end{aligned}$$

for all $(b,s) \in W_0 \oplus F$ and all $(a,t) \in X_0 \oplus K$. □

Proposition 22.8. $\nu(F^*) = \mathbb{Z}$ *or* $2\mathbb{Z}$, *where* ν *is the valuation chosen in 22.1.*

Proof. This holds by 22.2. □

Notation 22.9. Let $\nu_K = \nu$ and let ν_F denote the restriction of $\nu/\tilde{\delta}$ to F, where $\tilde{\delta} := |\mathbb{Z}/\nu(F^*)|$. Thus $\tilde{\delta} \le 2$ by 22.8 and ν_F is a valuation of F.

Notation 22.10. Let δ_K be the ν_K-index of Λ_K and let δ_F be the ν_F-index of Λ_F (as defined in 19.17).

Proposition 22.11. *Let* $\tilde{\delta}$ *be as in 22.9 and let* δ_K *and* δ_F *be as in 22.10. Then* $\delta_K \le \tilde{\delta} \le \tilde{\delta}\delta_F \le 2$. *In particular,* $\tilde{\delta}\delta_F/\delta_K = 1$ *or* 2.

Proof. By 19.17 and 22.8, δ_K, δ_F and $\tilde{\delta}$ are each equal to 1 or 2. Suppose that $\tilde{\delta} = 1$. Then $\nu(F^*) = \mathbb{Z}$. Since $q_K(0, s) = s$ for all $s \in F$, we have $\nu(q_K(0, F^*)) = \mathbb{Z}$ and hence $\delta_K = 1$. Now suppose that $\tilde{\delta} = 2$. Since $\nu_K(K^*) = \mathbb{Z}$ and $q_F(0, t) = t^2$ for all $t \in K$, we have $\nu_F(q_F(0, K^*)) = \nu_K(q_F(0, K^*))/2 = \mathbb{Z}$ and therefore $\delta_F = 1$. □

Notation 22.12. Let $\hat{K}$ be the completion of K with respect to ν_K and let $\hat{F}$ denote the closure of F in $\hat{K}$. We regard $\hat{F}$ as a vector space over $\hat{K}$ with respect to the scalar multiplication defined in 14.10 of [36]. The field $\hat{F}$ is also the completion of F with respect to ν_F. Let $W_{\hat{K}} = W_0 \otimes_K \hat{K}$, let $\hat{q}_1$ be the natural extension of q_1 to $W_{\hat{K}}$ (as defined in 19.1), let $Q_{\hat{K}}$ denote the quadratic form on $W_{\hat{K}} \oplus \hat{F}$ given by

$$Q_{\hat{K}}(\hat{b}, \hat{s}) = \hat{q}_1(\hat{b}) + \hat{s}$$

for all $(\hat{b}, \hat{s}) \in W_{\hat{K}} \oplus \hat{F}$ and let

$$\Lambda^{\hat{K}} = (\hat{K}, W_{\hat{K}} \oplus \hat{F}, Q_{\hat{K}}).^2$$

Let $X_{\hat{F}} = X_0 \otimes_F \hat{F}$, let $\hat{q}_2$ be the natural extension of q_2 to $X_{\hat{F}}$, let $Q_{\hat{F}}$ denote the quadratic form on $X_{\hat{F}} \oplus \hat{K}$ given by

$$Q_{\hat{F}}(\hat{a}, \hat{t}) = \hat{q}_2(\hat{a}) + \hat{t}^2$$

for all $(\hat{a}, \hat{t}) \in X_{\hat{F}} \oplus \hat{K}$ and let

$$\Lambda^{\hat{F}} = (\hat{F}, X_{\hat{F}} \oplus \hat{K}, Q_{\hat{F}}).$$

Proposition 22.13. *Suppose that $\Lambda_K = (K, L, q)$ is ν_K-compatible (as defined in 19.17), let $\hat{\Lambda} = (\hat{K}, \hat{L}, \hat{q})$ be as in 19.24.iii and let $\Lambda^{\hat{K}}$ be as in 22.12. We identify W_0 and F with their images in $L = W_0 \oplus F$ under the inclusions $b \mapsto (b, 0)$ and $s \mapsto (0, s)$. Let ρ_1 and ρ_2 be the extensions of these inclusions by continuity to $\hat{K}$-linear maps from $W_{\hat{K}}$ and $\hat{F}$ to $\hat{L}$ and let ρ be the map from $W_{\hat{K}} \oplus \hat{F}$ to $\hat{L}$ given by*

$$\rho(\hat{b}, \hat{s}) = \rho_1(\hat{b}) + \rho_2(\hat{s})$$

for all $(\hat{b}, \hat{s}) \in W_{\hat{K}} \oplus \hat{F}$. Then the following hold:

(i) *The map ρ is injective*[3] *and*

$$\hat{q}\big(\rho(\hat{b}, \hat{s})\big) = Q_{\hat{K}}(\hat{b}, \hat{s}) \qquad (22.14)$$

for all $(\hat{b}, \hat{s}) \in W_{\hat{K}} \oplus \hat{F}$.

(ii) *$\Lambda^{\hat{K}}$ is a quadratic space of type F_4.*

(iii) *The quadratic space Λ_F (introduced in 22.4) is ν_F-compatible.*

---

[2]The quadratic space $\Lambda^{\hat{K}}$ is not necessarily the same as the quadratic space $\Lambda_{\hat{K}}$ obtained by applying 19.1 with $E = \hat{K}$.

[3]In 22.30, we will show that ρ is, in fact, an isomorphism.

Proof. Let $\hat{W}$ be the closure of W_0 in $\hat{L}$ and let $\hat{q}_{\hat{W}}$ be the restriction of $\hat{q}$ to $\hat{W}$. By parts (iii) and (v) of 19.29 applied to (K, W_0, q_1), ρ_1 is an isomorphism from $(\hat{K}, W_{\hat{K}}, \hat{q}_1)$ to $(\hat{K}, \hat{W}, \hat{q}_{\hat{W}})$. Since $\rho_2(\hat{F}) \subset \hat{L}^\perp$, it follows that $\rho_1(W_{\hat{K}}) \cap \rho_2(\hat{F}) = 0$. Thus (i) holds. By 19.24.iii, $\hat{\Lambda}$ is anisotropic. From (i) it follows that $\Lambda^{\hat{K}}$ is anisotropic. Therefore $\Lambda^{\hat{K}}$ is a quadratic space of type F_4 by 14.1 in [36]. Thus (ii) holds. By 14.8 of [36] and 22.12, $\Lambda^{\hat{F}}$ is anisotropic. Hence $\Lambda^{\hat{F}}$ satisfies 19.4.iv. Thus (iii) holds (by 19.17). □

Corollary 22.15. *The quadratic space Λ_K is ν_K-compatible if and only if the quadratic space Λ_F is ν_F-compatible.*

Proof. This holds by 22.6 and 22.13.iii. □

Theorem 22.16. *Let*

$$\Lambda = (K, L, q)$$

be a quadratic space of type F_4, let ν be a discrete valuation of K, let

$$\mathcal{Q}_{\mathcal{F}}(\Lambda) = (U_+, U_1, \ldots, U_4)$$

and $x_1, \ldots, x_4$ be as in 30.8.vii, let (Δ, Σ, d) be the canonical realization of $\mathcal{Q}_{\mathcal{F}}(\Lambda)$ as defined in 30.6 and let $(G^\dagger, \xi)$ be the root datum of Δ based at Σ. Let $\Lambda_K = \Lambda$, let $q_K = q$, let $\nu_K = \nu$, let L be identified with $X_0 \oplus F$ as in 22.1, let $\Lambda_F = (F, X_0 \oplus K, q_F)$ be as in 22.4 and let $\tilde{\delta}$, ν_F, δ_K and δ_F be as in 22.9 and 22.10. Suppose that δ is a positive integer such that

$$\phi_1(x_1(0, t)) = \delta\nu_K(t) \tag{22.17}$$

for all $t \in K^*$. Then the following hold:

(i) *The map ϕ_1 extends to a valuation of $(G^\dagger, \xi)$ if and only if the quadratic space Λ_K is ν-compatible and $\delta = 2/(\tilde{\delta}\delta_F)$.*
(ii) *Suppose that ϕ_1 extends to a valuation ϕ of $(G^\dagger, \xi)$ and let ι be the associated root map of Σ (with target B_2). Then the root map ι is long if $\tilde{\delta}\delta_F/\delta_K = 1$ and short if $\tilde{\delta}\delta_F/\delta_K = 2$; and, after replacing ϕ by an equipollent valuation if necessary,*

$$\phi_1(x_1(a, t)) = \nu_F(q_F(a, t))/\delta_F \tag{22.18}$$

for all $(a, t) \in X_0 \oplus K$; and

$$\phi_4(x_4(b, s)) = \nu_K(q_K(b, s))/\delta_K \tag{22.19}$$

for all $(b, s) \in W_0 \oplus F$.

Proof. Let Ξ_d and Ξ_d° be as in 3.5. The four roots $a_1, \ldots, a_4$ in Ξ_d can be ordered so that $U_i = U_{a_i}$ for all $i \in [1, 4]$. We then have $\Xi_d^\circ = \{a_1, a_4\}$ and $[a_1, a_4] = (a_1, a_2, a_3, a_4)$, where $[a_1, a_4]$ is as defined in 3.1.

Suppose first that Λ_K is ν-compatible and that $\delta = 2/(\tilde{\delta}\delta_F)$. Let

$$\phi_i(x_i(a, t)) = \nu_F(q_F(a, t))/\delta_F$$

for $i = 1$ and 3 and for all $(a, t) \in X_0 \oplus K$, let

$$\phi_i(x_i(b, s)) = \nu_K(q_K(b, s))\delta_K$$

for $i = 2$ and 4 and for all $(b, s) \in W_0 \oplus F$ and let

$$\phi = \{\phi_1, \phi_2, \phi_3, \phi_4\}.$$

Then

$$\begin{aligned} \phi_1(x_1(0,t)) &= \nu_F(t^2)/\delta_F \\ &= \nu_K(t^2)/(\tilde{\delta}\delta_F) \quad \text{by 22.9} \\ &= 2\nu_K(t)(\tilde{\delta}\delta_F) \\ &= \delta\nu_K(t) \qquad (22.20) \end{aligned}$$

for all $t \in K^*$, so 22.17 holds. By 22.10, $\phi_i(U_i^*) = \mathbb{Z}$ for each $i \in [1, 4]$. Since Λ_K is ν_K-compatible, we have

$$\nu_K\big(q_K((b,s) + (b',s'))\big) \geq \min\{\nu_K(q_K(b,s)), \nu_K(q_K(b',s'))\} \qquad (22.21)$$

and

$$\nu_K\big(f_1((b,s),(b',s'))\big) \geq \big(\nu_K(q_K(b,s)) + \nu_K(q_K(b',s'))\big)/2 \qquad (22.22)$$

for all $(b, s), (b', s') \in W_0 \oplus F$ by 19.4. By 22.13.iii, the quadratic space Λ_F is ν_F-compatible. By 19.4 again, it follows that

$$\nu_F\big(q_F((a,t) + (a',t'))\big) \geq \min\{\nu_F(q_F(a,t)), \nu_F(q_F(a',t'))\} \qquad (22.23)$$

and

$$\nu_F\big(f_2((a,t),(a',t'))\big) \geq \big(\nu_F(q_F(a,t)) + \nu_F(q_F(a',t'))\big)/2 \qquad (22.24)$$

for all $(a, t), (a', t') \in X_0 \oplus K$. By 22.21 and 22.23, ϕ satisfies 15.1.i.

Let ι be a root map of Σ (with target B_2) that is long (as defined in 16.12) if $\tilde{\delta}\delta_F/\delta_K = 1$ and short if $\tilde{\delta}\delta_F/\delta_K = 2$. Thus U_1 is a long root group (i.e. the root a_1 is long) if and only if ι is long. By 16.7 of [36], 22.7, 22.22 and 22.24, ϕ satisfies 15.1.ii with respect to ι, ϕ is exact as defined in 15.24, the (a_1, a_4)-coefficient of ϕ at a_3 (as defined in 15.24) is $(1, 2\delta_K/(\delta_F\tilde{\delta}))$ and the (a_4, a_1)-coefficient of ϕ at a_2 is $(1, \delta_F\tilde{\delta}/\delta_K)$. By 15.25, 22.7 and 30.36.vii, it follows that ϕ satisfies 15.4.i. By 30.38.vii, ϕ satisfies 15.4.ii. We conclude that ϕ is a viable partial valuation of $(G^\dagger, \xi)$. By 15.21, ϕ extends to a valuation of $(G^\dagger, \xi)$.

Suppose, conversely, that ϕ is a valuation of $(G^\dagger, \xi)$ with respect to a root map ι such that 22.17 holds (with $\phi_1 = \phi_{a_1}$). Let $\phi_i = \phi_{a_i}$ for all $i \in [2, 4]$. By 2.50.i and 3.41.i, we have

$$\phi_4(x_4(b,s)) = -\Big(\phi_1\big(x_3(0,1)^{m_\Sigma(x_4(b,s))}\big) - \phi_3(x_3(0,1))\Big)/\delta_0$$

for all $(b, s) \in (W_0 \oplus F)^*$, where

$$\delta_0 = \begin{cases} 2 & \text{if } \iota \text{ is long and} \\ 1 & \text{if } \iota \text{ is short.} \end{cases}$$

By 32.11 in [36],

$$x_3(0,1)^{m_\Sigma(x_4(b,s))} = x_1(0, 1/q_K(b,s))$$

for all $(b,s) \in (W_0 \oplus F)^*$. Thus

$$\phi_4(x_4(b,s)) = \big(\delta\nu_K(q_K(b,s)) + \phi_3(x_3(0,1))\big)/\delta_0$$

for all $(b,s) \in (W_0 \oplus F)^*$. By 3.21, $\phi_i(U_i^*) = \mathbb{Z}$ for all $i \in [1,4]$. By 19.17, $\nu_K(q_K(L^*)) = \delta_K\mathbb{Z}$. It follows that

$$\delta_0 = \delta\delta_K \tag{22.25}$$

and $\phi_3(x_3(0,1))/\delta_0 \in \mathbb{Z}$. By 3.43, therefore, we can assume that 22.19 holds. Since ϕ satisfies condition (V1), 19.4.ii holds. Thus (K, L, q_K) is ν_K-compatible.

By 2.50.i and 3.41.i again, we have

$$\phi_1(x_1(a,t)) = -\delta_0\Big(\phi_4\big(x_2(0,1)^{m_\Sigma(x_1(a,t))}\big) - \phi_2(x_2(0,1))\Big)/2$$

By 32.11 in [36] as well as 22.19 and 22.25, it follows that

$$\begin{aligned}\phi_1(x_1(a,t)) &= -\delta_0\Big(\phi_4\big(x_4(0,1/q_F(a,t))\big) - \phi_2(x_2(0,1))\Big)/2\\ &= \delta\nu_K(q_F(a,t))/2 + \delta_0\phi_2(x_2(0,1))/2 \qquad (22.26)\end{aligned}$$

for all $(a,t) \in (X_0 \oplus K)^*$. By 22.17, therefore,

$$\begin{aligned}\delta\nu_K(t) &= \phi_1(x_1(0,t))\\ &= \delta\nu_K(q_F(0,t))/2 + \delta_0\phi_2(x_2(0,1))/2\\ &= \delta\nu_K(t^2)/2 + \delta_0\phi_2(x_2(0,1))/2\\ &= \delta\nu_K(t) + \delta_0\phi_2(x_2(0,1))/2\end{aligned}$$

for all $t \in K^*$. It follows that $\phi_2(x_2(0,1)) = 0$. By 22.26, therefore,

$$\phi_1(x_1(a,t)) = \delta\nu_K(q_F(a,t))/2 = \delta\tilde{\delta}\nu_F(q_F(a,t))/2$$

for all $(a,t) \in (X_0 \oplus K)^*$. Since $\phi_1(U_1^*) = \mathbb{Z}$, it follows that

$$\delta = 2/(\tilde{\delta}\delta_F) \tag{22.27}$$

and 22.18 holds. Thus (i) holds. By 22.25, we conclude that

$$\delta_0 = 2\delta_K/(\tilde{\delta}\delta_F).$$

Hence ι is long if and only if $\delta_0 = 2$ if and only if $\tilde{\delta}\delta_F/\delta_K = 1$. Thus (ii) holds. □

Notation 22.28. Let Δ, Λ, ν, etc. be as in 22.16, suppose that Λ is ν-compatible and let ϕ be the unique valuation of $(G^\dagger, \xi)$ satisfying 22.18 and 22.19. We denote by

$$\tilde{\mathsf{B}}_2^{\mathcal{F}}(\Lambda,\nu) \text{ or } \tilde{\mathsf{C}}_2^{\mathcal{F}}(\Lambda,\nu)$$

the pair $(\Delta, \mathcal{A})$ obtained by applying 14.47 to these data.

Thus the pair $\tilde{\mathsf{B}}_2^{\mathcal{F}}(\Lambda,\nu)$ defined in 22.28 exists precisely when the quadratic space Λ is ν-compatible. Note that we do not distinguish between $\tilde{\mathsf{B}}_2^{\mathcal{F}}(\Lambda,\nu)$ and $\tilde{\mathsf{C}}_2^{\mathcal{F}}(\Lambda,\nu)$.

Theorem 22.29. *Every Bruhat-Tits pair whose building at infinity is*

$$\mathsf{B}_2^{\mathcal{F}}(\Lambda) = \mathsf{C}_2^{\mathcal{F}}(\Lambda)$$

for some quadratic space (K, L, q) of type F_4 is of the form $\tilde{\mathsf{B}}_2^{\mathcal{F}}(\Lambda, \nu) = \tilde{\mathsf{C}}_2^{\mathcal{F}}(\Lambda, \nu)$ for some valuation ν of K such that Λ is ν-compatible.

Proof. This holds by 3.41.ii, 16.4, 22.16 and 22.28. □

We turn now to completions.

Proposition 22.30. *Let $\Lambda = (K, L, q)$ be a quadratic space of type F_4, let ν be a discrete valuation of K and suppose that Λ is ν-compatible (as defined in 19.17). Let*

$$\hat{\Lambda} = (\hat{K}, \hat{L}, \hat{q})$$

be as in 19.24.iii and let

$$\Lambda^{\hat{K}} = (\hat{K}, W_{\hat{K}} \oplus \hat{F}, Q_{\hat{K}})$$

be as in 22.12. Then the injection ρ from $W_{\hat{K}} \oplus \hat{F}$ to $\hat{L}$ given in 22.13 is, in fact, an isomorphism from $\Lambda^{\hat{K}}$ to $\hat{\Lambda}$.

Proof. We identify $W_{\hat{K}} \oplus \hat{F}$ with its image in $\hat{L}$. By 22.13.i, it suffices to show that $\hat{L} = W_{\hat{K}} \oplus \hat{F}$. Let $\hat{f}$ be the bilinear form belonging to $\hat{q}$. The restriction of $\hat{f}$ to $W_{\hat{K}}$ is non-degenerate, so

$$\hat{L} = W_{\hat{K}} \oplus W_{\hat{K}}^{\perp}. \tag{22.31}$$

We choose an ordered basis $v_1, \ldots, v_4$ of W_0 over K such that $f_1(v_1, v_2) = f_1(v_3, v_4) = 1$ and $\langle v_1, v_2 \rangle = \langle v_3, v_4 \rangle^{\perp}$, where f_1 is the bilinear form associated with q_1. Let

$$\hat{x} \in W_{\hat{K}}^{\perp}. \tag{22.32}$$

Then there exist sequences $(w_k)_{k \geq 1}$ of elements in W_0 and $(s_k)_{k \geq 1}$ of elements in F such that

$$\hat{x} = \lim_{k \to \infty} (w_k + s_k).$$

Let $t_1^{(k)}, \ldots, t_4^{(k)} \in K$ be the coordinates of w_k with respect to the ordered basis $v_1, \ldots, v_4$ (for each k). Then

$$\lim_{k \to \infty} \hat{f}(w_k, v_i) = \lim_{k \to \infty} \hat{f}(w_k + s_k, v_i) = \hat{f}(\hat{x}, v_i) = 0$$

for each $i \in [1, 4]$ by 22.32. By the choice of the basis $v_1, \ldots, v_4$, it follows that

$$\lim_{k \to \infty} t_i^{(k)} = 0$$

for each $i \in [1, 4]$ and thus

$$\lim_{k \to \infty} w_k = 0.$$

Hence $\hat{x} \in \hat{F}$. Therefore $\hat{L} = W_{\hat{K}} \oplus \hat{F}$ by 22.31. □

Theorem 22.33. *Let* $\Lambda = (K, L, q)$ *be a quadratic space of type* F_4*, let* ν *be a valuation of* K *such that* Λ *is* ν*-compatible, let*

$$(\Delta, \mathcal{A}) = \tilde{\mathsf{X}}_2^{\mathcal{F}}(\Lambda, \nu)$$

for $\mathsf{X} = \mathsf{B}$ *or* C *be as in 22.28 and let* $\Lambda^{\hat{K}}$ *be as in 22.12. Then* $\Lambda^{\hat{K}}$ *is a* ν*-compatible quadratic space of type* F_4*,* Δ *is completely Bruhat-Tits and its completion is* $\tilde{\mathsf{X}}_2^{\mathcal{F}}(\Lambda^{\hat{K}}, \nu)$*.*[4]

Proof. Since $\hat{K}$ is complete, $\Lambda^{\hat{K}}$ satisfies 19.4.iv and is thus ν-compatible. By 30.16, the building $\mathsf{X}_2^{\mathcal{F}}(\Lambda)$ is a subbuilding of $\mathsf{X}_2^{\mathcal{F}}(\Lambda^{\hat{K}})$. By 22.13.ii, the quadratic space $\Lambda^{\hat{K}}$ is of type F_4. Let $\Lambda_K = \Lambda$, let $\hat{\Lambda}_F$ be the quadratic space obtained by applying 19.24 to Λ_F and let $\Lambda^{\hat{F}}$ be as in 22.12. By 22.13.iii, Λ_F is ν_F-compatible. Hence by 22.30, $\Lambda^{\hat{F}} = \hat{\Lambda}_F$ as well as $\Lambda^{\hat{K}} = \hat{\Lambda}_K$. The claim holds, therefore, by 17.4 and 17.9. □

We do not determine the structure of the gems of a building Δ such that $(\Delta, \mathcal{A}) = \tilde{\mathsf{X}}_2^{\mathcal{F}}(\Lambda, \nu)$ for some Λ and ν in this book. We plan instead to publish these results separately at a later date; see 21.43.

[4]We are using the usual convention here (see 9.20) that the unique extension of ν to a valuation of the completion $\hat{K}$ of K is denoted by the same letter ν.

Chapter Twenty Three

Quadrangles of Involutory Type

In this chapter, we consider the case $X = \mathcal{Q}_\mathcal{I}$ of the problem framed at the end of Chapter 16. Our main result is 23.3. We use this result to give the classification of Bruhat-Tits pairs whose building at infinity is $\mathsf{B}_\ell^\mathcal{I}(\Lambda)$ for some involutory set Λ in 23.10. See 11.1 in [36] for the definition of an involutory set and 30.15 for the definition of the building $\mathsf{B}_\ell^\mathcal{I}(\Lambda)$.

We include honorary involutory sets and root group sequences

$$\mathcal{Q}_\mathcal{I}(\Lambda)$$

for honorary involutory sets Λ (as defined in 30.22) in this chapter. It will be convenient in some places, therefore, to refer to involutory sets as *genuine* involutory sets to distinguish them from honorary involutory sets.

Proposition 23.1. *Let (K, K_0, σ) be an involutory set, genuine or honorary, and let ν be a valuation of K. Suppose that ν is σ-invariant, i.e. that*

$$\nu(u^\sigma) = \nu(u)$$

for all $u \in K^*$. Then $\nu(K_0^*) = \mathbb{Z}$ or $2\mathbb{Z}$.

Proof. By 11.1 of [36], $1 \in K_0$ and $u^\sigma K_0 u \subset K_0$ for all $u \in K$. Therefore $0 = \nu(1) \in \nu(K_0^*)$ and for each $v \in K_0^*$, the set $\nu(K_0^*)$ contains all integers having the same parity as $\nu(v)$ (since ν is σ-invariant). □

Definition 23.2. Let $\Lambda = (K, K_0, \sigma)$ be an involutory set, genuine or honorary, and let ν be a discrete valuation of K. We say that Λ is *ν-compatible* if ν is σ-invariant. Suppose that Λ is ν-compatible. Then $\nu(K_0^*) = \mathbb{Z}$ or $2\mathbb{Z}$ by 23.1. Let $\delta = |\mathbb{Z}/\nu(K_0^*)|$. We will say that (K, K_0, σ) is *ramified* (respectively, *unramified*) *at* ν if $\delta = 2$ (respectively, $\delta = 1$).[1] We call δ the *ν-index* of Λ.

Theorem 23.3. *Let*

$$\Lambda = (K, K_0, \sigma)$$

be an involutory set, genuine or honorary, let ν be a discrete valuation of K, let

$$\mathcal{Q}_\mathcal{I}(\Lambda) = (U_+, U_1, \ldots, U_4)$$

[1]These notions coincide with the usual notions of a (totally) ramified and unramified extension in the special case that $F := K_0$ is a subfield of K, K/F is a separable quadratic extension, σ is the non-trivial element in $\mathrm{Gal}(K/F)$, K is complete with respect to ν and $\bar{K}/\bar{F}$ is separable. See 23.5.

and $x_1, \dots, x_4$ be as in 30.8.ii and 30.22, let (Δ, Σ, d) be the canonical realization of $\mathcal{Q}_{\mathcal{I}}(\Lambda)$ as defined in 30.6 and let $(G^\dagger, \xi)$ be the root datum of Δ based at Σ. Let δ be the ν-index of Λ as defined in 23.2 and let

$$\phi_4(x_4(t)) = \nu(t)$$

for all $t \in K^*$. Then the following hold:

(i) *The map ϕ_4 extends to a valuation of $(G^\dagger, \xi)$ if and only if (K, K_0, σ) is ν-compatible as defined in 23.2.*
(ii) *Suppose that ϕ_4 extends to a valuation of $(G^\dagger, \xi)$ and let ι be the associated root map from of Σ (with target B_2). Then ι is long if and only if $\delta = 2$; and, after replacing ϕ by an equipollent valuation if necessary,*

$$\phi_1(x_1(t)) = \nu(t)/\delta \tag{23.4}$$

for all $t \in K_0$.

Proof. Let Ξ_d and Ξ_d° be as in 3.5. The four roots $a_1, \dots, a_4$ in Ξ_d can be ordered so that $U_i = U_{a_i}$ for all $i \in [1, 4]$. We then have $\Xi_d^\circ = \{a_1, a_4\}$ and $[a_1, a_4] = (a_1, a_2, a_3, a_4)$, where $[a_1, a_4]$ is as defined in 3.1.

Suppose first that ν is σ-invariant. Let $\phi_2(x_2(t)) = \nu(t)$ for all $t \in K^*$, let

$$\phi_i(x_i(t)) = \nu(t)/\delta$$

for $i = 1$ and 3 and for all $t \in K_0^*$ and let

$$\phi = \{\phi_1, \phi_2, \phi_3, \phi_4\}.$$

By 23.2, $\phi_i(U_i^*) = \mathbb{Z}$ for all $i \in [1, 4]$. By 9.17.ii, ϕ satisfies 15.1.i. Let ι be a root map of Σ (with target B_2) that is long if $\delta = 2$ and short if $\delta = 1$ (as defined in 16.12). Thus U_4 is a long root group (i.e. the root a_4 is long) if and only if ι is long. By 16.2 of [36], 9.17.i and the σ-invariance of ν, ϕ satisfies 15.1.ii with respect to ι, ϕ is exact as defined in 15.24 and the (a_1, a_4)-index of ϕ at a_2 as defined in 15.24 is $(\delta, 1)$. By 15.25 and 30.36.ii, it follows that ϕ satisfies 15.4.i. By 30.38.ii, ϕ satisfies 15.4.ii. Thus ϕ is a viable partial valuation of $(G^\dagger, \xi)$. By 15.21, ϕ extends to a valuation of $(G^\dagger, \xi)$.

Suppose, conversely, that ϕ is a valuation of $(G^\dagger, \xi)$ with respect to a root map ι of Σ (with target B_2) that extends the map ϕ_4 (so $\phi_{a_4} = \phi_4$). Let $\phi_i = \phi_{a_i}$ for all $i \in [1, 3]$. By 30.38.ii,

$$x_4(v)^{m_\Sigma(x_1(1))} = x_2(v)$$

for all $v \in K$. By 3.41.i, therefore, there exists a constant C such that

$$\phi_2(x_2(v)) = \phi_4(x_4(v)) + C = \nu(v) + C$$

for all $v \in K^*$. By 30.38.ii, we also have

$$x_2(v)^{m_\Sigma(x_4(1))} = x_2(-v^\sigma)$$

for all $v \in K$. By 3.41.i again, therefore,

$$\phi_2(x_2(v)) = \phi_2(x_2(-v^\sigma))$$

for all $v \in K$ since $a_2 \cdot a_4 = 0$. We conclude that $\nu(v) = \nu(-v^\sigma) = \nu(v^\sigma)$ for all $v \in K$. In other words, ν is σ-invariant. Thus (i) holds.

By 2.50.i and another application of 3.41.i, we have

$$\phi_1(x_1(u)) = -\Big(\phi_4\big(x_2(1)^{m_\Sigma(x_1(u))}\big) - \phi_2(x_2(1))\Big)/\delta_0$$

for all $t \in K_0$, where

$$\delta_0 = \begin{cases} 2 & \text{if } \iota \text{ is long and} \\ 1 & \text{if } \iota \text{ is short.} \end{cases}$$

By 32.6 in [36],

$$x_2(1)^{m_\Sigma(x_1(u))} = x_4(-u^{-1})$$

for all $u \in K_0^*$ (since σ acts trivially on K_0). Thus

$$\begin{aligned} \phi_1(x_1(u)) &= -\big(\phi_4(x_4(-u^{-1})) - \phi_2(x_2(1))\big)/\delta_0 \\ &= \big(\nu(u) + \phi_2(x_2(1))\big)/\delta_0 \end{aligned}$$

for all $u \in K_0^*$. By 3.21, $\phi_i(U_1^*) = \mathbb{Z}$, and by 23.2, $\nu(K_0^*) = \delta\mathbb{Z}$. It follows that $\delta_0 = \delta$ and $\phi_2(x_2(1))/\delta \in \mathbb{Z}$. By 3.43, therefore, we can assume that 23.4 holds. Thus (ii) holds. □

Remark 23.5. Suppose that (K, K_0, σ) is a quadratic involutory set as defined in 30.21 or an honorary involutory set. Then

$$\mathcal{Q}_\mathcal{I}(K, F, \sigma) = \mathcal{Q}_\mathcal{Q}(F, K, q)$$

by 30.22, where $F = K_0$ and q is the norm of the composition algebra (K, F) as defined in 30.17. By 26.9, the results of 23.3 coincide with the results of 19.18 in this case.

Theorem 23.6. *Let (Δ, Σ, d) be a canonical realization of type 30.14.iv–vi. Thus*

$$\Delta = \mathsf{B}_\ell^\mathcal{I}(\Lambda) = \mathsf{C}_\ell^\mathcal{I}(\Lambda)$$

for some involutory set $\Lambda = (K, K_0, \sigma)$, proper with $\ell \geq 2$, quadratic with $\ell \geq 3$ or honorary with $\ell = 3$. Let $(G^\dagger, \xi)$ be the root datum of Δ based at Σ and let ν be a valuation of K. Let ι_B be a long root map of Σ and let ι_C be a short root map of Σ as defined in 16.12.[2] *Let a and x_a be as in 16.2 and 16.3 and let ϕ_a be given by*

$$\phi_a(x_a(t)) = \nu(t)$$

for all $t \in K^*$. Then the following hold:

(i) *If Λ is not ν-compatible, then ϕ_a does not extend to a valuation of $(G^\dagger, \xi)$.*
(ii) *If Λ is ν-compatible and Λ is ramified at ν, then ϕ_a extends to a valuation ϕ of $(G^\dagger, \xi)$ with respect to ι_B (but not with respect to ι_C).*

---

[2]Thus by 2.14.v, ι_X has target X_ℓ for $\mathsf{X} = \mathsf{B}$ and C.

(iii) *If Λ is ν-compatible and Λ is unramified at ν, then ϕ_a extends to a valuation ϕ of $(G^\dagger, \xi)$ with respect to ι_{C} (but not with respect to ι_{B}).*

In cases (ii) and (iii), the extension of ϕ is unique up to equipollence.

Proof. This holds by 16.14.ii, 23.2 and 23.3. □

Notation 23.7. Let Δ, Λ, ν, etc. be as in 23.6, suppose that Λ is ν-compatible and let ϕ be as in (ii) or (iii) of 23.6. We denote the pair $(\Delta^{\text{aff}}, \mathcal{A})$ obtained by applying 14.47 to these data by $\tilde{\mathsf{B}}_\ell^{\mathcal{I}}(\Lambda, \nu)$ in case (ii) and by $\tilde{\mathsf{C}}_\ell^{\mathcal{I}}(\Lambda, \nu)$ in case (iii).

Thus the pair $\tilde{\mathsf{X}}_\ell^{\mathcal{I}}(\Lambda, \nu)$ for $\mathsf{X} = \mathsf{B}$ or C defined in 23.7 exists precisely when the involutory set Λ is ν-compatible.

Remark 23.8. Suppose that (K, F, σ) and q are as in 23.5 and that (K, F, σ) is ν-compatible for some valuation ν of K (as defined in 23.2). Let δ be as in 23.2 and let ω be the restriction of ν/δ to F. We will observe in 26.9 that (F, K, q) is ω-compatible as defined in 19.17 and that (K, F, σ) is ramified at ν as defined in 23.2 if and only if (F, K, q) is ramified at ω as defined in 19.17. By 19.22, 23.5 and 23.7, it follows that

$$\tilde{\mathsf{B}}_2^{\mathcal{I}}\big((K, F, \sigma), \nu\big) = \tilde{\mathsf{C}}_2^{\mathcal{Q}}\big((F, K, q), \omega\big)$$

if (K, F, σ) is ramified at ν and

$$\tilde{\mathsf{C}}_2^{\mathcal{I}}\big((K, F, \sigma), \nu\big) = \tilde{\mathsf{B}}_2^{\mathcal{Q}}\big((F, K, q), \omega\big)$$

if (K, F, σ) is unramified at ν.

Remark 23.9. Suppose that $\Lambda = (K, F, \sigma)$ is a quadratic involutory set of type (i) or (ii) (so $\sigma = 1$) and that $\text{char}(K) = 2$ also if the type of Λ is (ii), let q be its norm and let ν be a discrete valuation of K. Then both Λ and the anisotropic space (K, K, q) are ν-compatible and unramified. By 30.14 and 30.27, we have

$$\mathsf{B}_\ell^{\mathcal{Q}}(K, K, q) = \mathsf{B}_\ell^{\mathcal{I}}(\Lambda).$$

By 19.22 and 23.7, it follows that

$$\tilde{\mathsf{C}}_\ell^{\mathcal{I}}(\Lambda, \nu) = \tilde{\mathsf{B}}_\ell^{\mathcal{Q}}\big((K, K, q), \nu\big).$$

Theorem 23.10. *Every Bruhat-Tits pair whose building at infinity is*

$$\mathsf{B}_\ell^{\mathcal{I}}(\Lambda) = \mathsf{C}_\ell^{\mathcal{I}}(\Lambda)$$

for some $\Lambda = (K, K_0, \sigma)$ and some $\ell \geq 2$ as in (iv)–(vi) of 30.14 is of the form $\tilde{\mathsf{B}}_\ell^{\mathcal{I}}(\Lambda, \nu)$ (if Λ is ramified at ν) or $\tilde{\mathsf{C}}_\ell^{\mathcal{I}}(\Lambda, \nu)$ (if Λ is unramified at ν) for some valuation ν of K such that Λ is ν-compatible.

Proof. This holds by 16.4, 23.6 and 23.7. □

We turn next to completions.

Notation 23.11. Let $\Lambda = (K, K_0, \sigma)$ be an involutory set, genuine or honorary, let ν be a σ-invariant discrete valuation of K and let $\hat{K}$ be the completion of K with respect to ν. Since σ is additive and ν is σ-invariant, σ maps Cauchy sequences to Cauchy sequences and hence has a unique extension by continuity to an involution of $\hat{K}$. We denote this involution by $\hat{\sigma}$. Let $\hat{K}_0$ be the closure of K_0 in $\hat{K}$.

Proposition 23.12. *Let $\Lambda = (K, K_0, \sigma)$ be an involutory set, genuine or honorary, let ν be a σ-invariant discrete valuation of K, let $\hat{K}$ be the completion of K with respect to ν and let $\hat{\sigma}$ and $\hat{K}_0$ be as in 23.11. Then*

$$\hat{\Lambda} := (\hat{K}, \hat{K}_0, \hat{\sigma})$$

is a ν-compatible involutory set, genuine if Λ is genuine, honorary if Λ is honorary.

Proof. By 11.1 in [36] and 23.11, $\hat{\Lambda}$ is a ν-compatible involutory set, genuine or honorary. The field K is associative if and only if $\hat{K}$ is associative. Therefore $\hat{\Lambda}$ is genuine if and only if Λ is genuine.[3] □

Theorem 23.13. *Let $\Lambda = (K, K_0, \sigma)$ be an involutory set, genuine or honorary, let ν be a σ-invariant discrete valuation of K, let*

$$\hat{\Lambda} = (\hat{K}, \hat{K}_0, \hat{\sigma})$$

be as in 23.12 and let $(\Delta, \mathcal{A}) = \tilde{\mathsf{X}}_\ell^{\mathcal{I}}(\Lambda, \nu)$ be as in 23.7 for $\mathsf{X} = \mathsf{B}$ or C. Then $\hat{\Lambda}$ is ν-compatible, Δ is completely Bruhat-Tits and its completion is $\tilde{\mathsf{X}}_\ell^{\mathcal{I}}(\hat{\Lambda}, \nu)$.

Proof. By 30.16, $\mathsf{X}_\ell^{\mathcal{I}}(\Lambda)$ is a subbuilding of $\mathsf{X}_\ell^{\mathcal{I}}(\hat{\Lambda})$. By 9.18.iii, Λ is ramified at ν if and only if $\hat{\Lambda}$ is. The claim holds, therefore, by 17.4 and 17.9. □

Proposition 23.14. *Let K be a field or a skew field or an octonion division algebra, let F be the center of K, let ν be a discrete valuation of K and suppose that K is finite-dimensional over F (which is automatic if K is octonion). Then the following hold:*

(i) *There exists a positive integer m such that the restriction ω of ν/m is a valuation of F.*
(ii) *K is complete with respect to ν if and only if F is complete with respect to ω.*

Proof. The restriction of ν to F^* is a subgroup of $\mathbb{Z}$. Since every element of K is algebraic over F, this subgroup cannot be the 0-subgroup. Thus $\nu(F^*) = m\mathbb{Z}$ for some positive integer m. Let ω be the restriction of ν/m to F. Then $\omega(F^*) = \mathbb{Z}$ and hence ω is a valuation of F. Thus (i) holds. Suppose that K is complete with respect to ν. Since F is the center of K, it is closed with respect to ν and hence complete with respect to ω. Suppose, conversely, that F is complete with respect to ω. By [20] and the identity 2.7

---

[3] See also 26.13.iii.

in [39], $\nu(u) = \omega(N(u))/n$ for all $u \in K$, where N is the reduced norm of K and n is the degree of K over F. Hence K is complete with respect to ν by 17.13. Thus (ii) holds □

The proof of the next observation is due to A. Rapinchuk.

Proposition 23.15. *Let K be a field, a skew field (not necessarily finite-dimensional over its center) or an octonion division algebra, let ν be a discrete valuation of K as defined in 9.17 and suppose that K is complete with respect to ν. Then ν is the only discrete valuation of K.*

Proof. Suppose that ω is a second valuation of K. By Statement B in Section 2.1 of [23],[4] we can choose $x \in K^*$ such that $\nu(x) < 0$ and $\omega(x) \geq 0$ or $\nu(x) \geq 0$ and $\omega(x) < 0$. Thus $\nu(x)\omega(x) \leq 0$. Let

$$S := \{u \in K^* \mid \nu(u)\omega(u) < 0\}.$$

Suppose that $x \notin S$. Then $\nu(x)\omega(x) = 0$. If $\nu(x) = 0$ (and hence $\omega(x) < 0$), choose $y \in K^*$ such that $\nu(y) > 0$. If $\omega(x) = 0$ (and hence $\nu(x) < 0$), choose $y \in K^*$ such that $\omega(y) > 0$. Then $x^n y \in S$ for all $n \in \mathbb{Z}$ sufficiently large. We conclude that S is not empty. Choose $u \in S$.

Let E be a subfield of K containing u. Then $\nu(E^*)$ is a non-zero additive subgroup of $\mathbb{Z}$. Hence the restriction of ν to E is a positive multiple of a unique valuation of E. If E is complete with respect to this valuation, we will say that E is complete with respect to ν.

We claim now that there exists a commutative subfield E of K containing u that is complete with respect to ν. Suppose first that K is octonion, let F be its center and let $E = F(u)$. Then E is a commutative subfield of degree at most 2 over F (by 20.9 in [36]) and by 17.14, E is complete with respect to this valuation. Next suppose that K is associative and let D be the centralizer of u in K. Then D is a skew field that is closed and hence complete with respect to ν. Therefore the center E of D is a commutative field containing u that is closed and hence complete with respect to ν. This proves our claim.

Since $\omega(E^*)$ is also a non-zero additive subgroup of $\mathbb{Z}$, also the restriction of ω to E is a positive multiple of a unique valuation of E. By Statement S in Section 5.2 of [23], therefore, the restrictions of ν and ω to E differ only by a positive factor. This contradicts the choice of u. □

Corollary 23.16. *Let $\Lambda = (K, K_0, \sigma)$ be a involutory set, genuine or honorary, let ν be a valuation of K and suppose that K is complete with respect to ν. Then Λ is ν-compatible.*

Proof. Set ω be the composition of ν with σ. Then ω is also a discrete valuation of K. The claim holds, therefore, by 23.15. □

[4] The proof of this statement in [23] is valid verbatim also for skew fields and octonion division algebras.

Remark 23.17. Let $\Lambda = (K, K_0, \sigma)$ be an involutory set, genuine or honorary, let F be the center of K, let

$$K_\sigma = \{a + a^\sigma \mid a \in K\}, \tag{23.18}$$

let

$$K^\sigma = \{a \in K \mid a^\sigma = a\} \tag{23.19}$$

and let ν be a valuation of K. Suppose that K is complete with respect to ν and that Λ is ν-compatible. By 11.1 in [36], $K_\sigma \subset K_0 \subset K^\sigma$. Since ν is σ-invariant, both K_σ and K^σ are closed with respect to σ. By 11.2 in [36], $K_\sigma = K_0 = K^\sigma$ if $\text{char}(K) \neq 2$. If Λ is honorary, then $K_\sigma = F$ by 30.20 and $K_0 = F$ by 30.22, so $K_0 = K_\sigma$. Suppose that $\text{char}(K) = 2$ and that Λ is genuine. By 11.3 in [36], the quotient $V := K^\sigma/K_\sigma$ has the structure of a right vector space over K, where

$$(x + K_\sigma) \cdot u = u^\sigma x u + K_\sigma \tag{23.20}$$

for all $x \in K^\sigma$ and all $u \in K$, and K_1 is the inverse image in K of a subspace of V that contains 1 if and only if (K, K_1, σ) is an involutory set. By 11.4 in [36], the quotient K^σ/K_σ can be infinite-dimensional even when K is finite-dimensional over its center. We conclude that K_0 is closed if $K_0 = K_\sigma$ or $K_0 = K^\sigma$ (in particular, if $\text{char}(K) \neq 2$ or Λ is honorary), but in general, K_0 might not be closed even if K is finite-dimensional over its center.

Remark 23.21. Let $\Lambda = (K, K_0, \sigma)$ be an involutory set, proper, quadratic or honorary. By 41.16 in [36], the spherical building $\mathsf{B}_\ell^{\mathcal{I}}(\Lambda)$ is algebraic as defined in 30.31 if and only if K is finite-dimensional over its center and K_0 is either K_σ or K^σ. Thus by 23.16 and 23.17, if $\mathsf{B}_\ell^{\mathcal{I}}(\Lambda)$ is algebraic and K is complete with respect to a valuation ν, then Λ is ν-compatible and $\Lambda = \hat{\Lambda}$.

We turn now to gems.

Definition 23.22. Let $\Lambda = (K, K_0, \sigma)$ be an involutory set, genuine or honorary, let ν be a valuation of K and suppose that Λ is ν-compatible as defined in 23.2. Let $\bar{K}_0$ be the image in $\bar{K}$ of the intersection $K_0 \cap \mathcal{O}_K$ under the natural homomorphism from the ring of integers $\mathcal{O}_K$ with respect to ν to $\bar{K}$. Since ν is σ-invariant, σ induces an involution $\bar{\sigma}$ on the residue field $\bar{K}$ of K and $\bar{\Lambda} := (\bar{K}, \bar{K}_0, \bar{\sigma})$ is an involutory set, genuine or honorary. Note that $\bar{\sigma}$ can be trivial when σ is not. If $\bar{\sigma}$ is assumed to be non-trivial, then $\bar{\Lambda}$ is proper if Λ is proper (simply by definition) and $\bar{\Lambda}$ is quadratic or honorary if Λ is quadratic or honorary (by 26.15). We call $\bar{\Lambda}$ the *residue of* Λ *at* ν.

Definition 23.23. Let $\Lambda = (K, K_0, \sigma)$ be an involutory set, genuine or honorary, let ν be a σ-invariant valuation of K and suppose that Λ is unramified at ν as defined in 23.2, i.e. that $\nu(K_0^*) = \mathbb{Z}$. Let π be a uniformizer contained in K_0. By 11.6 in [36], π^{-1} is also in K_0. Let $\Lambda_{\pi^{-1}}$ be the translate of Λ with respect to π^{-1} as defined in 11.8 of [36]. If Λ is honorary, then K_0 is the center of K and $\Lambda_{\pi^{-1}} = \Lambda$. If Λ is genuine, then it is not necessarily true that $\Lambda_{\pi^{-1}} = \Lambda$, but by 11.10 in [36], $\Lambda_{\pi^{-1}}$ is, in any case, an involutory set. Let $\overline{\Lambda_1}$ denote the residue of $\Lambda_{\pi^{-1}}$ at ν as defined in 23.22 (whether or not Λ is genuine). The involutory set $\overline{\Lambda_1}$ is independent of the choice of π up to similarity (as defined in 11.9 of [36]).[5] By 35.16 in [36] and 3.6, therefore, also the building $\mathsf{C}_\ell^{\mathcal{I}}(\overline{\Lambda_1})$ is independent of the choice of π.

Proposition 23.24. *Let $\Lambda = (K, K_0, \sigma)$ be a genuine involutory set, let ν be a σ-invariant valuation of K, let $\bar{\Lambda} = (\bar{K}, \bar{K}_0, \bar{\sigma})$ be the residue of Λ with respect to ν as defined in 23.22 and suppose that Λ is ramified at ν as defined in 23.2, i.e. that $\nu(K_0^*) = 2\mathbb{Z}$. Then $\bar{K}$ is commutative.*

Proof. If $\nu(u) = 1$ for some $u \in K^*$, then $\nu(u + u^\sigma) \geq 1$ (by 24.28); since $u + u^\sigma \in K_0$ (by 11.1.i in [36]) and $\nu(K_0^*) = 2\mathbb{Z}$, it follows that, in fact, $\nu(u + u^\sigma) \geq 2$.

Now choose $v, t \in K^*$ such that $\nu(v) = 0$ and $\nu(t) = 1$, let $w = tv + (tv)^\sigma$ and let $z = t + t^\sigma$. Then $\nu(w) \geq 2$ and $\nu(z) \geq 2$ (by the observation in the previous paragraph) and hence also $\nu(v^\sigma z) \geq 2$ by 9.17.i. From

$$v^\sigma t^\sigma = (tv)^\sigma = w - tv$$

and

$$v^\sigma t^\sigma = v^\sigma(z - t) = v^\sigma z - v^\sigma t,$$

it follows that $v^\sigma t - tv = v^\sigma z - w$. By 9.17.ii, therefore,

$$\nu(v^\sigma t - tv) \geq 2$$

and hence $\nu(v^\sigma - tvt^{-1}) \geq 1$, again by 9.17.i. We conclude that

$$\bar{v}^{\bar{\sigma}} = \overline{v^\sigma} = \overline{tvt^{-1}} \tag{23.25}$$

in $\bar{K}$ for all $v \in \mathcal{O}_K$ and all $t \in K^*$ such that $\nu(t) = 1$, where $\bar{\sigma}$ is as defined in 23.22.

By 23.25 and the assumption that K is associative, it follows that the map $\bar{\sigma}$ is multiplicative. Since $\bar{\sigma}$ is also an involution (and hence anti-multiplicative), it follows that $\bar{K}$ is commutative. □

Remark 23.26. In 26.15 we will show that if Λ is honorary and ramified at ν, then either $\bar{K}$ is commutative and $\bar{T}$ is identically zero or $\bar{K}$ is a quaternion division algebra. Thus the claim in 23.24 is not, in general, valid if Λ is not genuine.

Theorem 23.27. *Let*

$$(\Delta, \mathcal{A}) = \tilde{\mathsf{X}}_\ell^{\mathcal{I}}(\Lambda, \nu)$$

for some involutory set $\Lambda = (K, K_0, \sigma)$, genuine or honorary, and for some valuation ν of K such that Λ is ν-compatible (as defined in 23.7), where

[5] Here are the details. Suppose π' is a second uniformizer in K_0 and let $s = (\pi')^{-1}\pi$. Then $\nu(s) = 0$ and $\pi^{-1}\pi' \in \pi^{-1}K_0$. By 11.6 and 11.8 in [36], the set $\pi^{-1}K_0$ is closed under inverses. Therefore $s \in \pi^{-1}K_0$. Since $s\pi^{-1}K_0 = (\pi')^{-1}K_0$, the involutory set $\Lambda_{(\pi')^{-1}}$ is the translate of the involutory set $\Lambda_{\pi^{-1}}$ with respect to s as defined in 11.8 in [36].

$\mathsf{X} = \mathsf{B}$ if Λ is ramified at ν and $\mathsf{X} = \mathsf{C}$ if Λ is unramified at ν. Let $\bar{\Lambda}$ be the residue of Λ at ν as defined in 23.22, let the apartment A, the gem R_0 and the group G° be as in 18.1 and let the gem R_1 be as in 18.28. Then the following hold:

(i) *If $\mathsf{X} = \mathsf{B}$, then $R_0 \cong \mathsf{X}_\ell^{\mathcal{I}}(\bar{\Lambda})$ and all gems are in the same G°-orbit as R_0.*
(ii) *If $\mathsf{X} = \mathsf{C}$, then $R_0 \cong \mathsf{X}_\ell^{\mathcal{I}}(\bar{\Lambda})$, $R_1 \cong \mathsf{C}_\ell^{\mathcal{I}}(\overline{\Lambda_1})$, where $\overline{\Lambda_1}$ is as in 23.23, and every gem is in the same G°-orbit as R_0 or R_1 (or both).*
(iii) *If $\mathsf{X} = \mathsf{C}$ and Λ is quadratic (as defined in 30.21) or honorary (as defined in 30.22), then all gems are in the same G°-orbit as R_0.*

Proof. Let $\Sigma = A^\infty$, let d be a chamber of Σ, let $(G^\dagger, \xi)$ be the root datum of $\Delta_{\mathcal{A}}^\infty$ based at Σ and let $\phi = \phi_{R_0}$ be the valuation of $(G^\dagger, \xi)$ described in 13.8. Let c be the unique root in Ξ_d that is long (with respect to the root map ι_C) if $\mathsf{X} = \mathsf{C}$ and short (with respect to the root map ι_B) if $\mathsf{X} = \mathsf{B}$. By 19.22, we can assume that $(\Delta_{\mathcal{A}}^\infty, \Sigma, d)$ is the canonical realization of one of the root group labelings described in cases (iv)–(vi) of 30.14, that $\phi_a(x_a(t)) = \nu(t)$ for all $a \in \Xi_d^\circ \backslash \{c\}$ and all $t \in K^*$ and that $\phi_c(x_c(u)) = \nu(u)/\delta$ for all $u \in K_0^*$, where δ is the ν-index of Λ as defined in 23.2. By 18.27,

$$R_0 \cong \mathsf{X}_\ell^{\mathcal{I}}(\bar{\Lambda}).$$

Let $b \in \Xi_d^\circ$ be as in 18.28. Since b is long, we have $b = c$ if $\mathsf{X} = \mathsf{C}$. Let H be as in 18.1.

Suppose that $\mathsf{X} = \mathsf{B}$ and let $\pi \in K$ be a uniformizer. Suppose too that (K, K_0, σ) is genuine, i.e. that K is associative. By 18.9, there exists an element $\tau \in H$ centralizing U_a for all $a \in \Xi_d^\circ \backslash \{b\}$ such that

$$x_b(t)^\tau = x_b(t\pi)$$

for all $t \in K$ if $\ell = 2$ and

$$x_b(t)^\tau = x_b(\pi t)$$

for all $t \in K$ if $\ell \geq 3$.[6] Hence (for all values of $\ell \geq 2$) the map ϕ_a is

---

[6]Here are the details. Let

$$\mathcal{Q}_{\mathcal{I}}(\Lambda)^{\text{op}} = (U_+, U_1, U_2, U_3, U_4),$$

where $\mathcal{Q}_{\mathcal{I}}(\Lambda)$ is as in 30.8.ii. Thus x_i is an isomorphism from the additive group of K to U_i for $i = 1$ and 3, x_j is an isomorphism from K_0 to U_j for $j = 2$ and 4,

$$[x_1(u), x_4(t)] = x_2(-u^\sigma tu)x_3(tu)$$

and

$$[x_1(u), x_3(v)] = x_2(u^\sigma v + v^\sigma u)^{-1}$$

for all $u, v \in K$ and all $t \in K_0$ (as well as $[U_2, U_4] = 1$ and $[U_i, U_{i+1}] = 1$ for all $i \in [1, 3]$). There therefore exists an automorphism of U_+ extending the maps $x_i(u) \mapsto x_i(u\pi)$ for $i = 1$ and 3, $x_2(t) \mapsto x_2(\pi^\sigma t\pi)$ and $x_4(t) \mapsto x_4(t)$. Thus the automorphism τ exists by 18.9 if $\ell = 2$ (in which case $U_b = U_1$). Suppose that $\ell \geq 3$ and let

$$\mathcal{T}(K) = (U_+, U_1, U_2, U_3)$$

be the root group sequence in 30.8.i. Then there exists an automorphism of U_+ extending the maps $x_i(t) \mapsto x_i(\pi t)$ for $i = 1$ and 2 and $x_3(t) \mapsto x_3(t)$. Thus the automorphism τ exists by 18.9 and 30.14.iii.

τ-invariant for all $a \in \Xi_d^\circ \backslash \{b\}$ and

$$\phi_b(x_b(t)^\tau) = \phi_b(x_b(t)) + 1$$

for all $t \in K*$. By 18.6, there is an element $\rho \in G_A$ inducing τ on $\Delta_{\mathcal{A}}^\infty$. By 18.28.i, it follows that ρ maps R_0 to R_1. Thus (i) holds by 18.29 if K is associative.

Suppose now that (K, K_0, σ) is honorary. Thus we are in case 30.14.vi and $\ell = 3$. Let $F = K_0$, let q and T be the norm and trace of the corresponding composition algebra (K, F) as defined in 30.17, let λ be as in 18.10 with π in place of w and let $\beta(u) = \lambda(u)\pi^{-1}$ for all $u \in K$. Then β is an F-linear automorphism of K and both q and ν are β-invariant. Since

$$q(u + v) = q(u) + q(v) + T(u, v)$$

for all $u, v \in K$, also T is β-invariant. Let $a \in \Xi_d^\circ$ be as in 16.3 and let c be the other root in $\Xi_d^\circ \backslash \{b\}$. By 37.12 in [36], 18.9 and 30.14.vi, there exists an element τ in H centralizing c such that

$$x_b(u)^\tau = x_b(\lambda(u))$$

and

$$x_a(u)^\tau = x_a(\beta(u))$$

for all $u \in K$. Hence $\phi_b(x_b(u)^\tau) = \phi_b(x_b(u)) + 1$ for all $u \in K$ and ϕ_a is τ-invariant for $a \in \Xi_d^\circ \backslash \{b\}$. By 18.6, there is an element $\rho \in G_A$ inducing τ on $\Delta_{\mathcal{A}}^\infty$. By 18.28.i, it follows that ρ maps R_0 to R_1. Thus (i) holds by 18.29 also if K is non-associative.

Next suppose that $\mathsf{X} = \mathsf{C}$, so $\delta = 1$. We can thus assume that π is contained in K_0 and that π is the uniformizer used in the construction of $\overline{\Lambda_1}$ in 23.23. By 3.6 and 18.28.i, the maps $x_a(u) \mapsto x_a(u)$ for all $a \in \Xi_d^\circ \backslash \{b\}$ and $x_b(t) \mapsto x_b(\pi^{-1}t)$ induce an isomorphism from the residue R_1 to the building $\mathsf{C}_\ell^{\mathcal{I}}(\overline{\Lambda_1})$. By 18.29, therefore, (ii) holds.

Suppose, finally, that (K, K_0, σ) quadratic or honorary. In this case $F := K_0$ is contained in the center of K. By 18.9, there thus exists an element $\tau \in H$ centralizing U_a for all $a \in \Xi_d^\circ \backslash \{b\}$ such that

$$x_b(t)^\tau = x_b(\pi t)$$

and hence

$$\phi_b(x_b(t)^\tau) = \phi_b(x_b(t)) + 1$$

for all $t \in F$. By 18.6, there is an element $\rho \in G_A$ inducing τ on $\Delta_{\mathcal{A}}^\infty$. By 18.28.i, ρ maps R_1 to R_0. By 18.29, therefore, (iii) holds. $\square$

Chapter Twenty Four

Pseudo-Quadratic Quadrangles

In this chapter, we consider the case $X = \mathcal{Q}_{\mathcal{P}}$ of the problem framed at the end of Chapter 16. Our main result is 24.31. We use this result to give the classification of Bruhat-Tits pairs whose building at infinity is $\mathsf{B}_\ell^{\mathcal{P}}(\Lambda)$ for some anisotropic pseudo-quadratic space Λ in 24.37. See 11.17 in [36] for the definition of an anisotropic pseudo-quadratic space and 30.15 for the definition of the building $\mathsf{B}_\ell^{\mathcal{P}}(\Lambda)$.

Proposition 24.1. *Let*

$$\Lambda = (K, K_0, \sigma, L, q)$$

be a pseudo-quadratic space as defined in 11.17 of [36] (so Λ is not necessarily anisotropic), let f be the associated skew-hermitian form and suppose that $K_0 \neq K$. Let E be a skew field containing K that has an involution σ_E extending σ. Suppose that (E, E_0, σ_E) is an involutory set such that $K_0 \subset E_0 \cap K$. Let $L_E = L \otimes_K E$. Let f_E denote the unique sesquilinear map (over E) from $L_E \times L_E$ to E such that

$$f_E(u \otimes s, v \otimes t) = s^{\sigma_E} f(u, v)t \tag{24.2}$$

for all $u, v \in L$ and all $s, t \in E$. There there exists a unique map q_E from L_E to E (unique modulo E_0) such that

$$q_E(\sum_{i=1}^{d} v_i \otimes t_i) = \sum_{i=1}^{d} t_i^{\sigma_E} q(v_i) t_i + \sum_{i<j} t_i^{\sigma_E} f(v_i, v_j) t_j \tag{24.3}$$

for all $v_1, \ldots, v_d \in L$, for all $t_1, \ldots, t_d \in E$ and for all d. Moreover,

$$\Lambda_E := (E, E_0, \sigma_E, L_E, q_E)$$

is a pseudo-quadratic space and f_E is the associated skew-hermitian form.

Proof. Choose a basis B of L over K and define q_E using 24.3 for all finite subsets $v_1, \ldots, v_d$ of B. By 11.19 in [36], $f(v, v) = q(v) - q(v)^\sigma$ for all $v \in L$, and by 11.1.i in [36], K_0 contains the set of traces K_σ (as defined in 23.18). It follows (by a bit of calculation) that Λ_E is a pseudo-quadratic space and f_E is the associated skew-hermitian form. The identity 24.3 then holds in general by induction with respect to d. □

Definition 24.4. Let

$$\Lambda = (K, K_0, \sigma, L, q)$$

be a pseudo-quadratic space as defined in 11.17 of [36]. Then Λ is *anisotropic* if $q(u) \in K_0$ if and only if $u = 0$ and *proper* if $\sigma \neq 1$ (which implies that $K_0 \neq K$), $L \neq 0$ and the associated skew-hermitian form is non-degenerate. If Λ is a proper anisotropic pseudo-quadratic space, then by 23.23 in [36], the involutory set (K, K_0, σ) is either proper as defined in footnote 10 in Appendix B or quadratic of type (iii) or (iv) as defined in 30.21.

Notation 24.5. Let

$$(K, K_0, \sigma, L, q)$$

be an anisotropic pseudo-quadratic space (as defined in 11.17 in [36] and 24.4), let f be the skew-hermitian form associated with q and let T be the group defined in 11.24 of [36]. Thus

$$T = \{(u, t) \in L \times K \mid q(u) - t \in K_0\}$$

and

$$(u, t) \cdot (v, s) = \big(u + v, t + s + f(v, u)\big) \tag{24.6}$$

as well as

$$(u, t)^{-1} = (-u, -t^\sigma) \tag{24.7}$$

for all $(u, t), (v, s) \in T$. By 11.19 in [36], we also have

$$f(u, u) = t - t^\sigma \tag{24.8}$$

for all $(u, t) \in T$ since σ acts trivially on K_0.

The following result was first proved in Theorem 10.1.15 of [6]; see 19.4.

Proposition 24.9. *Let*

$$\Lambda = (K, K_0, \sigma, L, q)$$

be a proper anisotropic pseudo-quadratic space, let f and T be as in 24.5 and let ν be a valuation of K such that

$$\nu(u^\sigma) = \nu(u)$$

for all $u \in K$.[1] *Let $(\hat{K}, \hat{K}_0, \hat{\sigma})$ be as in 23.12 and let*

$$\Lambda_{\hat{K}} = (\hat{K}, \hat{K}_0, \hat{\sigma}, L_{\hat{K}}, q_{\hat{K}})$$

be the pseudo-quadratic space and $f_{\hat{K}}$ the skew-hermitian form obtained from Λ by applying 24.1 with $(E, E_0, \sigma_E) = (\hat{K}, \hat{K}_0, \hat{\sigma})$. Then the following are equivalent.

(i) $\nu(f(u, v)) \geq \min\{\nu(s), \nu(t)\}$ *for all* $(u, t), (v, s) \in T^*$.
(ii) $\nu(f(u, v)) \geq \big(\nu(s) + \nu(t)\big)/2$ *for all* $(u, t), (v, s) \in T^*$.
(iii) $\Lambda_{\hat{K}}$ *is anisotropic, i.e.* $\{\hat{u} \in L_{\hat{K}} \mid q_{\hat{K}}(\hat{u}) \in \hat{K}_0\} = \{0\}$.

[1] We observe that the statement and proof of 24.9 require only that ν be real-valued.

Proof. Suppose that (ii) does not hold. Then there exist $(u,t),(v,s) \in T$ such that

$$\eta := \nu(s) + \nu(t) - 2\nu(f(u,v)) > 0. \tag{24.10}$$

In particular, $f(u,v) \neq 0$. Let $v_0 = v$, $s_0 = s$ and

$$\lambda_1 = f(u,v)^{-\sigma}s. \tag{24.11}$$

Since $f(u,v)^\sigma = -f(v,u)$, we have $\nu(f(u,v)) = \nu(f(v,u))$. By 24.8,

$$f(u,u) = t - t^\sigma.$$

Hence

$$\nu(f(u,u)) \geq \nu(t) \tag{24.12}$$

and thus

$$\begin{aligned} \nu(f(u,u\lambda_1)) &= \nu(f(u,u)\lambda_1) \\ &\geq \nu(t) + \nu(\lambda_1) \\ &= -\nu(f(u,v)) + \nu(s) + \nu(t) \\ &= \eta + \nu(f(u,v)). \end{aligned}$$

Hence $\nu(f(u,u\lambda_1)) > \nu(f(u,v))$ since $\eta > 0$. By 9.23.ii, therefore,

$$\nu(f(u,v_1)) = \nu(f(u,v)) \tag{24.13}$$

for $v_1 := u\lambda_1 + v$. Let

$$s_1 = \lambda_1^\sigma t\lambda_1. \tag{24.14}$$

By 24.11, we have $s_1 = s^\sigma f(u,v)^{-1}tf(u,v)^{-\sigma}s$ and thus

$$\nu(s_1) = 2\nu(s) + \nu(t) - 2\nu(f(u,v)) = \eta + \nu(s). \tag{24.15}$$

The elements $q(u)-t$ and $q(v)-s$ are in contained in K_0 (by 24.5) and thus

$$\lambda_1^\sigma q(u)\lambda_1 \equiv \lambda_1^\sigma t\lambda_1 \pmod{K_0}$$

(by 11.1.ii in [36]). Hence $\lambda_1^\sigma q(u)\lambda_1 \equiv s_1 \pmod{K_0}$ by 24.14. Therefore

$$\begin{aligned} q(v_1) &= q(u\lambda_1 + v) \\ &\equiv \lambda_1^\sigma q(u)\lambda_1 + q(v) + \lambda_1^\sigma f(u,v) \pmod{K_0} && \text{by 11.16 in [36]} \\ &\equiv s_1 + s + s^\sigma \pmod{K_0} && \text{by 24.11} \\ &\equiv s_1 \pmod{K_0} && \text{by 11.1.i in [36]} \end{aligned}$$

Thus

$$q(v_1) - s_1 \in K_0; \tag{24.16}$$

i.e. $(v_1, s_1) \in T$.

Next we define inductively

$$\lambda_k = f(u, v_{k-1})^{-\sigma}s_{k-1} \tag{24.17}$$

$$v_k = u\lambda_k + v_{k-1} \tag{24.18}$$

and

$$s_k = \lambda_k^\sigma t \lambda_k \tag{24.19}$$

for all $k \geq 2$. (If we set $v_0 = v$ and $s_0 = s$, then 24.17 and 24.18 hold also for $k = 1$.) To show that λ_k is well defined, we need to show that $f(u, v_k) \neq 0$ for each $k \geq 1$. We will show by induction on k that, in fact,

$$\nu(f(u, v_k)) = \nu(f(u, v)), \tag{24.20}$$

so indeed $f(u, v_k) \neq 0$,

$$\nu(\lambda_k) = (2^{k-1} - 1)\eta + \nu(s) - \nu(f(u, v)) \tag{24.21}$$

and

$$\nu(s_k) = (2^k - 1)\eta + \nu(s) \tag{24.22}$$

as well as

$$q(v_k) - s_k \in K_0 \tag{24.23}$$

for all $k \geq 1$. These identities hold for $k = 1$ by 24.11, 24.13, 24.15 and 24.16. Let $k > 1$ and assume that all four identities hold for $k - 1$. Then

$$\begin{aligned} \nu(\lambda_k) &= -\nu(f(u, v_{k-1})) + \nu(s_{k-1}) && \text{by 24.17} \\ &= -\nu(f(u, v_{k-1})) + (2^{k-1} - 1)\eta + \nu(s) && \text{by 24.22} \end{aligned}$$

and thus 24.21 holds for k by 24.20. Hence

$$\begin{aligned} \nu(s_k) &= \nu(t) + 2\nu(\lambda_k) && \text{by 24.19} \\ &= (2^k - 2)\eta + 2\nu(s) + \nu(t) - 2\nu(f(u, v)) && \text{by 24.21} \\ &= (2^k - 1)\eta + \nu(s) && \text{by 24.10,} \end{aligned}$$

so 24.22 holds for k, and

$$\begin{aligned} \nu(f(u, u\lambda_k)) &\geq \nu(t) + \nu(\lambda_k) && \text{by 24.12} \\ &= -\nu(f(u, v)) + (2^{k-1} - 1)\eta + \nu(s) + \nu(t) && \text{by 24.21} \\ &= \nu(f(u, v)) + 2^{k-1}\eta > \nu(f(u, v)) && \text{by 24.10.} \end{aligned}$$

We have $f(u, v_k) = f(u, u\lambda_k) + f(u, v_{k-1})$ by 24.18 and $\nu(f(u, v_{k-1})) = \nu(f(u, v))$ by 24.20. Thus 24.20 holds for k by 9.23.ii. Finally, we observe that $\lambda_k^\sigma q(u)\lambda_k \equiv \lambda_k^\sigma t \lambda_k \equiv s_k$ by 11.1.ii in [36] and 24.19 (since $q(u) - t \in K_0$) and $q(v_{k-1}) \equiv s_{k-1} \pmod{K_0}$ by 24.23. It follows that

$$\begin{aligned} q(v_k) &= q(u\lambda_k + v_{k-1}) && \text{by 24.18} \\ &\equiv \lambda_k^\sigma q(u)\lambda_k + q(v_{k-1}) + \lambda_k^\sigma f(u, v_{k-1}) \pmod{K_0} \\ &\equiv s_k + s_{k-1} + s_{k-1}^\sigma \pmod{K_0} && \text{by 24.17} \\ &\equiv s_k \pmod{K_0} && \text{by 11.1.i in [36],} \end{aligned}$$

so 24.23 holds for k. Thus all four claims 24.20–24.23 hold for all $k \geq 2$.

By 24.20 and 24.22, we can choose N such that $\nu(f(u, v_N)) = \nu(f(u, v))$ and $\nu(s_N) > \nu(f(u, v))$ and let

$$\eta' = \nu(t) + \nu(s_N) - 2\nu(f(u, v_N)).$$

By 24.22, $\nu(s_N) \geq \nu(s)$. Thus $\eta' \geq \eta > 0$ by 24.10. By 24.23, $(v_N, s_N) \in T$. Thus we can replace (v, s) by (u, t) and (u, t) by (v_N, s_N) and begin the whole argument over again to conclude that there exists a sequence $(u_k, t_k)_{k\geq 1}$ of elements of T such that

$$\nu(f(u_k, v_N)) = \nu(f(u, v_N)) = \nu(f(u, v))$$

and

$$\nu(t_k) = (2^k - 1)\eta' + \nu(t)$$

for all $k \geq 1$. Hence we can choose M such that $\nu(t_M) > \nu(f(u, v))$. Thus $(u_M, t_M), (v_N, s_N) \in T$ and

$$\nu(f(u_M, v_M)) = \nu(f(u, v)) < \min\{\nu(t_M), \nu(s_N)\}.$$

Therefore (i) implies (ii).

We continue to assume that 24.10 holds and keep the previous notation. Let

$$\xi_k = \lambda_k + \lambda_{k-1} + \cdots + \lambda_1$$

for each $k \geq 1$. By 24.18, $v_k = u\xi_k + v$ for all $k \geq 1$. By 24.21, $(\xi_k)_{k\geq 1}$ is a Cauchy sequence. Hence there exists $\hat{\xi} \in \hat{K}$ such that

$$\lim_{k\to\infty} \xi_k = \hat{\xi}.$$

Let $\hat{v} = u\hat{\xi} + v$. Then

$$\lim_{k\to\infty} f_{\hat{K}}(u, v_k - \hat{v}) = \lim_{k\to\infty} f(u, u)(\xi_k - \hat{\xi}) = 0.$$

By 24.20, therefore, $f_{\hat{K}}(u, \hat{v}) \neq 0$. In particular, $\hat{v} \neq 0$. By 24.3,

$$\lim_{k\to\infty} q(v_k) = q_{\hat{K}}(\hat{v}).$$

By 24.22 and 24.23, it follows that $q_{\hat{K}}(\hat{v}) \in \hat{K}_0$. We conclude that also (iii) implies (ii).

Since the average of two numbers is at least as large as their minimum, (ii) implies (i). It thus remains only to show that (ii) implies (iii). Suppose that there exists a non-zero vector $\hat{u} \in \hat{L}_{\hat{K}}$ such that

$$q_{\hat{K}}(\hat{u}) \in \hat{K}_0. \tag{24.24}$$

Since Λ is proper, the form f is non-degenerate. Thus there exists $w \in L$ such that

$$f_{\hat{K}}(\hat{u}, w) \neq 0. \tag{24.25}$$

Let $B := (v_1, v_2, \ldots, v_d)$ be an ordered finite subset of a basis of L such that both $\hat{u}$ and w lie in the subspace of $L_{\hat{K}}$ spanned by the vectors in B. Let $\hat{t}_1, \ldots, \hat{t}_d$ be the coordinates of $\hat{u}$ with respect to B. Choose sequences $(t_{i,k})_{k\geq 1}$ of elements in K such that

$$\lim_{k\to\infty} t_{i,k} = \hat{t}_i$$

for all $i \in [1, d]$ and let

$$u_k = v_1 t_{1,k} + \cdots + v_d t_{d,k}$$

for all $k \geq 1$. Then

$$\lim_{k \to \infty} u_k = \hat{u}. \tag{24.26}$$

By 24.3, therefore,

$$\lim_{k \to \infty} q(u_k) = q_{\hat{K}}(\hat{u}).$$

By 24.24, there exist a sequence $(r_k)_{k \geq 1}$ of elements in K_0 such that

$$\lim_{k \to \infty} r_k = q_{\hat{K}}(\hat{u}).$$

Let $r'_k = q(u_k) - r_k$ for all $k \geq 1$. Then $(u_k, r'_k) \in T^*$ for all $k \geq 1$ and

$$\lim_{k \to \infty} \nu(r'_k) = \infty. \tag{24.27}$$

Let $r = q(w)$. Thus $(w, r) \in T^*$. By 24.26,

$$\lim_{k \to \infty} f(u_k, w) = f_{\hat{K}}(\hat{u}, w).$$

By 9.18.iii and 24.25, it follows that $\nu(f(u_k, w))$ is a constant for all k sufficiently large. By 24.27, therefore,

$$\nu(f(u_k, w)) < \big(\nu(r'_k) + \nu(r)\big)/2$$

for all k sufficiently large. We conclude that (iii) implies (iv). □

Definition 24.28. Let $\Lambda = (K, K_0, \sigma, L, q,)$ be a proper anisotropic pseudo-quadratic space and let ν be a discrete valuation of K. We will say that Λ is *ν-compatible* if the involutory set (K, K_0, σ) is ν-compatible as defined in 23.2 (i.e. ν is σ-invariant) and the three equivalent conditions in 24.9 hold.

Proposition 24.29. *Let*

$$\Lambda = (K, K_0, \sigma, L, q,)$$

be an anisotropic pseudo-quadratic space, let ν be a σ-invariant valuation of K and let T be as in 24.5. Then

$$\{\nu(t) \mid (a, t) \in T^*\} = \delta\mathbb{Z}$$

for $\delta = 1$ or 2.

Proof. Let $B = \{\nu(t) \mid (a, t) \in T^*\}$. Since $(0, K_0^*) \subset T^*$, we have $2\mathbb{Z} \subset B$ by 23.1. If $(a, t) \in T^*$, then $(as, s^\sigma ts) \in T^*$ for all $s \in K$. Thus if $\nu(t)$ is odd for some $(a, t) \in T^*$, then $B = \mathbb{Z}$. □

Definition 24.30. Let $\Lambda = (K, K_0, \sigma, L, q,)$ be an anisotropic pseudo-quadratic space, let ν be a discrete valuation of K that we assume to be σ-invariant and let δ be as in 24.29. We will say that Λ is *ramified* (respectively, *unramified*) *at* ν if $\delta = 2$ (respectively, $\delta = 1$).[2] We call δ the *ν-index* of Λ. Note that if Λ is ramified at ν, then the involutory set (K, K_0, σ) is also ramified at ν as defined in 23.2 since $(0, K_0) \subset T$. If Λ is unramified at ν, however, the involutory set (K, K_0, σ) can be ramified or unramified at ν.

---

[2]If $L = 0$, then Λ is just an involutory set. Note that this definition is compatible with 23.2 in this case.

Theorem 24.31. *Let*

$$\Lambda = (K, K_0, \sigma, L, q)$$

be a proper anisotropic pseudo-quadratic space, let ν be a discrete valuation of K, let

$$\mathcal{Q}_{\mathcal{P}}(\Lambda) = (U_+, U_1, \ldots, U_4)$$

and $x_1, \ldots, x_4$ be as in 30.8.v, let (Δ, Σ, d) be the canonical realization of $\mathcal{Q}_{\mathcal{P}}(\Lambda)$ as defined in 30.6 and let $(G^\dagger, \xi)$ be the root datum of Δ based at Σ. Let δ be the ν-index of Λ as defined in 24.30 and let

$$\phi_4(x_4(t)) = \nu(t) \tag{24.32}$$

for all $t \in K^*$. Then the following hold:

(i) *The map ϕ_4 extends to a valuation of $(G^\dagger, \xi)$ if and only if Λ is ν-compatible as defined in 24.28.*
(ii) *Suppose that ϕ_4 extends to a valuation ϕ of $(G^\dagger, \xi)$ and let ι be the associated root map of Σ (with target B_2). Then ι is long if and only $\delta = 2$; and after replacing ϕ by an equipollent valuation if necessary,*

$$\phi_1(x_1(a,t)) = \nu(t)/\delta \tag{24.33}$$

for all $(a,t) \in T^*$, where T is the group defined in 24.5 in terms of Λ.

Proof. Let Ξ_d and Ξ_d° be as in 3.5. The four roots $a_1, \ldots, a_4$ in Ξ_d can be ordered so that $U_i = U_{a_i}$ for all $i \in [1,4]$. We then have $\Xi_d^\circ = \{a_1, a_4\}$ and $[a_1, a_4] = (a_1, a_2, a_3, a_4)$, where $[a_1, a_4]$ is as defined in 3.1.

Suppose first that Λ is ν-compatible. Let $\phi_2(x_2(t)) = \nu(t)$ for all $t \in K^*$ and let

$$\phi_i(x_i(a,t)) = \nu(t)/\delta$$

for $i = 1$ and 3 and all $(a,t) \in T^*$. Then $\phi_i(U_i^*) = \mathbb{Z}$ for all $i \in [1,4]$. Let

$$\phi = \{\phi_1, \phi_2, \phi_3, \phi_4\}.$$

By 24.6 and 24.9.i, ϕ satisfies 15.1.i. Let ι be a root map of Σ (with target B_2) that is long if $\delta = 2$ and short if $\delta = 1$ (as defined in 16.12). Thus U_4 is a long root group if and only if ι is long. By 16.5 of [36] and 24.9.ii, ϕ satisfies 15.1.ii with respect to ι, ϕ is exact as defined in 15.24 and the (a_1, a_4)-coefficient of a_2 as defined in 15.24 is $(\delta, 1)$. By 15.25 and 30.36.v, it follows that ϕ satisfies 15.4.i. By 30.38.v, ϕ satisfies 15.4.ii. Thus ϕ is a viable partial valuation. By 15.21, ϕ extends to a valuation of $(G^\dagger, \xi)$.

Suppose, conversely, that ϕ is a valuation of $(G^\dagger, \xi)$ with respect to a root map ι of Σ (with target B_2) that extends the map ϕ_4 (so $\phi_{a_4} = \phi_4$). Let $\phi_i = \phi_{a_i}$ for all $i \in [1,3]$. By 30.38, we have

$$x_4(t)^{m_\Sigma(x_1(0,1))} = x_2(t)$$

for all $t \in K$. By 3.41.i, therefore, there exists a constant C such that

$$\phi_2(x_2(v)) = \phi_4(x_4(v)) + C = \nu(v) + C \tag{24.34}$$

for all $v \in K$. By 30.38.v, we also have

$$x_2(v)^{m_\Sigma(x_4(1))} = x_2(-v^\sigma)$$

for all $v \in K$. By 3.41.i again, therefore,

$$\phi_2(x_2(v)) = \phi_2(x_2(-v^\sigma))$$

for all $v \in K$ since $a_2 \cdot a_4 = 0$. We conclude that $\nu(v) = \nu(-v^\sigma) = \nu(v^\sigma)$ for all $v \in K$. In other words, ν is σ-invariant.

By 2.50.i and another application of 3.41.i, we have

$$\phi_1(x_1(a,t)) = -\Big(\phi_4\big(x_2(1)^{m_\Sigma(x_1(a,t))}\big) - \phi_2(x_2(1))\Big)/\delta_0$$

for all $(a,t) \in T^*$, where

$$\delta_0 = \begin{cases} 2 & \text{if } \iota \text{ is long and} \\ 1 & \text{if } \iota \text{ is short.} \end{cases}$$

By 32.9 in [36],

$$x_2(1)^{m_\Sigma(x_1(a,t))} = x_4(-t^{-\sigma})$$

for all $(a,t) \in T$. Thus

$$\phi_1(x_1(a,t)) = \big(\nu(t) + \phi_2(x_2(1))\big)/\delta_0$$

for all $(a,t) \in T^*$. By 3.21, $\phi_1(U_1^*) = \mathbb{Z}$, and by 24.30,

$$\{\nu(t) \mid (a,t) \in T^*\} = \delta\mathbb{Z}.$$

It follows that $\delta_0 = \delta$ and $\phi_2(x_2(1))/\delta \in \mathbb{Z}$. By 3.43, therefore, we can assume that 24.33 holds. Thus (ii) holds. Since ϕ satisfies condition (V1), 24.9.i follows from 24.6. Hence Λ is ν-compatible (since we have already seen that ν is σ-invariant). Thus (i) holds. □

Theorem 24.35. *Let (Δ, Σ, d) be a canonical realization of type 30.14.vii. Thus*

$$\Delta = \mathsf{B}_\ell^{\mathcal{P}}(\Lambda) = \mathsf{C}_\ell^{\mathcal{P}}(\Lambda)$$

for some anisotropic pseudo-quadratic space

$$\Lambda = (K, K_0, \sigma, L, q)$$

and some $\ell \geq 2$. Suppose that Λ is proper, let $(G^\dagger, \xi)$ be the root datum of Δ based at Σ and let ν be a valuation of K. Let ι_{B} be a long root map of Σ and let ι_{C} be a short root map of Σ as defined in 16.12.[3] *Let a and x_a be as in 16.2 and 16.3 and let ϕ_a be given by*

$$\phi_a(x_a(t)) = \nu(t)$$

for all $t \in K^*$. Then the following hold:

[3]Thus by 2.14.v, ι_{X} has target X_ℓ for $\mathsf{X} = \mathsf{B}$ and C.

(i) *If Λ is not ν-compatible, then ϕ_a does not extend to a valuation of $(G^\dagger, \xi)$.*
(ii) *If Λ is ν-compatible and Λ is ramified at ν, then ϕ_a extends to a valuation ϕ of $(G^\dagger, \xi)$ with respect to ι_{B} (but not with respect to ι_{C}).*
(iii) *If Λ is ν-compatible and Λ is unramified at ν, then ϕ_a extends to a valuation ϕ of $(G^\dagger, \xi)$ with respect to ι_{C} (but not with respect to ι_{B}).*

In cases (ii) and (iii), the extension of ϕ is unique up to equipollence.

Proof. This holds by 16.14.ii, 24.30 and 24.31. □

Notation 24.36. Let Δ, Λ, ν, etc. be as in 24.35, suppose that Λ is ν-compatible and let ϕ be as in (ii) or (iii) of 24.35. We denote the pair $(\Delta^{\rm aff}, \mathcal{A})$ obtained by applying 14.47 to this data by $\tilde{\mathsf{B}}_\ell^{\mathcal{P}}(\Lambda, \nu)$ in case (ii) and by $\tilde{\mathsf{C}}_\ell^{\mathcal{P}}(\Lambda, \nu)$ in case (iii).

Thus the pair $\tilde{\mathsf{X}}_\ell^{\mathcal{P}}(\Lambda, \nu)$ for $\mathsf{X} = \mathsf{B}$ or C defined in 24.36 exists if and only if the anisotropic pseudo-quadratic space Λ is ν-compatible.

Theorem 24.37. *Every Bruhat-Tits pair whose building at infinity is*

$$\mathsf{B}_\ell^{\mathcal{P}}(\Lambda) = \mathsf{C}_\ell^{\mathcal{P}}(\Lambda)$$

for some proper anisotropic pseudo-quadratic space

$$\Lambda = (K, K_0, \sigma, L, q)$$

and some $\ell \geq 2$ is of the form $\tilde{\mathsf{B}}_\ell^{\mathcal{Q}}(\Lambda, \nu)$ (if Λ is ramified at ν) or $\tilde{\mathsf{C}}_\ell^{\mathcal{Q}}(\Lambda, \nu)$ (if Λ is unramified at ν) for some valuation ν of K such that Λ is ν-compatible.

Proof. This holds by 16.4, 24.35 and 24.36. □

We turn now to completions.

Definition 24.38. Let

$$\Lambda = (K, K_0, \sigma, L, q)$$

be a proper anisotropic pseudo-quadratic space, let T be the group defined in 24.5, let ν be a valuation of K, let

$$\omega(a, t) = \nu(t) \tag{24.39}$$

for all $(a, t) \in T^*$ and let ∂_ν denote the map from $T \times T$ to $\mathbb{R}^*$ defined in 9.19 with ω in place of ν. Suppose that Λ is ν-compatible as defined in 24.28. By 24.9.i, we have

$$\omega\big((a, t) \cdot (b, s)\big) \geq \min\{\omega(a, t), \omega(a, s)\}$$

for all $(a, t), (b, s) \in T$, and by 24.7 and 24.28, we have

$$\omega\big((a, t)^{-1}\big) = \omega(-a, -t^\sigma) = \omega(a, t)$$

for all $(a, t) \in T$. By 9.18, therefore, ∂_ν is a metric on T.

Proposition 24.40. *Let*

$$\Lambda = (K, K_0, \sigma, L, q)$$

be a proper anisotropic pseudo-quadratic space, let f and T be as in 24.5 and let ν be a valuation of K. Suppose that Λ is ν-compatible (as defined in 24.28) and that K is complete with respect to ν.[4] *Let $\hat{K}_0$ be the closure of K_0 in K, let let ω be as in 24.39, let ∂_ν be as in 24.38 and let $\hat{T}$ denote the completion of T with respect to ∂_ν. Then the following hold:*

(i) *There is a unique group structure on $\hat{T}$ with respect to which T is a dense subgroup of $\hat{T}$.*
(ii) *The set $\hat{K}_0$ is a normal subgroup of $\hat{T}$, the map $(a,t)u = (au, u^\sigma tu)$ from $T \times K$ to T has a unique extension by continuity to a map from $\hat{T} \times K$ to $\hat{T}$ and this map endows $\hat{L} := \hat{T}/\hat{K}_0$ with the structure of a right vector space over K containing L as a subspace.*
(iii) *There exist unique maps $\hat{f}\colon \hat{L} \times \hat{L} \to K$ and $\hat{q}_0\colon \hat{T} \to K$ such that*

$$\hat{f}(a,b) = \lim_{k\to\infty} f(a_k, b_k)$$

whenever $(a_k, t_k)_{k\geq 1}$ and $(b_k, s_k)_{k\geq 1}$ are sequences in T that converge to elements of $\hat{T}$ whose images in $\hat{L} = \hat{T}/\hat{K}_0$ are a and b and

$$\hat{q}_0(x) = \lim_{k\to\infty} t_k$$

whenever $(a_k, t_k)_{k\geq 1}$ is a sequence in T that converges to an element $x \in \hat{T}$. The map $\hat{f}$ is a skew-symmetric form on $\hat{L}$ with respect to $(K, \hat{K}_0, \sigma)$.
(iv) *$\hat{q}_0(\hat{K}_0) \subset \hat{K}_0$ and*

$$\hat{\Lambda} := (K, \hat{K}_0, \sigma, \hat{L}, \hat{q})$$

is an anisotropic pseudo-quadratic space, where $\hat{q}$ is the map from $\hat{L}$ to K/K_0 induced by $\hat{q}_0$.
(v) *The map $x \mapsto \big(x + \hat{K}_0, \hat{q}_0(x)\big)$ is an isomorphism from $\hat{T}$ to the group $T_{\hat{\Lambda}}$ obtained by applying 24.5 to $\hat{\Lambda}$.*

Proof. Since we are working with valuations, it will be convenient to write the multiplication in T additively even though it is not abelian.

We first define an addition on the set $\hat{T}$ that extends the addition on T. Let $x, y \in \hat{T}$ and suppose that $x_k = (a_k, t_k)$ and $y_k = (b_k, s_k)$ are elements of T for all $k \geq 1$ such that

$$x = \lim_{k\to\infty} x_k \quad \text{and} \quad y = \lim_{k\to\infty} y_k.$$

Choose $k, l \geq 0$. Then

$$(x_k + y_k) - (x_l + y_l) = (x_k - x_l) + [-x_l, y_l - y_k] + (y_k - y_l),$$

---

[4]See 23.16, 24.48 and 24.49.

where

$$\begin{aligned}[-x_l, y_l - y_k] &= x_l + y_k - y_l - x_l + y_l - y_k \\ &= (a_l, t_l) + \big(b_k - b_l, s_k - s_l^\sigma - f(b_l, b_k)\big) \\ &\quad + (-a_l, -t_l^\sigma) + \big(b_l - b_k, s_l - s_k^\sigma - f(b_k, b_l)\big).\end{aligned}$$

By 24.8 and a bit of calculation, we conclude that

$$\begin{aligned}(x_k + y_k) - (x_l + y_l) &= (x_k - x_l) \\ &\quad + \big(0, f(a_l, b_l - b_k) + f(a_l, b_l - b_k)^\sigma\big) + (y_k - y_l).\end{aligned} \tag{24.41}$$

Since $(x_k)_{k\geq 1}$ converges, the sequence $(\omega(x_k))_{k\geq 1}$ is bounded below. Since $(y_k)_{k\geq 1}$ is a Cauchy sequence, it follows by 24.9.ii that for all N there exists M such that

$$\nu(f(a_l, b_l - b_k)) > N \tag{24.42}$$

for all $k, l \geq M$. By 24.41, we conclude that $(x_k + y_k)_{k\geq 1}$ is also a Cauchy sequence. We define $x + y$ to be the limit of this sequence. This yields a well-defined associative addition on $\hat{T}$ that extends the addition in T. By 24.7, ω is invariant under inverses. Thus if the sequence $(x_k)_{k\geq 1}$ of elements of T converges to the element x of $\hat{T}$, then the sequence $(-x_k)_{k\geq 1}$ is also a Cauchy sequence; its limit in $\hat{T}$ is then an additive inverse of x. Thus $\hat{T}$ with this addition is, in fact, a group. Thus (i) holds.

The map $t \mapsto (0, t)$ from K_0 to T extends to a map from $\hat{K}_0$ into the center of $\hat{T}$. Thus the subgroup $\hat{K}_0$ is normal in $\hat{T}$ and $[\hat{T}, \hat{T}] \subset \hat{K}_0$ (since $[T, T] \subset K_0$), so the quotient group $\hat{L} = \hat{T}/\hat{K}_0$ is abelian. Let $u \in K$, let $x \in \hat{T}$, let $x_k = (a_k, t_k) \in T$ for all $k \geq 1$ and suppose that the sequence $(x_k)_{k\geq 1}$ converges to x. Then

$$(a_k u, u^\sigma t_k u)_{k\geq 1}$$

is a Cauchy sequence; let xu denote its limit in $\hat{T}$. The map $(x, u) \mapsto xu$ gives $\hat{L}$ the structure of a right vector space over K that extends the structure of L as a right vector space over K. Thus (ii) holds.

Again let $x, y \in \hat{T}$ and let $x_k = (a_k, t_k)$ and $y_k = (b_k, s_k)$ for all $k \geq 1$ be elements of T such that the sequences $(x_k)_{k\geq 1}$ and $(y_k)_{k\geq 1}$ are sequences converging to x and y. We have

$$f(a_k, b_k) - f(a_l, b_l) = f(a_k - a_l, b_k) + f(a_l, b_k - b_l)$$

for all $k, l \geq 0$. By 24.42, therefore, $f(a_k, b_k)_{k\geq 1}$ is a Cauchy sequence. Let $\hat{f}_0(x, y)$ denote its limit in K. This defines a map $\hat{f}_0$ from $\hat{T} \times \hat{T}$ to K such that $\hat{f}_0(\hat{K}_0, \hat{T}) = \hat{f}_0(\hat{T}, \hat{K}_0) = 0$. The map $\hat{f}_0$ thus induces a form $\hat{f}$ on $\hat{L}$ that is skew-symmetric with respect to $(K, \hat{K}_0, \sigma)$. By 24.6 and 24.7,

$$x_k - x_l = \big(a_k - a_l, t_k - t_l + f(a_l, a_l - a_k)\big)$$

for all $k, l \geq 0$. By 24.42 (with x in place of y), it follows that $(t_k)_{k\geq 1}$ is a Cauchy sequence in K. Let

$$\hat{q}_0(x) = \lim_{k\to\infty} t_k. \tag{24.43}$$

This defines a map from $\hat{T}$ to K. Hence (iii) holds.

Since $(a_k, t_k)u = (a_k u, u^\sigma t_k u)$ for all $k \geq 1$ and all $u \in K$, we conclude that $\hat{q}_0(xu) = u^\sigma \hat{q}_0(x)u$ for all $x \in \hat{T}$ and all $u \in K$. We also have

$$\begin{aligned}\hat{q}_0(x+y) &= \hat{q}_0\big(\lim_{k\to\infty}(x_k + y_k)\big)\\ &= \hat{q}_0\Big(\lim_{k\to\infty}\big(a_k + b_k, t_k + s_k + f(b_k, a_k)\big)\Big)\\ &= \lim_{k\to\infty}\big(t_k + s_k + f(b_k, a_k)\big)\\ &= \hat{q}_0(x) + \hat{q}_0(y) + \hat{f}_0(y, x) \qquad (24.44)\end{aligned}$$

for all $x, y \in \hat{T}$. Suppose that

$$\hat{q}_0(x) \in \hat{K}_0$$

for some $x \in \hat{T}$. Let $(t_k)_{k\geq 1}$ be a sequence of elements of K_0 that converges to $\hat{q}_0(x)$ and let $z_k = x - (0, t_k)$ for each $k \geq 1$. Then $(z_k)_{k\geq 1}$ is a Cauchy sequence in $\hat{T}$; let z be its limit. By 24.44, $\hat{q}_0(z) = 0$. Let $(c_k, r_k)_{k\geq 1}$ be a sequence in T that converges to z. By 24.43,

$$\hat{q}_0(z) = \lim_{k\to\infty} r_k.$$

If $z \neq 0$, then $\nu(r_k)$ equals $\omega(z)$ for all k sufficiently large (by 9.18.iii). Hence $z = 0$, so $x \in \hat{K}_0$. Thus (iv) holds.

Let π denote the map $x \mapsto \big(x + \hat{K}_0, \hat{q}_0(x)\big)$ from $\hat{T}$ to the group $T_{\hat{\Lambda}}$. By 24.44, π is a homomorphism. By (iii),

$$\hat{q}_0(z) = z \qquad (24.45)$$

for all $z \in \hat{K}_0$. Thus if $x \in \hat{K}_0$ and $\hat{q}_0(x) = 0$, then $x = 0$. Hence π is injective. Let $(a, t) \in T_{\hat{\Lambda}}$. Then $a = x + \hat{K}_0$ for some $x \in \hat{T}$ such that

$$\hat{q}_0(x) \equiv t \pmod{\hat{K}_0}.$$

Let $y = t - \hat{q}_0(x)$. Then $\hat{q}_0(x+y) = \hat{q}_0(x) + \hat{q}_0(y)$ by 24.44. By 24.45, therefore, $\hat{q}_0(x+y) = t$. Hence $\pi(x+y) = (a, t)$. We conclude that π is surjective. Thus (v) holds. □

Theorem 24.46. *Let*

$$\Lambda = (K, K_0, \sigma, L, q)$$

be a proper anisotropic pseudo-quadratic space, let ν be a discrete valuation of K and suppose that Λ is ν-compatible (as defined in 24.28). Let $\hat{K}$ be the completion of K with respect to ν, let $\Lambda_{\hat{K}}$ be as in 24.9 and let

$$\hat{\Lambda} = (\hat{K}, \hat{K}_0, \hat{\sigma}, \hat{L}, \hat{q})$$

be the anisotropic quadratic space obtained by applying 24.40 to $\Lambda_{\hat{K}}$. Let $(\Delta, \mathcal{A}) = \tilde{\mathsf{X}}_\ell^{\mathcal{P}}(\Lambda, \nu)$ for $\mathsf{X} = \mathsf{B}$ or C be as in 24.36. Then $\hat{\Lambda}$ is ν-compatible, Δ is completely Bruhat-Tits and its completion is $\tilde{\mathsf{X}}_\ell^{\mathcal{P}}(\hat{\Lambda}, \nu)$.

Proof. By 30.16, $\mathsf{X}_\ell^{\mathcal{P}}(\Lambda)$ is a subbuilding of $\mathsf{X}_\ell^{\mathcal{P}}(\hat{\Lambda})$. Since $\hat{K}$ is complete, $\hat{\Lambda}$ satisfies 24.9.iii. Thus $\hat{\Lambda}$ is ν-compatible. It follows from 9.18.iii that Λ is ramified at ν if and only if $\hat{\Lambda}$ is. The claim holds, therefore, by 17.4, 17.9 and 24.40. □

Proposition 24.47. *With all the hypotheses and notation of 24.40, suppose that* $\dim_K L$ *is finite and that* K_0 *is a closed subset of* K. *Then* $\hat{\Lambda} = \Lambda$.

Proof. It will suffice to show that T is complete with respect to ω. Let $v_1, v_2, \ldots, v_m$ be a basis of L over K, let $(a_k, t_k)_{k\geq 1}$ be a Cauchy sequence in T and let s_{ij} be elements of K such that

$$a_k = v_1 s_{k1} + v_2 s_{k2} + \cdots + v_m s_{km}$$

for all $k \geq 1$. Since Λ is proper, the skew-hermitian form f is non-degenerate. Thus there exist elements $w_1, w_2, \ldots, w_m \in L$ such that

$$f(v_i, w_j) = \delta_{ij}$$

for all $i, j \in [1, m]$. Therefore $f(a_k, w_j) = s_{kj}$ for all $k \geq 1$ and all $j \in [1, m]$. By 24.40.iii, it follows that $(s_{kj})_{k\geq 1}$ is a Cauchy sequence for each $j \in [1, m]$; let $s_j \in K$ be its limit. By 24.40.iii, we also know that the sequence $(t_k)_{k\geq 1}$ converges; let $t \in K$ be its limit. Let

$$a = \sum_{j=1}^{m} v_k s_k.$$

Since $q(a_k) - t_k \in K_0$ for all $k \geq 1$, we have

$$\sum_{i=1}^{n} s_{ki}^\sigma q(v_i) s_{ki} + \sum_{i\neq j} s_{ki}^\sigma f(v_i, v_j) s_{kj} - t_k \in K_0$$

for all $k \geq 1$ and hence $q(a) = t \in K_0$ since K_0 is closed in K. Since

$$(a_k, t_k) - (a, t) = \big(a_k - a, t_k - t + f(a, a - a_k)\big)$$

and

$$f(a, a - a_k) = \sum_{j=1}^{m} f(a, v_j)(s_j - s_{kj})$$

for each $k \geq 1$, we have

$$\lim_{k\to\infty} f(a, a - a_k) = 0.$$

It follows that

$$\lim_{k\to\infty} (a_k, t_k) = (a, t).$$

□

Proposition 24.48. *Let*

$$\Lambda = (K, K_0, \sigma, L, q)$$

be a proper anisotropic pseudo-quadratic space, let ν *be a valuation of* K, *let* K_σ *and* K^σ *be as in 23.18 and 23.19 and suppose that the following hold:*

(i) *K is complete with respect to ν.*
(ii) *L is finite-dimensional over K.*
(iii) *K_0 equals K_σ or K^σ.*

Then Λ is ν-compatible and Λ equals the anisotropic pseudo-quadratic space $\hat{\Lambda}$ obtained by applying 24.40 to Λ.

Proof. By (i) and 23.16, the involutory set (K, K_0, σ) is ν-compatible. By (iii), it follows that K_0 is closed in K, and by (i), 24.9.iii and 24.28, Λ is ν-compatible. Therefore $\Lambda = \hat{\Lambda}$ by (ii) and 24.47. □

Remark 24.49. Let

$$\Lambda = (K, K_0, \sigma, L, q)$$

be a proper anisotropic pseudo-quadratic space. By 30.32.iv, the spherical building $\mathsf{B}_\ell^{\mathcal{P}}(\Lambda)$ is algebraic as defined in 30.31 if and only if Λ satisfies the conditions (ii)–(iv) in 24.48. Thus (by 24.48) if $\mathsf{B}_\ell^{\mathcal{P}}(\Lambda)$ is algebraic and K is complete with respect to a valuation ν, then Λ is ν-compatible and $\Lambda = \hat{\Lambda}$.

We now consider gems.

Definition 24.50. Let

$$\Lambda = (K, K_0, \sigma, L, q)$$

be a proper anisotropic pseudo-quadratic space, let T be as in 24.5, let ν be a valuation of K and suppose that Λ is ν-compatible. Let $\mathcal{O}_K$ be the ring of integers in K with respect to ν, let $\bar{K}$ be the residue field of K, let

$$T_i = \{(a, t) \in T \mid \nu(t) \geq i\}$$

and let $Z_i = T_i \cap (0, K_0)$ for all $i \in \mathbb{Z}$. If $(u, t) \in T_{i+1}$ and $(v, s) \in T_i$ for some i, then $\nu(f(u, v)) \geq i + 1$ by 24.9.ii. By 24.6–24.8 and 24.9.i, it follows that T_i is a subgroup of T, T_{i+1} is a normal subgroup of T_i, Z_i is a subgroup of the center of T_i and the quotient group

$$\bar{L}_i := T_i/(T_{i+1} + Z_i) \tag{24.51}$$

is abelian for each i.[5] Choose $i \in \mathbb{Z}$ and let $u \mapsto \bar{u}$ be the natural homomorphism from T_i to $\bar{L}_i$. By 24.6–24.8, we also have

$$\begin{aligned}&\big(a(s + s_1), (s + s_1)^\sigma t(s + s_1)\big)\\ &\quad - (as, s^\sigma ts) = (as_1, s^\sigma t^\sigma s_1 + s_1^\sigma ts + s_1^\sigma ts_1)\end{aligned} \tag{24.52}$$

for all $(a, t) \in T$ and all $s, s_1 \in K$, from which it follows that the map

$$(\overline{(a, t)}, \bar{s}) \mapsto \overline{(as, s^\sigma ts)}$$

from $\bar{L}_i \times \bar{K}$ to $\bar{L}_i$ is well defined and gives $\bar{L}_i$ the structure of a right vector space over $\bar{K}$. Now let $i = 0$, let $\bar{K}_0$ and $\bar{\sigma}$ be as in 23.22, let $\bar{L} = \bar{L}_0$ and let $\bar{q}$ be the map from $\bar{L}$ to $\bar{K}$ modulo $\bar{K}_0$ given by

$$\bar{q}\big(\overline{(a, t)}\big) = \bar{t}$$

[5] In 24.50 and 24.55 we are using additive notation for the group T even though T is not, in general, abelian.

for all $(a,t) \in \bar{T}_0$. Then

$$\bar{\Lambda} := (\bar{K}, \bar{K}_0, \bar{\sigma}, \bar{L}, \bar{q}) \tag{24.53}$$

is an anisotropic pseudo-quadratic space.[6] We call $\bar{\Lambda}$ the *residue of* Λ *at* ν. Let $\bar{T}$ be the group defined by applying 24.5 to $\bar{\Lambda}$ (so $\bar{T}$ is, in particular, a subset of $\bar{L} \times \bar{K}$) and let ρ be the map from $\bar{T}_0$ to $\bar{T}$ induced by the map

$$(a,t) \to (\overline{(a,t)}, \bar{t}).$$

If $(\overline{(a,t)}, \bar{s}) \in \bar{T}$ for some $(a,t) \in T_0$ and some $s \in \mathcal{O}_K$, then $\bar{t} - \bar{s} \in \bar{K}_0$ (by 24.5) and hence $(a,s) \in T_0$ and ρ maps $\overline{(a,s)}$ to $(\overline{(a,t)}, \bar{s}) \in \bar{T}$. It follows that ρ is an isomorphism.

Definition 24.54. Let

$$\Lambda = (K, K_0, \sigma, L, q)$$

be a proper anisotropic pseudo-quadratic set and let ν be a valuation of K. Suppose that Λ is ν-compatible and that $\nu(K_0^*) = \mathbb{Z}$. Thus, in particular, the ν-index of Λ is 1. Let π be a uniformizer contained in K_0. By 11.6 in [36], π^{-1} is also in K_0. Let $\Lambda_{\pi^{-1}}$ be the translate of Λ with respect to π^{-1} as defined in 11.8 of [36] and let $\overline{\Lambda_1}$ denote the residue of $\Lambda_{\pi^{-1}}$ at ν as defined in 24.50. The pseudo-quadratic set $\overline{\Lambda_1}$ is independent of the choice of π up to similarity (as defined in 11.26 of [36]); compare 23.23. By 35.19 in [36] and 3.6, therefore, also the building $\mathsf{C}_\ell^{\mathcal{P}}(\overline{\Lambda_1})$ is independent of the choice of the uniformizer π.

Lemma 24.55. *Let*

$$\Lambda = (K, K_0, \sigma, L, q)$$

be a proper anisotropic pseudo-quadratic space and let ν *be a valuation of* K. *Suppose that* Λ *is* ν*-compatible and that the* ν*-index* δ *of* Λ *is 1 but that* $\nu(K_0^*) = 2\mathbb{Z}$. *Let* π *be a uniformizer of* K. *Then the following hold (with all the notation as in 24.50):*

(i) *The residue field* $\bar{K}$ *is commutative.*
(ii) $Z_1 \subset T_2$, *so* $\bar{L}_1 = T_1/T_2$.
(iii) $\bar{L}_1 \neq 0$ *and the map* $q_1 \colon \bar{L}_1 \to \bar{K}$ *given by*

$$q_1\big((a,t) + T_2\big) = \overline{\pi^{-1}t} \tag{24.56}$$

for all $(a,t) \in T_1$ *is an anisotropic quadratic form on* $\bar{L}_1$ *with bilinear form* f_1 *given by*

$$f_1\big((a,t) + T_2, (b,s) + T_2\big) = \overline{\pi^{-1}f(a,b)} \tag{24.57}$$

for all $(a,t), (b,s) \in T_1$.

---

[6]The anisotropic pseudo-quadratic space $\bar{\Lambda}$ need not, however, be proper; in fact, it can happen that $\bar{L} = 0$, in which case

$$\mathsf{B}_\ell^{\mathcal{P}}(\bar{\Lambda}) = \mathsf{B}_\ell^{\mathcal{I}}(\bar{K}, \bar{K}_0, \bar{\sigma}).$$

Proof. Assertion (i) holds by 23.24. If $(0,t) \in Z_1$, then t is an element of K_0 such that $\nu(t) \geq 1$. Since $\nu(K_0^*) = 2\mathbb{Z}$, it follows that $Z_1 \subset T_2$. Thus (ii) holds. Since $T_1 \neq T_2$, $\bar{L}_1$ is non-trivial.

Let $(a,t) + T_2$ be a non-trivial element of $\bar{L}_1$ and let q_1 be as in 24.56 (which is well defined by 24.9.ii). Then $\nu(t) = 1$, so

$$q_1\big((a,t) + T_2\big) \neq 0,$$

and

$$\begin{aligned} q_1\big((av, v^\sigma tv) + T_2\big) &= \overline{\pi^{-1}v^\sigma tv} \\ &= \overline{\pi^{-1}tv^2} && \text{by 23.25} \\ &= \overline{\pi^{-1}t}\bar{v}^2 \\ &= q_1\big((a,t) + T_2\big)\bar{v}^2 \end{aligned}$$

for all $v \in \mathcal{O}_K$. Now choose $(b,s) \in T_1$. Then

$$\begin{aligned} q_1\big(((a,t) + T_2) + ((b,s) + T_2)\big) &= q_1\big((a+b, t+s+f(b,a)) + T_2\big) \\ &= \overline{\pi^{-1}t} + \overline{\pi^{-1}s} + \overline{\pi^{-1}f(b,a)} \end{aligned}$$

by 24.6. Since $\bar{L}_1$ is abelian, it follows that

$$\overline{\pi^{-1}f(b,a)} = \overline{\pi^{-1}f(a,b)}.$$

Since $f(b,as) = f(b,a)s$ for all $s \in K$, it follows that the map f_1 defined in 24.57 is bilinear over $\bar{K}$. We conclude that $(\bar{K}, \bar{L}_1, q_1)$ is an anisotropic quadratic space. Thus (iii) holds. □

Theorem 24.58. *Let $(\Delta, \mathcal{A}) = \tilde{\mathsf{X}}_\ell^{\mathcal{P}}(\Lambda, \nu)$ for $\mathsf{X} = \mathsf{B}$ or C, some proper $\Lambda = (K, K_0, \sigma, L, q)$ and some valuation ν of K such that Λ is ν-compatible, where $\mathsf{X} = \mathsf{B}$ if Λ is ramified at ν and $\mathsf{X} = \mathsf{C}$ if Λ is unramified at ν. Let $\bar{\Lambda}$ be the residue of Λ at ν as defined in 24.50, let the apartment A, the gem R_0 and the group G° be as in 18.1 and let the gem R_1 be as in 18.28. Then the following hold:*

(i) *If $\mathsf{X} = \mathsf{B}$, then $R_0 = \mathsf{B}_\ell^{\mathcal{P}}(\bar{\Lambda})$ and all gems are in the same G°-orbit as R_0.*
(ii) *If $\mathsf{X} = \mathsf{C}$ and $\nu(K_0^*) = \mathbb{Z}$, then $R_0 = \mathsf{B}_\ell^{\mathcal{P}}(\bar{\Lambda})$, $R_1 \cong \mathsf{B}_\ell^{\mathcal{P}}(\overline{\Lambda_1})$, where $\overline{\Lambda_1}$ is as in 24.54, and every gem is in the same G°-orbit as R_0 or R_1 (or both).*
(iii) *If $\mathsf{X} = \mathsf{C}$ and $\nu(K_0^*) = 2\mathbb{Z}$, then $R_0 = \mathsf{B}_\ell^{\mathcal{P}}(\bar{\Lambda})$, $R_1 \cong \mathsf{B}_\ell^{\mathcal{Q}}(\bar{K}, \bar{L}_1, q_1)$, where $(\bar{K}, \bar{L}_1, q_1)$ is as in 24.55, and every gem is in the same G°-orbit as R_0 or R_1 (or both).*
(iv) *If $\mathsf{X} = \mathsf{C}$, (K, K_0, σ) is a quadratic involutory set, $\nu(\pi) = 1$ for some $\pi \in K_0^*$ and there exists a K-linear automorphism ψ of L such that $q(\psi(x)) = \pi q(x)$ for all $x \in L$, then every gem is in the same G°-orbit as R_0.*

Proof. Let $\Sigma = A^\infty$, let d be a chamber of Σ, let $(G^\dagger, \xi)$ be the root datum of $\Delta_{\mathcal{A}}^\infty$ based at Σ and let $\phi = \phi_{R_0}$ be the valuation of $(G^\dagger, \xi)$ described in

13.8. Let c be the unique root in Ξ_d that is long (with respect to the root map ι_C) if $\mathsf{X} = \mathsf{C}$ and short (with respect to the root map ι_B) if $\mathsf{X} = \mathsf{B}$. By 19.22, we can assume that $(\Delta_{\mathcal{A}}^{\infty}, \Sigma, d)$ is the canonical realization of one of the root group labelings described in 30.14.vii, that $\phi_a(x_a(t)) = \nu(t)$ for all $a \in \Xi_d^{\circ}\backslash\{c\}$ and all $t \in K^*$ and that $\phi_c(x_c(a,t)) = \nu(t)/\delta$ for all non-trivial elements (a,t) in the group T defined in 24.5, where δ is the ν-index of Λ as defined in 24.30. By 18.27,

$$R_0 \cong \mathsf{X}_\ell^{\mathcal{P}}(\bar{\Lambda}).$$

Let $b \in \Xi_d^{\circ}$ be as in 18.28. Since b is long, we have $b = c$ if $\mathsf{X} = \mathsf{C}$. Let H be as in 18.1.

Suppose that $\mathsf{X} = \mathsf{B}$ and let $\pi \in K$ be a uniformizer. By 18.9, there exists an element $\tau \in H$ centralizing U_a for all $a \in \Xi_d^{\circ}\backslash\{b\}$ such that

$$x_b(t)^\tau = x_b(\pi t)$$

for all $t \in K$. Hence ϕ_a is τ-invariant for all $a \in \Xi_d^{\circ}\backslash\{b\}$ and

$$\phi_b(x_b(t)^\tau) = \phi_b(x_b(t)) + 1$$

for all $t \in K*$. By 18.6, there is an element $\rho \in G_A$ inducing τ on $\Delta_{\mathcal{A}}^{\infty}$. By 18.28.i, it follows that ρ maps R_0 to R_1. Thus (i) holds by 18.29.

Now suppose that $\mathsf{X} = \mathsf{C}$. Suppose, too, that $\nu(K_0^*) = \mathbb{Z}$. Let π be a uniformizer of K in K_0 and suppose that π is the uniformizer used to construct $\overline{\Lambda_1}$ in 24.54. By 3.6 and 18.28.i, the maps $x_a(u) \mapsto x_a(u)$ for all $a \in \Xi_d^{\circ}\backslash\{b\}$ and $x_b(a,t) \mapsto x_b(a, \pi^{-1}t)$ induce an isomorphism from the residue R_1 to $\mathsf{B}_\ell^{\mathcal{P}}(\overline{\Lambda_1})$. Thus (ii) holds by 18.29.

Next suppose that $\mathsf{X} = \mathsf{C}$ but $\nu(K_0^*) = 2\mathbb{Z}$ and let $(\bar{K}, \bar{L}_1, q_1)$ be as in 24.55. By 3.6 and 18.28.i again, the maps $x_a(u) \mapsto x_a(u)$ for all $a \in \Xi_d^{\circ}\backslash\{b\}$ and $x_b(a,t) \mapsto x_b(a,t)$ induce an isomorphism from R_1 to $\mathsf{B}_\ell^{\mathcal{Q}}(\bar{K}, \bar{L}_1, q_1)$. Thus (iii) holds by 18.29.

Suppose, finally, that all the hypotheses in (iv) hold. Thus, in particular, $F := K_0$ is contained in the center of K and $\pi F = F$. By 18.9, there thus exists a map $\tau \in H$ centralizing U_a for $a \in \Xi_d^{\circ}\backslash\{b\}$ such that

$$x_b(a,t)^\tau = x_b(\psi(a), \pi t)$$

for all $(a,t) \in T$. By 18.6, there is an element $\rho \in G_A$ inducing τ on $\Delta_{\mathcal{A}}^{\infty}$, and by 18.28, ρ maps R_1 to R_0. Thus (iv) holds by 18.29. □

The following result will be needed in the proof of 28.20.ii and in 28.36. The proof we give is a modification of the proof of 18.34.

Proposition 24.59. *Let*

$$\Lambda = (K, K_0, \sigma, L, q)$$

be a proper anisotropic quadratic space and suppose that K is complete with respect to a valuation ν (so Λ is ν-compatible by 24.9). Let T, T_i and Z_i be as in 24.50 (for all i). Then the following hold, where all the notation is as in 24.50:

(i) *$T_2 + Z_0$ is a normal subgroup of T_0.*[7]

(ii) *Let $V = T_0/(T_2 + Z_0)$ and let $(a,t) \mapsto \overline{(a,t)}$ denote the natural map from T_0 to V. Then the map*

$$(\overline{(a,t)}, \bar{s}) \mapsto \overline{(as, s^\sigma ts)}$$

from $V \times \bar{K}$ to V is well defined and gives V the structure of a right vector space over $\bar{K}$.

(iii) $\dim_{\bar{K}} V = \dim_{\bar{K}} \bar{L}_0 + \dim_{\bar{K}} \bar{L}_1$, *where $\bar{L}_0$ and $\bar{L}_1$ are as in 24.51.*

(iv) *If $\dim_{\bar{K}} \bar{L}_0$ and $\dim_{\bar{K}} \bar{L}_1$ are both finite, then*

$$\dim_{\bar{K}} \bar{L}_0 + \dim_{\bar{K}} \bar{L}_1 = \dim_K L.$$

(v) *If $\dim_{\bar{K}} \bar{L}_0$ is finite and Λ is ramified, then*

$$\dim_{\hat{K}} \bar{L}_0 = \dim_K L.$$

Proof. Since $[T_0, T_0] \subset Z_0$, (i) holds. By 11.1.i in [36], $K_\sigma \subset K_0$. By 24.52, therefore, (ii) holds and $(T_1 + Z_0)/(T_2 + Z_0)$ is a $\bar{K}$-subspace of V. Let ψ be the group homomorphism from T_1 to $(T_1 + Z_0)/(T_2 + Z_0)$ given by

$$\psi(x) = x + T_2 + Z_0$$

for all $x \in T_1$. Then ψ is surjective and its kernel is $T_1 \cap (T_2 + Z_0) = T_2 + Z_1$. Hence ψ induces an isomorphism from $\bar{L}_1 = T_1/(T_2 + Z_1)$ to $(T_1 + Z_0)/(T_2 + Z_0)$ that is linear over $\bar{K}$. Therefore

$$\begin{aligned}\dim_{\bar{K}} V &= \dim_{\bar{K}} T_0/(T_2 + Z_0) \\ &= \dim_{\bar{K}} T_0/(T_1 + Z_0) + \dim_{\bar{K}} (T_1 + Z_0)/(T_1 + Z_0) \\ &= \dim_{\bar{K}} \bar{L}_0 + \dim_{\bar{K}} \bar{L}_1.\end{aligned}$$

Thus (iii) holds.

It will be useful now to set

$$(a,t)s = (as, s^\sigma ts)$$

for all $(a,t) \in T$ and all $s \in K$. Note that by 24.6–24.8,

$$(a,t)r - (a,t)s = \big(a(r-s), (r-s)^\sigma tr + s^\sigma t^\sigma (r-s)\big)$$

for all $(a,t) \in T$ and all $r, s \in K$. Thus if $(a,t) \in T$ and $(s_i)_{i \geq 1}$ is a sequence of elements in K that converges to an element $s \in K$, then

$$\lim_{i \to \infty} (a,t)s_i = (a,t)s \tag{24.60}$$

with respect to the metric ∂_ν on T defined in 24.38.

Let $m = \dim_{\bar{K}} V$ and suppose from now on that that m is finite. We claim that

$$\dim_{\bar{K}} V = \dim_K L. \tag{24.61}$$

Let $\omega(a,t) = \nu(t)$ for all $(a,t) \in T^*$ and let

$$B := \{(a_i, s_i) \mid i \in [1, m]\}$$

[7] As in 24.50, we are using additive notation for the group T.

be the inverse image of a basis of V under the natural map from T_0 to V. The set B is linearly independent modulo $(0, K_0)$ (i.e. the image of B in $L = T/(0, K_0)$ is linearly independent) since otherwise we could choose scalars $t_1, \dots, t_m \in K$ such that

$$\sum_{i=1}^{m} (a_i, s_i) t_i \equiv 0 \ (\mathrm{mod}(0, K_0))$$

and $\min\{\nu(t_i) \mid i \in [1, m]\} = 0$. To prove that 24.61 holds, it remains only to show that B spans T modulo $(0, K_0)$.

Let $(b, u) \in T^*$ and let π be a uniformizer of K. Then $\omega(b, u) = 2k_1$ or $2k_1 + 1$ for some integer k_1. Thus $(c, v) := (b, u)\pi^{-k_1}$ is contained in T_0. Hence there exist elements $t_i^{(1)}$ in the ring of integers $\mathcal{O}_K$ for $i \in [1, m]$ such that

$$\overline{(c, v)} = \overline{(a_1, s_1)} \bar{t}_1^{(1)} + \cdots + \overline{(a_m, s_m)} \bar{t}_m^{(1)}.$$

There thus exists an element $(b_1, u_1) \in T$ such that

$$(b, u) = \sum_{i=1}^{m} (a_i, s_i) t_i^{(1)} \pi^{k_1} + (b_1, u_1) \ (\mathrm{mod}\ (0, K_0))$$

such that $\omega(b_1, u_1) > \omega(b, u)$. If $(b_1, u_1) = 0$, then (b, u) is in the span of B modulo $(0, K_0)$. We can thus assume that $(b_1, u_1) \neq 0$.

By induction and the conclusion of the previous paragraph, we can assume that there exist elements $k_1, k_2, k_3, \dots$ of $\mathbb{Z}$, elements

$$(b_1, u_1), (b_2, u_2), (b_3, u_3), \dots$$

of T^* and elements

$$t_i^{(1)}, t_i^{(2)}, t_i^{(3)}, \dots$$

of $\mathcal{O}_K$ for all $i \in [1, m]$ such that

$$(b, u) \equiv \sum_{i=1}^{m} (a_i, s_i) \big(t_i^{(1)} \pi^{k_1} + t_i^{(2)} \pi^{k_2} + \cdots + t_i^{(r)} \pi^{k_r} \big)$$

$$+ (b_r, u_r) \ (\mathrm{mod}\ (0, K_0)) \tag{24.62}$$

as well as

$$\omega(b, u) < \omega(b_1, u_1) < \omega(b_2, u_2) < \cdots < \omega(b_r, u_r) \tag{24.63}$$

and

$$\omega(b_{r-1}, u_{r-1}) = 2k_r \text{ or } 2k_r + 1 \tag{24.64}$$

for all $r \geq 1$ (where $b_0 = b$ and $u_0 = u$). By 24.63 and 24.64,

$$k_1 \leq k_2 \leq k_3 \leq \cdots \tag{24.65}$$

and no value is repeated more than once in the sequence 24.65. Hence

$$\lim_{r \to \infty} k_r = \infty. \tag{24.66}$$

By 24.63,

$$\lim_{r\to\infty}(b_r, u_r) = 0$$

with respect to the metric ∂_ν. For each $i \in [1, m]$ and each $r \geq 1$, let $z_i^{(r)}$ denote the coefficient of (a_i, s_i) in the sum on the right-hand side of 24.62. Then $(z_i^{(r)})_{r\geq 1}$ is a Cauchy sequence for each $i \in [1, m]$ (by 24.66) and hence has a limit in K that we denote by f_i. Thus

$$\lim_{r\to\infty}(a_i, s_i)z_i^{(r)} = (a_i, s_i)f_i$$

for each $i \in [1, m]$ by 24.60. Hence

$$(b, u) = (a_1, s_1)f_1 + \cdots + (a_m, s_m)f_m \pmod{(0, K_0)}$$

since the subgroup $(0, K_0)$ of T is closed with respect to ∂_ν.[8] Therefore (b, u) is in the space spanned by B modulo $(0, K_0)$. We conclude that 24.61 holds as claimed. By (iii), therefore, (iv) and (v) hold (since $\bar{L}_1 = 0$ if Λ is ramified). □

* * *

By 30.14 and the results of Chapters 19–24, we have now concluded the classification of Bruhat-Tits pairs of type $\tilde{\mathsf{B}}_\ell$ and $\tilde{\mathsf{C}}_\ell$. The precise results are listed in Table 27.2. We close this chapter with one more observation.

Proposition 24.67. *Let $(\Delta, \mathcal{A})$ be a Bruhat-Tits pair of type $\tilde{\mathsf{B}}_\ell$ for some $\ell \geq 3$, let K be the defining field of the building at infinity $\Delta_{\mathcal{A}}^\infty$ (as defined in 30.15) and let R be a maximal irreducible residue of Δ that is not contained in a gem. Then the following hold:*

(i) *$R \cong \mathsf{A}_3(\bar{K})$ if $\ell = 3$ and $R \cong \mathsf{D}_\ell(\bar{K})$ if $\ell \geq 4$.*
(ii) *If $\bar{K}$ is not commutative, then $(\Delta, \mathcal{A}) = \tilde{\mathsf{B}}_3^{\mathcal{I}}(\Lambda, \nu)$ for some honorary involutory set $\Lambda = (K, K_0, \sigma)$ and $\bar{K}$ is a quaternion division algebra.*

Proof. A maximal connected proper subdiagram of the Coxeter diagram $\tilde{\mathsf{B}}_\ell$ that contains both special vertices is isomorphic to A_3 if $\ell = 3$, to D_ℓ if $\ell \geq 4$. By 30.14, therefore, $R \cong \mathsf{A}_3(F)$ for some field or skew field F if $\ell = 3$ and $R \cong \mathsf{D}_\ell(F)$ for some field F if $\ell \geq 4$. In particular, every irreducible rank 2 residue of R is isomorphic to $\mathsf{A}_2(F)$. By 19.35, 23.27 and 24.58 (as well as 30.14), every irreducible rank 2 residue of a gem is isomorphic to $\mathsf{A}_2(\bar{K})$. Since R has irreducible rank 2 residues that are also residues of a gem, it follows by 35.6 in [36] that $F \cong \bar{K}$. Thus (i) holds.

Suppose now that $\bar{K}$ is not commutative. Since F must be commutative if $\ell \geq 4$, it follows that $\ell = 3$. By 19.35, 23.27 and 24.58 again as well as 23.6 and 24.35, we have either

$$(\Delta, \mathcal{A}) = \tilde{\mathsf{B}}_3^{\mathcal{I}}(\Lambda, \nu)$$

[8] If $u, v \in S$ and $(v_k)_{k\geq 1}$ is a sequence of elements of S converging to v, then the sequence $(uv_k)_{k\geq 1}$ converges to uv and the sequence $(v_k u)_{k\geq 1}$ converges to vu (by 24.9.ii). By 38.10 in [36], $(0, K_0)$ is the center of T. It follows from this that $(0, K_0)$ is closed with respect to ∂_ν.

for some involutory set Λ, genuine or honorary, or

$$(\Delta, \mathcal{A}) = \tilde{\mathsf{B}}_3^{\mathcal{P}}(\Lambda, \nu)$$

for some anisotropic pseudo-quadratic space Λ, where ν is a valuation of K such that Λ is ν-compatible (as defined in 23.2 and 24.28) and Λ is ramified at ν (as defined in 23.2 and 24.30). By 23.24 and 23.26, it follows that Λ is honorary and $\bar{K}$ is a quaternion division algebra. $\square$

Let $\Lambda = (K, K_0, \sigma)$ be an involutory set and suppose that ν is a valuation of K such that Λ is ν-compatible and ramified at ν. Let $\ell \geq 4$, let

$$(\Delta, \mathcal{A}) = \tilde{\mathsf{B}}_\ell^{\mathcal{I}}(\Lambda, \nu)$$

and let R be as in 24.67. As we have just shown, $R \cong \mathsf{D}_\ell(\bar{K})$, and the proof of this we gave does not use 23.24. Since (by 30.14) buildings of type D_ℓ exist only over commutative fields, we obtain from this observation an alternative, more geometrical (but much less elementary!) proof of 23.24.

Chapter Twenty Five

Hexagons

In this chapter, we consider the case $X = \mathcal{H}$ of the problem framed at the end of Chapter 16. Our main results are 25.21 and 25.25 in which we give the classification of Bruhat-Tits pairs of type $\tilde{\mathsf{G}}_2$.

We begin with the analogue to 19.1, 21.8 and 24.1:

Proposition 25.1. *Let $\Lambda = (J, K, N, \#, T, \times, 1)$ be an hexagonal system as defined in 15.15 in [36],*[1] *let E be a field containing K and let $J_E = J \otimes_K E$. Let T_E be the unique bilinear form on J_E and let $\times_E$ be the unique bilinear map from $J_E \times J_E$ to J_E such that*

$$T_E(u \otimes s, v \otimes t) = T(u, v)st \tag{25.2}$$

and

$$(u \otimes s) \times_E (v \otimes t) = (u \times v) \otimes st$$

for all $u, v \in J$ and all $s, t \in E$. Then there are unique maps $\#_E$ from J_E to itself and N_E from J_E to E such that

$$\Big(\sum_{i=1}^{d} v_i \otimes t_i\Big)^{\#_E} = \sum_{i=1}^{d} t_i^2 (v_i)^\# + \sum_{i<j} (v_i \times v_j) \otimes t_i t_j \tag{25.3}$$

and

$$\begin{aligned} N_E\Big(\sum_{i=1}^{d} v_i \otimes t_i\Big) &= \sum_{i=1}^{d} t_i^3 N(v_i) \\ &\quad + \sum_{i<j} \big(t_i^2 t_j T(v_i^\#, v_j) + t_i t_j^2 T(v_i, v_j^\#)\big) \\ &\qquad + \sum_{i<j<k} t_i t_j t_k T(v_i \times v_j, v_k) \end{aligned} \tag{25.4}$$

for all $v_1, \dots, v_d \in L$, for all $t_1, \dots, t_d \in E$ and for all d. Moreover,

$$\Lambda_E := (J_E, E, N_E, \#_E, T_E, \times_E, 1)$$

is a cubic norm structure; that is to say, Λ_E satisfies conditions (i)–(xi), but perhaps not the last condition (xii), in 15.15 of [36].[2]

[1] Hexagonal systems are a class of Jordan division algebras. Jordan division algebras over a field with valuation were first investigated, to our knowledge, by Holger Petersson [19], [20].

[2] Let $\Lambda = (J, K, N, \#, T, \times, 1)$ be a cubic norm structure. Then Λ satisfies condition 15.15.xii in [36] if and only if it is *anisotropic*. Here anisotropic means that $N(w) = 0$ for $w \in J$ if and only if $w = 0$. Thus an hexagonal system and an anisotropic cubic norm structure are the same thing.

Proof. Choose a basis B of J over K and define $\#_E$ and N_E using 25.3 and 25.4 for all finite subsets $v_1, \ldots, v_d$ of B. It follows by a bit of calculation that Λ_E is a cubic norm structure. The identities 25.3 and 25.4 then hold in general by induction with respect to d. □

Proposition 25.5. *Let*

$$\Lambda = (J, K, N, \#, T, \times, 1)$$

be an hexagonal system that is not of type $1/K$ as defined in 15.20 of [36], let ν be a valuation of K,[3] *let $\hat{K}$ be the completion of K with respect to K and let $\Lambda_{\hat{K}}$ be as in 25.1. Let $\hat{J} = J_{\hat{K}}$, $\hat{N} = N_{\hat{K}}$, $\hat{T} = T_{\hat{K}}$, etc. Then the following assertions are equivalent:*

(i) *The cubic norm structure $\Lambda_{\hat{K}}$ is, in fact, an hexagonal system; equivalently, $\hat{N}(\hat{u}) = 0$ for $\hat{u} \in \hat{J}$ if and only if $\hat{u} = 0$.*
(ii) $\nu(T(u, v)) \geq \big(\nu(N(u)) + \nu(N(v))\big)/3$ *for all $u, v \in J$.*

Proof. Suppose first that (i) holds. We want to show that (ii) holds. Since $\hat{T}$ and $\hat{N}$ are extensions of T and N, it suffices to assume that $K = \hat{K}$. Let $w \in J^*$ and let

$$\Lambda_w = (J, K, N_w, \#_w, T_w, \times_w, w)$$

be the translate of Λ with respect to w as defined in 29.36 of [36]. Then Λ_w is also an hexagonal system and by 29.37 in [36],

$$T_w(u, v) = T(U_{w^\#/N(w)}(u), v) \tag{25.6}$$

for all $u, v \in J$, where U is as in 15.42 of [36], and

$$N_w(u) = N(u)/N(w) \tag{25.7}$$

for all $u \in J$.

Choose $u, v \in J^*$ and suppose that

$$\nu(T_w(U_w(u), v)) \geq \big(\nu(N_w(U_w(u))) + \nu(N_w(v))\big)/3. \tag{25.8}$$

By 29.38 of [36], we have

$$N(U_w(u)) = N(w)^2 N(u), \tag{25.9}$$

and by 29.39 of [36], $U_{w^\#/N(w)}(U_w(u)) = u$. Therefore,

$$\begin{aligned} \nu(T(u, v)) &= \nu(T_w(U_w(u), v)) && \text{by 25.6} \\ &\geq \big(\nu(N_w(U_w(u))) + \nu(N_w(v))\big)/3 && \text{by 25.8} \\ &= \big(\nu(N(U_w(u))) + \nu(N(v)) - 2\nu(N(w))\big)/3 && \text{by 25.7} \\ &= \big(\nu(N(u)) + \nu(N(v))\big)/3 && \text{by 25.9.} \end{aligned}$$

It thus suffices to show that for some choice of $w \in J^*$, 25.8 holds. Setting $w = v$ and replacing $U_v(u)$ by u and Λ by Λ_v, we conclude that it suffices to show that

$$\nu(T(u, 1)) \geq \nu(N(u))/3 \tag{25.10}$$

[3] We observe that the statement and proof of 25.5 require only that ν be real-valued.

for all $u \in J$ (since $N_v(v) = 1$ by 25.7).

We can assume that $T(u,1) \neq 0$. Let J_0 be the subspace of J spanned by $1, u, u^\#$. By the proof of 30.6 of [36], J_0 is contained in a substructure of J of type $3/K$ (as defined in 15.21 and 30.2 of [36]). We can thus assume that Λ itself is of type $3/K$. This means that there exists a separable cubic extension E/K such that $\Lambda = (E/K)^+$ as defined in 15.21 of [36]. In particular, N is the norm of the extension E/K and the map $T\colon E \to K$ given by $T(u) = T(u,1)$ for all $u \in E$ is its trace. Let L be the normal closure of E/K, let $N_{L/K}$ be the norm of the extension L/K and let

$$\omega(z) = \nu(N_{L/K}(z))$$

for all $z \in L$. By 17.14.i, ω satisfies 9.17.ii. Let σ be an element of order 3 in $\mathrm{Gal}(L/K)$. Then

$$T(u) = u + u^\sigma + u^{\sigma^2} \text{ and } N(u) = u \cdot u^\sigma \cdot u^{\sigma^2}.$$

Since ω is σ-invariant, it follows by 9.17 that

$$\omega(T(u)) \geq \omega(u) = \omega(N(u))/3.$$

Dividing both sides of this inequality by the degree $[L : K]$, we obtain 25.10. Thus (i) implies (ii).

To show that (ii) implies (i), we assume that there exists an element $\hat{u} \in \hat{J}^*$ such that $\hat{N}(\hat{u}) = 0$. By 15.38, 15.41 and 17.6 in [36], the bilinear form T is non-degenerate. Hence also the bilinear form $\hat{T}$ is non-degenerate. There thus exists $z \in J$ such that

$$\hat{T}(\hat{u}, z) \neq 0. \tag{25.11}$$

By 17.6 in [36] again, J is finite-dimensional over K.[4] Let $B = (v_1, \ldots, v_d)$ be an ordered basis of J and let $\hat{t}_1, \ldots, \hat{t}_d$ be the coordinates of $\hat{u}$ with respect to B. For each $i \in [1, d]$, choose a sequence $(t_{i,k})_{k \geq 1}$ of elements in K converging to $\hat{t}_i$ with respect to ν and let

$$u_k = t_{1,k}v_1 + \cdots + t_{d,k}v_d$$

for each $k \geq 1$. By 25.4,

$$\lim_{k \to \infty} N(u_k) = \hat{N}(\hat{u}) = 0$$

and thus

$$\lim_{k \to \infty} \nu(N(u_k)) = \infty.$$

By 25.2, it follows similarly that

$$\lim_{k \to \infty} T(u_k, z) = \hat{T}(\hat{u}, z).$$

Thus $\nu(T(u_k, z))$ is a constant for all k sufficiently large (by 9.18.iii and 25.11). Therefore

$$\nu(T(u_k, z)) < \big(\nu(N(u_k)) + \nu(N(z))\big)/3$$

for k sufficiently large. Hence (ii) implies (i). □

---

[4]In fact, the dimension of J over K divides 27.

Proposition 25.12. *Let*

$$\Lambda = (J, K, N, \#, T, \times, 1)$$

be an hexagonal system and let ν *be a discrete valuation of* K. *Then*

$$\nu(N(J^*)) = \delta\mathbb{Z}$$

for $\delta = 1$ *or* 3.

Proof. By (vii) and (x) of 15.15 in [36], $N(1) = 1$. By 30.4.i in [36],

$$N(a^\#) = N(a)^2 \tag{25.13}$$

for all $a \in J$, and by 15.15.ii of [36], $N(ta) = t^3N(a)$ for all $a \in J$ and all $t \in K$. Thus $3\mathbb{Z} \subset \nu(N(J^*))$ and if $\nu(N(a))$ is not divisible by 3 for some $a \in J^*$, then every integer is of the form $\nu(N(t \cdot 1))$, $\nu(N(ta))$ or $\nu(N(ta^\#))$ for some $t \in K^*$. □

Definition 25.14. Let

$$\Lambda = (J, K, N, \#, T, \times, 1)$$

be an hexagonal system, let ν be a discrete valuation of K and let δ be as in 25.12. Then δ is called the *ν-index* of Λ. We will say that Λ is *ramified* (respectively, *unramified*) *at* ν if $\delta = 1$ (respectively, $\delta = 3$).[5] If Λ is not of type $1/K$, we will say that Λ is *ν-compatible* if it fulfills the two equivalent conditions of 25.5. We then simply declare that *all* hexagonal systems of type $1/K$ are ν-compatible.

Proposition 25.15. *Let*

$$\Lambda = (J, K, N, \#, T, \times, 1)$$

be an hexagonal system, let ν *be a valuation of* K *and suppose that* Λ *is* ν*-compatible as defined in 25.14. Then*

(i) $\nu(T(u, v)) \geq \big(\nu(N(u)) + \nu(N(v))\big)/3$;
(ii) $\nu(N(u + v)) \geq \min\{\nu(N(u)), \nu(N(v))\}$; *and*
(iii) $\nu(N(u \times v) \geq \nu(N(u)) + \nu(N(v))$

for all $u, v \in J$.

Proof. The inequality (i) holds by 15.20 of [36] if Λ is of type $1/K$ and by 25.5.ii if it is not. Let $u, v \in J$. By 15.15.v of [36],

$$N(u + v) = N(u) + T(u^\#, v) + T(u, v^\#) + N(v). \tag{25.16}$$

By 25.13 and (i), we have

$$\begin{aligned} \nu(T(u^\#, v)) &\geq \big(2\nu(N(u)) + \nu(N(v))\big)/3 \\ &\geq \min\{\nu(N(u)), \nu(N(v))\} \end{aligned} \tag{25.17}$$

---

[5]This coincides with the usual meaning of the terms (totally) "ramified" and "unramified" in the case that $\Lambda = (E/K)^+$ for some separable cubic extension E/K (as defined in 15.21 of [36]) such that K is complete and $\bar{E}/\bar{K}$ is separable (where $\bar{E}$ is defined with respect to the valuation ω in 17.14). Compare footnote 5 in Chapter 19.

and

$$\nu(T(u, v^\#)) \geq \big(\nu(N(u)) + 2\nu(N(v))\big)/3 \geq \min\{\nu(N(u)), \nu(N(v))\}. \tag{25.18}$$

By 9.17.ii and 25.16, it follows that (ii) holds.

By 30.4.ii in [36],

$$N(u \times v) = T(u^\#, v)T(u, v^\#) - N(u)N(v). \tag{25.19}$$

Adding 25.17 and 25.18, we obtain

$$\nu(T(u^\#, v)) + \nu(T(u, v^\#)) \geq \nu(N(u)) + \nu(N(v)). \tag{25.20}$$

Therefore

$$\begin{aligned} \nu(T(u^\#, v)T(u, v^\#)) &= \nu(T(u^\#, v)) + \nu(T(u, v^\#)) && \text{by 9.17.i} \\ &\geq \nu(N(u)) + \nu(N(v)) && \text{by 25.20} \\ &= \nu(N(u)N(v)) && \text{by 9.17.i.} \end{aligned}$$

By 9.17.ii and 25.19, we conclude that (iii) holds. □

Theorem 25.21. *Let*

$$\Lambda = (J, K, N, \#, T, \times, 1)$$

be an hexagonal system, let ν be a discrete valuation of K, let

$$\mathcal{H}(\Lambda) = (U_+, U_1, \ldots, U_6)$$

and $x_1, \ldots, x_6$ be as in 30.8.viii, let (Δ, Σ, d) be the canonical realization of $\mathcal{H}(\Lambda)$ as defined in 30.6 and let $(G^\dagger, \xi)$ be the root datum of Δ based at Σ. Let δ be the ν-index of Λ as defined in 25.14 and let

$$\phi_6(x_6(t)) = \nu(t) \tag{25.22}$$

for all $t \in K^*$. Then the following hold:

(i) *The map ϕ_6 extends to a valuation of $(G^\dagger, \xi)$ if and only if Λ is ν-compatible as defined in 25.14.*
(ii) *Suppose that ϕ_6 extends to a valuation of $(G^\dagger, \xi)$ and let ι be the associated root map from of Σ (with target G_2). Then ι is long if $\delta = 3$ and short if $\delta = 1$; and, after replacing ϕ by an equipollent valuation if necessary,*

$$\phi_1(x_1(u)) = \nu(N(u))/\delta \tag{25.23}$$

for all $u \in J^*$.

Proof. Let Ξ_d and Ξ_d° be as in 3.5. The six roots $a_1, \ldots, a_6$ in Ξ_d can be ordered so that $U_i = U_{a_i}$ for all $i \in [1, 6]$. We then have $\Xi_d^\circ = \{a_1, a_6\}$ and $[a_1, a_6] = (a_1, a_2, \cdots, a_6)$, where $[a_1, a_6]$ is as defined in 3.1.

Suppose first that Λ is ν-compatible. Let

$$\phi_i(x_i(t)) = \nu(t)$$

for $i = 2$ and 4 and all $t \in K$, let

$$\phi_i(x_i(u)) = \nu(N(u))/\delta$$

for $i = 1$, 3 and 5 and all $u \in J$ and let $\phi = \{\phi_1, \phi_2, \ldots, \phi_6\}$. By 25.14, $\phi_i(U_i^*) = \mathbb{Z}$ for all $i \in [1, 6]$. By 25.15.ii, ϕ satisfies 15.1.i. Let ι be the root map of Σ (with target G_2) that is long (as defined in 16.12) if $\delta = 3$ and short if $\delta = 1$ Thus U_6 is a long root group if and only if ι is long. By 16.8 in [36] as well as 25.13 and (i)–(ii) of 25.15, ϕ satisfies 15.1.ii with respect to ι, ϕ is exact as defined in 15.24 and the (a_1, a_6)-coefficient of ϕ at a_4 as defined in 15.24 is $(\delta, 2)$. By 15.25 and 30.36.viii, it follows that ϕ satisfies 15.4.i. By 30.38.viii, ϕ satisfies 15.4.ii. We conclude that ϕ is a viable partial valuation. By 15.21, ϕ extends to a valuation of $(G^\dagger, \xi)$.

Suppose, conversely, that ϕ is a valuation of $(G^\dagger, \xi)$ with respect to a root map ι of Σ (with target G_2) that extends the map ϕ_6 (so $\phi_{a_6} = \phi_6$). Let $\phi_i = \phi_{a_i}$ for all $i \in [1, 5]$. By 2.50.ii and 3.41.i, we have

$$\phi_1(x_1(u)) = -\Big(\phi_6\big(x_2(1)^{m_\Sigma(x_1(u))}\big) - \phi_2(x_2(1))\Big)/\delta_0$$

for all $u \in J^*$, where

$$\delta_0 := \begin{cases} 3 & \text{if } \iota \text{ is long and} \\ 1 & \text{if } \iota \text{ is short.} \end{cases}$$

By 32.12 in [36],

$$x_2(1)^{m_\Sigma(x_1(u))} = x_6(-1/N(u))$$

for all $u \in J^*$. Thus

$$\phi_1(x_1(u)) = \big(\nu(N(u)) + \phi_2(x_2(1))\big)/\delta_0$$

for all $u \in J^*$. By 3.21, $\phi_i(U_i^*) = \mathbb{Z}$ for all $i \in [1, 6]$. By 25.12, $\nu(N(J^*)) = \delta\mathbb{Z}$. It follows that $\delta_0 = \delta$ and $\phi_2(x_2(1))/\delta \in \mathbb{Z}$. By 3.43, therefore, we can assume that 25.23 holds. Thus (ii) holds. Since ϕ satisfies condition (V2), 25.5.ii holds. Hence Λ is ν-compatible. Thus (i) holds. □

Notation 25.24. Let Δ, Λ, ν, etc. be as in 25.21, suppose that Λ is ν-compatible and let ϕ be the unique valuation of $(G^\dagger, \xi)$ satisfying 25.22 and 25.23. We denote by

$$\tilde{\mathsf{G}}_2(\Lambda, \nu)$$

the pair $(\Delta^{\text{aff}}, \mathcal{A})$ obtained by applying 14.47 to these data.

Thus the pair $\tilde{\mathsf{G}}_2(\Lambda, \nu)$ exists precisely when the hexagonal system Λ is ν-compatible.

Theorem 25.25. *Every Bruhat-Tits pair whose building at infinity is $\mathsf{G}_2(\Lambda)$ for some hexagonal system*

$$\Lambda = (J, K, N, \#, T, \times, 1)$$

is of the form $\tilde{\mathsf{G}}_2(\Lambda, \nu)$ for some valuation ν of K such that Λ is ν-compatible.

Proof. This holds by 3.41.ii, 16.4, 25.21 and 25.24. □

Next we consider completions.

Notation 25.26. Let

$$\Lambda = (J, K, N, \#, T, \times, 1)$$

be an hexagonal system that is ν-compatible for some discrete valuation ν of K and let $\hat{K}$ denote the completion of K with respect to ν. Suppose that Λ is of type $1/K$, i.e. that $\Lambda = (E/K)^\circ$ as defined in 15.20 of [36] for some field E containing K such that $E^3 \subset K$ (so $E = K$ if $\text{char}(K) \neq 3$). If $E = K$, let $\hat{E} = \hat{K}$. If $E \neq K$, let F denote the closure of the subfield E^3 in $\hat{K}$ and then let $\hat{E} = F^{1/3}$. Let $\hat{\Lambda}$ be the hexagonal system $(\hat{E}/\hat{K})^\circ$ (whether or not $E = K$). By 25.14, $\hat{\Lambda}$ is ν-compatible. Now suppose that Λ is not of type $1/K$ and let $\hat{\Lambda} = \Lambda_{\hat{K}}$ be as defined in 25.1. By 25.14, $\hat{\Lambda}$ is a ν-compatible hexagonal system also in this case.

Theorem 25.27. *Let $\Lambda = (J, K, \#)$ be an hexagonal system, let ν be a valuation of K such that Λ is ν-compatible, let $\hat{\Lambda}$ be as in 25.26 and let $(\Delta, \mathcal{A}) = \tilde{\mathsf{G}}_2(\Lambda, \nu)$ be as in 25.24. Then $\hat{\Lambda}$ is ν-compatible, Δ is completely Bruhat-Tits and its completion is $\tilde{\mathsf{G}}_2(\hat{\Lambda}, \nu)$.*

Proof. By 30.16, $\mathsf{G}_2(\Lambda)$ is a subbuilding of $\mathsf{G}_2(\hat{\Lambda})$. The claim holds, therefore, by 17.4, 17.9, 17.13 and 25.26. □

We turn now to gems.

Definition 25.28. Let

$$\Lambda = (J, K, N, \#, T, \times, 1)$$

be an hexagonal system, let ν be a discrete valuation of K and suppose that Λ is ν-compatible. Let

$$J_i = \{u \mid \nu(N((u)) \geq 0\}$$

for each $i \in \mathbb{Z}$. By 25.15.ii, J_i is an additive subgroup of J for each i. Let $\bar{J} = J_0/J_1$ and let $u \mapsto \bar{u}$ be the natural homomorphism from J_0 to $\bar{J}$. The quotient $\bar{J}$ has the structure of a vector space over $\bar{K}$ where $\bar{t} \cdot \bar{u} = \overline{tu}$ for all $t \in \mathcal{O}_K$ and all $u \in J_0$. The map $\bar{u} \mapsto \overline{N(u)}$ from $\bar{J}$ to $\bar{K}$ is well defined and anisotropic (i.e. it sends only 0 to 0). We call this map $\bar{N}$. By 25.5.ii, 25.15.iii and 25.13, also the maps T, $\times$ and $\#$ induce maps $\bar{T}$, $\bar{\#}$ and $\bar{\times}$ such that

$$\bar{\Lambda} := (\bar{J}, \bar{K}, \bar{N}, \bar{\#}, \bar{T}, \bar{\times}, \bar{1})$$

satisfies all the properties in 15.15 of [36] defining an hexagonal system. We call $\bar{\Lambda}$ the *residue* of Λ (with respect to ν).

Theorem 25.29. *Let*

$$\Lambda = (J, K, N, \#, T, \times, 1)$$

be an hexagonal system, let ν be a discrete valuation of K and suppose that Λ is ν-compatible. Let $(\Delta, \mathcal{A}) = \tilde{\mathsf{G}}_2(\Lambda, \nu)$ be as in 25.24, let $\bar{\Lambda}$ be as in 25.28 and let the group $G^\dagger$ be as in 18.1. Then $G^\dagger$ acts transitively on the set of gems of Δ and every gem is isomorphic to $\mathsf{G}_2(\bar{\Lambda})$.

Proof. This holds by 18.5 and 18.27. □

Using 18.34 (and 25.29), we can say a bit more about gems:

Proposition 25.30. *Let*

$$\Lambda = (J, K, N, \#, T, \times, 1)$$

be an hexagonal system not of type $1/K$,[6] *let ν be a valuation of K and suppose K is complete with respect to ν.*[7] *Let $\bar{\Lambda}$ be as in 25.28. Then the following hold:*

(i) $\dim_{\bar{K}} \bar{J} = \dim_K J$ *if Λ is unramified at ν (as defined in 25.14).*
(ii) $\dim_{\bar{K}} \bar{J} = \big(\dim_K J\big)/3$ *if Λ is ramified at ν.*
(iii) *If Λ is ramified at ν, then $\bar{\Lambda}$ is not of type $9^*/K$.*
(iv) *If Λ is ramified at ν and $\bar{\Lambda}$ is of type $9/K$, then Λ is of type $27/K$ (i.e. Λ is a first Tits construction).*

Proof. Let J_i be as in 25.28 for each i. Since Λ is not of type $1/K$, the dimension of J over K divides 27 (by 17.6 in [36]). Let

$$k_i = \dim_{\bar{K}} J_i/J_{i+1}$$

for all i. Thus k_0 is the dimension of $\bar{J}$ over $\bar{K}$. By 15.15.ii in [36], $N(tu) = t^3N(u)$ for all $t \in K$ and all $u \in J$. By 25.15.ii, the norm N satisfies 18.36. Thus by 18.34 with $Q = N$ and $m = 3$, the quotient J_0/J_3 has the structure of a vector space over $\bar{K}$ (where $\bar{t} \cdot \bar{u} = \overline{tu}$ for all $t \in \mathcal{O}_K$ and all $u \in J_0$) containing J_1/J_3 and J_2/J_3 as subspaces and

$$\dim_K J = \dim_{\bar{K}} J_0/J_3 = k_0 + k_1 + k_2. \tag{25.31}$$

If Λ is unramified at ν, then $k_1 = k_2 = 0$, i.e. $J_0/J_3 = \bar{J}$. Thus (i) holds. Suppose that Λ is ramified at ν. Thus $k_i > 0$ for all i. Choose i and then choose $w \in J^*$ such that $\nu(N(w)) = i$. Let Λ_w be the translate of Λ with respect to w as defined in 29.36 of [36] and let $\overline{\Lambda_w}$ be its residue as defined in 25.28. By 29.37 of [36], the underlying vector space of $\overline{\Lambda_w}$ is J_i/J_{i+1}. Thus k_i is a power of 3.[8] Since i is arbitrary, we conclude that k_0, k_1 and k_2 are all powers of 3. By 25.31, therefore, they are equal. Thus (ii) holds. Assertions (iii) and (iv) are proved in Proposition 6 of [20].[9] □

---

[6]See 15.20–15.22, 15.29, 15.31 and 15.34 in [36] for the definition of the six types of hexagonal systems. In 17.6 of [36] it is proved that these are, in fact, the only hexagonal systems.

[7]Hence 25.5.i holds and thus Λ is ν-compatible.

[8]By 25.31, k_i is finite. Thus by 15.20 in [36], k_i is a power of 3 even if $\overline{\Lambda_w}$ is of type $1/\bar{K}$.

[9]There is a misprint in the statement of Proposition 6 in [20]: the word "unramified" in the second line should read "ramified."

Remark 25.32. Let

$$\Lambda = (J, K, N, \#, T, \times, 1)$$

be an hexagonal system and let ν be a valuation of K. Suppose that K is complete with respect to ν and that Λ is ramified at ν. By (ii)–(iv) of 25.30 and 17.6 in [36], exactly one of the following holds: [10]

(i) $\bar{T}$ is identically zero, $\text{char}(\bar{K}) = 3$, $\bar{\Lambda} \cong (\bar{E}/\bar{K})^\circ$ for some field $\bar{E}$ containing $\bar{K}$ such that $\bar{E}^3 \subset \bar{K}$ and $\bar{E} \neq \bar{K}$.
(ii) Λ is of type $3/K$ and $\bar{\Lambda} = (\bar{E}/\bar{K})^\circ$ for $\bar{E} = \bar{K}$.
(iii) Λ is of type $9/K$ or $9^*/K$ and $\bar{\Lambda} = (\bar{E}/\bar{K})^+$ for some field $\bar{E}$ such that $\bar{E}/\bar{K}$ is a separable cubic extension.
(iv) Λ is of type $27/K$ and $\bar{\Lambda} = \bar{D}^+$ for some skew field $\bar{D}$ of degree 3 over $\bar{K}$ (i.e. $\bar{K}$ is the center of $\bar{D}$ and the dimension of $\bar{D}$ over $\bar{K}$ is 9).

Bruhat-Tits buildings of type $\tilde{\mathsf{G}}_2$ are the only Bruhat-Tits buildings that have residues that are contained neither in a gem nor in a proper residue of rank at least 3; see 18.19. We use the rest of this chapter to determine the structure of these residues.

Proposition 25.33. *Let $(\Delta, \mathcal{A}) = \tilde{\mathsf{G}}_2(\Lambda, \nu)$ for some hexagonal system*

$$\Lambda = (J, K, N, \#, T, \times, 1)$$

and some valuation ν of K and suppose that K is complete with respect to ν. Let R be an $\{x, y\}$-residue of Δ, where $\{x, y\}$ is the unique edge of $\tilde{\Pi} = \tilde{\mathsf{G}}_2$ with label 3. Then R is Moufang and the following hold:

(i) *If Λ is unramified at ν, then*

$$R \cong \mathsf{A}_2(\bar{K}),$$

where $\bar{K}$ is the residue field of K.

(ii) *If Λ is ramified at ν, then*

$$R \cong \mathsf{A}_2(\bar{E}) \text{ or } \mathsf{A}_2(\bar{D}),$$

where $\bar{E}$ and $\bar{D}$ are as in 25.32.

Proof. Let $A \in \mathcal{A}$, let $\Sigma = A^\infty$, let d be a chamber of Σ, let $(G^\dagger, \xi)$ be the root datum of $\Delta_{\mathcal{A}}^\infty$ based at Σ, let R_0 be a gem cut by A, let $\phi = \phi_{R_0}$ be the valuation of $(G^\dagger, \xi)$ described in 13.8, let ι be the root map of Σ described in 13.13 and let δ be the ν-index of Λ as defined in 23.2. By 25.24, we can assume that $(\Delta_{\mathcal{A}}^\infty, \Sigma, d)$ is the canonical realization of

$$\mathcal{H}(\Lambda) = (U_+, U_1, U_2, \ldots, U_6)$$

as defined in 30.6, that

$$\phi_1(x_1(u)) = \nu(N(u))/\delta$$

for all $u \in J^*$, that $\phi_6(x_6(t)) = \nu(t)$ for all $t \in K^*$ and that ι is long if and only if $\delta = 3$.

[10] The hexagonal systems $(E/K)^\circ$, $(E/K)^+$ and D^+ are defined in 15.20–15.22 of [36].

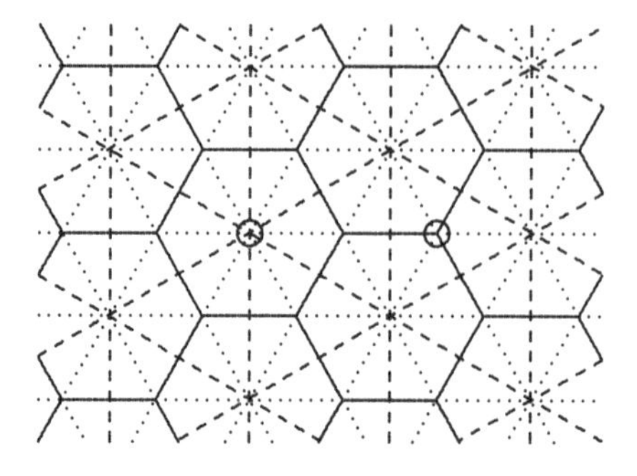

Figure 25.1 Alcoves for $\tilde{\mathsf{G}}_2$

We identify the apartment A with the chamber system $\tilde{\Sigma}_{\mathsf{G}_2}$ defined in 2.29 and pictured in Figure 25.1. This figure is a reproduction of Figure 2.3, but with two points circled; we call the one on the left p_1 and the one on the right p_2. We can assume that p_1 is the center of the residue $A \cap R_0$ of A and that the residue $A \cap R$ consists of the six alcoves whose closure contains p_2. Let x be the alcove of $A \cap R_0$ that goes from three o'clock to four o'clock in Figure 2.3 and let S be the sector $\sigma_A(R_0, x)$. We can assume as well that $d = S^\infty$.

Let α_1 be the root of A whose wall is the horizontal line through p_1 and p_2 that contains all the alcoves below this line. Then α_1 is long (in the sense of 2.35). Rotating around p_1 in the counterclockwise direction, we denote by $\alpha_2, \ldots, \alpha_6$ (in order) the remaining *long* roots of A cutting R_0. Let $a_i = \alpha_i^\infty$ and

$$\breve{U}_i = U_{a_i, k_i} \tag{25.34}$$

for all $i \in [1, 6]$, where

$$(k_1, k_2, \ldots, k_6) = (0, -2, -2, 0, 2, 2). \tag{25.35}$$

Then for each $i \in [1, 6]$, the group $\breve{U}_i$ fixes the residue R and acts trivially on a unique root β_i of $R \cap A$. Let $\bar{U}_i$ denote the group induced by $\breve{U}_i$ on R. By 13.5 and 29.14.iii, $\bar{U}_i$ acts transitively on the set of apartments of R containing the root β_i for each $i \in [1, 6]$.

Let b_i be the unique root in the open interval (a_i, a_{i+1}) for each $i \in [1, 6]$ (where $a_7 = a_1$). The roots b_i are all short. We claim next that for each $i \in [1, 6]$, the commutator group $[\bar{U}_i, \bar{U}_{i+1}]$ is trivial (where $U_7 := U_1$). Let $i \in [1, 6]$ and let g_j be an element of U_{a_j} mapping R to itself for $j = i$ and $j = i + 1$. We have $[U_{a_i}, U_{a_{i+1}}] \subset U_{b_i}$ (by 3.2.iii). Thus $g := [g_i, g_{i+1}]$ is an element of U_{b_i} that maps R to itself. Since b_i is short, it follows that g acts trivially on R (since R is not cut by any short roots). This proves the claim. By 29.57 (and 29.15), it follows that R is Moufang and that $\bar{U}_i$ is the root group associated with the root β_i of R for each i.

To identify R, it suffices to identify the root group sequence

$$\Omega_1 := (W, \bar{U}_1, \bar{U}_2, \bar{U}_3),$$

where $W = \bar{U}_1\bar{U}_2\bar{U}_3$.

Let Ω_2 be the root group sequence

$$(X, U_{a_1}, U_{b_1}, U_{a_2}, U_{b_2}, U_{a_3}, U_{b_3}),$$

where

$$X = U_{a_1}U_{b_1}U_{a_2}U_{b_2}U_{a_3}U_{b_3}.$$

The roots $a_1, b_1, a_2, b_2, a_3, b_4$ are precisely the roots of Σ that contain d. Hence

$$\Omega_2 = \mathcal{H}(\Lambda) \text{ or } \mathcal{H}(\Lambda)^{\text{op}}.$$

Suppose that U_6 is a long root group. This means that $\Omega_2 = \mathcal{H}(\Lambda)^{\text{op}}$. In 16.8 of [36], we have

$$[x_6(t), x_2(s)] = x_4(-st)$$

for all $s, t \in K$ in $\mathcal{H}(\Lambda)$. By 25.35, we have, with respect to the identification of Ω_2 with $\mathcal{H}(\Lambda)^{\text{op}}$,

$$\breve{U}_1 = \{x_6(t) \mid \nu(t) \geq 0\}$$

and

$$\breve{U}_j = \{x_j(t) \mid \nu(t) \geq -2\}$$

for $j = 2$ and 3 (where $\breve{U}_i$ is as in 25.34). Let π be a uniformizer of K. Then the maps $x_6(\bar{t}) \mapsto x_1(\bar{t})$, $x_4(t) \mapsto x_2(\overline{\pi^2 t})$ and $x_2(t) \mapsto x_3(\overline{-\pi^2 t})$ extend to an isomorphism from the root group sequence Ω_1 to the root group sequence $\mathcal{T}(\bar{K})$ defined in 16.1 of [36]. Hence $R \cong \mathsf{A}_2(\bar{K})$.

Suppose now that U_6 is short, i.e. that Λ is ramified at ν. This time $\Omega_2 = \mathcal{H}(\Lambda)$. In particular, the root group $\bar{U}_{a_1,R_0}$ is parametrized by the additive group of A, where $A := E$ in cases (i), (ii) and (iii) of 25.32 and $A := D$ in case (iv).

Let P be the panel that contains the two alcoves (i.e. chambers) of A whose closure contains both p_1 and the midpoint between p_1 and p_2. We have $U_{a_1,R_0} = U_{a_1,R}$ and an element of the group $U_{a_1,R}$ acts trivially on R_0 if and only if it acts trivially on P if and only if it acts trivially on R. We can thus identify canonically the group $\bar{U}_{a_1,R_0}$ with the group $\bar{U}_{a_1,R}$. Similarly, we can identify canonically the group $\bar{U}_{-a_1,R_0}$ with the group $\bar{U}_{-a_1,R}$.

Let Σ_1 be the apartment $A \cap R_0$ of R_0 and let Σ_2 be the apartment $A \cap R$ of R. Let g be an element of $U_{a_1,R}$ whose image in $\bar{U}_{a_1,R}$ is non-trivial and let $m_\Sigma(g) = ugv$ for $u, v \in U^*_{-a_1}$ be as in 3.8. By 13.28, $u, v \in U_{-a_1,R}$ both act non-trivially on P and the product

$$\bar{u}\bar{g}\bar{v} \in \bar{U}_{-a_1,R}\bar{U}_{a_1,R}\bar{U}_{-a_1,R}$$

acts like $m_{\Sigma_1}(\bar{g})$ on R_0 and $m_{\Sigma_2}(\bar{g})$ on R.

By 30.14, the root sequence Ω_1 is isomorphic to $\mathcal{T}(F)$ for some alternative division ring F. This means, in particular, that the group $\bar{U}_{a_1,R}$ is isomorphic to the additive group of F. We can thus assume that $F = \bar{A}$ as additive groups. By 37.12 in [36], we can assume, in fact, that they have the same multiplicative identity 1. Let a be an element in the valuation ring $\mathcal{O}_A$ whose image $\bar{a}$ in $\bar{A}$ is non-trivial and let

$$h = m_\Sigma(x_{a_1}(\bar{1}))m_\Sigma(x_{a_1}(\bar{u})).$$

By 15.2.iv, 15.6.iv and 32.12 in [36], h induces the map

$$x_{a_1}(\bar{v}) \mapsto x_{a_1}(\bar{u} \cdot \bar{v} \cdot \bar{u})$$

on $\bar{U}_{a_1,R_0}$, where $\cdot$ is multiplication in $\bar{A}$. By 32.5 in [36], h induces the map

$$x_{a_1}(\bar{v}) \mapsto x_{a_1}(\bar{u} * \bar{v} * \bar{u})$$

on $\bar{U}_{a_1,R}$, where $*$ is multiplication in F. Since $\bar{U}_{a_1,R_0} = \bar{U}_{a_1,R}$ and this group acts faithfully on both R_0 and R, it follows that $\bar{u} \cdot \bar{v} \cdot \bar{u} = \bar{u} * \bar{v} * \bar{u}$ for all u, v in the additive group $F = \bar{A}$. By the result at the bottom of page 2 in [18], it follows that $F = \bar{A}$ or $\bar{A}^{\text{op}}$ as fields (or skew fields). We conclude that $R \cong \mathsf{A}_2(\bar{A})$ or $\mathsf{A}_2(\bar{A}^{\text{op}})$. By 35.15 in [36], there is a (non-type-preserving) isomorphism from $\mathsf{A}_2(\bar{A})$ to $\mathsf{A}_2(\bar{A}^{\text{op}})$. □

In the situation described in 25.33 we can ask a more subtle question when $R \cong \mathsf{A}_2(D)$ and D is not isomorphic to D^{op}, in which case there is no type-preserving isomorphism from $\mathsf{A}_2(D)$ to $\mathsf{A}_2(D^{\text{op}})$. The residue R, on the other hand, has a natural "orientation" due to its embedding in Δ: some of its panels are contained in gems and some are not. More precisely, let x_0 be an arbitrary chamber of $R \cap A$. Then there is unique gallery

$$(x_0, x_1, x_2, x_3, x_4, x_5)$$

in $R \cap A$ such that all the x_i are distinct and the panel containing x_0 and x_1 is not contained in a gem. Let α_i be the root of $R \cap A$ containing x_i but not x_{i-1}, let U_i be the corresponding root group of R for each $i \in [1, 3]$ and let Ω_R denote the root group sequence

$$(U_+, U_1, U_2, U_3),$$

where $U_+ = U_1U_2U_3$. Then Ω_R is, up to isomorphism, independent of the choice of R and x_0. We can thus assume that Ω_R is the root group sequence called Ω_1 in the proof of 25.33. Since $R \cong \mathsf{A}_2(D)$, we have $\Omega_R \cong \mathcal{T}(D)$ or $\mathcal{T}(D^{\text{op}})$. Which is it? We answer this question below using a result due to H. P. Petersson (personal communication):

Lemma 25.36. *Suppose that Λ and ν are as in (iii) of 25.32. Let $\omega(u) = \nu(N(u))$ for all $u \in J$, let J_i for all $i \in \mathbb{Z}$ be as in 25.28 and let π be a uniformizer of K. Then the following hold:*

(i) *There exists an unramified division ring D of degree 3 over K and a scalar $s \in K^*$ such that $\nu(s) = 1$ or 2 such that Λ is isomorphic to the hexagonal system $J(D, K, s)$ described in 15.23 of [36]. In particular, $J = D \oplus D \oplus D$.*

(ii) *We identify* Λ *with* $J(D,K,s)$ *and* D *with its image in* J *under the map* $a \mapsto (v,0,0)$, *so the restriction of* $\eta := \omega/3$ *to* D *is a valuation of* D, *and set*

$$D_i = \pi^i \mathcal{O}_D$$

for each i, *where* $\mathcal{O}_D$ *is the valuation ring of* D *with respect to* η. *Then*

$$J_0 = D_0 \oplus D_0 \oplus D_1 \tag{25.37}$$

as well as

$$J_1 = D_1 \oplus D_0 \oplus D_1$$

and

$$J_3 = \pi J_0 = D_1 \oplus D_1 \oplus D_2.$$

(iii) *Let* $\phi_0(\overline{v_0, v_1, v_2}) = \bar{v}_0$ *for all* $(v_1, v_2, v_3) \in J_0$. *Then* ϕ_0 *is an isomorphism of* $\bar{K}$*-vector spaces from* $\bar{J} = J_0/J_1$ *to* $\bar{D}$.

(iv) *Suppose* $\nu(s) = 1$. *Then*

$$J_2 = D_1 \oplus D_1 \oplus D_1. \tag{25.38}$$

Let ϕ_2 *be the map from* $\bar{J}_2 = J_2/J_3$ *to* $\bar{D}$ *given by*

$$\phi_2(\overline{v_0, v_1, v_2}) = \overline{s^{-1}v_2}$$

for all $(v_0, v_1, v_2) \in J_2$. *Then* ϕ_2 *is a* $\bar{K}$*-linear isomorphism such that*

$$\phi_2(\overline{x \times y}) = -\phi_2(\bar{y}) \cdot \phi_0(\bar{x}) \tag{25.39}$$

for all $x \in J_0$ *and all* $y \in J_2$, *where* $\cdot$ *denotes multiplication in* $\bar{D}$.

(v) *Suppose* $\nu(s) = 2$. *Then*

$$J_2 = D_1 \oplus D_0 \oplus D_2. \tag{25.40}$$

Let ϕ_2 *be the map from* $\bar{J}_2 = J_2/J_3$ *to* $\bar{D}$ *given by*

$$\phi_2(\overline{v_0, v_1, v_2}) = \bar{v}_1$$

for all $(v_0, v_1, v_2) \in J_2$. *Then* ϕ_2 *is a* $\bar{K}$*-linear isomorphism such that*

$$\phi_2(\overline{x \times y}) = -\phi_0(\bar{x}) \cdot \phi_2(\bar{y})$$

for all $x \in J_0$ *and all* $y \in J_2$.

Proof. Assertion (i) is proved in Proposition 6 in [21] and assertion (ii) in Lemma 12 in [21]. The assertion (iii) follows from (ii). If $\omega(w) = 1$ for some $w \in J$, then $J_2 = U_w(J_0)$ by 25.9. Thus 25.38 and 25.40 also hold by Lemma 12 in [19].

Let $x = (v_0, v_1, v_2) \in J_0$ and $y = (z_0, z_1, z_2) \in J_2$. Note that by 25.15.iii, $x \times y \in J_2$, so the expression $\phi_2(\overline{x \times y})$ in (iv) and (v) makes sense.

Suppose that $\nu(s) = 1$. By Equation (13) in [21],

$$\phi_2(\overline{x \times y}) = \overline{v_1 \times z_1 - s^{-1}v_2 z_0 - s^{-1}z_2 v_0} \in \bar{D}. \tag{25.41}$$

By 25.15.iii again as well as 25.37 and 25.38, we have $v_1 \times z_1 \in D_0 \times D_1 \subset D_1$. By 25.37 and 25.38, we also have $s^{-1}v_2z_0 \in D_1$. Hence by 25.41, we have

$$\phi_2(\overline{x \times y}) = \overline{-s^{-1}z_2v_0} = -\phi_2(y) \cdot \phi_0(x).$$

Thus (iv) holds.

Suppose now that $\nu(s) = 2$. By Equation (13) in [21],

$$\phi_2(\overline{x \times y}) = \overline{s^{-1}v_2 \times z_2 - v_0z_1 - z_0v_1} \in \bar{D}. \tag{25.42}$$

By 25.15.iii, 25.37 and 25.39, we have $s^{-1}v_2 \times z_2 \in D_2$. By 25.37 and 25.39, we also have $z_0v_1 \in D_1$. Hence by 25.42, we have

$$\phi_2(\overline{x \times y}) = \overline{-v_0z_1} = -\phi_0(\bar{x}) \cdot \phi_2(\bar{y}).$$

Thus (v) holds. □

We now return to the point in the proof of 25.33 (and to all the notation in this proof) where we assume that U_6 is a short root group (i.e. that Λ is ramified at ν) and that the root group sequence Ω_2 equals $\mathcal{H}(\Lambda)$. We have

$$[x_1(v), x_5(z)] = x_3(v \times z)$$

for all $v, z \in J$ in 16.8 in [36]. By 25.35, we have, with respect to the identification of Ω_2 with $\mathcal{H}(\Lambda)$,

$$\breve{U}_1 = \{x_1(u) \mid u \in J_0\}$$

and

$$\breve{U}_j = \{x_j(u) \mid u \in J_2\}$$

for $j = 2$ and 3.

Let D, ϕ_0 and ϕ_2 be as in 25.36. Then the maps $x_1(\bar{x}) \mapsto x_1(\phi_0(\bar{x}))$, $x_3(\bar{x}) \mapsto x_2(\phi_2(\bar{x}))$ and $x_5(\bar{x}) \mapsto x_3(-\phi_2(\bar{x}))$ extend to an isomorphism from Ω_1 to $\mathcal{T}(\bar{D}^{\text{op}})$ in case 25.36.iv, to $\mathcal{T}(\bar{D})$ in case 25.36.v. Thus the answer to our question is given by the value $\nu(s)$. The element s is not an invariant of Λ, but the root group sequence $\Omega_1 = \Omega_R$ is (up to isomorphism) and hence the value $\nu(s)$ is an invariant as well.

Suppose that we have an hexagonal system $J(D, K, s)$ that gives rise to the situation 25.32.ii. Then $J(D, K, s^2)$ is also an hexagonal system and $\{\nu(s), \nu(s^2)\} = \{1, 2\}$ modulo 3. Hence we have *two* Bruhat-Tits buildings, one with $\Omega_R = \mathcal{T}(D)$ and the other with $\Omega_R = \mathcal{T}(D^{\text{op}})$.

Chapter Twenty Six

Assorted Conclusions

We have now finished answering the question posed at the end of Chapter 16. To complete the classification of Bruhat-Tits pairs, it remains only to say a few more words about Bruhat-Tits pairs of type $\tilde{\mathsf{F}}_4$. We do this in the first part of this chapter. In the second part of this chapter, we describe the classification of algebraic Bruhat-Tits buildings (as defined in 26.2) and in the third part we prove a few more facts about the automorphism group of a Bruhat-Tits pair.

The principal results in this chapter are 26.12, 26.25–26.27, 26.37 and 26.39.

* * *

Before going on, we observe that the result 26.12 will be the last step in our proof of the classification of Bruhat-Tits pairs. Here is the conclusion:

Theorem 26.1. *Every Bruhat-Tits pair (as defined in 13.1) is isomorphic to one of the Bruhat-Tits pairs in the fourth column of Table 27.2.*

Proof. By the classification of Moufang spherical buildings (as summarized in 30.14), the building at infinity of a Bruhat-Tits pair is isomorphic to one of the buildings in the second column of Table 27.1. The claim holds, therefore, by 16.18, 19.23, 20.6, 21.35, 22.29, 23.10, 24.37, 25.25 and 26.12 (which we prove below). □

* * *

We begin this chapter with a few definitions.

Definition 26.2. A Bruhat-Tits pair is *algebraic* if its building at infinity is algebraic as defined in 30.31. An affine building Δ is *algebraic* if it has a system of apartments $\mathcal{A}$ such that $(\Delta, \mathcal{A})$ is an algebraic Bruhat-Tits pair. (Thus an algebraic affine building is, in particular, a Bruhat-Tits building as defined in 13.1.)

Definition 26.3. A Bruhat-Tits pair is *exceptional* if its building at infinity is exceptional as defined in 30.31, i.e. if its building at infinity is in one of the cases (i)–(vi) of 30.32. An affine building Δ is *exceptional* if it has a system of apartments $\mathcal{A}$ such that $(\Delta, \mathcal{A})$ is an exceptional Bruhat-Tits pair. Note that the term "exceptional" implies "algebraic."

Definition 26.4. A Bruhat-Tits pair is *mixed* or *of mixed type* if its building at infinity is mixed as defined in 30.24. An affine building Δ is *mixed* or *of mixed type* if it has a system of apartments $\mathcal{A}$ such that $(\Delta, \mathcal{A})$ is a mixed Bruhat-Tits pair.

Exceptional Bruhat-Tits pairs whose building at infinity is in one of the cases (iii), (iv) or (vi) of 30.32 were studied in Chapters 16, 21 and 25. In the first part of this chapter, we examine the remaining cases. The Bruhat-Tits pairs in these remaining cases are all parametrized by composition algebras.

Proposition 26.5. *Let (K, F) be a composition algebra with standard involution σ and norm q as defined in 30.17. Let ν be a map from K^* to $\mathbb{Z}$, let ω be a map from F^* to $\mathbb{Z}$ and let $\delta \in \mathbb{N}$. Then the following are equivalent:*

(i) *ν is a valuation of K, the involutory set (genuine or honorary)*
$$(K, F, \sigma)$$
is ν-compatible as defined in 23.2, δ is its ν-index as defined in 23.2 and
$$\omega(t) = \nu(t)/\delta$$
for all $t \in F^*$.

(ii) *ω is a valuation of F, the anisotropic quadratic space*
$$(F, K, q)$$
is ω-compatible as defined in 19.17, $2/\delta$ is its ω-index as defined in 19.17 and
$$\nu(u) = \delta\omega(q(u))/2 \qquad (26.6)$$
for all $u \in K^*$.

Proof. Suppose that (i) holds. By 23.2, ω is a surjective map from F^* to $\mathbb{Z}$ (so ω is a valuation of F) and ν is σ-invariant. By 9.17.i and 30.18, we have
$$\nu(q(u)) = \nu(uu^\sigma) = 2\nu(u) \qquad (26.7)$$
for all $u \in K^*$. Thus 26.6 holds and q satisfies 19.4.ii with ω in place of ν (since ν satisfies 9.17.ii). By 19.17, therefore, the anisotropic quadratic space (F, K, q) is ω-compatible. By 26.7, we also have $\nu(q(K^*)) = 2\mathbb{Z}$ and hence $\omega(q(K^*)) = (2/\delta)\mathbb{Z}$. Hence by 19.17 again, the ω-index of (F, K, q) is $2/\delta$. Thus (i) implies (ii).

Suppose, conversely, that (ii) holds. Since ω is a valuation of F, we have $\omega(F^*) = \mathbb{Z}$ and hence $\nu(F^*) = \delta\mathbb{Z}$. By 30.19, ν satisfies 9.17.i, and by 19.4.ii, ν satisfies 9.17.ii. Thus ν is a valuation of K. By 30.18, q is σ-invariant. Hence ν is also σ-invariant. By 23.2, therefore, the involutory set (K, F, σ), genuine or honorary, is ν-compatible. Thus (ii) implies (i). □

Definition 26.8. Let (K, F) be a composition algebra with standard involution σ and norm q and let ν be a valuation of K. We will say that the composition algebra (K, F) is *ν-compatible* if the two conditions in 26.5 are satisfied. If (K, F) is ν-compatible, we call the number δ in 26.5 the *ν-index* of (K, F). We call (K, F) *ramified* (respectively, *unramified*) at ν if $\delta = 2$ (respectively, $\delta = 1$).

Proposition 26.9. *Let (K, F) be a composition algebra with standard involution σ and norm q, let ν be a valuation of K and let ω be a valuation of F. Then the following are equivalent:*

(i) *The composition algebra is ν-compatible with ν-index δ and ω is the restriction of ν/δ to F.*
(ii) *The involutory set (K, F, σ) (honorary or genuine) is ν-compatible with ν-index δ and ω is the restriction of ν/δ to F.*
(iii) *The anisotropic quadratic space (F, K, q) is ω-compatible with ω-index δ and $\nu(u) = \omega(q(u))/\delta$ for all $u \in K^*$.*

Proof. This holds by 26.5 and 26.8. □

Proposition 26.10. *Let (K, F) be a composition algebra with standard involution σ and norm q, let ν be a valuation of K and let ω be a valuation of F. Suppose that the three equivalent assertions in 26.9 hold. Then the following are equivalent:*

(i) *(K, F) is ramified at ν as defined in 26.8.*
(ii) *The involutory set (K, F, σ) (honorary or genuine) is ramified at ν as defined in 23.2.*
(iii) *The quadratic space (F, K, q) is ramified at ω as defined in 19.17.*

Proof. Also this holds by 26.5 and 26.8. □

Notation 26.11. Let $\Lambda = (K, F)$ be a composition algebra and let (Δ, Σ, d) be the corresponding canonical realization of type 30.14.xiii. Let q be the norm and let σ be the standard involution of (K, F). Let $(G^\dagger, \xi)$ be the root datum of Δ based at Σ. We label the vertices of the Coxeter diagram F_4 so that (x_1, x_2), (x_2, x_3) and (x_3, x_4) are the three positively oriented edges as defined in 30.11 and let $a_i := a_{x_i}$ for all $i \in [1, 4]$ (as defined in 3.5). Let ν be a valuation of K and suppose that Λ is ν-compatible as defined in 26.8. Let δ and ω be as in 26.5. By 16.14 and 23.3 (or 19.17), there exists a root map ι of Σ (with target F_4) and a valuation ϕ of the root datum $(G^\dagger, \xi)$ with respect to ι such that

$$\phi_{a_i}(x_{a_i}(u)) = \nu(u)$$

for $i = 1$ and 2 and all $u \in K^*$ and

$$\phi_{a_i}(x_{a_i}(t)) = \omega(t)$$

for $i = 3$ and 4 and all $t \in F^*$. We denote the pair $(\Delta^{\rm aff}, \mathcal{A})$ obtained by applying 14.47 to $(G^\dagger, \xi)$ and ϕ by

$$\tilde{\mathsf{F}}_4(\Lambda, \nu).$$

Note that by 26.6, ν is uniquely determined by ω, so we could equally well have called this Bruhat-Tits pair $\tilde{\mathsf{F}}_4(\Lambda, \omega)$.

Thus the pair $\tilde{\mathsf{F}}_4(\Lambda, \nu)$ defined in 26.11 exists precisely when the composition algebra Λ is ν-compatible.

Theorem 26.12. *Every Bruhat-Tits pair whose building at infinity is*

$$\mathsf{F}_4(\Lambda)$$

for some composition algebra $\Lambda = (K, F)$ *is of the form* $\tilde{\mathsf{F}}_4(\Lambda, \nu)$ *for some valuation* ν *of* K *such that* Λ *is* ν*-compatible.*

Proof. This holds by 3.41.iii, 16.4, 16.14.ii and 21.27. □

We turn next to completions of Bruhat-Tits pairs of the form $\tilde{\mathsf{F}}_4(\Lambda, \nu)$.

Theorem 26.13. *Let* $\Lambda = (K, F)$ *be a composition algebra, let* ν *be a valuation of* K*, suppose that* (K, F) *is* ν*-compatible as defined in 26.8 and let* $(\hat{K}, \hat{F}, \hat{\sigma})$ *be as in 23.12. Then the following hold:*

(i) $\hat{\Lambda} := (\hat{K}, \hat{F})$ *is a* ν*-compatible composition algebra with standard involution* $\hat{\sigma}$*.*

(ii) *The Bruhat-Tits pair*

$$\tilde{\mathsf{F}}_4(\Lambda, \nu)$$

is completely Bruhat-Tits (as defined in 17.1) and its completion is $\tilde{\mathsf{F}}_4(\hat{\Lambda}, \nu)$*.*

(iii) *If* Λ *is of type (i) (as defined in 30.17), then* $\hat{\Lambda}$ *is of type (i) or (ii) and if* Λ *is of type (k) for k not equal to i, then* $\hat{\Lambda}$ *is also of type (k).*

Proof. Let q be the norm and let T be the trace of Λ (as defined in 30.20). Let $\hat{T}(\hat{u}) = \hat{u} + \hat{u}^{\hat{\sigma}}$ and $\hat{q}(\hat{u}) = \hat{u}^{\hat{\sigma}}\hat{u}$ for all $\hat{u} \in \hat{K}$. Thus $T(u) \in F$, $q(u) \in F$ and

$$u^2 - T(u)u + q(u) = u^2 - (u + u^\sigma)u + u^\sigma u = 0$$

for all $u \in K$ and hence $\hat{T}(\hat{u}) \in \hat{F}$, $\hat{q}(\hat{u}) \in \hat{F}$ and

$$\hat{u}^2 - \hat{T}(\hat{u})\hat{u} + \hat{q}(\hat{u}) = \hat{u}^2 - (\hat{u} + \hat{u}^\sigma)\hat{u} + \hat{u}^{\hat{\sigma}}\hat{u} = 0$$

for all $\hat{u} \in \hat{K}$ by 23.11. Since $\hat{F}$ is a subfield of the center of $\hat{K}$, it follows by 20.3 in [36] that $(\hat{K}, \hat{F})$ is a composition algebra with standard involution $\hat{\sigma}$. By 23.12 and 26.9, this composition algebra is ν-compatible. Thus (i) holds. By 30.16, $\mathsf{F}_4(\Lambda)$ is a subbuilding of $\mathsf{F}_4(\hat{\Lambda})$. By 17.4 and 17.9, therefore, (ii) holds.

Let (k) be the type of Λ. If $K^2 \subset F$, then $\hat{K}^2 \subset \hat{F}$, and if $K = F$, then $\hat{K} = \hat{F}$. To prove (iii), it thus suffices to assume that k is at least iii. Hence the bilinear form associated with q is non-degenerate. By 19.29, it follows that the quadratic spaces $(\hat{F}, \hat{K}, \hat{q})$ and $(\hat{F}, K \otimes_F \hat{F}, q_{\hat{F}})$ are isomorphic. In particular,

$$\dim_{\hat{F}} \hat{K} = \dim_F K.$$

Since the bilinear form associated with $\hat{q}$ is also non-degenerate, it follows that $\hat{\Lambda}$ is also of type (k). Thus (iii) holds. □

Proposition 26.14. *Let* $\Lambda = (K, F)$ *be a composition algebra and suppose that* K *is complete with respect to a valuation* ν*. Then* Λ *is* ν*-compatible as defined in 26.8.*

Proof. This holds by 23.16. □

We turn now to residues of Bruhat-Tits pairs of the form $\tilde{\mathsf{F}}_4(\Lambda, \nu)$.

Proposition 26.15. *Let $\Lambda = (K, F)$ be a composition algebra of type (k) as defined in 30.17 (so k is i, ii, iii, iv or v). Let σ be the standard involution of Λ, let T be the trace of Λ (as defined in 30.20) and let ν be a valuation of K. Suppose that K is complete with respect to ν (so Λ is ν-compatible by 26.14). Let ω be as in 26.5, let $\bar{K}$ be as defined in 9.22 with respect to ν and let $\bar{F}$ be the image of the ring of integers in F (with respect to ω) in $\bar{K}$. By 23.2, ν is σ-invariant, so there are well-defined maps $\bar{\sigma}$ from $\bar{K}$ to itself induced by σ and $\bar{T}$ from $\bar{K}$ to $\bar{F}$ induced by T. Then $(\bar{K}, \bar{F})$ is a composition algebra, $\bar{\sigma}$ is its standard involution and $\bar{T}$ is its trace. If k equals i, then $(\bar{K}, \bar{F})$ is of type (i) or (ii). If k is not equal to i, then*

$$\dim_{\bar{F}} \bar{K} = \dim_F K$$

if (K, F) is unramified at ν and

$$\dim_{\bar{F}} \bar{K} = \big(\dim_F K\big)/2$$

if (K, F) is ramified at ν.[1] *If, in addition, $\bar{T}$ is not identically zero, then:*

(i) *$(\bar{K}, \bar{F})$ is of type (k) if Λ is unramified at ν and*
(ii) *$(\bar{K}, \bar{F})$ is of type (k−1) if Λ is ramified at ν.*

Thus, for example, if K is an octonion division algebra with center F, then $\bar{K}$ is either an octonion division algebra with center $\bar{F}$, a quaternion division algebra with center $\bar{F}$ or (if $\bar{T}$ is identically zero) a purely inseparable extension of $\bar{F}$. We call $\bar{\Lambda} := (\bar{K}, \bar{F})$ the residue of Λ at ν.

Proof. For all $u \in K$, both $u + u^\sigma$ and $u^\sigma u$ are in F. Let u be an element of the ring of integers of K. Then

$$u^2 - (u + u^\sigma)u + u^\sigma u = 0.$$

Since ν is σ-invariant (by 26.8), the elements $u + u^\sigma$ and $u^\sigma u$ are in the ring of integers of F. It follows that every element of $\bar{K}$ is the root of a monic quadratic polynomial with coefficients in $\bar{F}$. By 20.3 of [36], it follows that $(\bar{K}, \bar{F})$ is a composition algebra of type (m) for some m, that $\bar{\sigma}$ is its standard involution and that $\bar{T}$ is its trace.

If k equals i, then T is identically zero, so $\bar{T}$ is also identically zero; therefore, m equals i or ii. Suppose that k is not equal to i. Then (F, K, q) is a finite-dimensional anisotropic quadratic space, where q is the norm of Λ. If Λ is unramified at ν, then $\dim_{\bar{F}} \bar{K} = \dim_F K$ by 19.37. Suppose that Λ is ramified at ν and let w be an element of K^* such that $\nu(w) = 1$. Then multiplication by w induces an isomorphism from $\bar{\Lambda}_0$ to $\bar{\Lambda}_1$, where $\bar{\Lambda}_0$ and $\bar{\Lambda}_1$ are the anisotropic quadratic spaces obtained by applying 19.33 to (F, K, q). By 19.36, it follows that

$$\dim_{\bar{F}} \bar{K} = \big(\dim_F K\big)/2.$$

[1]If k equals ii, then Λ is automatically unramified at ν.

By 30.17, we conclude that if k is not equal to i and $\bar{T}$ is not identically zero, then m equals k if Λ is unramified at ν and m equals k − 1 if Λ is ramified at ν. □

Theorem 26.16. *Let $\Lambda = (K, F)$ be a composition algebra and let ν be a valuation of K. Suppose that $\Lambda = (K, F)$ is ν-compatible as defined in 26.8, let*

$$(\Delta, \mathcal{A}) = \tilde{\mathsf{F}}_4(\Lambda, \nu)$$

and let $\bar{\Lambda} = (\bar{K}, \bar{F})$ be as in 26.15. Then the following hold:

(i) *The group $G^\dagger$ acts transitively on the set of gems in Δ.*
(ii) *Every gem of Δ is isomorphic to $\mathsf{F}_4(\bar{\Lambda})$ (in the notation of 30.15).*

Proof. The first assertion holds by 18.5. The second follows, therefore, by 18.27, 26.15 and 30.14.xiii. □

Proposition 26.17. *Let $(\Delta, \mathcal{A}) = \tilde{\mathsf{F}}_4(\Lambda, \nu)$ for some composition algebra $\Lambda = (K, F)$ and some valuation ν of K, let σ be the standard involution of (K, F) and q the norm of Λ. Let $\bar{\Lambda} = (\bar{K}, \bar{F})$ and $\bar{\sigma}$ be as in 26.15 and let $\bar{q}$ be the norm of $\bar{\Lambda}$. Finally, let R be a residue of Δ of type B_4. Then the following hold:*

(i) *If Λ is unramified at ν as defined in 26.8, then*

$$R \cong \mathsf{B}_4^{\mathcal{Q}}(\bar{F}, \bar{K}, \bar{q}). \quad (26.18)$$

(ii) *If Λ is ramified at ν, then*

$$R \cong \mathsf{B}_4^{\mathcal{I}}(\bar{K}, \bar{F}, \bar{\sigma}). \quad (26.19)$$

Proof. Let $(\Delta_{\mathcal{A}}^\infty, \Sigma, d)$, ι and ϕ be as in 26.11 (where $\Delta_{\mathcal{A}}^\infty$ is called Δ) and let A be the unique apartment in $\mathcal{A}$ such that $\Sigma = A^\infty$. We can assume that R is cut by A. Let R_0 be as in 18.1, let R_1 be a residue of R of type A_3 and let R_2 be a residue of R_0 of type B_3. Let Φ be the root system F_4; we identify A with $\tilde{\Sigma}_\Phi$ (as defined in 2.29) via ι. By 2.47, all the roots of A that cut R_1 are long (as defined in 2.35). Hence all the roots of A that cut $R_1 \cap R_2$ are long.

Let o be the unique special vertex of the Coxeter diagram $\tilde{\Pi} := \tilde{\mathsf{F}}_4$ and let I be the vertex set of the Coxeter diagram $\Pi := \tilde{\Pi}_o$ of $\Delta_{\mathcal{A}}^\infty$. Let x be the element of I chosen in 16.3 and let $a = a_x$ be the corresponding root of Σ as defined in 3.5. Let J and J' be the two subsets of I such that the Coxeter diagrams Π_J and $\Pi_{J'}$ are both isomorphic to B_3. We order these two sets so that x is contained in an edge with label 3 of Π_J. Let R_3 be a J-residue of the gem R_0 and let R_3' be a J'-residue of R_0. By 26.16.ii (or, more precisely, 18.27) and the choice of the vertex x,

$$R_3 \cong \mathsf{B}_3^{\mathcal{I}}(\bar{K}, \bar{F}, \bar{\sigma})$$

and

$$R_3' \cong \mathsf{B}_3^{\mathcal{Q}}(\bar{F}, \bar{K}, \bar{q}).$$

We can assume that either $R_2 = R_3$ or $R_2 = R_3'$. Since all the roots of A that cut $R_1 \cap R_2$ are long, we have $R_2 = R_3$ if a is long and $R_2 = R_3'$ if a is short. By 23.6, 23.27 and 26.10, a is long if and only if Λ is ramified at ν. □

The result 26.15 can be applied also to exceptional Bruhat-Tits pairs whose building at infinity is in case (i) or case (ii) of 30.32:

Proposition 26.20. *Let $(\Delta, \mathcal{A}) = \tilde{\mathsf{A}}_2(K, \nu)$ for some octonion division algebra K and some valuation ν of K such that K is complete with respect to ν. Let F be the center of K, let Λ denote the composition algebra (K, F), let $\bar{K}$ be the residue field of K and let $(\bar{K}, \bar{F})$ and $\bar{T}$ be as in 26.15. Then the gems of Δ are all isomorphic to $\mathsf{A}_2(\bar{K})$, where*

(i) *$\bar{K}$ is an octonion division algebra if $\bar{T}$ is not identically zero and Λ is unramified.*
(ii) *$\bar{K}$ is a quaternion division algebra if $\bar{T}$ is not identically zero and Λ is ramified.*
(iii) *$\bar{K}$ is a commutative field and $\bar{K}/\bar{F}$ is an inseparable extension of exponent 1 and of degree m, where $m = 4$ if Λ is ramified and $m = 8$ if it is not.*

Proof. This holds by 18.30 and 26.15. □

Proposition 26.21. *Let $(\Delta, \mathcal{A}) = \tilde{\mathsf{X}}_3^{\mathcal{I}}(\Lambda, \nu)$ for some honorary involutory set $\Lambda = (K, F, \sigma)$ and some valuation ν of K such that Λ is ν-compatible and K is complete with respect to ν, where $\mathsf{X} = \mathsf{B}$ if Λ is unramified at ν and $\mathsf{X} = \mathsf{C}$ if Λ is ramified at ν. Thus K is an octonion division algebra and F is its center. Let Λ denote the composition algebra (K, F) and let $(\bar{K}, \bar{F})$, $\bar{\sigma}$ and $\bar{T}$ be as in 26.15. Then the gems of Δ are all isomorphic to $\mathsf{X}_3^{\mathcal{I}}(\bar{K}, \bar{F}, \bar{\sigma})$, where*

(i) *$\bar{K}$ is an octonion division algebra if $\bar{T}$ is not identically zero and Λ is unramified.*
(ii) *$\bar{K}$ is a quaternion division algebra if $\bar{T}$ is not identically zero and Λ is ramified.*
(iii) *$\bar{K}$ is a commutative field and $\bar{K}/\bar{F}$ is an inseparable extension of exponent 1 and of degree m, where $m = 4$ if Λ is ramified and $m = 8$ if it is not.*

Proof. This holds by 23.27.iii and 26.15. □

While we are on the subject of quaternions, we record here a result that we will need in Chapter 28:

Proposition 26.22. *Suppose that K is a quaternion division algebra. Let F be the center of K, let N be the reduced norm of K and let σ be the standard involution of K. For each involution τ of K, let K^τ be as in 23.19. For each $w \in K^*$, let ι_w be the inner automorphism of K given by $x^{\iota_w} = wxw^{-1}$ for all $x \in K$. For each F-subspace X of K, let $X^\perp$ denote the F-subspace* *perpendicular to* X *with respect to the reduced trace of* K.[2]

(i) *Let* E *be a subfield of* K *containing* F *such that* E/F *is a separable quadratic extension and let* e *be an element in* $K^* \cap E^\perp$. *Then*

$$K = E \oplus eE,$$

all the relations in 20.17 of [36] hold (where σ *is denoted by* $x \mapsto \bar{x}$*) and*

$$N(u + ev) = N(u) - \beta N(v)$$

for all $u, v \in E$, *where* $\beta = -N(e)$.

(ii) *For every pair* (E, e) *as in (i),*

$$(u + ev)^\sigma = u^\sigma - ev \qquad (26.23)$$

for all $u, v \in E$ *and the restriction of* σ *to* E *is the unique non-trivial element of* $\mathrm{Gal}(E/F)$.

(iii) *If* τ *is a non-standard involution of* K *acting trivially on* F, *then* $\tau = \sigma\iota_e$ *for some* $e \in K^*$ *such that* $e^\tau = -e$.

(iv) *If* τ *and* e *are as in (iii), then there exists a subfield* E *of* K *such that the pair* (E, e) *is as in (i), and for all such* E,

$$(u + ev)^\tau = u - ev^\sigma$$

for all $u, v \in E$.

(v) *If* $\mathrm{char}(K) = 2$ *and* τ *is a non-standard involution of* K *acting trivially on* F, *then the involutory set* (K, K^τ, τ) *is the translate of* (K, K^σ, σ) *with respect to* e.[3]

(vi) *If* $\mathrm{char}(K) \neq 2$ *and* τ *and* ρ *are two non-standard involutions of* K *acting trivially on* F, *then the involutory sets* (K, K^τ, τ) *and* (K, K^ρ, ρ) *are translates of each other.*

Proof. Assertions (i) and (ii) hold by 20.20 in [36]. Let τ be an involution of K distinct from σ such that τ acts trivially on F. Then the product $\sigma^{-1}\tau$ is an automorphism of K acting trivially on F. By the Skolem-Noether theorem (see, for example, 2.27 in [36]), it follows that $\tau = \sigma\iota_e$ for some $e \in K^*$. Since $\tau \neq \sigma$, we have $e \notin F$. Since $\tau^2 = 1$, the product $t := e^\sigma e^{-1}$ is contained in the center F of K. Since $\sigma^2 = 1$, it then follows that $t^2 = 1$. In other words, $e^\sigma = \pm e$. If $\mathrm{char}(K) \neq 2$, then $K^\sigma = F$ (by 26.23) and hence $e^\sigma = -e$ since $e \notin F$. Thus (iii) holds and $F \subset e^\perp$.

We claim now that there exists a subfield E of K containing F and contained in $e^\perp$ such that E/F is a separable quadratic extension. By 9.6 in [36], every element of $K\backslash F$ is quadratic over F. If $\mathrm{char}(K) \neq 2$, it suffices, therefore, to choose an arbitrary element v in $e^\perp\backslash F$ and set $E = F(v)$. Suppose that $\mathrm{char}(K) = 2$. If $e^\perp \subset K^\sigma$, then these two F-spaces must, in fact,

---

[2] In other words, $X^\perp = \{u \in K \mid u^\sigma v + v^\sigma u = 0 \text{ for all } v \in X\}$.

[3] An involutory set $\Lambda' = (K', K_0', \sigma')$ is a *translate* of another involutory set $\Lambda = (K, K_0, \sigma)$ (according to the definition in 11.8 of [36]) if there exists an element $w \in K_0^*$ such that $K' = K$, $K_0' = wK_0$ and $\sigma' = \sigma\iota_w$ (in which case we say that Λ' is the translate of Λ *with respect to* w).

be equal since they have the same dimension. This implies, however, that $e \in (K^\sigma)^\perp = F$. We conclude that there exists an element $v \in e^\perp$ that is not contained in K^σ. Then the subfield $E = F(v)$ has the desired properties also in this case. This proves our claim (in all characteristics). Since $e^\sigma = -e$, we have $e^2 = -ee^\sigma = -N(e) \in F$. Let $\beta = e^2$. By (i) and (ii), we now calculate that

$$(u + ev)^{\sigma\iota_e} = e(u^\sigma - ev)e/\beta = u - ev^\sigma \tag{26.24}$$

for all $u, v \in E$. Thus (iv) holds.

If $\mathrm{char}(K) = 2$, then $e^\sigma = e$ by (iii). Thus (v) holds. Now suppose that $\mathrm{char}(K) \neq 2$ and let ρ be a second non-standard involution of K acting trivially on F. By (iii), we have $\rho = \sigma\iota_w$ for some $w \in K^*$ such that $w^\sigma = -w$. Let $z = we^{-1}$. Then $\tau\iota_z = \rho$ and

$$\begin{aligned} z^\tau = z^{\sigma\iota_e} &= \big((e^{-1})^\sigma \cdot w^\sigma\big)^{\iota_e} \\ &= \big((-e^{-1}) \cdot (-w)\big)^{\iota_e} = we^{-1} = z. \end{aligned}$$

Therefore (K, K^ρ, ρ) is the translate of (K, K^τ, τ) with respect to z. Thus (vi) holds. □

* * *

This concludes the first part of this chapter. We turn now to the classification of algebraic Bruhat-Tits buildings, in other words, of complete Bruhat-Tits pairs

$$(\Delta, \mathcal{A}) = \tilde{\mathsf{X}}_\ell^*(\Lambda, \nu)$$

whose building at infinity is algebraic as defined in 30.31 and of rank at least 2. The classification of algebraic spherical buildings of rank at least 2 is summarized in 30.33. Since $(\Delta, \mathcal{A})$ is complete, the defining field K of $\Delta_{\mathcal{A}}^\infty$ is complete (by the results cited in the fifth column of Table 27.2).

Theorem 26.25. *Let*

$$\Delta = \mathsf{X}_\ell^*(\Lambda)$$

be an algebraic spherical building with $\ell \geq 2$ (see 30.31 and 30.33), let K be the defining field of Δ (as defined in 30.15), let ν be a discrete valuation of K and suppose that K is complete with respect to ν. Suppose as well that 21.23 holds if $\mathsf{X}_\ell^* = \mathsf{B}_2^{\mathcal{E}}$. Then the following hold:

(i) *There exists a unique Bruhat-Tits pair whose building at infinity is Δ.*
(ii) *The Bruhat-Tits pair in (i) is complete.*

Proof. By 23.15, ν is the only discrete valuation of K. If X_ℓ is simply laced, then the claim in (i) holds by 16.17 and 16.18 (even if Δ is not algebraic and K is not complete). Suppose that X_ℓ is not simply laced. Since Δ is algebraic, K is finite-dimensional over its center and hence 24.9 applies. By 19.1, 23.16, 24.9 and 25.5, the parameter system Λ is, in every case, ν-compatible (respectively, exceptionally ν-compatible, given our extra hypothesis, when

$\mathsf{X}_\ell^* = \mathsf{B}_2^{\mathcal{E}}$). The claim in (i) holds, therefore, by 19.21, 21.27, 22.16, 23.6, 24.35, 25.21 and 26.11. Assertion (ii) holds by 19.29, 23.21, 24.49, 30.33 and the results cited in the fifth column of Table 27.2. □

We can now formulate the classification of algebraic Bruhat-Tits buildings. See also 26.37.

Corollary 26.26. *Algebraic Bruhat-Tits buildings are classified (modulo 21.23) by the algebraic spherical buildings of rank at least 2 as described in 30.33 whose defining field is complete with respect to a discrete valuation.*

Proof. This holds by 26.25. □

Here is another formulation of the classification:

Theorem 26.27. *Algebraic Bruhat-Tits buildings are classified (modulo 21.23) by pairs (G, F), where F is a field that is complete with respect to a discrete valuation ν and G is an absolutely simple algebraic group of F-rank at least 2.*

Proof. Let F be a field, let G be an absolutely simple algebraic group of F-rank ℓ at least 2, let Δ be the spherical building associated with the pair (G, F) and let K be the defining field of Δ (as defined in 30.15). By 30.32, F is contained in the center of K and K is finite-dimensional over F. The claim holds, therefore, by 23.14.ii and 26.26. □

* * *

We come now to the third and final part of this chapter. Our goal is to prove various results about the automorphism group of a Bruhat-Tits pair.

Notation 26.28. Let

$$(\Delta, \mathcal{A}) = \tilde{\mathsf{X}}_\ell^*(\Lambda, \nu)$$

for some parameter system Λ and some valuation ν of the defining field K. We adopt all the notation in 18.1. Thus A is an apartment in $\mathcal{A}$, H is the pointwise stabilizer of $\Sigma := A^\infty$ in $\mathrm{Aut}(\Delta_{\mathcal{A}}^\infty)$ and G_A is the subgroup of $G = \mathrm{Aut}(\Delta, \mathcal{A})$ consisting of all the elements that map A to itself and induce elements of H on A. Let d be a chamber of Σ. As in 18.9, we can assume that $(\Delta_{\mathcal{A}}^\infty, \Sigma, d)$ is the canonical realization of the corresponding root group labeling of the Coxeter diagram $\Pi := \mathsf{X}_\ell$ as described in 30.15. Let ϕ be the valuation of the root datum of $\Delta_{\mathcal{A}}^\infty$ described in the result in the column of Table 27.2 to the left of $\tilde{\mathsf{X}}_\ell(\Lambda, \nu)$. Let Ξ_d° be as in 3.5.

In 18.6, we gave a necessary and sufficient condition for an element of H to be induced by an element of G_A. In fact, *most* elements in H are induced by elements of G_A. We make this comment more concrete in a series of observations. See also 26.38.

In 26.29–26.36, we assume that τ is an arbitrary element of H.

Example 26.29. Suppose first that X_ℓ^* has edges with label 3. Equivalently, we assume that either $\ell \geq 3$ or $\mathsf{X}_\ell^* = \mathsf{A}_2$. Thus if a is the root in Ξ_d° fixed in 16.3, then $\phi_a(x_a(t)) = \nu(t)$ for all $t \in K^*$. If K is associative (which is necessarily the case unless $\mathsf{X}_\ell^* = \mathsf{A}_2$, C_3 or F_4), then by 37.13 in [36], there exist a root $a \in \Xi_d^\circ$ such that the root group U_a is parametrized by K, an automorphism ρ of K and an element $w \in K^*$ such that

$$x_a(t)^\tau = x_a(\rho(t)w)$$

for all $t \in K$. Suppose that K is an octonion division algebra, let F be the center of K and let N be the norm of the composition algebra (K, F). By 37.12 in [36], there exist a root $a \in \Xi_d^\circ$ such that the root group U_a is parametrized by K, an element κ in the group X defined in 37.9 in [36] and an element $t \in K^*$ such that

$$x_a(u)^\tau = x_a(\kappa(u)t)$$

for all $u \in K$.[4] Let $w = \kappa(1)$, let λ be the element of the group X obtained by applying 18.10 to w and let ρ denote the composition $\lambda^{-1} \circ \kappa$. By 37.17 of [36] and 18.10.i, ρ is an automorphism of K. By 18.6 and 18.10.ii, we conclude (whether K is associative or not) that τ is induced by an element of G_A if and only if ν is ρ-invariant.

In the next four examples (26.30–26.36), $\ell = 2$, so $(\Delta_{\mathcal{A}}^\infty, \Sigma, d)$ is the canonical realization of one of the root group sequences

$$(U_+, U_1, \ldots, U_n)$$

described in 30.8.

Example 26.30. Let $\mathsf{X}_\ell^* = \mathsf{X}_2^{\mathcal{Q}}$, $\mathsf{X}_2^{\mathcal{E}}$ or $\mathsf{X}_2^{\mathcal{F}}$ for $\mathsf{X} = \mathsf{B}$ or C. The parameter system Λ is an anisotropic quadratic space (K, L, q) that we can assume is unitary. By 37.30 in [36], there exist an automorphism ρ of K, a ρ-linear automorphism ψ of L and elements $\gamma, w \in K^*$ such that $\rho(q(u)) = \gamma q(\psi(u))$ and

$$x_4(u)^\tau = x_4(\psi(u)) \tag{26.31}$$

for all $u \in L$ as well as

$$x_1(t)^\tau = x_1(w\rho(t)) \tag{26.32}$$

for all $t \in K$ if $\mathsf{X}_\ell^* = \mathsf{X}_2^{\mathcal{Q}}$ and

$$x_1(0,t)^\tau = x_1(0, w\rho(t)) \tag{26.33}$$

for all $t \in K$ if $\mathsf{X}_\ell^* = \mathsf{X}_2^{\mathcal{E}}$ or $\mathsf{X}_2^{\mathcal{F}}$. By 18.6 and 26.32, τ is induced by an element of G_A if and only if ν is ρ-invariant when $\mathsf{X}_\ell^* = \mathsf{X}_2^{\mathcal{Q}}$. Suppose that $\mathsf{X}_\ell^* = \mathsf{X}_2^{\mathcal{E}}$ or $\mathsf{X}_2^{\mathcal{F}}$. It follows from 18.6 and 26.33 that if τ is induced by an element of G_A, then ν is ρ-invariant. By 21.30 and 22.19,

$$\phi_4(x_4(u)) = \nu(q(u))/\delta$$

for all $u \in L^*$, where δ is the ν-index of Λ as defined in 19.17. It follows from 18.6, 26.31 and this last equation that if ν is ρ-invariant, then τ extends to an element of G_A. We conclude in all three cases that ν is ρ-invariant if and only if τ extends to an element of G_A.

[4] See 18.10 and 18.13.

Example 26.34. Let $\mathsf{X}_\ell^* = \mathsf{X}_2^{\mathcal{D}}$ for $\mathsf{X} = \mathsf{B}$ or C, so Λ is an indifferent set (K, K_0, L_0). By 37.32 of [36], there exist an automorphism ρ of K and an element $w \in K^*$ such that

$$x_1(t)^\tau = x_1(\rho(t)w)$$

for all $t \in K_0$. By 20.2, $\phi_1(x_1(t)) = \nu(t)/\delta_K$ for all $t \in K_0^*$, where (δ_K, δ_L) is the ν-index of Λ. By 18.6, therefore, τ is induced by an element of G_A if and only if $\nu(\rho(t)) = \nu(t)$ for all $t \in K_0$. Since $K^2 \subset K_0$, this last condition holds if and only if ν is ρ-invariant.

Example 26.35. Let $\mathsf{X}_\ell^* = \mathsf{X}_2^{\mathcal{I}}$ or $\mathsf{X}_2^{\mathcal{P}}$ for $\mathsf{X} = \mathsf{B}$ or C, so Λ is an involutory set (K, K_0, σ) in the first case and an anisotropic pseudo-quadratic space in the second. By 37.33 in [36], there exist (in both cases) an automorphism ρ of K and an element $w \in K^*$ such that

$$x_4(t)^\tau = x_4(\rho(t)w)$$

for all $t \in K$. By 23.3 and 24.31, $\phi_4(x_4(t)) = \nu(t)$ for all $t \in K^*$. By 18.6, it follows that τ is induced by an element of G_A if and only if ν is ρ-invariant.

Example 26.36. Let $\mathsf{X}_\ell^* = \mathsf{G}_2$, so Λ is an hexagonal system

$$(J, K, N, \#, T, \times, 1).$$

By 37.41 of [36], there exist an automorphism ρ of K and an element $w \in K^*$ such that

$$x_6(t)^\tau = x_6(\rho(t)w)$$

for all $t \in K$. By 25.21, $\phi_6(x_6(t)) = \nu(t)$ for all $t \in K^*$. By 18.6, therefore, τ is induced by an element of G_A if and only if ν is ρ-invariant.

Assembling these observations, we have the following conclusion.

Theorem 26.37. *Let*

$$(\Delta, \mathcal{A}) = \tilde{\mathsf{X}}_\ell(\Lambda, \nu)$$

be a Bruhat-Tits pair, let K be the defining field of $\Delta_{\mathcal{A}}^\infty$ (as defined in 30.15) and suppose that K is complete with respect to ν. Then every type-preserving automorphism of $\Delta_{\mathcal{A}}^\infty$ is induced by an element of $\mathrm{Aut}(\Delta, \mathcal{A})$.

Proof. Let $\tau \in H$. By 26.29–26.36, there is an automorphism ρ of K such that ν is ρ-invariant if and only if τ is induced by an element of G_A. By 23.15, it follows that every element of H is induced by an automorphism of Δ. The claim holds, therefore, by 12.31 and 18.2. □

Example 26.38. We give an example of an element $\tau \in H$ that is not induced by an element of G_A. Suppose that $(\Delta, \mathcal{A}) = \mathsf{A}_\ell(K, \nu)$ and let ϕ be as in 16.14.i. If ρ is an arbitrary automorphism of K, then there exists an element τ of H such that $x_a(u)^\tau = x_a(\rho(u))$ and hence $\phi_a(x_a(u)^\tau) = \nu(\rho(u))$ for all $a \in \Xi_d^\circ$ and all $u \in K$. If ν is *not* ρ-invariant, then τ is *not* induced by an element of G_A (by 18.6). To obtain a concrete example, let $K = \mathbb{Q}(i)$, let ρ be the restriction of complex conjugation to K and let ν be the unique valuation of K such that $\nu(2+i)=1$. Then

$$\nu\big((2+i)^\rho\big)=\nu(2-i)=0,$$

so ν is not ρ-invariant.

Our next result is a strengthened version of 18.15:

Theorem 26.39. *Let $(\Delta,\mathcal{A})$ be a Bruhat-Tits pair, let $\tilde{\mathsf{X}}_\ell$ be the Coxeter diagram of Δ and let G° be as in 18.1. Suppose that $\tilde{\mathsf{X}}_\ell\neq\tilde{\mathsf{C}}_\ell$ (so, in particular, $\tilde{\mathsf{X}}_\ell\neq\tilde{\mathsf{B}}_2$). Then the group G° acts transitively on the set of gems of Δ.*

Proof. If $\mathsf{X}=\mathsf{B}$, then $\Delta_{\mathcal{A}}^\infty$ is as in (iii)–(vii) of 30.14 and the claim holds by 19.35.i, 23.27.i and 24.58.i.[5] The remaining cases are covered by 18.5 and 18.15. □

To conclude this chapter, we consider elements of $\mathrm{Aut}(\Delta,\mathcal{A})$ that do not induce type-preserving automorphisms of $\Delta_{\mathcal{A}}^\infty$.

Proposition 26.40. *Let $(\Delta,\mathcal{A})=\tilde{\mathsf{X}}_\ell^*(\Lambda,\nu)$ be a Bruhat-Tits pair, let K be the defining field of $\Delta_{\mathcal{A}}^\infty$ (as defined in 30.15) and suppose that one of the following holds:*

(i) $\mathsf{X}_\ell^*=\mathsf{A}_\ell$ *and there exists an anti-isomorphism κ of K such that ν is κ-invariant.*
(ii) $\mathsf{X}_\ell^*=\mathsf{D}_\ell$ *for $\ell\geq 4$ or* E_6.

Let o be a special vertex of the Coxeter diagram $\Pi:=\tilde{\mathsf{X}}_\ell$ and let σ_o be a non-trivial automorphism of Π of order 2 that fixes o.[6] *Then there exist σ_o-automorphisms in* $\mathrm{Aut}(\Delta,\mathcal{A})$.

Proof. Let Σ, d, ϕ, Ξ_d°, etc. be as in 26.28, let R_0 be as in 18.27, let the subdiagram Π_o (as defined in 8.21) be identified with the Coxeter diagram of $\Delta_{\mathcal{A}}^\infty$ as explained in 18.24 and let σ be the restriction of σ_o to Π_o. We thus have a permutation $a_x\mapsto a_{\sigma(x)}$ of Ξ_d° (in the notation introduced in 3.5) which we also denote by σ. By 7.5 in [36] and 3.6, there exists an element τ of $\mathrm{Aut}(\Delta_{\mathcal{A}}^\infty)$ mapping the pair (Σ,d) to itself such that

$$x_a(u)^\tau=x_{\sigma(a)}(\kappa(u))$$

for all $a\in\Xi_d^\circ$ and all $u\in K$ in case (i) and

$$x_a(u)^\tau=x_{\sigma(a)}(u)$$

for all $a\in\Xi_d^\circ$ and all $u\in K$ in case (ii). By 13.31.ii with $(\Delta',\mathcal{A}')=(\Delta,\mathcal{A})$, τ is induced by an element ρ of G_A. We have $U_{a,0}^\rho=U_{\sigma(a),0}$ for all $a\in\Xi_d^\circ$, where $U_{a,0}$ and $U_{\sigma(a),0}$ are as in condition (V1) in 3.21. By 1.12, therefore, ρ maps R_0 to itself. By 8.22, it follows that ρ is a σ_o-automorphism of Δ. □

---

[5] In fact, the claim holds also if $(\Delta,\mathcal{A})$ is of the form $\tilde{\mathsf{B}}_2^{\mathcal{X}}(\Delta,\nu)$ for $\mathcal{X}=\mathcal{Q}$, $\mathcal{I}$ or $\mathcal{P}$. These are exactly the three cases in which we distinguish between $\tilde{\mathsf{B}}_2^{\mathcal{X}}(\Delta,\nu)$ and $\tilde{\mathsf{C}}_2^{\mathcal{X}}(\Delta,\nu)$; see 19.22, 23.7 and 24.36, but also 20.5, 21.34 and 22.28.

[6] Note that σ_o is uniquely determined by o unless $\mathsf{X}_\ell^*=\mathsf{D}_4$.

Theorem 26.41. *Let*

$$(\Delta, \mathcal{A}) = \tilde{\mathsf{X}}_\ell^*(\Lambda, \nu)$$

be a Bruhat-Tits pair and let K *be the defining field of* $\Delta_{\mathcal{A}}^\infty$ *(as defined in 30.15). Suppose that* $\mathsf{X} \neq \mathsf{C}$ *and that* K *is isomorphic to its opposite if* $\mathsf{X} = \mathsf{A}$. *Then* $\mathrm{Aut}(\Delta, \mathcal{A})$ *induces the full automorphism group on the Coxeter diagram* $\tilde{\mathsf{X}}_\ell$.

Proof. This holds by 26.39 and 26.40. □

Proposition 26.42. *Let* $(\Delta, \mathcal{A}) = \tilde{\mathsf{A}}_\ell(K, \nu)$ *for some skew field* K *that is not isomorphic to its opposite. Then the subgroup of the automorphism group of the Coxeter diagram* $\tilde{\mathsf{A}}_\ell$ *induced by* $\mathrm{Aut}(\Delta, \mathcal{A})$ *is cyclic of order* $\ell + 1$.

Proof. By 35.6 in [36], the root group sequence determined by applying 16.1 in [36] to K is not isomorphic to its opposite. It follows that no automorphism of the diagram $\tilde{\mathsf{A}}_\ell$ interchanging the two vertices of an edge is induced by an element of $\mathrm{Aut}(\Delta, \mathcal{A})$. The claim holds, therefore, by 18.16. □

If $(\Delta, \mathcal{A}) = \tilde{\mathsf{C}}_\ell^*(\Lambda, \nu)$ for some Λ and ν, there are cases when $\mathrm{Aut}(\Delta, \mathcal{A})$ acts non-trivially on the Coxeter diagram $\tilde{\mathsf{C}}_\ell$ and cases when it does not. We will show, for example, that if Δ is a building in one of the Tables 28.4, 28.5 or 28.6 and $\mathcal{A}$ is its complete system of apartments, then $\mathrm{Aut}(\Delta, \mathcal{A})$ acts transitively on the set of special vertices of the Coxeter diagram of Δ if and only if there is only one entry in the last column to the right of Δ in the table (Table 28.4, 28.5 or 28.6) in which Δ appears.

Chapter Twenty Seven

Summary of the Classification

Looking back, we can observe that the classification of Bruhat-Tits pairs falls naturally into three parts. Part I is covered in Chapters 1–2 and 4–12. It includes the construction (starting with an arbitrary irreducible affine building) of the building at infinity and, in particular, the key result 7.24[1] that makes this construction possible, as well as the construction of the wall and panel trees $(T_m, \mathcal{A}_m)$ and $(T_F, \mathcal{A}_F)$. Part I culminates in the result 12.3 that an affine building is uniquely determined by its building at infinity together with its tree structure. This part of the classification is valid for arbitrary affine buildings. Part I also includes the various results about trees and projective valuations in Chapter 9.

In Part II, covered in Chapters 3 and 13–14, we assume that the building at infinity, which is a spherical building, satisfies the Moufang property. This means that the building at infinity is uniquely determined by the collection of root groups associated with the roots of a fixed apartment. This is the root datum of the building at infinity. The two main results of Part II are 13.31, in which the tree structure in 12.3 is translated into a valuation of the root datum of the building at infinity, and 14.47, which says that for each root datum with valuation, there exists, in fact, an affine building with the corresponding building at infinity and the corresponding tree structure. Thus Part II leaves us with the conclusion that Bruhat-Tits buildings, by which we mean affine buildings whose building at infinity is Moufang, are classified by root data with valuation (up to equipollence).

In Part III, covered in Chapters 15–26, the classification of Moufang spherical buildings is invoked. This means that we have a list, summarized in 30.14, of all possible root data.[2] These come in various families, each determined by a certain class of algebraic structures defined over a field K (which is, in some cases, a skew field or octonion division algebra). In 3.41.iii and 16.4, it is observed that a valuation of the root datum is uniquely determined by a discrete valuation of the field K.[3] The main goal of Part III is the determination of necessary and sufficient conditions, one family of root data at a time, for a valuation of the field K to give rise to a valuation of

[1] "For every two sectors there is an apartment containing subsectors of each."

[2] The root data of Moufang octagons are omitted from this list simply because there is no affine diagram having $\mathsf{I}(8)$ as a subdiagram (from which it follows that the building at infinity of an affine building cannot be a generalized octagon).

[3] Thus the word "valuation" is being used to mean both a valuation of a root datum as defined in 3.21 and a valuation of a field as defined in 9.17.

30.14	*Name*	*Nature of* Λ
i	$\mathsf{A}_\ell(\Lambda)$	Field or skew field
ii	$\mathsf{A}_2(\Lambda)$	Octonion division algebra
iii	$\mathsf{B}_\ell^{\mathcal{Q}}(\Lambda) = \mathsf{C}_\ell^{\mathcal{Q}}(\Lambda)$	Anisotropic quadratic space
iv	$\mathsf{B}_\ell^{\mathcal{I}}(\Lambda) = \mathsf{C}_\ell^{\mathcal{I}}(\Lambda)$	Proper involutory set
v	$\mathsf{B}_\ell^{\mathcal{I}}(\Lambda) = \mathsf{C}_\ell^{\mathcal{I}}(\Lambda)$	Quadratic involutory set
vi	$\mathsf{B}_3^{\mathcal{I}}(\Lambda) = \mathsf{C}_3^{\mathcal{I}}(\Lambda)$	Honorary involutory set
vii	$\mathsf{B}_\ell^{\mathcal{P}}(\Lambda) = \mathsf{C}_\ell^{\mathcal{P}}(\Lambda)$	Proper anisotropic pseudo-quadratic space
viii	$\mathsf{B}_2^{\mathcal{E}}(\Lambda) = \mathsf{C}_2^{\mathcal{E}}(\Lambda)$	Quadratic space of type E_6, E_7 or E_8
ix	$\mathsf{B}_2^{\mathcal{F}}(\Lambda) = \mathsf{C}_2^{\mathcal{F}}(\Lambda)$	Quadratic space of type F_4
x	$\mathsf{B}_2^{\mathcal{D}}(\Lambda) = \mathsf{C}_2^{\mathcal{D}}(\Lambda)$	Proper indifferent set
xi	$\mathsf{D}_\ell(\Lambda)$	Field
xii	$\mathsf{E}_\ell(\Lambda)$	Field
xiii	$\mathsf{F}_4(\Lambda)$	Composition algebra
xiv	$\mathsf{G}_2(\Lambda)$	Hexagonal system

Table 27.1 Moufang Spherical Buildings

the corresponding root datum.[4]

* * *

We use the rest of the chapter to recapitulate the conclusions of the classification of Bruhat-Tits buildings, first in Tables 27.1 and 27.2 and then in 27.1–27.5.

The starting point of our recapitulation is the classification of Moufang spherical buildings. This is described in 30.8–30.15 in terms of root group labelings of spherical Coxeter diagrams. We have summarized the classification of Moufang spherical buildings in Table 27.1, where the entry in the first column is a reference to the corresponding case of 30.14 and the names in the second column are those assigned in 30.15.[5] These are the candidates for the building at infinity of a Bruhat-Tits pair.

Let $\Delta = \mathsf{X}_\ell^*(\Lambda)$ be one of the buildings in Table 27.1 (where $*$ has the meaning indicated in 30.15), let Σ be an apartment of Δ, let Ξ be the set of roots of Σ, let $(G^\dagger, \xi)$ be the root datum of Δ based at Σ and let K denote

[4]Note that Part III is, except for Chapter 3, essentially independent of Parts I and II.
[5]Note, too, that 30.28 applies to the descriptions of Λ in the third column.

30.14	ν-*Compat.*	ϕ	*Name*	$\hat{\Lambda}$	*Gems*
i–ii	16.18	16.17	$\tilde{\mathsf{A}}_\ell(\Lambda,\nu)$	17.11	18.30
iii	19.17	19.22	$\tilde{\mathsf{B}}_\ell^{\mathcal{Q}}(\Lambda,\nu)$ or $\tilde{\mathsf{C}}_\ell^{\mathcal{Q}}(\Lambda,\nu)$	19.27	19.35
iv–vi	23.2	23.7	$\tilde{\mathsf{B}}_\ell^{\mathcal{I}}(\Lambda,\nu)$ or $\tilde{\mathsf{C}}_\ell^{\mathcal{I}}(\Lambda,\nu)$	23.13	23.27
vii	24.30	24.36	$\tilde{\mathsf{B}}_\ell^{\mathcal{P}}(\Lambda,\nu)$ or $\tilde{\mathsf{C}}_\ell^{\mathcal{P}}(\Lambda,\nu)$	24.46	24.58
viii	21.22	21.34	$\tilde{\mathsf{B}}_2^{\mathcal{E}}(\Lambda,\nu)=\tilde{\mathsf{C}}_2^{\mathcal{E}}(\Lambda,\nu)$	21.42	–
ix	22.15	22.28	$\tilde{\mathsf{B}}_2^{\mathcal{F}}(\Lambda,\nu)=\tilde{\mathsf{C}}_2^{\mathcal{F}}(\Lambda,\nu)$	22.33	–
x	20.6	20.5	$\tilde{\mathsf{B}}_2^{\mathcal{D}}(\Lambda,\nu)=\tilde{\mathsf{C}}_2^{\mathcal{D}}(\Lambda,\nu)$	20.8	20.10
xi	16.18	16.17	$\tilde{\mathsf{D}}_\ell(\Lambda,\nu)$	17.11	18.30
xii	16.18	16.17	$\tilde{\mathsf{E}}_\ell(\Lambda,\nu)$	17.11	18.30
xiii	26.8	26.11	$\tilde{\mathsf{F}}_4(\Lambda,\nu)$	26.13	26.16, etc.
xiv	25.14	25.24	$\tilde{\mathsf{G}}_2(\Lambda,\nu)$	25.27	25.30, etc.

Table 27.2 Bruhat-Tits Pairs

the defining field of Δ (as defined in 30.15).

Bruhat-Tits pairs having Δ as the building at infinity are classified by valuations of the root datum $(G^\dagger,\xi)$ up to equipollence. This is 14.54.

Let

$$\phi=\{\phi_a \mid a\in\Xi\}$$

be a valuation of $(G^\dagger,\xi)$. By 16.4, we can pick out a root a in Ξ such that ϕ_a has a simple description in terms of a valuation ν of the field K. By 3.41.iii, ϕ is uniquely determined up to equipollence by ϕ_a and hence by ν. The fundamental question of Bruhat-Tit theory is then the following: Given a discrete valuation ν of K and thus the map ϕ_a defined by the formulas in 16.4, when does ϕ_a extend to a valuation of $(G^\dagger,\xi)$?

The answer is "always" if the Coxeter diagram X_ℓ is simply laced (by 16.14) or if $\Delta=\mathsf{B}_2^{\mathcal{D}}(\Lambda)$ for some indifferent set Λ (by 20.2) or if $\Delta=\mathsf{G}_2(\Lambda)$ for some hexagonal system Λ of type $1/K$ (by 25.14). In every other case, however, the map ϕ_a extends to a valuation of $(G^\dagger,\xi)$ if and only if the parameter system Λ is ν-*compatible* (or *exceptionally* ν-*compatible* in case 30.14.viii). The relevant definitions and results are referenced in the first four columns of Table 27.2.[6] Here are some explanations:

[6]We extend the notion of ν-compatibility (which has already been defined in every other case) by declaring from now on that Λ is *always* ν-compatible if $\Lambda=K$ (which can happen precisely when the diagram X_ℓ is simply laced), if Λ is an indifferent set or if Λ is an hexagonal system Λ of type $1/K$. With this extension of the notion of ν-compatibility,

The entry in the first column of Table 27.2 refers to the case or cases of 30.14 that describe the building at infinity. The entry in the second column refers to the suitable definition of ν-compatibility. Note that by these definitions, Δ is always ν-compatible if K is complete with respect to the valuation ν.

The entry in the third column refers to the result in which an extension of ϕ_a to a valuation ϕ of the root datum $(G^\dagger, \xi)$ is described. In the fourth column, we give the name we have assigned to the Bruhat-Tits pair determined by Δ and ϕ. This name is of the form $\tilde{\mathsf{X}}_\ell^*(\Lambda, \nu)$. In each case $\tilde{\mathsf{X}}_\ell$ is the Coxeter diagram associated with the affine building and $\mathsf{X}_\ell^*(\Lambda)$ is (up to isomorphism) the building at infinity. The ambiguity in the name in the second to seventh rows of Table 27.2 is explained in 19.21, 23.6 and 24.35.

In each case there is a parameter system $\hat{\Lambda}$ of the same type as Λ defined over the completion $\hat{K}$ of the field K with respect to ν such that the pair $\tilde{\mathsf{X}}_\ell^*(\hat{\Lambda}, \nu)$ is the completion of the pair $\tilde{\mathsf{X}}_\ell^*(\Lambda, \nu)$ as defined in 17.1. In the fifth column of Table 27.2, we give the result in which the parameter system $\hat{\Lambda}$ is determined. Since each Bruhat-Tits building is part of a unique *complete* Bruhat-Tits pair, these are the parameter systems that classify Bruhat-Tits buildings (rather than Bruhat-Tits pairs).

In last column of Table 27.2 we indicate the principal results in which we prove something about the structure of the gems of Δ^{aff}. (These results play an essential role in the next chapter.)

This concludes our description of Tables 27.1 and 27.2. We now summarize the classification of Bruhat-Tits pairs and buildings in a sequence of results that expand on the formulation given in 26.1.

Here is the main uniqueness result:

Theorem 27.1. *Let $(\Delta, \mathcal{A})$ be a Bruhat-Tits pair of type $\tilde{\mathsf{X}}_\ell$ as defined in 13.1 (so $\ell \geq 2$) and let $\Delta_{\mathcal{A}}^\infty$ be its building at infinity as constructed in 8.9 and 8.24. Then the following hold:*

(i) *The building at infinity $\Delta_{\mathcal{A}}^\infty$ is, up to isomorphism, one of the buildings $\mathsf{X}_\ell^*(\Lambda)$ described in 30.15 (and listed in Table 27.1) for some suitable parameter system Λ.*[7]
(ii) *There exists a valuation ϕ of the root datum of $\Delta_{\mathcal{A}}^\infty$ satisfying 13.8 that is uniquely determined up to equipollence.*
(iii) *There exists a discrete valuation ν of the defining field K of $\Delta_{\mathcal{A}}^\infty$ that is uniquely determined by 16.4 applied to the equipollence class of the valuation ϕ.*

---

we can delete the phrase "in every other case" from the previous sentence. Note that if $\Lambda = K$ and K is non-commutative (which can happen precisely when $\mathsf{X}_\ell^* = \mathsf{A}_\ell$) and if we start only with a valuation of the center of K, then there *are* non-vacuous conditions that need to hold in order for this valuation to extend to a valuation of K.

[7]The parameter system Λ and the defining field K of $\mathsf{X}_\ell^*(\Lambda)$ are invariants of $\Delta_{\mathcal{A}}^\infty$ to the extent described in 30.29. See 27.6.

(iv) *The parameter system* Λ *is* ν*-compatible as defined in the results listed in the second column of Table 27.2* [8] *and the pair* $(\Delta, \mathcal{A})$ *is uniquely determined by the spherical building* $\Delta_{\mathcal{A}}^{\infty} = \mathsf{X}_{\ell}^{*}(\Lambda)$ *and the valuation* ν.[9] *In particular,*

$$(\Delta, \mathcal{A}) = \tilde{\mathsf{X}}_{\ell}^{*}(\Lambda, \nu),$$

where $\tilde{\mathsf{X}}_{\ell}^{*}(\Lambda, \nu)$ *is as defined in the results in the third column of Table 27.2.*

Proof. By 8.24, the building $\Delta_{\mathcal{A}}^{\infty}$ is irreducible and spherical, and by 13.1, it is Moufang. Assertion (i) holds, therefore, by the classification of Moufang spherical buildings as summarized in 30.14. Assertion (ii) holds by 13.30 and assertion (iii) is just 16.4. By 3.41.iii, ϕ is uniquely determined by ν up to equipollence. Thus assertion (iv) holds by 13.34. □

Here is the main existence result:

Theorem 27.2. *Let* $\Delta = \mathsf{X}_{\ell}^{*}(\Lambda)$ *be one of the spherical buildings described in 30.15 for a suitable parameter system* Λ, *let* K *be the defining field of* Δ, *let* ν *be a discrete valuation of* K *and suppose that* Λ *is* ν*-compatible as defined in the results in the second column of Table 27.2 (and modified in footnote 8). Then there exists a Bruhat-Tits pair* $\tilde{\mathsf{X}}_{\ell}(\Lambda, \nu)$ *whose building at infinity is* Δ *that satisfies 16.4 (with respect to the unique valuation* ϕ *defined in 13.8) with the following proviso: If* $\Delta = \mathsf{B}_{\ell}^{*}(\Lambda) = \mathsf{C}_{\ell}^{*}(\Lambda)$ *for* $* = \mathcal{Q}, \mathcal{I}$ *or* $\mathcal{P}$, *then* $\mathsf{X}_{\ell}^{*}(\Lambda, \nu)$ *exists, but only for one value of* X *in the set* $\{\mathsf{B}, \mathsf{C}\}$. *This value of* X *is determined in 19.22, 23.7 and 24.36 according to whether* Λ *is ramified at* ν *or not (as defined in 19.17, 23.2 and 24.30).*

Proof. These assertions hold by 14.47, 16.14, 19.21, 23.6, 24.35 and the results in the third column of Table 27.2. □

To conclude this chapter, we record several conclusions about complete Bruhat-Tits pairs.

Definition 27.3. Let Δ be a Bruhat-Tits building. We call $\Delta_{\mathcal{A}}^{\infty}$ the *complete building at infinity* of Δ if $\mathcal{A}$ is the complete system of apartments of Δ.

Theorem 27.4. *The complete building at infinity of a Bruhat-Tits building always satisfies the Moufang condition. Equivalently, every Bruhat-Tits building is completely Bruhat-Tits (as defined in 17.1).*

Proof. This holds by the results in the fifth column of Table 27.2. □

---

[8] If $\mathsf{X}_{\ell}^{*} = \mathsf{B}_{2}^{\mathcal{E}}$, then "$\nu$-compatible" is to be replaced by "exceptionally ν-compatible" as defined in 21.22.

[9] Note that by 30.29, this makes sense even in the few cases where K is not quite an invariant of $\Delta_{\mathcal{A}}^{\infty}$.

Theorem 27.5. *Let $\mathsf{X}_\ell^*(\Lambda)$ and K be as in 30.15 and let ν be a discrete valuation of K. Suppose that the parameter system Λ is ν-compatible (in the suitable sense that depends on X^*) and let*

$$(\Delta, \mathcal{A}) = \tilde{\mathsf{X}}_\ell^*(\Lambda, \nu).$$

Then the pair $(\Delta, \mathcal{A})$ is complete if and only if K is complete with respect to ν and the following hold:

(i) *If $\mathsf{X}_\ell^* = \mathsf{X}_\ell^{\mathcal{Q}}$ or $\mathsf{X}_\ell^{\mathcal{F}}$ for $\mathsf{X} = \mathsf{B}$ or C, so Λ is an anisotropic quadratic space (K, L, q), then L is complete with respect to the metric ∂_ν on L defined in 19.24.*
(ii) *If $\mathsf{X}_\ell^* = \mathsf{X}_\ell^{\mathcal{I}}$ for $\mathsf{X} = \mathsf{B}$ or C, so Λ is an involutory set (K, K_0, σ), then K_0 is closed in K (with respect to ν).*
(iii) *If $\mathsf{X}_\ell^* = \mathsf{X}_\ell^{\mathcal{P}}$ for $\mathsf{X} = \mathsf{B}$ or C, so Λ is a proper anisotropic pseudo-quadratic space*

$$(K, K_0, \sigma, L, q),$$

then the subset K_0 is closed in K (with respect to ν) and T is complete with respect to the metric ∂_ν, where T and ∂_ν are as in 24.38.
(iv) *If $\mathsf{X}_\ell^* = \mathsf{X}_2^{\mathcal{D}}$ for $\mathsf{X} = \mathsf{B}$ or C, so $\Lambda = (K, K_0, L_0)$ is an indifferent set, then both K_0 and L_0 are closed in K (with respect to ν).*
(v) *If $\mathsf{X}_\ell^* = \mathsf{X}_2^{\mathcal{F}}$ for $\mathsf{X} = \mathsf{B}$ or C, so Λ is a quadratic space of type F_4, then the subfield F of K defined in 14.3 of [36] is closed (with respect to ν).*
(vi) *If $\mathsf{X}_\ell^* = \mathsf{F}_4$ and Λ is a composition algebra (K, F) of type (i), then F is closed in K (with respect to ν).*
(vii) *If $\mathsf{X}_\ell^* = \mathsf{G}_2$, $\Lambda = (E/K)^\circ$ for some field extension E/K as defined in 15.20 of [36] and $\mathrm{char}(K) = 3$, then E^3 is closed in K (with respect to ν).*

Proof. These assertions hold by 17.9 and the descriptions of ϕ in the third column of Table 27.2. □

Theorem 27.6. *A Bruhat-Tits building is uniquely determined by its complete building at infinity (as defined in 27.3).*

Proof. Let Δ be a Bruhat-Tits building and let K be the defining field of its complete system of apartments (as defined in 30.15). By 27.5, K is complete with respect to ν. The claim holds, therefore, by 23.15 and 27.1.iii. □

Theorem 27.7. *Bruhat-Tits buildings are classified (modulo 21.23) by Moufang spherical buildings $\mathsf{X}_\ell^*(\Lambda)$ (in the notation of 30.15) such that*

(i) *the defining field K of $\mathsf{X}_\ell^*(\Lambda)$ is complete with respect to a discrete valuation ν and*
(ii) *the conditions in (i)–(vii) of 27.5 are met.*

If $\mathsf{X}_\ell^*(\Lambda)$ is algebraic (as defined in 30.31), then condition (i) implies condition (ii).

Proof. Let Δ be a Bruhat-Tits building. By 27.4 and the classification 30.14 of Moufang spherical buildings, its complete building at infinity (as defined in 27.3) is of the form $\mathsf{X}_\ell^*(\Lambda)$. By 27.6, Δ is uniquely determined by its complete building at infinity, and by 27.5, (i) and (ii) hold. Suppose, conversely, that the pair $(\mathsf{X}_\ell^*(\Lambda), \nu)$ satisfies (i) and (ii). By (i), 23.15 and 30.29, ν is uniquely determined by $\mathsf{X}_\ell^*(\Lambda)$, and by (i) as well as the results in the second column of Table 27.2, Λ is ν-compatible. By 27.2 and 27.5, there exists a Bruhat-Tits building whose complete building at infinity is $\mathsf{X}_\ell(\Lambda)$. The last claim holds by 26.25.ii. □

We recall that various results about the automorphism group of a Bruhat-Tits pair can be found in Chapter 18 and in the third part of Chapter 26. See especially 18.6 and 26.37–26.41

The structure of a maximal irreducible residue that is not a residue of a gem is determined in 18.31 when $\tilde{\Pi} = \tilde{\mathsf{E}}_7$ or $\tilde{\mathsf{E}}_8$, in 24.67 when $\tilde{\Pi} = \tilde{\mathsf{B}}_\ell$ for $\ell \geq 3$, in 25.33 when $\tilde{\Pi} = \tilde{\mathsf{G}}_2$ and in 26.17 when $\tilde{\Pi} = \tilde{\mathsf{F}}_4$; see also 18.19.

In the next chapter, we apply the results summarized in this chapter to work out the classification of locally finite Bruhat-Tits buildings.

Chapter Twenty Eight

Locally Finite Bruhat-Tits Buildings

A building is called *locally finite* if each chamber is adjacent to only finitely many other chambers.[1] In this chapter we apply the results summarized in the previous chapter to produce a list (in 28.33–28.38) of all locally finite Bruhat-Tits buildings and their principal features.

As we will see, almost all locally finite Bruhat-Tits buildings are algebraic as defined in 26.2.[2] That is to say, they are the affine buildings associated with pairs (G, F), where F is a local field (as defined in 28.1) and G is an absolutely simple algebraic group of F-rank $\ell \geq 2$. These are precisely the pairs whose classification (for arbitrary $\ell \geq 0$) is described in Tits's Corvallis notes [33] and summarized (for $\ell \geq 2$) in Tables 28.1–28.3.

As we produce our list of locally finite Bruhat-Tits buildings in 28.33–28.38, we also correlate our findings with the results in [33] and at the end of this chapter, we summarize our conclusions in Tables 28.4–28.6.

Suppose that $(\Delta, \mathcal{A})$ is a complete Bruhat-Tits pair such that the building Δ is locally finite and let K be the defining field of $\Delta_{\mathcal{A}}^{\infty}$. By 26.1, the pair $(\Delta, \mathcal{A})$ is of the form $\tilde{\mathsf{X}}_{\ell}^{*}(\Lambda, \nu)$, and by 27.5, K is complete with respect to the valuation ν. By 1.7, all the proper residues of Δ are spherical buildings. Since all the panels of Δ are finite, it follows that all the proper residues of Δ are, in fact, finite.

By the results in the last column of Table 27.2, there are panels of Δ on which the additive group of the residue field $\bar{K}$ of K acts faithfully.[3] Therefore $\bar{K}$ is finite. We conclude that K is a local field:

Definition 28.1. A *local field* is a field, a skew field or an octonion division algebra K having a discrete valuation ν with respect to which K is complete and the residue field $\bar{K}$ is finite.[4]

In the first part of this chapter, 28.2–28.27, we assemble the results about local fields and the algebraic structures defined over local fields — anisotropic quadratic spaces, involutory sets, hexagonal systems, etc. — that are needed to describe the classification of locally finite Bruhat-Tits buildings. A good number of these results are well-known facts about local fields.

[1] Since by definition each chamber is contained in only finitely many panels, this is equivalent to saying that all the panels are finite.

[2] The only exceptions are three small families of locally finite Bruhat-Tits buildings of mixed type as defined in 26.4.

[3] To see this when $(\Delta, \mathcal{A}) = \tilde{\mathsf{B}}_2^{\mathcal{D}}(\Lambda, \nu)$ for some indifferent set $\Lambda = (K, K_0, L_0)$, we need to observe that $K^2 \subset L_0$ and hence $\bar{K}^2 \subset \bar{L}_0$ by 10.2 in [36].

[4] Note that by 23.15, the valuation ν in 28.1 is unique. See also 28.4.i.

We begin with the structure of K itself:

Theorem 28.2. *If K is a commutative local field, then one of the following holds:*

(i) *K is a finite extension of a p-adic field $\mathbb{Q}_p$ for some prime p.*
(ii) *K is the field $D((t))$ of Laurent series over a finite field D.*

Proof. This is proved in Chapter 2, Sections 4 and 5, of [28]. □

Remark 28.3. If K is a commutative local field of characteristic $p > 0$, then by 28.2, K/K^p is an extension of degree p.

Theorem 28.4. *Suppose that K is a non-commutative local field with valuation ν. Let F be its center. Then the following hold:*

(i) *K is not octonion.*
(ii) *The residue field $\bar{K}$ is commutative.*
(iii) *K is of finite dimension n^2 over F.*
(iv) *$\nu(F^*) = n\mathbb{Z}$ and the degree of the extension $\bar{K}/\bar{F}$ is n.*
(v) *If $n > 2$, then K does not have any involutions, and if $n = 2$, then every involution of K acts trivially on F.*

Proof. There are no inseparable finite extensions of finite fields, no finite non-commutative fields and no finite octonion division algebras.[5] By 26.15, therefore, (i) and (ii) hold.

By Theorem 56.12 in [26],[6] K is locally compact with respect to the metric in 9.18.ii. By Theorem 58.9 in [26], a locally compact skew field is finite dimensional over its center. Thus (iii) holds. Hence (iv) holds by Theorem 14.3 in [24] and (v) holds by Theorem 2.2 in Chapter 10 of [27]. □

Our proof of 28.4.iii is due to T. Grundhöfer. He also observed that 28.4.iii can be shown to follow from Propositon 3.1.4 in [12].

Remark 28.5. Let F be a commutative local field, let $n > 1$ and let ϕ be the Euler function. By Theorem 31.8 in [24], there are exactly $\phi(n)$ isomorphism types of skew fields of degree n over F and they are all cyclic. In particular, there is a unique quaternion division algebra over F.

Remark 28.6. Let E/K be a separable extension of finite degree n and let T be trace of this extension. If E is a local field with valuation ω, then ω is invariant under every automorphism of E (by 23.15), and since $K = T(E)$, it follows that K is closed with respect to ω and hence complete. By 17.14.ii, it follows that E is a local field if and only if K is a local field.

Remark 28.7. Let K be a commutative local field. By Theorem 2 in Chapter 3, Section 5, of [28], there exists a unique field E containing K such

[5] The non-existence of finite octonion division algebras was first proved by E. Artin; see, for example, 34.5 in [36].

[6] Theorem 56.12 in [26] is stated only for commutative fields, but the proof is valid also for skew fields.

that E/K is an unramified separable quadratic extension (unique up to a K-linear isomorphism). In general, there are, however, many fields E such that E/F is a ramified separable quadratic extension over F.

Remark 28.8. Let K be a commutative local field and let M be the unique quaternion algebra over K. As observed in the Appendix to Chapter VI, Section 1, of [11], it follows from Theorem 3 in Chapter VI, Section 1.1, of [11] that every field E containing K such that E/K is a quadratic extension is K-linearly isomorphic to a subfield of M containing K.

Proposition 28.9. *Suppose that K is a local quaternion division algebra with center F, let $N\colon K \to F$ be its reduced norm, let (E, e) be a pair as in 26.22.i and let $\beta = -N(e)$. Then $F = N(E) \cup \beta N(E)$. In particular, N is surjective.*

Proof. For each $\gamma \in F^*$, let (E, γ) be the quaternion algebra defined in 9.3 of [36]. By 20.20 in [36], $(E, \beta) \cong K$. Now suppose that γ is an element of F not in $N(E) \cup \beta N(E)$. By 9.4 in [36], (E, γ) and $(E, \beta/\gamma)$ are both division algebras. By 23.14.ii, F is a local field, so by 28.5, there is only one quaternion division algebra over F up to isomorphism. Hence (E, γ) and $(E, \beta/\gamma)$ are isomorphic. By 9.7 in [36], it follows that $\beta = \gamma \cdot \beta/\gamma \in N(E)$. Thus 9.4 in [36] implies that (E, β) is not a division algebra. This contradicts, however, our observation that $(E, \beta) \cong K$. We conclude that $F = N(E) \cup \beta N(E)$. Since $N(u + ev) = N(u) - \beta N(v)$ for all $u, v \in E$ (by 26.22.i), it follows that N is surjective. □

Proposition 28.10. *Let K and N be as in 28.9 and let ν be the unique valuation of K. Then $\nu(u) = \nu(N(u))/2$ for all $u \in K^*$.*

Proof. This holds by 30.18 since ν is invariant with respect to the standard involution of K (by 23.15). □

Proposition 28.11. *Let $\Lambda = (K, L, q)$ be a unitary anisotropic quadratic space and suppose that K is a local field. Let $m = \dim_K L$ and for each i, let*

$$\bar{\Lambda}_i = (\bar{K}, \bar{L}_i, \bar{q}_i)$$

be as in 19.33 and let $m_i = \dim_{\bar{K}} \bar{L}_i$. Then the following hold:

(i) *$m_i > 0$ for all even i.*
(ii) *$m_i \le 2$ for all i.*
(iii) *$m = m_0 + m_1 \le 4$.*
(iv) *$m_1 > 0$ if and only if Λ is ramified.*[7]
(v) *Λ is unramified if $m = 1$ and ramified if $m = 3$ or 4.*
(vi) *If $m = 2$, then there is a quadratic extension E/K with norm N such that $\Lambda \cong (K, E, N)$. If E/K is inseparable, then $q(L) = K$ and Λ is*

[7] Properly speaking, we should say that Λ is ramified *at* ν, where ν is the unique valuation of K. Because of the uniqueness of ν, however, we will generally omit such references to ν in this chapter.

ramified. If E/K is separable, then Λ is ramified if and only if the extension E/K is ramified.

(vii) *If $m = 3$ or 4, then the bilinear form associated with q is not identically zero.*

(viii) *Λ is not of type E_6, E_7 or E_8 nor is Λ of type F_4.*

Proof. Since Λ is unitary, (i) holds; (ii) holds by 34.3 in [36]. By 19.36, we have $m = m_0 + m_1$, so (iii) follows from (ii); (iv) holds by 19.17 and (v) is a consequence of (i)–(iv). Suppose $m = 2$. By 34.2.ii in [36], there exists E such that E/K is a quadratic extension with norm N such that $\Lambda \cong (K, E, N)$. If E/K is inseparable, then $N(x) = x^2$ for each $x \in E$. By 28.3, $E = K^2$, so $N(E) = K$. Thus $q(L) = K$, so Λ is ramified. By 19.17 (and its footnote), therefore, (vi) holds. Suppose that $m \geq 2$ and that the bilinear form associated with q is identically zero. By (vi), there exists a two-dimensional subspace E of L such that $q(E) = K$. This implies that $m = 2$ (since q is anisotropic). Thus (vii) holds. Quadratic forms of type E_6, E_7, E_8 and F_4 are anisotropic and have dimension greater than 4. By (iii), therefore, (viii) holds. □

Remark 28.12. If K is a local field and M is the unique quaternion division algebra over K, then, up to similarity, (K, M, N) is the unique anisotropic quadratic space of dimension four, where N is the reduced norm of M. This is proved in 4.1 in Chapter 6 of [27] when $\mathrm{char}(K) \neq 2$, but the proof is valid also when $\mathrm{char}(K) = 2$ since there is only one two-dimensional anisotropic quadratic space over $\bar{K}$ and it is non-degenerate.

Remark 28.13. Suppose that K is a local field and that $\Lambda = (K, L, q)$ is an anisotropic quadratic space of dimension 3. By 28.11.vii, the bilinear form f associated with q is not identically zero. We can therefore choose a subspace E of L of dimension 2 such that the restriction q_E of q to E is non-degenerate. Let $u \in E^*$. After replacing q by $q/q(u)$, we can assume that E has the structure of a field containing F such that q_E is the norm of this extension (by 28.11.vi). Let e be a non-zero element of L perpendicular to E with respect to f and let $\beta = -q(e)$. Then $\beta \notin q_E(E)$ because q is anisotropic. Thus $M := (E, \beta)$ is a quaternion division algebra, where (E, β) is as in the proof of 28.9. Moreover, M contains $L = \langle E, e \rangle$ as a K-subspace and the restriction N_L of the reduced norm N of M to L is precisely q. Now suppose that $\Lambda_1 = (K, L_1, q_1)$ is a second anisotropic quadratic space of dimension 3. Since there is only one quaternion algebra over K (by 28.5), it follows that there is a K-subspace of M that we can identify with L_1 so that restriction N_{L_1} of N to L_1 is similar to q_1. Since the reduced trace T of M is non-degenerate, we can choose non-zero elements $z, z_1 \in M$ such that $z^\perp = L$ and $z_1^\perp = L_1$ (where $\perp$ is as in 26.22). Let $s = z_1 z^{-1}$. Since $T(su, z_1) = T(su, sz) = N(s)T(u, z)$ for all $u \in M$, we have $L_1 = sL$. Thus the map $x \mapsto sx$ is a K-linear similarity from the quadratic space (K, L, N_L) to the quadratic space (K, L_1, N_{L_1}). We conclude that (K, L, q) is unique up to similarity.

Theorem 28.14. *There do not exist any proper indifferent sets* (K, K_0, L_0) *for* K *a local field.*

Proof. Let $\Lambda = (K, K_0, L_0)$ be an indifferent set (as defined in 10.1 in [36]), let $F = K^2$ and let L be the subring of K generated by L_0. By 10.2 in [36],

$$F \subset L_0 \subset L \subset K_0 \subset K$$

and $FK_0 \subset K_0$. Thus, in particular, K_0 is an F-subspace of K containing F. Now suppose that K is a local field. By 28.3, K/F is a quadratic extension. Therefore either $K_0 = K$ or $F = L_0 = L = K_0$. Thus Λ is not proper (as defined in 38.8 of [36]). □

Theorem 28.15. *Let* $\Lambda = (K, K_0, \sigma)$ *be an involutory set with* $\sigma \neq 1$ *and suppose that* K *is a local field. Then exactly one of the following holds:*

(i) Λ *is a quadratic involutory set of type (iii) as defined in 30.17, i.e.* K *is commutative,* $F := K_0$ *is a subfield of* K *and* K/F *is a separable quadratic extension. Furthermore,* Λ *is ramified (as defined in 23.2) if and only if the extension* K/F *is ramified.*

(ii) Λ *is a ramified quadratic involutory set of type (iv).*

(iii) K *is a quaternion division algebra,* $K_0 = K^\sigma$ *(as defined in 23.19),* K_0 *contains the center* F *of* K*,* K_0 *is a vector space over* F *of dimension three and* Λ *is both proper and unramified.*

Proof. Let ν be the valuation of K. If K is commutative, then (K, K_0, σ) is a quadratic involutory set of type (iii) by 11.3 in [36], and by 23.2, the field extension E/F is ramified if and only if the involutory set Λ is ramified. Thus (i) holds.

Suppose now that K is not commutative and let F denote its center. By 28.4.v, K is a quaternion division algebra and σ acts trivially on F. Suppose that $F = K_0$. By 11.1.i in [36], $K_\sigma \subset F$, and by 11.2 in [36], $K_\sigma = F$ if $\mathrm{char}(K) \neq 2$. By 26.22.iv, it follows that σ is the standard involution of K. By 28.4.iv, $\nu(F^*) = 2\mathbb{Z}$. Hence Λ is ramified. Thus (ii) holds.

Suppose that $K_0 \neq F$. If $\mathrm{char}(K) \neq 2$, then $K_0 = K_\sigma = K^\sigma$ (by 11.2 in [36]). Suppose that $\mathrm{char}(K) = 2$. Suppose, too, that σ is the standard involution of K. Let E, e and β be as in 26.22.i. By 28.2, we have

$$E = \{a^2 + \beta b^2 \mid a, b \in E\}.\text{[8]} \tag{28.16}$$

As in 23.16, we consider $V := K^\sigma/K_\sigma$ as a right vector space of K with scalar multiplication given by 23.20. By (i) and (ii) of 26.22, we have

$$(a + eb)^\sigma e(a + eb) \equiv e(a^2 + \beta b^2) \pmod{K_\sigma}$$

for all $a, b \in E$. By 28.16, therefore, $\dim_K V = 1$. Since $K_0 \neq F = K_\sigma$, we conclude that $K_0 = K^\sigma$. By 26.22.v, it follows that $K_0 = K^\sigma$ even if σ is not the standard involution.

---

[8]Here are the details. By 28.2, $F = D((t))$ for some finite subfield D and some $t \in F$. Because the reduced norm of K is anisotropic, we have $\beta \notin F^2$. Hence there exist $u, v \in F$ such that $u^2 + \beta v^2 = t +$ terms of degree > 1. It follows that $F = \{x^2 + \beta y^2 \mid x, y \in F\}$. Since E/F is separable, we can choose $c \in E$ such that $c^2 \notin F$ and hence $E = F + c^2F$.

Now suppose that $\mathrm{char}(K)$ is arbitrary and let N be the reduced norm of K. By 26.22.ii and 26.22.iv, $\dim_F K^\sigma = \dim_F K_0 = 3$. By 23.23 in [36], Λ is proper. By 28.11.v, the anisotropic quadratic space (F, K_0, N) is ramified. By 28.4.iv, the restriction of $\nu/2$ to F is the unique valuation of F. Hence $\nu(N(K_0^*))/2 = \mathbb{Z}$ (by 19.17). By 28.10, therefore, $\nu(K_0^*) = \mathbb{Z}$. Hence Λ is unramified (by 23.2). Thus (iii) holds. □

Proposition 28.17. *Let K be a local quaternion division algebra with valuation ν, let F be the center of K, let σ be the standard involution of K, let E be a subfield of K containing F such that E/F is a separable quadratic extension and let $e \in K^* \cap E^\perp$. Then the following hold:*

(i) *If E/F is ramified, then $\nu(E^*) = \mathbb{Z}$.*
(ii) *If E/F is unramified, then $\nu(E^*) = 2\mathbb{Z}$ and $\nu(ev)$ is odd for all $v \in E^*$.*

Proof. By 28.4.iv, $\nu(F^*) = 2\mathbb{Z}$. Therefore $\nu(E^*) = \delta\mathbb{Z}$ for $\delta = 1$ or 2 and $\nu(F^*)/\delta = (2/\delta)\mathbb{Z}$. The restriction of ν/δ to E is the unique valuation of E. By 23.2, therefore, E/F is ramified if and only if $2/\delta = 2$. Suppose that E/F is unramified, so $\delta = 2$, and let N be the reduced norm of K. By 28.10, we have $\nu(N(E^*)) = 4\mathbb{Z}$. Since $\nu(F^*) = 2\mathbb{Z}$, it follows by 28.9 that $\nu(N(e))$ is twice an odd integer. Therefore $\nu(ev)$ is odd for all $v \in E^*$ (again by 28.10). □

Proposition 28.18. *Let K, F, ν and σ be as in 28.17, let E be a subfield of K containing F such that E/F is an unramified separable quadratic extension, let $e \in K^* \cap E^\perp$, let τ be the involution of K given by*

$$(u + ev)^\tau = u - ev^\sigma$$

for all $u, v \in E$, let $K_0 = K^\tau$ and let $\Lambda = (K, K_0, \tau)$. Suppose that $\mathrm{char}(K) \neq 2$. Then the following hold:

(i) *The residue $\bar{\Lambda}$ of Λ is a quadratic involutory set of type (ii).*
(ii) *There exists an element $\pi \in K_0^*$ such that $\nu(\pi) = 1$ and the residue $\overline{\Lambda_\pi}$ of the translate Λ_π of Λ with respect to π is a quadratic involutory set of type (iii).*

Proof. By 28.4.iv, $\bar{K}/\bar{F}$ is a quadratic extension. Since E/F is unramified, also $\bar{E}/\bar{F}$ is a quadratic extension. Since $E \subset K^\tau = K_0$, it follows that $\bar{E} = \bar{K}_0 = \bar{K}$. Thus (i) holds.

By 28.17.ii, $\nu(ev)$ is odd for all $v \in E^*$. Moreover, $eE = E^\perp$ and $\dim_F K_0 = 3$. We can thus choose $w \in E$ such that $ew \in K_0^* \cap E^\perp$ and $\nu(ew) = 1$. Let $\beta = (ew)^2 = -N(ew)$ and let $\rho = \tau\iota_{ew}$, where ι is as in 26.22. Thus (K, K^ρ, ρ) is the translate of Λ with respect to ew. We have

$$\begin{aligned} u^\rho &= ew \cdot u \cdot ew/\beta \\ &= u^\sigma \end{aligned}$$

for all $u \in E$ (by 26.22.i). By 26.15, $\bar{\sigma}$ acts non-trivially on $\bar{K}$. Since $\bar{E} = \bar{K}$, we conclude that $\bar{\rho} \neq 1$. By 11.3 in [36], it follows that $\bar{\Lambda}$ is a quadratic involutory set of type (iii). Thus (ii) holds with $\pi = ew$. □

Proposition 28.19. *Let K, F, ν and σ be as in 28.17, let $K_0 = K^\sigma$ and let $\Lambda = (K, K_0, \sigma)$. Suppose that* $\text{char}(K) = 2$. *Then the following hold:*

(i) *The residue $\bar{\Lambda}$ of Λ is a quadratic involutory set of type (iii).*
(ii) *There exists an element $\pi \in K_0^*$ such that $\nu(\pi) = 1$ and the residue $\overline{\Lambda_\pi}$ of the translate Λ_π of Λ with respect to π is a quadratic involutory set of type (ii).*

Proof. By 26.15 and 28.4.iv, $(\bar{K}, \bar{F}, \bar{\sigma})$ is a quadratic involutory set of type (iii). In particular, $\bar{\sigma} \neq 1$. Since $K_0 = K^\sigma$, $\bar{K}_0$ is contained in the set of fixed points of $\bar{\sigma}$. Hence $\bar{K}_0 = \bar{F}$. Thus (i) holds.

By 28.7 and 28.8, there exists a subfield E of K containing F such that E/F is an unramified quadratic extension. By 26.22.i and 28.17.ii, we can choose e in $K^* \cap E^\perp$ such that $\nu(e) = 1$. By 26.22.ii, $K_0 = K^\sigma = F + eE$. The translate Λ_e of Λ with respect to e is $(K, eK_0, \sigma\iota_e)$, where ι is as in 26.22. Since $E \subset eK_0$ and $\bar{E} = \bar{K}$, it follows that $\overline{\Lambda_e}$ is of type (ii). Thus (ii) holds with $\pi = e$. □

Theorem 28.20. *Let*

$$\Lambda = (K, K_0, \sigma, L, q)$$

be an anisotropic pseudo-quadratic space and let $m := \dim_K L$. Suppose that Λ is proper (as defined in 35.5 in [36]) and that K is a local field. Then one of the following holds:

(i) *(K, K_0, σ) is a quadratic involutory set of type (iii), $m \leq 2$ and Λ is unramified.*
(ii) *(K, K_0, σ) is a quadratic involutory set of type (iv), $m \leq 3$ and Λ is unramified if $m = 2$ or 3.*
(iii) *(K, K_0, σ) is as in case (iii) of 28.15, $m = 1$ and Λ is unramified.*

Proof. Let ν be the valuation of K. Since Λ is proper, σ is non-trivial. We can thus apply 28.15. Suppose that $K_0 = K^\sigma$ (as defined in 23.19), let $d = m$ if $m < \infty$ and let d be an arbitrary positive integer otherwise. If $v \in L^*$, then $q(v) \notin K_0$ (by 11.16.iii in [36]) and hence $f(v, v) \neq 0$ by 11.19 in [36]. Hence we can choose non-zero vectors $v_1, v_2, \ldots, v_d$ that are pairwise orthogonal with respect to f. By 11.16 in [36], there exist scalars $\alpha_1, \ldots, \alpha_d \in K \backslash K_0$ such that

$$q(v_1 t_1 + \cdots + v_d t_d) \equiv t_1^\sigma \alpha_1 t_1 + \cdots + t_d^\sigma \alpha_d t_d \pmod{K_0}. \tag{28.21}$$

Suppose now that (K, K_0, σ) is as in case (i) of 28.15. Thus, in particular, $K_0 = K^\sigma$. Let $F = K_0$, let N be the norm of the extension K/F, let α be an arbitrary element of K not in F if K/F is unramified and let α be an element of K^* such that $\nu(\alpha) = 1$ if K/F is ramified. Since $\nu(F^*) = 2\mathbb{Z}$ if K/F is ramified, we have $\alpha \notin F$ in both cases. We can thus choose elements $s_1, w_1, \ldots, s_d, w_d \in F$ such that $\alpha_i = \alpha s_i + w_i$ for each $i \in [1, d]$. By 28.21, therefore,

$$q(v_1 t_1 + \cdots + v_d t_d) \equiv \alpha\big(s_1 N(t_1) + \cdots + s_d N(t_d)\big) \pmod{F} \tag{28.22}$$

for all $t_1, \dots, t_d \in K$. Since $q(v) \notin K_0 = F$ for all $v \in L^*$, it follows that the quadratic space (F, V, Q) is anisotropic, where $V = E^d$ and

$$Q(t_1, \dots, t_m) = s_1 N(t_1) + \cdots + s_d N(t_d)$$

for all $(t_1, \dots, t_d) \in V$. By 28.11.iii, $d \le 2$. Therefore $m = d \le 2$ by the choice of d. By 24.30, Λ is unramified if (K, F, σ) is unramified. If (K, F, σ) is ramified, then $\nu(\alpha s_1)$ is odd by the choice of α. Since $(v_1, \alpha s_1)$ is an element of the group T^* defined in 11.24 of [36], it follows that Λ is unramified also in this case (by 24.30). Thus (i) holds.

Suppose next that (K, K_0, σ) is as in case (ii) of 28.15, so (K, K_0, σ) is ramified (but $K_0 = K^\sigma$ only if the characteristic of K is not 2). Let $F = K_0$ and let $\bar{L}_i$ for $i \in \mathbb{Z}$ be as in 24.51. An anisotropic pseudo-quadratic space over a finite field with non-trivial involution cannot have dimension greater than 1; this is shown in the middle of page 378 in [36]. Since $\bar{\sigma} \ne 1$ (by 26.15) and $\bar{\Lambda}$ as defined in 24.53 is an anisotropic pseudo-quadratic space, it follows that $\dim_{\bar{K}} \bar{L}_0 \le 1$. Thus $m \le 1$ if Λ is ramified by 24.59.v. Suppose that Λ is unramified. By 34.3 in [36] and 24.55, $\dim_{\bar{K}} \bar{L}_1 \le 2$. Hence $m \le 3$ by 24.59.iv. Thus (ii) holds.

Suppose, finally, that (K, K_0, σ) is as in case (iii) of 28.15. Then $K_0 = K^\sigma$ again, so 28.21 holds. By the formulas in (ii) and (iv) of 26.22 and some calculation, we have

$$t^\sigma \alpha_1 t \equiv N(t)\alpha_1 \pmod{K_0} \tag{28.23}$$

for all $t \in K$, where N denotes the reduced norm of K. By 28.9, N is surjective. By 28.21 and 28.23 (and the hypothesis that Λ is anisotropic as defined in 11.16.iii in [36]), it follows that $m = 1$. By 28.15.iii, (K, K_0, σ) is unramified. By 24.30, it follows that Λ is unramified. Thus (iii) holds. □

A completely different proof of the bound $m \le 3$ in 28.20.ii for the case that $\mathrm{char}(K) \ne 2$ can be found in Theorem 3.6.ii in Chapter 10 of [27].

Remark 28.24. Suppose that $\Lambda = (K, K_0, \sigma, L, q)$ and m are as in case (i) of 28.20 and let $F = K_0$. Suppose, too, that the extension K/F is unramified and that $m = 2$. By 28.22 (where now $d = 2$), the vector space L can be endowed with the structure of a quaternion division algebra over F containing K as a subfield so that

$$q(x) \equiv \beta N(x) \pmod{F}$$

for some $\beta \in K \backslash F$ and for all $x \in L$, where N is the reduced norm of this quaternion division algebra. Since a pseudo-quadratic form is really only defined modulo K_0, we can assume that, in fact, $q(x) = \beta N(x)$ for all $x \in L$. Since K/F is unramified, we have $\nu(F^*) = \mathbb{Z}$ (by 23.2).[9] By 28.9, we have $F^* = N(L^*)$. We can therefore choose $w \in L^*$ such that $\nu(N(w)) = 1$. Let $\psi(x) = wx$ for all $x \in L$. Then ψ is a K-linear automorphism of L such that $q(\psi(x)) = N(w)q(x)$ for all $x \in L$. This means that we will be able

[9] Note that if ω is the unique valuation of L, then by 28.17.ii, its restriction to K is 2ν, not ν.

to apply 24.58.iv in this case when we want to determine the gems of the corresponding affine building (in 28.36).

Proposition 28.25. *Let*

$$\Lambda = (K, K_0, \sigma, L, q)$$

be a proper anisotropic pseudo-quadratic space and suppose that the residue $(\bar{K}, \bar{K}_0, \bar{\sigma})$ *is a quadratic involutory set of type (k). Let* $\bar{L} = \bar{L}_0$ *be as in 24.51. Then the following hold:*

(i) *If k equals ii, then* $\bar{L} = 0$.
(ii) *If k equals iii and* Λ *is as in case (i) or (iii) of 28.20, then* $\bar{L} \neq 0$.

Proof. Let T_0, T_1 and Z_0 be as in 24.50. Suppose that k equals ii and let $(a,t) \in T_0 \backslash T_1$. Since $\bar{K} = \bar{K}_0$, there exists $s \in K_0^*$ such that $\nu(s) = 0$ and $\nu(t - s) \geq 1$. Therefore

$$(a, t) = (a, t - s) + (0, s) \in T_1 + Z_0.$$

Thus (i) holds.

Now suppose that k equals iii and that Λ is as in case (i) or (iii) of 28.20 (so $K_0 = K^\sigma$). If Λ is as in case (i) of 28.20, let $F = K_0$ and let N be the norm of the extension K/F. If Λ is as in case (iii) of 28.20, let F be the center of K and let N be the reduced norm of K. In both cases N is surjective (by 28.9). It follows from 28.21–28.23 that for each $t \in K$, there exists an element $a \in L$ such that $(a,t) \in T$. Since k equals iii, we have $\bar{K}_0 = \bar{F}$. Hence we can choose $(a,t) \in T$ such that $\nu(t) = 0$ and $\bar{t} \notin \bar{K}_0$ and hence $(a, t) \notin T_1 + Z_0$. Thus (ii) holds. □

Proposition 28.26. *Let* $\Lambda = (K, F)$ *be a composition algebra and suppose that* K *is a local field. Then the following hold:*

(i) Λ *is not of type (v).*
(ii) *If* Λ *is of type (i), then* $F = K^2$.
(iii) F *is also a local field.*
(iv) *If* Λ *is of type (i), (ii) or (iv) as defined in 30.17, then* Λ *is uniquely determined by* F.

Proof. By 28.4.i, (i) holds. Suppose that Λ is of type (i). By 28.3, K/K^2 is a quadratic extension. Since $K^2 \subset F$, it follows that $F = K^2$. Thus (ii) holds, F is a local field and Λ is uniquely determined by F. If Λ is of type (iii), then F is a local field by 28.6 and if Λ is of type (iv), then F is a local field by 23.14.ii. Thus (iii) holds. By 28.5, we conclude that (iv) holds; see also 28.7. □

Theorem 28.27. *Let*

$$\Lambda = (J, K, N, \#, T, \times, 1)$$

be an hexagonal system and suppose that K *is a local field. Then there exists a local field* E *containing* K *such that exactly one of the following holds:*

(i) *Either* $E = K$ *or* $\text{char}(K) = 3$ *and* $K = E^3$ *and (in both cases)* $\Lambda \cong (E/K)^\circ$*; or*
(ii) E/K *is a separable cubic extension and* $\Lambda \cong (E/K)^+$*; or*
(iii) E *is a skew field of degree* 3 *over* K *(so the dimension of* E *over* K *is* 9*) and* $\Lambda \cong E^+$,

where $(E/K)^\circ$, $(E/K)^+$ *and* E^+ *are as defined in 15.20–15.22 of [36].*

Proof. By 28.3, if $\text{char}(K) = 3$ and E is a field containing K properly such that $E^3 \subset K$, then, in fact, $E^3 = K$. By 28.5, there are two skew fields of degree 3 over K, but by 28.4.v, neither of them has an involution. Thus Λ cannot be of type $9^*/K$ as defined in 15.31 of [36]. Suppose that $\dim_K J = 27$ and let $\bar{\Lambda}$ and $\bar{J}$ be as in 25.28. Then $\bar{\Lambda}$ is an hexagonal system and by (i) and (ii) of 25.30, $\dim_{\bar{K}} \bar{J}$ is either 9 or 27. By 15.20 in [36], the only hexagonal system of type $1/\bar{K}$ is one-dimensional (since $\bar{K}$ is perfect). By 30.5 in [36], therefore, $\bar{T}$ is not identically zero. By 30.6 and 30.17 (as well as 15.22 and 15.31) in [36], it follows that there exists a skew field of degree 3 over $\bar{K}$. Since there are no finite skew fields, $\dim_K J$ cannot, in fact, be 27. By 17.6 in [36], we conclude that Λ is as in (i), (ii) or (iii). □

* * *

This concludes the first part of this chapter. Equipped with the results 28.2–28.27 and the classification 30.14, we can now make a list of all Bruhat-Tits buildings whose building at infinity is defined over a local field. It will turn out that all of these Bruhat-Tits buildings are, in fact, locally finite. As we make the list, we also determine the structure of the gems (and thus, implicitly, of all the irreducible residues) and correlate our findings with the classification of absolutely simple algebraic groups defined over a local field given in [33]. More precisely, we determine where (and whether) each family of locally finite Bruhat-Tits buildings belongs in Tables 4.2 and 4.3 in [33].[10]

Each row in these tables is uniquely specified by the "name" that appears in the first column. This name refers to the *local index* of a family of forms of algebraic groups, *local* because it refers to algebraic groups defined over fields that are local in the sense of 28.1. Given the subject of this book, we prefer to use the term *affine index* in place of local index.

Let G be an absolutely simple algebraic group defined over a local field F and let F_1 be the maximal unramified extension of F. The affine index of the pair (G, F) is the triple (Ω_1, Θ, S), where Ω_1 is the *absolute local Dynkin diagram* of the given algebraic group (as defined in Section 1.11 of [33][11]), Θ is the subgroup of $\text{Aut}(\Omega_1)$ induced by $\text{Gal}(F_1/F)$ and S is the vertex set of Ω_1. In the tables in [33], Ω_1 is the first entry in the next-to-last column. The diagram Ω_1 is drawn according to the convention that vertices of Ω_1

[10] When we refer to the tables in [33], we will always ignore those rows in which the affine Coxeter diagram has fewer than three vertices. In other words, we ignore those algebraic groups whose relative rank (the second subscript in the index) is less than 2.

[11] This diagram is called Δ_1 in [33]; we prefer to reserve the letter Δ in this book for buildings.

are in the same Θ-orbit if and only if they are drawn "near" and one above the other. The order of Θ is the superscript (to the left) in the name of the index unless the order of Θ is 1, in which case there is no superscript.[12] The indices where Θ is trivial are precisely those in Table 4.2 in [33].

Let $\Omega = \Omega_1$ if Ω_1 is in Table 4.2 in [33] and let Ω be the diagram directly below Ω_1 if Ω_1 is in Table 4.3 in [33]. In each case, there is a visible correspondence between the orbits of Θ and the vertices of the diagram Ω. The diagram Ω is called the *relative local Dynkin diagram* of the pair (G, F).[13]

Let Π_1 denote Ω_1 stripped of the arrows on its multiple edges (so $\Pi_1 = \Omega_1$ if Ω_1 has no multiple edges) and let Π denote the diagram Ω stripped of the arrows on its multiple edges (and all its other labels). Then Π_1 is the Coxeter diagram of the Bruhat-Tits building Δ_1 corresponding to the pair (G, F_1) and Π is the Coxeter diagram of the Bruhat-Tits building Δ associated with the pair (G, F).

The entry in the last column of the tables in [33] is the *index* of (G, F) as defined in Tits's Boulder notes [31] (and in the notation introduced there). We will refer to the index as the *spherical index* of (G, F) since it is really an invariant of the spherical building $\Delta_{\mathcal{A}}^{\infty}$.

We have reproduced most of Tables 4.2 and 4.3 of [33] in Tables 28.1–28.3 below:[14] We have omitted only the second column of Table 4.3 of [33] and the rows of Tables 4.2 and 4.3 of [33] where the Coxeter diagram Π has fewer than three vertices. We have also applied 28.32 below, so ℓ always denotes the number of vertices of the Coxeter diagram Π minus 1. The vertices of the Coxeter diagram Π that are special in the sense of 1.1 are labeled either "s" (for *special*) or "hs" (for *hyperspecial*) in Tables 4.2 and 4.3 of [33]. In our tables we have included the "hs" but omitted the "s" since it is clear by inspection which vertices are special. All the other "decorations" of the Coxeter diagram Π have been reproduced faithfully as well as the arrows of both Ω and Ω_1.

We will say a few words about the meaning of these crosses, arrows and numbers as well as the label "hs" in 28.40 below.

Note 28.28. The affine index A_ℓ in Table 28.1 is the triple (Ω_1, Θ, S), where Ω_1 is the Coxeter diagram $\tilde{\mathsf{A}}_\ell$ and $\Theta = 1$. The affine index ${}^d\mathsf{A}_{(\ell+1)d-1}$ in Table 28.2 is the triple (Ω_1, Θ, S), where Ω_1 is the Coxeter diagram $\tilde{\mathsf{A}}_{(\ell+1)d-1}$ and Θ is the group of rotations of Ω_1 of order d. The Coxeter diagram Π is $\tilde{\mathsf{A}}_\ell$ in both cases. In the first case, every vertex of Π is hyperspecial and in the second case every vertex of Π is labeled with a "d."

[12] It would be more consistent with the notation in [31] to add ℓ as a second subscript to the name of each affine index to indicate the number of orbits of Θ minus 1. It also might make sense to put a tilde over the name of an affine index to distinguish it more clearly from the index of the corresponding pair (G, F) in the last column of the tables in [33]. Neither of these modifications is necessary, however, and we do not make them.

[13] In Table 4.2 in [33], where the relative and absolute local Dynkin diagrams coincide, the adjectives "relative" and "absolute" are omitted.

[14] Tits's Corvallis notes [33] (as well as Tits's Boulder notes [31]) can be downloaded from the AMS website.

Name	*Affine Index*	*Spherical Index*
A_ℓ, $\ell \geq 2$	See 28.28.	${}^1\mathsf{A}^{(1)}_{\ell,\ell}$
B_ℓ, $\ell \geq 3$	 hs hs	$\mathsf{B}_{\ell,\ell}$
B-C_ℓ, $\ell \geq 3$		${}^2\mathsf{A}^{(1)}_{2\ell-1,\ell}$
C_ℓ, $\ell \geq 2$	 hs hs	$\mathsf{C}^{(1)}_{\ell,\ell}$
C-B_ℓ, $\ell \geq 2$		${}^2\mathsf{D}^{(1)}_{\ell+1,\ell}$
C-BC_ℓ, $\ell \geq 2$		${}^2\mathsf{A}^{(1)}_{2\ell,\ell}$
D_ℓ, $\ell \geq 4$	 hs hs hs hs	${}^1\mathsf{D}^{(1)}_{\ell,\ell}$
E_6	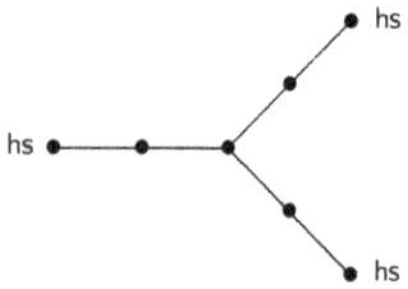 hs hs hs	${}^1\mathsf{E}^0_{6,6}$
E_7	 hs hs	$\mathsf{E}^0_{7,7}$
E_8	 hs	$\mathsf{E}^0_{8,8}$
F_4	 hs	$\mathsf{F}^0_{4,4}$
$\mathsf{F}^{\mathrm{I}}_4$		${}^2\mathsf{E}^2_{6,4}$
G_2	 hs	$\mathsf{G}_{2,2}$
$\mathsf{G}^{\mathrm{I}}_2$		${}^3\mathsf{D}_{4,2}$ or ${}^6\mathsf{D}_{4,2}$

Table 28.1 Affine Indices with Trivial Θ

Name	*Affine Index*	*Spherical Index*
${}^d\mathsf{A}_{(\ell+1)d-1}$, $d \geq 2$, $\ell \geq 2$	See 28.28.	${}^1\mathsf{A}^{(d)}_{(\ell+1)d-1,\ell}$
${}^2\mathsf{A}'_{2\ell-1}$, $\ell \geq 2$		${}^2\mathsf{A}^{(1)}_{2\ell-1,\ell}$
${}^2\mathsf{A}''_{2\ell+1}$, $\ell \geq 2$		${}^2\mathsf{A}^{(1)}_{2\ell+1,\ell}$
${}^2\mathsf{A}'_{2\ell}$, $\ell \geq 2$		${}^2\mathsf{A}^{(1)}_{2\ell,\ell}$
${}^2\mathsf{B}_{\ell+1}$, $\ell \geq 2$		$\mathsf{B}_{\ell+1,\ell}$
${}^2\mathsf{B}\text{-}\mathsf{C}_{\ell+1}$, $\ell \geq 2$		${}^2\mathsf{A}^{(1)}_{2\ell+1,\ell}$
${}^2\mathsf{C}_{2\ell+1}$, $\ell \geq 2$		$\mathsf{C}^{(2)}_{2\ell+1,\ell}$
${}^2\mathsf{C}_{2\ell}$, $\ell \geq 2$		$\mathsf{C}^{(2)}_{2\ell,\ell}$
${}^2\mathsf{C}\text{-}\mathsf{B}_{2\ell+1}$, $\ell \geq 2$		${}^2\mathsf{D}^{(2)}_{2\ell+2,\ell}$
${}^2\mathsf{C}\text{-}\mathsf{B}_{2\ell}$, $\ell \geq 2$		${}^2\mathsf{D}^{(2)}_{2\ell+1,\ell}$
${}^2\mathsf{D}_{\ell+1}$, $\ell \geq 3$		${}^2\mathsf{D}^{(1)}_{\ell+1,\ell}$
${}^2\mathsf{D}'_{\ell+2}$, $\ell \geq 2$		${}^1\mathsf{D}^{(1)}_{\ell+2,\ell}$

Table 28.2 Affine Indices with Non-Trivial Θ, I

Name	*Affine Index*	*Spherical Index*
${}^2\mathsf{D}''_{2\ell}$, $\ell \geq 3$		${}^1\mathsf{D}^{(2)}_{2\ell,\ell}$
${}^2\mathsf{D}''_{2\ell+1}$, $\ell \geq 3$		${}^2\mathsf{D}^{(2)}_{2\ell+1,\ell}$
${}^2\mathsf{D}''_5$		${}^2\mathsf{D}^{(2)}_{5,2}$
${}^3\mathsf{D}_4$		${}^3\mathsf{D}_{4,2}$
${}^4\mathsf{D}_{2\ell+2}$, $\ell \geq 2$		${}^2\mathsf{D}^{(2)}_{2\ell+2,\ell}$
${}^4\mathsf{D}_{2\ell+3}$, $\ell \geq 2$		${}^1\mathsf{D}^{(2)}_{2\ell+3,\ell}$
${}^2\mathsf{E}_6$		${}^2\mathsf{E}^2_{6,4}$
${}^3\mathsf{E}_6$		${}^1\mathsf{E}^{16}_{6,2}$
${}^2\mathsf{E}_7$		$\mathsf{E}^5_{7,4}$

Table 28.3 Affine Indices with Non-Trivial Θ, II

We can now explain how we determine the affine index for each complete Bruhat-Tits pair $\tilde{\mathsf{X}}_\ell(\Lambda, \nu)$ whose defining field K is local.[15] For each family of these pairs, we can determine the spherical index from 41.16 in [36] and the indications on page 67 of [33].[16] The Coxeter diagram Π is, of course, just $\tilde{\mathsf{X}}_\ell$. This is enough information to determine the affine index in most cases, since most of the affine indices in Tables 28.1–28.3 are uniquely determined by the spherical index and Π. In fact, there are only four pairs of affine indices that are not distinguished by these data: $\mathsf{C}\text{-}\mathsf{BC}_\ell$ and ${}^2\mathsf{A}'_{2\ell}$, ${}^2\mathsf{B}\text{-}\mathsf{C}_{\ell+1}$ and ${}^2\mathsf{A}''_{2\ell+1}$, ${}^2\mathsf{C}\text{-}\mathsf{B}_{2\ell+1}$ and ${}^4\mathsf{D}_{2\ell+2}$ as well as $\mathsf{G}_2^{\mathrm{I}}$ and ${}^3\mathsf{D}_4$. To match these affine indices to the corresponding Bruhat-Tits pairs, we need to consider also the structure of gems and interpret a few of the decorations of Π in the tables. In fact, it will be enough to know that there is a cross next to a vertex in these tables if and only if there exist root groups of gems that are non-abelian[17] and that there is an integer m next to a vertex that does not have a cross if and only if there exist root groups of gems that are parametrized by vector spaces of dimension m over $\bar{F}$, where F is the subfield of K defined in 28.40.[18]

Notation 28.29. For the rest of this chapter, we let

$$(\Delta, \mathcal{A}) = \tilde{\mathsf{X}}_\ell^*(\Lambda, \nu)$$

be a complete Bruhat-Tits pair such that the defining field K of $\Delta_{\mathcal{A}}^\infty$ (as defined in 30.15) is a local field (as defined in 28.1) with valuation ν. Let

$$G := \mathrm{Aut}(\Delta, \mathcal{A}).$$

Since the pair $(\Delta, \mathcal{A})$ is complete, G is, in fact, equal to $\mathrm{Aut}(\Delta)$. Let R_0 be the gem described in 18.27 and, if $\mathsf{X} = \mathsf{C}$, let R_1 be the gem described in 18.28 (with respect to a fixed apartment A of Δ and all the notation in 18.1). In our list of locally finite Bruhat-Tits buildings, we will include an indication of the structure of R_0 and, if $\mathsf{X} = \mathsf{C}$, also R_1. By 26.39, the group G acts transitively on the set of gems of Δ if $\mathsf{X} \neq \mathsf{C}$, and by 18.29, every gem is in the same G-orbit as R_0 or R_1 if $\mathsf{X} = \mathsf{C}$. (Sometimes R_0 and R_1 are in the same G-orbit even when $\mathsf{X} = \mathsf{C}$. We will indicate precisely when this happens.)

---

[15]If a Bruhat-Tits building Δ is algebraic, it is natural to refer to the affine and spherical indices of the corresponding algebraic group as the affine and spherical indices of the building. (That these indices are invariants of Δ follows from the Theorem in Section 5.8 of [32].) As we observed at the beginning of this chapter, there are locally finite Bruhat-Tits buildings that are not algebraic and therefore do not have an affine (or a spherical) index. These exceptions, which are all of mixed type, are displayed in Table 28.4.

[16]See also 11.18 and 16.18 in [36]. Note, too, that the forms, quadratic and skew-hermitian, that appear in cases (iii) and (vii) of 30.14 are the *anisotropic parts* of the corresponding forms on page 67 of [33]. The dimension of one of these forms is always $m + 2\ell$, where ℓ is the rank of the corresponding spherical building and either m is the dimension of the anisotropic part or $m = 0$ and the corresponding spherical building is of involutory type.

[17]See Section 1.8 of [33]

[18]See Section 1.6 of [33], where F is called K. That these dimensions are invariants of the gems is a consequence of 35.7–35.13 in [36].

Convention 28.30. The gems of Δ are all finite Moufang spherical buildings (by 18.18). It will turn out that in every case $\bar{K}$ is the defining field of every gem (as defined in 30.15). In other words, every gem is, up to isomorphism, one of the buildings in 30.34 with $\bar{K}$ in place of K. We will always use the notation in 30.35 to refer to these buildings.

Convention 28.31. In light of 23.15, it is safe to write $\tilde{\mathsf{X}}_\ell^*(\Lambda)$ rather than $\tilde{\mathsf{X}}_\ell^*(\Lambda, \nu)$ to specify the pair $(\Delta, \mathcal{A})$, and we will do this for the rest of this chapter. In fact, since we are only considering complete Bruhat-Tits pairs in this chapter, we can think of $\tilde{\mathsf{X}}_\ell^*(\Lambda)$ as the name for Δ itself rather than for the pair $(\Delta, \mathcal{A})$.

Convention 28.32. The second subscript of the spherical index in the last column of the tables in [33] is always equal to the parameter we have been calling ℓ. In [33], this subscript is sometimes n, $n-1$ or $n-2$ and sometimes m or $m-1$. In Tables 28.1–28.3 (and everywhere below), we set this subscript equal to ℓ and replace n and m accordingly. Thus, for example, we write ${}^2\mathsf{A}''_{2\ell+1}$ for the affine index called ${}^2\mathsf{A}''_{2m-1}$ in Table 4.3 in [33].

We now make our list of Bruhat-Tits buildings whose defining field is local. We organize this list in 28.33–28.38 according to the different possibilities for $\mathsf{X}_\ell^*(\Lambda)$ (which we consider in roughly the same order in which they appear in 30.14). In each case, we determine the structure of the gems (that is to say, R_0 and, when relevant, R_1) and the affine index (for those buildings that are algebraic).

28.33. Suppose that the Coxeter diagram X_ℓ is simply laced, i.e. $\mathsf{X}_\ell^* = \mathsf{A}_\ell$, D_ℓ or E_ℓ. By 28.4.i and 30.14, either $\Lambda = K$ is commutative or $\mathsf{X} = \mathsf{A}$ and K is a skew field. By 18.30, $R_0 \cong \mathsf{X}_\ell(\bar{K})$. If K is commutative, then the affine index of Δ is X_ℓ. If instead K is a cyclic algebra of degree $d > 1$ over its center F (so $\bar{K}/\bar{F}$ is an extension of degree d by 28.4.iv), the affine index of Δ is ${}^d\mathsf{A}_{(\ell+1)d-1}$.

28.34. Suppose that $\mathsf{X}_\ell^* = \mathsf{B}_\ell^{\mathcal{Q}}$ or $\mathsf{C}_\ell^{\mathcal{Q}}$. Thus Λ is an anisotropic quadratic space (K, L, q) which can be assumed to be unitary by 19.15 and 19.16. Let m and m_i for each i be as in 28.11. By 28.11, we have $m_0 > 0$, $m_i \leq 2$ for $i = 0$ and 1, $m = m_0 + m_1$ and $m_1 > 0$ if and only if $\mathsf{X} = \mathsf{C}$. By 19.35, $R_0 \cong \mathsf{B}_\ell(\bar{K}, m_0)$ and, if $\mathsf{X} = \mathsf{C}$, then $R_1 \cong \mathsf{B}_\ell(\bar{K}, m_1)$. Thus R_0, R_1 and X are determined by m_0 and m_1. If $m = 1$, then $m_0 = 1$ and $m_1 = 0$, so $\mathsf{X} = \mathsf{B}$; in this case the affine index of Δ is B_ℓ if $\ell \geq 3$ and C_2 if $\ell = 2$.[19] Suppose next that $m = 2$. If $m_0 = m_1 = 1$, then $\mathsf{X} = \mathsf{C}$. If the bilinear form f associated with q is identically zero, then Δ is of mixed type (as defined in 26.4) and does not have an index. If f is not identically zero, then the affine index of Δ is $\mathsf{C}\text{-}\mathsf{B}_\ell$. If $m_0 = 2$ and $m_1 = 0$, then $\mathsf{X} = \mathsf{B}$. In this case, the affine index of Δ is ${}^2\mathsf{D}_{\ell+1}$ if $\ell \geq 3$ and ${}^2\mathsf{A}'_3$ if $\ell = 2$.[20] If $m = 3$, then either

---

[19] Note that by 23.8, $(\Delta, \mathcal{A}) \cong \tilde{\mathsf{C}}_2^{\mathcal{I}}(K, K, \mathrm{id})$ when $m = 1$ and $\ell = 2$.

[20] Note that by 23.8, $(\Delta, \mathcal{A}) \cong \tilde{\mathsf{C}}_2^{\mathcal{I}}(E, K, \sigma)$ when $m = m_0 = 2$ and $\ell = 2$, where E is as in 28.11.vi and σ is the non-trivial element of $\mathrm{Gal}(E/K)$.

$m_0 = 1$ and $m_1 = 2$ or vice versa, and hence $\mathsf{X} = \mathsf{C}$. In this case, the affine index of Δ is ${}^2\mathsf{B}_{\ell+1}$. If $m = 4$, then $m_0 = m_1 = 2$ and hence $\mathsf{X} = \mathsf{C}$. In this case, the affine index of Δ is ${}^2\mathsf{D}'_{\ell+2}$. Finally, we observe that if $m = 1$, 2 or 4, then by 19.35.iii (and its footnote), 28.11.vi and 28.12, all the residues are in the same G-orbit as R_0 (even if the bilinear form f is identically zero).

28.35. Suppose that $\mathsf{X}_\ell^* = \mathsf{B}_\ell^{\mathcal{I}}$ or $\mathsf{C}_\ell^{\mathcal{I}}$. Thus Λ is an involutory set (K, K_0, σ), genuine or honorary. By 28.4.i, Λ is in fact genuine, and by 30.14, we can assume either that Λ is proper (and hence $\sigma \neq 1$) or $\ell \geq 3$ and Λ is a quadratic involutory set of type (ii) with $\text{char}(K) \neq 2$, or of type (iii) or (iv). Let $\bar{\Lambda} = (\bar{K}, \bar{K}_0, \bar{\sigma})$ be as in 23.22. Suppose first that Λ is proper. Then Λ must be as in 28.15.iii and the affine index of Δ is ${}^2\mathsf{C}_{2\ell}$. By 35.16 in [36], we can replace Λ by a translate without changing Δ (up to isomorphism). By (v) and (vi) in 26.22, therefore, we can assume that σ is the standard involution of K if $\text{char}(K) = 2$ and that σ equals the involution called τ in 28.18 if $\text{char}(K) \neq 2$. By 23.27.ii, 28.18 and 28.19, it follows that $R_j \cong \mathsf{B}_\ell^{\mathcal{I}}(\bar{K})$ and $R_{1-j} \cong \mathsf{B}_\ell^{\mathcal{I}}(\bar{K}, 2)$ for $j = 0$ or 1. Now suppose that $\ell \geq 3$ and Λ is a quadratic involutory set of type (ii), (iii) or (iv). Let $F = K_0$. By 23.27 and 30.34, G acts transitively on gems even if $\mathsf{X} = \mathsf{C}$ and $R_0 \cong R_1 \cong \mathsf{B}_\ell^{\mathcal{I}}(\bar{K}, d)$ for $d = 1$ or 2. If Λ is of type (ii), then $\mathsf{X} = \mathsf{C}$ (by 23.2) and $d = 1$. In this case, the affine index of Δ is C_ℓ. Suppose that Λ is of type (iii). Then K/F is a separable quadratic extension and this extension is ramified if and only if Λ is ramified (by 28.15.i). Thus $\mathsf{X} = \mathsf{B}$ and $d = 1$ if K/F is ramified and $\mathsf{X} = \mathsf{C}$ and $d = 2$ if K/F is unramified. The affine index of Δ is $\mathsf{B}\text{-}\mathsf{C}_\ell$ if K/F is ramified and ${}^2\mathsf{A}'_{2\ell-1}$ if K/F is unramified. Suppose, finally, that Λ is of type (iv). By 28.4.iv, Λ is ramified and $d = 2$. Thus $\mathsf{X} = \mathsf{B}$ and the affine index of Δ is ${}^2\mathsf{D}''_{2\ell}$. Note that the affine indices C_2 and ${}^2\mathsf{A}'_3$ are accounted for in 28.34.

28.36. Suppose that $\mathsf{X}_\ell^* = \mathsf{B}_\ell^{\mathcal{P}}$ or $\mathsf{C}_\ell^{\mathcal{P}}$. Thus

$$\Lambda = (K, K_0, \sigma, L, q)$$

is a proper anisotropic pseudo-quadratic space. In particular, $\sigma \neq 1$ (by 35.5 in [36]). Let $m = \dim_K L$ and let $\bar{\Lambda} = (\bar{K}, \bar{K}_0, \bar{\sigma}, \bar{L}, \bar{q})$ be as in 24.50. Suppose first that Λ is as in case (i) of 28.20 and let $F = K_0$. Then Λ is unramified, so $\mathsf{X} = \mathsf{C}$ (by 24.36). Suppose that K/F is ramified. Then $(\bar{K}, \bar{F}, \bar{\sigma})$ is a quadratic involutory set of type (ii) by 26.15.ii, so $\bar{L} = 0$ by 28.25.i and therefore $R_0 \cong \mathsf{B}_\ell^{\mathcal{I}}(\bar{K})$ and $R_1 \cong \mathsf{B}_\ell^{\mathcal{Q}}(\bar{K}, m)$ by 24.58.iii and 24.59.iv. The affine index of Δ is $\mathsf{C}\text{-}\mathsf{BC}_\ell$ if $m = 1$ and ${}^2\mathsf{B}\text{-}\mathsf{C}_{\ell+1}$ if $m = 2$. Suppose, instead, that K/F is unramified. By 26.15.i, $(\bar{K}, \bar{F}, \bar{\sigma})$ is a quadratic space of type (iii). By 28.25.ii, it follows that $\bar{L} \neq 0$. By 24.58.ii and 24.59.iv, therefore, $R_0 \cong \mathsf{B}_\ell^{\mathcal{P}}(\bar{K})$ and $R_1 \cong \mathsf{B}_\ell^{\mathcal{I}}(\bar{K}, 2)$ if $m = 1$ and $R_0 \cong R_1 \cong \mathsf{B}_\ell^{\mathcal{P}}(\bar{K})$ if $m = 2$. Note that if $m = 2$, then by 24.58.iv and 28.24, R_0 and R_1 are in the same G-orbit.[21] The affine index of Δ is ${}^2\mathsf{A}'_{2\ell}$

[21] If K/F is unramified and $m = 1$, then R_0 and R_1 are not even isomorphic. It is an interesting puzzle to find the place where the argument in 28.24 breaks down when $m = 1$.

Δ	*Conditions*	*Gems*
$\tilde{\mathsf{C}}_\ell^{\mathcal{Q}}(K, L, q)$	$L = D((t))$, $\text{char}(D) = 2$, $K = L^2$, $q(x) = x^2$	$\mathsf{B}_\ell^{\mathcal{Q}}(D)$
$\tilde{\mathsf{F}}_4(K, F)$	$K = D((t))$, $\text{char}(D) = 2$, $F = K^2$	$\mathsf{F}_4(D)$
$\tilde{\mathsf{G}}_2((J/K)^\circ)$	$J = D((t))$, $\text{char}(D) = 3$, $K = J^3$	$\mathsf{G}_2(D)$

Table 28.4 Mixed Type

if $m = 1$ and ${}^2\mathsf{A}''_{2\ell+1}$ if $m = 2$. Suppose next that Λ is as in case (ii) of 28.20 and again let $F = K_0$. By 28.15.ii, (K, F, σ) is ramified. If $m = 1$ and Λ is ramified, then $\mathsf{X} = \mathsf{B}$, $R_0 \cong \mathsf{B}_\ell^{\mathcal{P}}(\bar{K})$ by 24.58.i and 24.59.v and the affine index of Δ is ${}^2\mathsf{D}''_{2\ell+1}$.[22] If $m = 1$ and Λ is unramified, then $\mathsf{X} = \mathsf{C}$, $R_0 \cong \mathsf{B}_\ell^{\mathcal{I}}(\bar{K}, 2)$ and $R_1 \cong \mathsf{B}_\ell^{\mathcal{Q}}(\bar{K})$ by 24.58.iii and 24.59.iv (since $\bar{L}_1 \neq 0$ by 24.55.iii) and the affine index of Δ is ${}^2\mathsf{C}\text{-}\mathsf{B}_{2\ell}$. Now let $m > 1$. By 28.20.ii, Λ is unramified. Thus 24.58.iii and 24.59.iv apply and $\mathsf{X} = \mathsf{C}$. If $m = 2$, then either $R_0 \cong \mathsf{B}_\ell^{\mathcal{P}}(\bar{K})$ and $R_1 \cong \mathsf{B}_\ell^{\mathcal{Q}}(\bar{K})$, in which case the affine index of Δ is ${}^2\mathsf{C}\text{-}\mathsf{B}_{2\ell+1}$, or $R_0 \cong \mathsf{B}_\ell^{\mathcal{I}}(\bar{K}, 2)$ and $R_1 \cong \mathsf{B}_\ell^{\mathcal{Q}}(\bar{K}, 2)$, in which case the affine index of Δ is ${}^4\mathsf{D}_{2\ell+2}$. If $m = 3$, then $R_0 \cong \mathsf{B}_\ell^{\mathcal{P}}(\bar{K})$, $R_1 \cong \mathsf{B}_\ell^{\mathcal{Q}}(\bar{K}, 2)$ and the affine index of Δ is ${}^4\mathsf{D}_{2\ell+3}$. Suppose, finally, that Λ is as in case (iii) of 28.20. Then Λ is unramified, so $\mathsf{X} = \mathsf{C}$. By 35.19 of [36], we can replace Λ by a translate as defined in 11.26 of [36]. By (v) and (vi) of 26.22, therefore, we can assume that σ is the standard involution of K if $\text{char}(K) = 2$ and that σ equals the involution called τ in 28.18 if $\text{char}(K) \neq 2$. By 24.58.ii, 28.18, 28.19 and 28.25.ii, it follows that $R_j \cong \mathsf{B}_\ell^{\mathcal{P}}(\bar{K})$ and $R_{1-j} \cong \mathsf{B}_\ell^{\mathcal{I}}(\bar{K})$ for $j = 0$ or 1. The affine index of Δ is ${}^2\mathsf{C}_{2\ell+1}$.

By 28.11.viii and 28.14, our list is now complete in the case that X is either B or C.[23]

28.37. Suppose that $\mathsf{X}_\ell^* = \mathsf{F}_4$. Thus Λ is a composition algebra (K, F), so 28.26 applies, and $\bar{\Lambda} := (\bar{K}, \bar{F})$ is a composition algebra whose type can be determined by 26.15 and 28.4.iv. By 26.16 and 30.34, we have $R_0 \cong \mathsf{F}_4(\bar{K}, d)$ for $d = 1$ or 2. If Λ is of type (i) (as defined in 30.17), then $d = 1$ and Δ is of mixed type, so it does not have an affine index. If Λ is of type (ii), then $d = 1$ and the affine index of Δ is F_4. If Λ is of type (iii), then $d = 1$ if E/F is ramified, in which case the affine index of Δ is $\mathsf{F}_4^{\mathrm{I}}$, and $d = 2$ if E/F is unramified, in which case the affine index is ${}^2\mathsf{E}_6$. If Λ is of type (iv), then $d = 2$ (by 28.4.iv) and the affine index of Δ is ${}^2\mathsf{E}_7$.

28.38. Suppose that $\mathsf{X}_\ell^* = \mathsf{G}_2$. Thus

$$\Lambda = (J, K, N, \#, T, \times, 1)$$

[22] Note that the affine index ${}^2\mathsf{D}''_{2\ell+1}$ occupies two rows both in Table 4.3 in [33] and in Table 28.3, one for the case $\ell \geq 3$ and one for the case $\ell = 2$.

[23] The restriction $\text{char}(K) \neq 2$ in row ${}^2\mathsf{C}_\ell$ of Table 28.5 is explained in 23.9.

Affine Index	Δ	*Conditions*	*Gems*
A_ℓ	$\tilde{\mathsf{A}}_\ell(K)$	–	$\mathsf{A}_\ell(\bar{K})$
B_ℓ	$\tilde{\mathsf{B}}_\ell^{\mathcal{Q}}(K, L, q)$	$m = 1$, $\ell \geq 3$	$\mathsf{B}_\ell^{\mathcal{Q}}(\bar{K})$
$\mathsf{B}\text{-}\mathsf{C}_\ell$	$\tilde{\mathsf{B}}_\ell^{\mathcal{I}}(K, F, \sigma)$	K/F ram., $\ell \geq 3$	$\mathsf{B}_\ell^{\mathcal{I}}(\bar{K})$
C_ℓ	$\tilde{\mathsf{C}}_\ell^{\mathcal{I}}(K, K, \mathrm{id})$	$\mathrm{char}(K) \neq 2$	$\mathsf{B}_\ell^{\mathcal{I}}(\bar{K})$
$\mathsf{C}\text{-}\mathsf{B}_\ell$	$\tilde{\mathsf{C}}_\ell^{\mathcal{Q}}(K, L, q)$	$m = 2$	$\mathsf{B}_\ell^{\mathcal{Q}}(\bar{K})$
$\mathsf{C}\text{-}\mathsf{BC}_\ell$	$\tilde{\mathsf{C}}_\ell^{\mathcal{P}}(K, F, \sigma, L, q)$	K/F ram., $m = 1$	$\mathsf{B}_\ell^{\mathcal{I}}(\bar{K})$ & $\mathsf{B}_\ell^{\mathcal{Q}}(\bar{K})$
D_ℓ	$\tilde{\mathsf{D}}_\ell(K)$	$\ell \geq 4$	$\mathsf{D}_\ell(\bar{K})$
E_ℓ	$\tilde{\mathsf{E}}_\ell(K)$	$\ell = 6, 7, 8$	$\mathsf{E}_\ell(\bar{K})$
F_4	$\tilde{\mathsf{F}}_4(K/K)$	–	$\mathsf{F}_4(\bar{K})$
$\mathsf{F}_4^{\mathrm{I}}$	$\tilde{\mathsf{F}}_4(E/K)$	E/K ram.	$\mathsf{F}_4(\bar{K})$
G_2	$\tilde{\mathsf{G}}_2((K/K)^\circ)$	–	$\mathsf{G}_2(\bar{K})$
$\mathsf{G}_2^{\mathrm{I}}$	$\tilde{\mathsf{G}}_2((J/K)^+)$	J/K ram.	$\mathsf{G}_2(\bar{K})$

Table 28.5 Residually Split Groups

is an hexagonal system. Let E be as in 28.27. By 25.29 and 30.34, we have $R_0 \cong \mathsf{G}_2(\bar{K}, d)$ for $d = 1$ or 3. If Λ is as in case (i) of 28.27, then $d = 1$ and either $E \neq K$, Δ is of mixed type and does not have an affine index or $E = K$ and the affine index of Δ is G_2. If Λ is as in case (ii) of 28.27, then $d = 1$ if the extension E/K is ramified, in which case the affine index of Δ is $\mathsf{G}_2^{\mathrm{I}}$, and $d = 3$ if the extension E/K is unramified, in which case the affine index of Δ is ${}^3\mathsf{D}_4$. If Λ is as in case (iii) of 28.27, then $d = 3$ (by 28.4.iv) and the affine index of Δ is ${}^3\mathsf{E}_6$.

This completes the list of Bruhat-Tits buildings whose defining field is local. We can observe now that all of these buildings are, in fact, locally finite. We thus have produced in 28.33–28.38 a complete list of locally finite Bruhat-Tits buildings.

* * *

We summarize the results of this chapter in Tables 28.4–28.6. In all three tables we use various conventions spelled out in 28.39. In the first table, we give the three families of locally finite Bruhat-Tits buildings that are mixed rather than algebraic and consequently do not have an affine index. The remaining locally finite Bruhat-Tits buildings, which are all algebraic, appear in Tables 28.5 and 28.6.

In the first column of these tables we have listed the affine indices (of groups of relative rank at least 2) in the same order that they occur in

Affine Index	Δ	*Conditions*	*Gems*
${}^d\mathsf{A}_{(\ell+1)d-1}$	$\tilde{\mathsf{A}}_\ell(K)$	$\deg_F K = d$	$\mathsf{A}_\ell(\bar{K})$
${}^2\mathsf{A}'_{2\ell-1}$	$\tilde{\mathsf{C}}_\ell^{\mathcal{I}}(K, F, \sigma)$	K/F unram.	$\mathsf{B}_\ell^{\mathcal{I}}(\bar{K}, 2)$
${}^2\mathsf{A}''_{2\ell+1}$	$\tilde{\mathsf{C}}_\ell^{\mathcal{P}}(K, F, \sigma, L, q)$	K/F unram., $m = 2$	$\mathsf{B}_\ell^{\mathcal{P}}(\bar{K})$
${}^2\mathsf{A}'_{2\ell}$	$\tilde{\mathsf{C}}_\ell^{\mathcal{P}}(K, F, \sigma, L, q)$	K/F unram., $m = 1$	$\mathsf{B}_\ell^{\mathcal{I}}(\bar{K}, 2)$ & $\mathsf{B}_\ell^{\mathcal{P}}(\bar{K})$
${}^2\mathsf{B}_{\ell+1}$	$\tilde{\mathsf{C}}_\ell^{\mathcal{Q}}(K, L, q)$	$m = 3$	$\mathsf{B}_\ell^{\mathcal{Q}}(\bar{K}, 2)$ & $\mathsf{B}_\ell^{\mathcal{Q}}(\bar{K})$
${}^2\mathsf{B}\text{-}\mathsf{C}_{\ell+1}$	$\tilde{\mathsf{C}}_\ell^{\mathcal{P}}(K, F, \sigma, L, q)$	K/F ram., $m = 2$	$\mathsf{B}_\ell^{\mathcal{I}}(\bar{K})$ & $\mathsf{B}_\ell^{\mathcal{Q}}(\bar{K}, 2)$
${}^2\mathsf{C}_{2\ell+1}$	$\tilde{\mathsf{C}}_\ell^{\mathcal{P}}(K, K_0, \tau, L, q)$	K quatern., $m = 1$	$\mathsf{B}_\ell^{\mathcal{P}}(\bar{K})$ & $\mathsf{B}_\ell^{\mathcal{I}}(\bar{K})$
${}^2\mathsf{C}_{2\ell}$	$\tilde{\mathsf{C}}_\ell^{\mathcal{I}}(K, K_0, \tau)$	K quatern.	$\mathsf{B}_\ell^{\mathcal{I}}(\bar{K}, 2)$ & $\mathsf{B}_\ell^{\mathcal{I}}(\bar{K})$
${}^2\mathsf{C}\text{-}\mathsf{B}_{2\ell+1}$	$\tilde{\mathsf{C}}_\ell^{\mathcal{P}}(K, F, \sigma, L, q)$	K quatern., $m = 2$	$\mathsf{B}_\ell^{\mathcal{P}}(\bar{K})$ & $\mathsf{B}_\ell^{\mathcal{Q}}(\bar{K})$
${}^2\mathsf{C}\text{-}\mathsf{B}_{2\ell}$	$\tilde{\mathsf{C}}_\ell^{\mathcal{P}}(K, F, \sigma, L, q)$	K quatern., $m = 1$	$\mathsf{B}_\ell^{\mathcal{I}}(\bar{K}, 2)$ & $\mathsf{B}_\ell^{\mathcal{Q}}(\bar{K})$
${}^2\mathsf{D}_{\ell+1}$	$\tilde{\mathsf{B}}_\ell^{\mathcal{Q}}(K, L, q)$	$m = 2$, $\ell \geq 3$	$\mathsf{B}_\ell^{\mathcal{Q}}(\bar{K}, 2)$
${}^2\mathsf{D}'_{\ell+2}$	$\tilde{\mathsf{C}}_\ell^{\mathcal{Q}}(K, L, q)$	$m = 4$	$\mathsf{B}_\ell^{\mathcal{Q}}(\bar{K}, 2)$
${}^2\mathsf{D}''_{2\ell}$	$\tilde{\mathsf{B}}_\ell^{\mathcal{I}}(K, F, \sigma)$	K quatern., $\ell \geq 3$	$\mathsf{B}_\ell^{\mathcal{I}}(\bar{K}, 2)$
${}^2\mathsf{D}''_{2\ell+1}$	$\tilde{\mathsf{B}}_\ell^{\mathcal{P}}(K, F, \sigma, L, q)$	K quatern., $m = 1$	$\mathsf{B}_\ell^{\mathcal{P}}(\bar{K})$
${}^3\mathsf{D}_4$	$\tilde{\mathsf{G}}_2((J/K)^+)$	J/K unram.	$\mathsf{G}_2(\bar{K}, 3)$
${}^4\mathsf{D}_{2\ell+2}$	$\tilde{\mathsf{C}}_\ell^{\mathcal{P}}(K, F, \sigma, L, q)$	K quatern., $m = 2$	$\mathsf{B}_\ell^{\mathcal{I}}(\bar{K}, 2)$ & $\mathsf{B}_\ell^{\mathcal{Q}}(\bar{K}, 2)$
${}^4\mathsf{D}_{2\ell+3}$	$\tilde{\mathsf{C}}_\ell^{\mathcal{P}}(K, F, \sigma, L, q)$	K quatern., $m = 3$	$\mathsf{B}_\ell^{\mathcal{P}}(\bar{K})$ & $\mathsf{B}_\ell^{\mathcal{Q}}(\bar{K}, 2)$
${}^2\mathsf{E}_6$	$\tilde{\mathsf{F}}_4(E/K)$	E/K unram.	$\mathsf{F}_4(\bar{K}, 2)$
${}^3\mathsf{E}_6$	$\tilde{\mathsf{G}}_2(J^+)$	$\deg_K J = 3$	$\mathsf{G}_2(\bar{K}, 3)$
${}^2\mathsf{E}_7$	$\tilde{\mathsf{F}}_4(K/F)$	K quatern.	$\mathsf{F}_4(\bar{K}, 2)$

Table 28.6 Non-Residually Split Groups

Tables 4.2 and 4.3 in [33] (and in Tables 28.1–28.3). We will refer to rows in the tables (both those in this chapter and those in [33]) by naming the entry in the first column.

If we regard 28.33–28.38 as a lexicon indicating the algebraic group corresponding to each algebraic locally finite Bruhat-Tits building, then Tables 28.5 and 28.6 can be seen as a lexicon for translating in the reverse sense.

Notes 28.39. Tables 28.4–28.6 are to be interpreted according to the following rules and observations:

(i) The names of the affine indices are the same as those in the tables in [33]; note, however, that we have applied 28.32 to the subscripts.

(ii) The entries in the columns labeled Δ are the names introduced in 16.17, 19.22, 23.7, 24.36, 25.24 and 26.11 to which we have applied 28.31.[24] In every case, K is the defining field of the building at infinity (as defined in 30.15). In the last columns we use the names introduced in 30.35.

(iii) In Table 28.4, D is an arbitrary finite field of the indicated characteristic.

(iv) K is a commutative field except in those rows where it is indicated that K is quaternion and in row ${}^d\mathsf{A}_{(\ell+1)d-1}$, where K is a skew field of degree d over its center.

(v) If K is commutative, then F always denotes a subfield such that K/F is a separable quadratic extension and σ always denotes the non-trivial element of $\mathrm{Gal}(K/F)$.

(vi) If K is quaternion, then σ always denotes its standard involution and F always denotes its center.

(vii) E is always a commutative field containing K such that E/K is a separable quadratic extension.

(viii) In Tables 28.5 and 28.6, J is always a commutative field containing K such that J/K is a separable cubic extension except in row ${}^3\mathsf{E}_6$, where J is a skew field over K of degree 3. The hexagonal systems $(K/K)^\circ$, $(J/K)^+$ and J^+ referred to in the tables are defined in 15.20–15.22 of [36].

(ix) The letter m always denotes $\dim_K L$.

(x) (K, K_0, τ) always denotes an involutory set such that K is quaternion and $\dim_F K_0 = 3$. This occurs only in rows ${}^2\mathsf{C}_{2\ell+1}$ and ${}^2\mathsf{C}_{2\ell}$.[25]

(xi) It is to be assumed that $\ell \geq 2$ except where indicated otherwise.[26]

---

[24]The building at infinity (in the notation in 30.15) is obtained by simply deleting the tilde (and substituting C for B or B for C if desired).

[25]As observed in 28.35, it can be assumed that τ is as in 28.18 if $\mathrm{char}(K) \neq 2$ and that τ is the standard involution of K if $\mathrm{char}(K) = 2$.

[26]Note that the buildings in the following pairs of rows are isomorphic for $\ell = 2$: $(\mathsf{B}_\ell, \mathsf{C}_\ell)$, $(\mathsf{B}\text{-}\mathsf{C}_\ell, \mathsf{C}\text{-}\mathsf{B}_\ell)$, $({}^2\mathsf{A}'_{2\ell-1}, {}^2\mathsf{D}_{\ell+1})$ and $({}^2\mathsf{D}'_{\ell+2}, {}^2\mathsf{D}''_{2\ell})$. It is for this reason that we require $\ell \geq 3$ in one of each of these pairs of rows. See also 23.8. Note, too, that the index ${}^2\mathsf{D}''_{2\ell+1}$, which occupies two rows in Table 28.3, occupies only one row in Table 28.6.

(xii) If there is only one item in the last column, then there is only one G-orbit of gems (where G is as in 28.29).

The buildings in our tables, it is to be emphasized, are *not*, in general, uniquely determined by K and the affine index. There can be, for example, many ramified separable quadratic extensions E of K, hence many pairwise non-similar anisotropic quadratic spaces of dimension 2 over K and therefore many different buildings defined over K whose affine index is $\mathsf{B}\text{-}\mathsf{C}_\ell$ (for each $\ell \geq 3$). Also the buildings defined over K with index ${}^3\mathsf{E}_6$ are not unique, since by 28.5, there are two non-isomorphic skew fields of degree 3 over K (which are opposites of each other). On the other hand, there are a number of locally finite buildings with defining field K that are uniquely determined by their affine index. For example, anisotropic quadratic spaces of dimension 1, 3 and 4 over K are unique up to similarity (by 28.12 and 28.13) and there is a unique field E such that E/K is an unramified quadratic extension (by 28.7); hence the buildings defined over K whose affine indices are B_ℓ, ${}^2\mathsf{B}_{\ell+1}$, ${}^2\mathsf{D}_{\ell+1}$ and ${}^2\mathsf{D}'_{\ell+2}$ are all unique. Also the buildings defined over K whose affine indices are ${}^2\mathsf{C}_{2\ell+1}$ and ${}^2\mathsf{C}_{2\ell}$ are both unique; this follows from (v) and (vi) of 26.22 and 28.23.

Final Comments 28.40. We conclude this chapter with a few more words about the crosses, arrows, integers and letters decorating the Coxeter diagrams in Tables 28.1–28.3.

Let $(\Delta, \mathcal{A}) = \tilde{\mathsf{X}}^*_\ell(\Lambda)$ be one of the buildings in Table 28.5 or 28.6 (so $\Pi := \tilde{\mathsf{X}}_\ell$ is its Coxeter diagram). In these tables K denotes the defining field of the building $\mathsf{X}^*_\ell(\Lambda)$ in the sense of 30.15. In some rows of these tables, K is commutative and F is the fixed point set of an involution σ of K, and in other rows, K is quaternion and F is its center. In all other cases (i.e. where there is no involution called σ), we let F denote the center of K (so $F = K$ when K is commutative and there is no involution called σ). Then in every case, there exists an absolutely simple algebraic group G that is defined over F such that Δ is the Bruhat-Tits building associated to the pair (G, F).

Let A be an apartment of Δ, let e be a chamber of A, let x be a vertex of the Coxeter diagram Π, let m be the integer next to the vertex x of Π in Tables 28.4 and 28.5 (which, it is to be understood, is equal to 1 if there is no integer next to x), let e_x be the unique chamber of A that is x-adjacent to e, let α be the unique root of A that contains e but not e_x and let $a = \alpha^\infty$. By 1.16, we can choose a gem R cut by α. Let $\bar{U}_{a,R}$ be as in 18.17. By 18.20, the group $\bar{U}_{a,R}$ is, up to isomorphism, independent of the choice of R.[27] If there is no cross next to x, then $\bar{U}_{a,R}$ is abelian and has canonically the structure of a vector space over the residue field $\bar{F}$ (by 35.6–35.13 in [36]), and the integer m is precisely the dimension of this vector space. If there is a cross next to x in the tables, then $\bar{U}_{a,R}$ is non-abelian, $m = 3$, the center of $\bar{U}_{a,R}$ is isomorphic to the additive group of $\bar{F}$ and $\bar{U}_{a,R}$ modulo its center has canonically the structure of a vector space of dimension 2 over $\bar{F}$.

[27]Note that we cannot choose R independently of x.

Let o be special vertex of Π and let R be an o-special gem of Δ. Unless the affine index of Δ is ${}^d\mathsf{A}_{(\ell+1)d-1}$ for some $d > 1$, there is an absolutely simple algebraic group $G_{\rm o}$ defined over the residue field $\bar{F}$ such that R is the spherical building associated with the pair $(G_{\rm o}, \bar{F})$.[28] The arrows and crosses decorating the diagram Π in Tables 28.5 and 28.6, which turn it into the relative local Dynkin diagram Ω, are arranged so that in every case the subdiagram Ω_o obtained from Ω by deleting the vertex o and all the edges containing o is precisely the *relative Dynkin diagram* associated with the relative root system of the pair $(G_{\rm o}, \bar{F})$ as defined in 2.5.2 in [31]. The arrows on the diagram Π_1, which turn it into the absolute local Dynkin diagram Ω_1, have an analogous description.

The special vertex o of the Coxeter diagram Π is *hyperspecial* (and labeled "hs" in the tables) exactly when the o-special gems of Δ are defined by algebraic data having all the same dimensions as the algebraic data defining Δ itself, where by "dimension" we mean the dimension over the field of definition F, respectively $\bar{F}$, of the corresponding algebraic group. Thus, for example, we have the following:

(i) In row ${}^2\mathsf{A}'_{2\ell-1}$ both special vertices of Π are hyperspecial since the building Δ and all the gems of Δ are defined by a quadratic involutory set of type (iii).
(ii) In row ${}^2\mathsf{A}''_{2\ell+1}$ neither special vertex is hyperspecial since Δ is defined by a two-dimensional anisotropic pseudo-quadratic space, whereas the gems are all defined by one-dimensional anisotropic pseudo-quadratic spaces.
(iii) In row ${}^2\mathsf{A}'_{2\ell}$ the building Δ is defined by a one-dimensional pseudo-quadratic space over a quadratic extension, so the special vertex on the left is not hyperspecial since the corresponding gems are defined by a zero-dimensional anisotropic pseudo-quadratic spaces (i.e. by involutory sets), whereas the special vertex on the right *is* hyperspecial since the corresponding gems are also defined by one-dimensional anisotropic pseudo-quadratic spaces over quadratic extensions.
(iv) In row ${}^2\mathsf{B}\text{-}\mathsf{C}_{\ell+1}$ we have $\dim_K L = \dim_{\bar{K}} \bar{L} = 2$ but nevertheless the vertex on the right is not hyperspecial since $\dim_F K \neq \dim_{\bar{F}} \bar{K}$.
(v) If K is non-commutative, then there are no hyperspecial vertices since $\dim_F K \neq \dim_{\bar{F}} \bar{K}$ by 28.4.iv.

[28] If the affine index of Δ is ${}^d\mathsf{A}_{(\ell+1)d-1}$ for some $d > 1$, then there is still an absolutely simple group $G_{\rm o}$ that gives rise to R, but in this case R is the spherical building associated with the pair $(G_{\rm o}, \bar{K})$ rather than the pair $(G_{\rm o}, \bar{F})$. At this point, we can observe that the captions of Tables 28.5 and 28.6 (which are borrowed from Tables 4.2 and 4.3 in [33]) refer to the fact that pairs $(G_{\rm o}, \bar{F})$ are all split for the buildings in Table 28.5 but not for the buildings in Table 28.6. See 1.10.2 and 3.5.2 in [33].

Chapter Twenty Nine

Appendix A

This appendix consists of two parts. In the first part (29.1–29.15), we review some of the definitions and results from [37] that are used most frequently in this book.[1] In the second, we assemble a variety of small results about Coxeter chamber systems (29.16–29.30) and buildings (29.32–29.63) that would have been included in [37] had we anticipated writing this sequel. These results are, for the most part, just variations on results in [37] and they could, in fact, serve as a set of exercises to [37].

We begin the first part. All the definitions in 29.1–29.3 are from Chapters 1 and 2 of [37].

29.1. Edge-colored graphs. Let Δ be an edge-colored graph and let I be the set of colors appearing on the edges of Δ. The set I will be called the *index set* of Δ. Suppose that Δ is connected and let J be a subset of the index set I. A *J-residue* of Δ is a connected component of the graph obtained from Δ by deleting all the edges whose color is *not* in J (but without deleting any vertices). A *residue* is a J-residue for some $J \subset I$. The *type* of a residue (including Δ itself) is the set of colors that appear on its edges (i.e. its index set) and the *rank* of a residue is the cardinality of its type. A *panel* is a residue of rank 1.

29.2. Chamber systems. A *chamber system* is a connected edge-colored graph Δ such that all the panels are complete graphs having at least two vertices. The vertices of a chamber system are usually called *chambers*. A chamber system is called *thick* (respectively, *thin*) if all its panels contain at least three chambers (respectively, exactly two chambers). Suppose Δ and Δ' are two chamber systems with index sets I and I' and suppose that σ is a bijection from I to I'. A *σ-homomorphism* from Δ to Δ' is a map from the chamber set of Δ to the chamber set of Δ' such that whenever two vertices of Δ are joined by an edge of color $i \in I$, their images are contained in a panel of color $\sigma(i) \in I'$ in Δ'. A *σ-isomorphism* from Δ to Δ' is a bijection from the chamber set of Δ to the chamber set of Δ' such that two vertices of Δ are joined by an edge of color $i \in I$ if and only if their images are joined by an edge of color $\sigma(i) \in I'$. A *homomorphism* is a σ-homomorphism for some bijection σ and an *isomorphism* is a σ-isomorphism for some bijection σ. A homomorphism (or an isomorphism) ϕ is called *special* (or *type-preserving*)

[1] It was our original intention to include in this appendix *all* the definitions and results from [37] that we require here, but this turned out to be infeasible.

if $I = I'$ and ϕ is a σ-homomorphism (or a σ-isomorphism) for $\sigma = \text{id}$. We denote the group of automorphisms of Δ by $\text{Aut}(\Delta)$ and the group of special automorphisms by $\text{Aut}^\circ(\Delta)$.

29.3. Galleries. A *gallery* in a chamber system Δ is a sequence

$$(x_0, x_1, \ldots, x_k)$$

of chambers such that for each $i \in [1, k]$, $\{x_{i-1}, x_i\}$ is an edge. Let $\gamma = (x_0, x_1, \ldots, x_k)$ be a gallery. The *length* of γ is k and the *type* of γ is the word $j_1 j_2 \cdots j_k$ in the free monoid M_I, where I is the type of Δ and j_i is the color of the edge $\{x_{i-1}, x_i\}$ for each $i \in [1, k]$. A gallery is called *minimal* if there is no shorter gallery having the same first and last vertices. The distance between two chambers x and y is the length of a minimal gallery from x to y. The distance between two chambers x and y is denoted by $\text{dist}(x, y)$. The *diameter* of Δ is the supremum of the set of lengths of minimal galleries in Δ; we denote this number by $\text{diam}(\Delta)$.

For the remainder of this appendix, we fix a Coxeter diagram Π and let I be its vertex set.

29.4. Coxeter chambers systems. Let $R = \{r_i \mid i \in I\}$ be a set in one-to-one correspondence with the vertex set I of Π and let

$$W_\Pi = \langle r_i \mid (r_i r_j)^{m_{ij}} = 1 \text{ for all } i, j \in I \rangle, \tag{29.5}$$

where $m_{ij} = 1$ if $i = j$ (so $r_i^2 = 1$ for each i); $m_{ij} = 2$ if i and j are distinct but not adjacent in Π (so $r_i r_j = r_j r_i$ in this case); and m_{ij} is the label on the edge $\{i, j\}$ of Π if i and j are distinct and adjacent in Π. A *Coxeter group of type* Π is any group isomorphic to W_Π. A *Coxeter system of type* Π is a pair (W, S), where W is a group and S is a subset of W, such that there is an isomorphism from W to W_Π mapping S to the set $R = \{r_i \mid i \in I\}$. The *Coxeter chamber system* Σ_Π belonging to Π is the edge-colored graphs whose index set (i.e. set of colors) is I and whose vertices are the elements of the Coxeter group W_Π, where two elements u and v are joined by an edge of color i if and only if $u = vr_i$. The Coxeter group W_Π acts sharply transitively on Σ_Π by left multiplication and (by 2.8 in [37]) every special automorphism of Σ_Π is given by left multiplication by an element of W_Π. A *Coxeter chamber system of type* Π is an edge-colored graph isomorphic to Σ_Π.

29.6. Roots. Let Σ be a Coxeter chamber system of type Π as defined in 29.4. Thus Σ is, in particular, a thin chamber system as defined in 29.2, i.e. the panels of Σ are precisely the edges. We identify W_Π with the group $\text{Aut}^\circ(\Sigma)$ of special automorphisms of Σ. A *reflection* is an element of W_Π that interchanges two adjacent chambers of Σ. A *wall* is the set M_s of edges fixed by a given reflection s. The graph $\Sigma \backslash M_s$ obtained from Σ by deleting the edges in a wall M_s (but without deleting any chambers) has two connected components (by 3.10 of [37]). A *root* is one of the two connected components of $\Sigma\backslash M_s$ for some reflection s (regarded simply as a set of chambers). The two roots associated with a given reflection are called *opposites*. Thus the root opposite a given root α is simply the complement of α in Σ; it is often denoted by $-\alpha$. Roots have the following properties:

(i) Every edge $\{x, y\}$ of Σ is in the wall of a unique reflection s, the two opposite roots associated with this reflection are

$$\{u \mid \text{dist}(x, u) < \text{dist}(y, u)\}$$

and

$$\{u \mid \text{dist}(y, u) < \text{dist}(x, u)\};$$

these are the only two roots that contain exactly one chamber in $\{x, y\}$ and they are interchanged by s.[2]

(ii) Roots are convex.[3]

(iii) The distance between two chambers x and y equals the number of roots that contain x but not y.

(iv) Let x and y be two chambers. Then a chamber u is contained in a minimal gallery from x to y if and only if it is contained in every root that contains both x and y (even when there is no such root).

The references in [37] for the assertions in 29.6 are as follows: (i) is proved in 3.11 and 3.15, (ii) in 3.19, (iii) in 3.20 and (iv) in 3.21.

29.7. Residues. Let Σ be a Coxeter chamber system of type Π as defined in 29.4. Residues of Σ (as defined in 29.1) have the following properties:

(i) For each residue R and each chamber u of Σ, there is a unique chamber $\text{proj}_R\, u$ of R that is nearest u. For every chamber u of Σ and every chamber v of R, there is a minimal gallery from u to v that passes through $\text{proj}_R\, u$.[4]

(ii) Residues are convex.

(iii) Each residue is a Coxeter chamber system of type Π_J, where J is its type and Π_J is the subdiagram of Π spanned by the set J.[5]

(iv) If R is a residue and α is a root, then either $R \subset \alpha$ or $R \subset -\alpha$ or $\alpha \cap R$ is a root of R.

The references in [37] for the assertions in 29.7 are as follows: (i) is proved in 3.22, (ii) in 3.24, (iii) in 4.9 and (iv) in 4.10.

---

[2]Note that if s' is another reflection interchanging these two opposite roots, then $M_{s'} \subset M_s$, from which it follows that $s' = s$.

[3]A set X of chambers is *convex* if every minimal gallery whose first and last chambers are contained in X lies entirely in X.

[4]The map $u \mapsto \text{proj}_R\, u$ is called the *projection map* of Π (with respect to R).

[5]The *subdiagram* Π_J *of* Π *spanned by* J is the subdiagram with vertex set J where two elements of J are joined by an edge with label m if and only if they are joined by an edge with label m in Π.

29.8. Reduced words. Let Σ be a Coxeter chamber system of type Π, let M_I be the free monoid on I (so the type of a gallery is a word in M_I) and let $f \mapsto r_f$ be the unique homomorphism from M_I to the Coxeter group

$$W_\Pi = \langle r_i \mid i \in I \rangle$$

extending the map $i \mapsto r_i$. A word in M_I is called *reduced* (with respect to Π) if it is *not* homotopic (as defined in 4.1 of [37]) to a word of the form $fiig$ for some $i \in I$ and some $f, g \in M_I$ (possibly empty). The following hold:

(i) Two minimal galleries in Σ having the same first and last chambers have homotopic types.
(ii) A gallery of Σ is minimal if and only if its type is reduced.
(iii) Two reduced words f and g in M_I have the same image $r_f = r_g$ in W if and only if they are homotopic.

The references in [37] for the assertions in 29.8 are as follows: (i) is proved in 4.2, (ii) in 4.3 and (iii) in 4.4.

29.9. Opposites. Let Σ be a Coxeter chamber system of type Π and suppose that Π is spherical.[6] Two chambers x and y of Σ are called *opposite* if the distance between them is $\text{diam}(\Sigma)$. The following hold:

(i) If x and y are opposite chambers, then
$$|\alpha \cap \{x, y\}| = 1$$
for every root α.
(ii) Σ has $2 \cdot \text{diam}(\Sigma)$ roots.
(iii) If x and y are opposite chambers, then every chamber of Σ is contained in a minimal gallery from x to y.
(iv) For each chamber u of Σ, there is a unique chamber $\text{op}_\Sigma u$ that is opposite u.[7]
(v) There exists a unique automorphism σ of Π such that the opposite map op_Σ defined in (iv) is a σ-automorphism of Σ (as defined in 29.2).

The references in [37] for the assertions in 29.9 are as follows: (i) is proved in 5.2, (ii) in 5.3, (iii) in 5.4, (iv) in 5.5 and (v) in 5.11.

29.10. Buildings. Let Δ be a building of type Π as defined in 7.1 of [37]. This means that Δ is a chamber system with index set I and there exists a map

$$\delta\colon \Delta \times \Delta \to W_\Pi, \quad (29.11)$$

called the *Weyl distance function* of Δ, such that for every reduced word f in M_I and every ordered pair of chambers (x, y),

$$\delta(x, y) = r_f \quad (29.12)$$

[6]See footnote 2 in Chapter 1.
[7]The map op_Σ is called the *opposite map* of Σ.

(where r_f is as in 29.8) if and only if there is a gallery of type f from x to y.[8] The following hold:

(i) A gallery is minimal if and only if its type is reduced.
(ii) Residues are convex.
(iii) Every J-residue (for $J \subset I$) is a building of type Π_J.
(iv) The Coxeter chamber system Σ_Π is the only thin building of type Π.
(v) For each chamber u of Δ and each residue R, there is a unique chamber $\text{proj}_R u$ of R that is nearest u. For every chamber u of Δ and every chamber v of a residue R, there is a minimal gallery from u to v that passes through $\text{proj}_R u$.[9]

The references in [37] for the assertions in 29.10 are as follows: (i) is proved in 7.7.ii, (ii) and (iii) in 7.20, (iv) in 8.11 and (v) in 8.21.

29.13. Apartments. Let Δ be a building of type Π and let Σ be a Coxeter chamber system of type Π. An *apartment* is the image of a special isomorphism from Σ into Δ. Apartments have the following properties:

(i) Every isometry (as defined in 8.1 of [37]) from a subset of Σ to Δ extends to a special isomorphism from Σ to Δ.
(ii) Every two chambers lie in an apartment.
(iii) Apartments are convex.
(iv) If A is an apartment, α is a root in A and R is a residue of Δ such that $R \cap A \neq \emptyset$, then $A \cap R$ is an apartment of R and either $\alpha \cap R$ is a root of $A \cap R$ or $A \cap R \subset \alpha$ or $R \cap \alpha = \emptyset$.
(v) If α is a root of an apartment A and A' is another apartment containing α, then α is also a root of A'.

The references in [37] for the assertions in 29.13 are as follows: (i) is proved in 8.2 and 8.5, (ii) in 8.6, (iii) in 8.9, (iv) in 8.13.i and (v) in 8.19.

29.14. Spherical buildings. A building is called *spherical* if its apartments are finite. If Δ is a spherical building, then the following hold:

(i) The diameter of Δ is finite and equals the diameter of each apartment of Δ.
(ii) If two chambers u and v of Δ are opposite, then there is a unique apartment containing them both; this apartment consists of all the chambers of Δ that lie on a minimal gallery from u to v.
(iii) If α is a root of Δ (i.e. a root of some apartment of Δ) and P is a panel of Δ containing exactly one chamber x of α, then for each $y \in P \backslash \{x\}$, there is a unique apartment A_y containing $\alpha \cup \{y\}$ and the map $y \mapsto A_y$ from $P \backslash \{x\}$ to the set of apartments containing α is a bijection.

---

[8] By 29.8.iii, this is equivalent to assuming that Δ is a chamber system with index set I such that for each ordered pair (x, y) of chambers, there are galleries of reduced type from x to y and the types of any two such galleries are homotopic.

[9] The map $u \mapsto \text{proj}_R u$ is called the *projection map* of Δ (with respect to R); compare 29.7.i.

(iv) Let Σ be an apartment of Δ and let u be a chamber in Σ. If g is an element of $\mathrm{Aut}(\Delta)$ acting trivially on Σ and on every panel containing u, then $g = 1$.

The references in [37] for the assertions in 29.14 are as follows: (i) is proved in 9.1, (ii) in 9.2, (iii) in 9.3 and (iv) in 9.7.

29.15. The Moufang property. Suppose that Δ is a spherical building. For each root α of Δ, the *root group* U_α is the pointwise stabilizer in $\mathrm{Aut}(\Delta)$ of the union of all panels of Δ that contain two chambers of α.[10] The spherical building Δ is called *Moufang* if

(a) it is thick and irreducible,
(b) its rank is at least 2 and
(c) for each root α, the root group U_α acts transitively on the set of all apartments containing α.

The following hold:

(i) If Δ is thick, irreducible and of rank at least 3, then Δ is Moufang.
(ii) If Δ is Moufang, then every irreducible[11] residue of Δ of rank at least 2 is also Moufang.
(iii) If Δ is Moufang, then the subgroup $G^\dagger$ of $\mathrm{Aut}(\Delta)$ generated by all the root groups of Δ acts transitively on the set of pairs (Σ, d) such that Σ is an apartment of Δ and d is a chamber of Σ.
(iv) Suppose that Δ is Moufang and let (Σ, d) be a pair as in (iii). Let N be the stabilizer of Σ in the group $G^\dagger$ defined in (iii) and let B be the stabilizer of d in $G^\dagger$. Let m_Σ be as in 3.8 and let Ξ_d° be the set of roots of Σ containing d but not some chamber of Σ adjacent to d (as defined in 3.5). For each $\alpha \in \Xi_d^\circ$, choose $m_\alpha \in m_\Sigma(U_\alpha^*)$ and let

$$S = \{m_\alpha \mid \alpha \in \Xi_d^\circ\}.$$

Then $(G^\dagger, B, N, S)$ is a Tits system as defined in 14.37.
(v) For each root α, the root group U_α acts sharply transitively on the set of apartments containing α.

The references in [37] for the assertions in 29.15 are as follows: (i) is proved in 11.6, (ii) in 11.8, (iii) in 11.12 and (iv) in 11.14. Assertion (v) holds by 9.3 and 11.4 in [37].

This concludes the first part of this appendix.

* * *

[10] By the comment following 11.1 in [37], U_α acts trivially on α.
[11] This makes sense by 29.10.iii.

We use the rest of this appendix to assemble an assortment of results about Coxeter chamber systems (29.16–29.30) and buildings (29.32–29.63) that are needed at various places in the text.

We begin with Coxeter chamber systems.

Proposition 29.16. *Let Σ be a Coxeter chamber system, let E be a residue of Σ, let α be a root such that $E \cap \alpha \neq \emptyset$ and let $u \in \alpha$. Then $\text{proj}_E\, u \in \alpha$, where proj_E is as defined in 29.7.i.*

Proof. Let $v = \text{proj}_E\, u$ and $x \in E \cap \alpha$. By 29.7.i, there is a minimal gallery from x to u that passes through v. Therefore $v \in \alpha$ since roots are convex (by 29.6.ii). □

Proposition 29.17. *Let Σ be a Coxeter chamber system, let E be a residue and let α and β be roots cutting E (as defined in 1.11) such that*

$$\alpha \cap E = \beta \cap E.$$

Then $\alpha = \beta$.

Proof. Since α cuts E and E is connected, there exists a panel $\{d, e\}$ in E such that $d \in \alpha$ and $e \in -\alpha$. Since $\beta \cap E = \alpha \cap E$, we have $d \in \beta$ and $e \in -\beta$. By 29.6.i, there is only one root that contains d but not e. □

Proposition 29.18. *Let Σ be a finite Coxeter chamber system and let α be a root of Σ. Then $|\alpha| = |\Sigma|/2$.*

Proof. By 29.6, α and $-\alpha$ form a partition of the chamber set of Σ and there is a reflection interchanging these two subsets. □

Proposition 29.19. *Let Σ be a Coxeter chamber system, let E be a residue of Σ and let X be the set of roots of Σ cutting E. Then the map $\alpha \mapsto \alpha \cap E$ is a bijection from X to the set of roots of E.*

Proof. Let $\psi(\alpha) = \alpha \cap E$ for each $\alpha \in X$. By 29.7.iv, $\psi(\alpha)$ is a root of E for each $\alpha \in X$, and by 29.17, the map ψ is injective. Suppose now that β is an arbitrary root of E. Choose a panel $\{x, y\}$ in E such that $x \in \beta$ and $y \notin \beta$ and let

$$\alpha = \{u \mid \text{dist}(x, u) < \text{dist}(y, u)\}.$$

By 29.7.iii, E is itself a Coxeter chamber system. By 29.6.i and 29.7.ii, therefore, $\alpha \in X$ and $\psi(\alpha) = \beta$. Thus ψ is surjective. □

Proposition 29.20. *Let Σ be a Coxeter chamber system, let $X \subset \Sigma$ and let W be the convex hull*[12] *of X. Then either $W = \Sigma$ or W is the intersection of all the roots of Σ that contain X.*

Proof. Let V denote the intersection of all the roots of Σ that contain X assuming there are such roots; if not, let $V = \Sigma$. By 29.6.ii, V is convex.

---

[12] The *convex hull* of X is the intersection of all the convex subsets of Σ that contain X; see footnote 3.

Therefore $W \subset V$ and V is connected. Thus if $W \neq V$, then we can choose a pair (x, y) of adjacent chambers such that $x \in W$ and $y \in V \backslash W$. Let

$$\alpha = \{u \mid \text{dist}(u, x) < \text{dist}(u, y)\}.$$

By 29.6.i, α is the unique root of Σ that contains x but not y and its opposite $-\alpha$ is

$$\{u \mid \text{dist}(u, x) > \text{dist}(u, y)\}.$$

Since $y \notin \alpha$ but $y \in V$, we have $X \not\subset \alpha$. Choose $u \in -\alpha \cap X$. Then $\text{dist}(u, x) > \text{dist}(u, y)$. Hence y lies on a minimal gallery from x to u. Since W is convex, this contradicts the choice of y. We conclude that $W = V$. □

Proposition 29.21. *Let Σ be a Coxeter chamber system, let E be a residue of Σ cut by a root α, let $\{d, e\}$ be a panel such that α contains d but not e, let $u = \text{proj}_E\, d$ and let $v = \text{proj}_E\, e$. Then α contains u but not v and the chambers u and v are adjacent.*

Proof. By 29.16, $u \in \alpha$ and $v \in -\alpha$. By 29.6.i,

$$\text{dist}(u, e) = \text{dist}(u, d) + 1$$

and

$$\text{dist}(v, d) = \text{dist}(v, e) + 1.$$

By 29.7.i applied to E and e,

$$\text{dist}(u, e) = \text{dist}(u, v) + \text{dist}(v, e),$$

and by 29.7.i applied to E and d,

$$\text{dist}(v, d) = \text{dist}(v, u) + \text{dist}(u, d).$$

Therefore

$$\text{dist}(u, d) + 1 = \text{dist}(u, v) + \text{dist}(v, e)$$

and

$$\text{dist}(v, e) + 1 = \text{dist}(v, u) + \text{dist}(u, d).$$

From these last two equations it follows first that $\text{dist}(v, e) = \text{dist}(u, d)$ and then that $\text{dist}(u, v) = 1$. □

Definition 29.22. Let Σ be a Coxeter chamber system. For each root α of Σ, the *wall of* α is the set of panels of Σ containing exactly one chamber in α (and thus exactly one in $-\alpha$). By 3.14 in [37], this is the same as the wall M_s as defined in 29.6, where s is the unique reflection of Σ interchanging α and its opposite. We denote the wall of a root α by $\mu(\alpha)$. A *wall* is a set of the form $\mu(\alpha)$ for some root α and the *chamber set* of a wall M is the set of chambers contained in some panel in M.

Proposition 29.23. *Let Σ be a Coxeter chamber system and let α be a root of Σ. Let Ω be the graph with vertex set the set of panels in the wall $\mu(\alpha)$ (as defined in 29.22), where two panels in $\mu(\alpha)$ are joined by an edge of Ω whenever they are contained in a residue of rank 2. Then Ω is connected.*

Proof. Let P and Q be two panels in $\mu(\alpha)$, let $d \in P \cap \alpha$, let $x \in Q \cap \alpha$ and let $k = \text{dist}(d, x)$. We proceed by induction with respect to k. Let y be the other chamber in Q. By 29.6.i, there is a minimal gallery from $(d, \dots, z, x, y)$ from d to y that passes through x. By 29.6, there is a reflection s that interchanges α and $-\alpha$ and fixes every panel in the wall $\mu(\alpha)$. Thus $P = \{d, d^s\}$ and $Q = \{x, x^s\}$. Hence if $k = 0$, then $P = Q$.

Suppose that $k > 0$, let E be the residue of rank 2 that contains x, y and z, let $Q_1 = \text{proj}_E P$ and let $x_1 = \text{proj}_E d$. By 29.7.i, $\text{dist}(d, x_1) < k$. By 29.21, Q_1 is a panel of E contained in $\mu(\alpha)$. Thus Q_1 is a vertex of Ω adjacent to Q. By induction, there is a path in Ω from P to Q_1. □

Proposition 29.24. *Let Σ be a Coxeter chamber system and let α and β be roots such that the four sets $\pm\alpha \cap \pm\beta$ are all non-empty. Then there exists a residue of rank 2 cut by both α and β.*

Proof. There exist chambers $x, y \in \beta$ such that $x \in \alpha$ but $y \in -\alpha$. Let γ be a minimal gallery from x to y. Then γ contains adjacent chambers x' and y' such that $x' \in \alpha$ and $y' \in -\alpha$. By 29.6.ii, x' and y' are both contained in β. Similarly, there exist adjacent chambers u and v contained in $-\beta$ such that $u \in \alpha$ and $v \in -\alpha$. The claim follows now by 29.23. □

Corollary 29.25. *Let Σ be a finite Coxeter chamber system and let α and β be two roots that are not opposite each other. Then there exists a residue of rank 2 cut by both α and β.*

Proof. Since $|\alpha| = |\Sigma|/2 = |\beta|$ (by 29.18) and $\alpha \cap \beta \neq \emptyset$, the sets $\pm\alpha \cap \pm\beta$ are all non-empty. The claim holds, therefore, by 29.24. □

Definition 29.26. Let Σ be a Coxeter chamber system. Two panels P and Q of Σ will be called *comural* if they belong to the same wall.

Proposition 29.27. *Let Σ be a Coxeter chamber system of type* Π. *Suppose that P and Q are comural panels of types i and j. Then i and j are connected by a path in Π all of whose edges have an odd label.*

Proof. Let i, j be distinct vertices of Π and let E be an $\{i, j\}$-residue. By 29.7.iii, $|E| = 2n$, where $n = 2$ if i and j are not adjacent in Π and n is the label on the edge $\{i, j\}$ if they are adjacent in Π. Furthermore, comural (i.e. opposite) panels in E have the same type if and only if n is even. The claim holds, therefore, by 29.23. □

Proposition 29.28. *Let E and E_1 be two residues of a Coxeter chamber system Σ and let $(d_0, d_1, \dots, d_k)$ be a gallery of minimal length such that $d_0 \in E$ and $d_k \in E_1$. For each $i \in [1, k]$, let α_i be the unique root of Σ containing d_i but not d_{i-1}. Then the roots $\alpha_1, \dots, \alpha_k$ are distinct and these are precisely the roots of Σ containing E_1 whose opposites contain E.*

Proof. We have $d_0 = \text{proj}_E d_i$ and $d_k = \text{proj}_{E_1} d_i$ for all $i \in [0, k]$. By 29.7.i, therefore, $d_0, \dots, d_{i-1}$ and all the chambers in E are nearer to d_{i-1} than to d_i, and $d_i, \dots, d_k$ and all the chambers of E_1 are nearer to d_i than to d_{i-1}

(for all $i \in [1,k]$). Thus by 29.6.i, the roots $\alpha_1,\ldots,\alpha_k$ are distinct, they all contain E_1 and their opposites all contain E. If α is another root containing E_1 whose opposite contains E, then α contains d_i but not d_{i-1} for some $i \in [1,k]$ and thus $\alpha = \alpha_i$ (again by 29.6.i). □

Proposition 29.29. *Let Σ be a Coxeter chamber system of type Π and let σ be an automorphism of the Coxeter diagram Π. Let J be a subset of the vertex set I of Π and let ϕ be a $\sigma|_J$-isomorphism (as defined in 29.2) from a J-residue E to a $\sigma(J)$-residue E' of Σ. Then ϕ has a unique extension to a σ-automorphism of Σ.*

Proof. Let the elements $r_i \in W$ for $i \in I$ be as in 29.4. Then the map

$$r_i \mapsto r_{\sigma(i)}$$

extends to an σ-automorphism ψ of Σ. Thus the composition $\psi^{-1}\phi$ is a special isomorphism from E to $E_1 := \psi^{-1}\phi(E)$. The group $\text{Aut}^\circ(\Sigma)$ of special automorphisms of Σ acts transitively on the chamber set of Σ. There thus exists $\pi \in \text{Aut}^\circ(\Sigma)$ such that $\pi^{-1}\psi^{-1}\phi$ is a special isomorphism from E to $\pi^{-1}(E_1)$ fixing a chamber u of E. Since E is the unique J-residue containing u, it follows that $\pi^{-1}(E_1) = E$. Since E is thin, the only special automorphism of E fixing a chamber is the identity. Thus $\psi\pi$ is a σ-automorphism of Π extending ϕ. Since Σ is thin, there is no other extension of ϕ to a σ-automorphism of Π. □

Proposition 29.30. *Let Σ be a Coxeter chamber system of type Π and let Γ be the underlying graph (i.e. Σ without its edge coloring). Suppose that every label of Π is finite. Then every automorphism of Γ is an automorphism of Σ in the sense of 29.2.*

Proof. Let d be a chamber of Σ and let $i \mapsto d_i$ be the bijection from I to the set of chambers adjacent to d such that i is the type of the panel $\{d, d_i\}$ for each $i \in I$. For each two-element subset $\{i,j\}$ of I, let $m_{ij} = |E_{ij}|/2$, where E_{ij} is the $\{i,j\}$-residue containing d, let x_{ij} be the chamber opposite d in E_{ij} (which exists because $m_{ij} < \infty$), let γ_{ij} be the minimal gallery in E_{ij} from d to x_{ij} that passes through d_i and let γ_{ji} be the minimal gallery in E_{ij} from d to x_{ij} that passes through d_j.

Let $\{i,j\}$ be a two-element subset of I. Suppose that $\gamma = (d, d_i, \ldots, w)$ and $\gamma' = (d, d_j, \ldots, w')$ are minimal galleries both of length m_{ij} such that $w = w'$. Let f and g be the types of γ and γ'. By 29.8.i, f and g are homotopic as defined in 4.1 of [37]. It follows that f and g are the two words $p(i,j)$ and $p(j,i)$ defined in 3.1 of [37]. Hence $\gamma = \gamma_{ij}$ and $\gamma' = \gamma_{ji}$, so the union of γ and γ' is the residue E_{ij}.

Let $\sigma \in \text{Aut}(\Gamma)$. We want to show that $\sigma \in \text{Aut}(\Sigma)$. Since $\text{Aut}(\Sigma)$ acts transitively on the chamber set of Σ (by 29.4), we can assume that σ fixes the chamber d. There thus exists a unique permutation $i \mapsto i'$ of I such that $d_i^\sigma = d_{i'}$ for each $i \in I$. By the conclusion of the previous paragraph,

$$E_{ij}^\sigma = E_{i'j'} \tag{29.31}$$

and hence $m_{ij} = |E_{ij}|/2 = |E_{i'j'}|/2 = m_{i'j'}$ for each two-element subset $\{i,j\}$ of I. This means that the permutation $i \mapsto i'$ lies in $\text{Aut}(\Pi)$. By 29.5, there is therefore an automorphism τ of Σ that fixes d and maps d_i to $d_{i'}$ for all $i \in I$. Replacing σ by $\sigma\tau^{-1}$, we can thus assume that σ fixes each neighbor of d. By 29.31, σ acts trivially on each rank 2 residue containing d.

Let e be a chamber adjacent to d and let w be a chamber adjacent to e distinct from d. Then σ acts trivially on the unique residue of rank 2 containing d, e and w. Thus, in particular, σ fixes w. We conclude that σ acts trivially on the set of chambers adjacent to e for each chamber e adjacent to d. Since Σ is connected, it follows that $\sigma = 1$. □

The conclusion of 29.30 is not, in general, true without the restriction on the labels of Π. If *all* the labels are infinite, for example, then Γ is a tree, so if $|I| > 2$, $\text{Aut}(\Gamma)$ is much larger than $\text{Aut}(\Sigma)$.

* * *

We now turn to buildings.

Definition 29.32. Let Δ be a building and let α be a root of Δ (i.e. a root of some apartment of Δ). Then $\mu(\alpha)$ denotes the set of panels P of Δ such that $|P \cap \alpha| = 1$. The set $\mu(\alpha)$ is called the *wall of the root* α. A *wall of* Δ is a set of panels of the form $\mu(\alpha)$ for some root α. We call the set of chambers contained in some panel of a wall M the *chamber set* of M.

Proposition 29.33. *Let Σ be an apartment in a building (Δ, δ) as defined in 29.10 and 29.13, let α be a root of Σ, let P be a panel in the wall $\mu(\alpha)$ as defined in 29.32, let x be the unique chamber in $P \cap \alpha$ and let u be any other chamber in P. Then there exists an apartment of Δ containing $\alpha \cup \{u\}$.*

Proof. Let y be the unique chamber in $P \cap \Sigma$ that is not in α and let v be an arbitrary chamber of α. By 29.6.i, there exists a minimal gallery γ from v to y that passes through x. Let f be the type of γ. Replacing y by u yields a gallery from v to u that is also of type f. By 29.10.i, f is reduced. By 29.12, therefore, $\delta(v,y) = \delta(v,u)$. The map from $\alpha \cup \{y\}$ to $\alpha \cup \{u\}$ that maps y to u and acts trivially on α is thus an isometry (as defined in 8.1 of [37]). By 29.13, this map extends to a special isomorphism from Σ to Δ and the image of this extension is an apartment containing $\alpha \cup \{u\}$. □

Proposition 29.34. *Let Σ and Σ' be two apartments of a building Δ. Then there exists a special isomorphism π from Σ to Σ' that acts trivially on $\Sigma \cap \Sigma'$, and π is the unique special isomorphism from Σ to Σ' that fixes any given chamber in $\Sigma \cap \Sigma'$.*

Proof. Let Π be the Coxeter diagram of Δ. All apartments of Δ are isomorphic to Σ_Π, so there is nothing to prove if $\Sigma \cap \Sigma' = \emptyset$. Suppose, therefore, that $\Sigma \cap \Sigma'$ contains a chamber x. Then $\{x\}$ is a J-residue for $J = \emptyset$. By 29.29, there exists a unique special isomorphism π from Σ to Σ' that fixes x. Let y be an arbitrary chamber in $\Sigma \cap \Sigma'$. By 29.13.iii, apartments are convex. We can thus choose a gallery from x to y contained in $\Sigma \cap \Sigma'$. By

2.5 of [37], γ is the only gallery of its type beginning at x that is contained in Σ. Since π is special, it follows that π fixes y. Thus π acts trivially on $\Sigma \cap \Sigma'$. □

The following characterization of buildings is borrowed from 3.11 in the book [25] by M. Ronan.

Proposition 29.35. *Let Δ be a chamber system whose index set is the vertex set I of the Coxeter diagram Π and let $\mathcal{A}$ be a collection of subgraphs*[13] *isomorphic to the Coxeter chamber system Σ_Π. Suppose that the following hold:*

(i) *For every two chambers x and y of Δ, there exists an element of $\mathcal{A}$ that contains them both.*
(ii) *For every two elements $A, A' \in \mathcal{A}$ and for every pair x, y of chambers in $A \cap A'$, there exists a special isomorphism from A to A' that fixes x and y.*
(iii) *For every two elements $A, A' \in \mathcal{A}$, for every chamber x in $A \cap A'$ and for every panel P such that $P \cap A$ and $P \cap A'$ are both non-empty, there exists a special isomorphism from A to A' that fixes x and maps $A \cap P$ to $A' \cap P$.*

Then Δ is a building of type Π.

Proof. Let $\Sigma = \Sigma_\Pi$, let $W = W_\Pi$ and let $\delta_W(x, y) = x^{-1}y$ for all $x, y \in W$. By 2.5 in [37], $\delta_W(x, y) = r_f$ for each ordered pair (x, y) in $W \times W$, where f is the type of an arbitrary gallery in Σ from x to y and r_f is as defined in 29.8. The map δ_W is invariant under the group $\text{Aut}^\circ(\Sigma)$ of special automorphisms of Σ (i.e. under left multiplication by elements of W).

Now let $A \in \mathcal{A}$, let π be a special isomorphism from A to Σ and let

$$\delta_A(x, y) = \delta_W(x^\pi, y^\pi)$$

for all chambers x and y of A. By 29.4, the group $\text{Aut}^\circ(\Sigma)$ of special automorphisms of Σ acts transitively on Σ. By 29.29, therefore, π is unique up to an element of $\text{Aut}^\circ(\Sigma)$. It follows that δ_A is independent of the choice of the special isomorphism π.

Now let x and y be chambers of Δ. By (i), there exists $A \in \mathcal{A}$ containing x and y. We set

$$\delta(x, y) = \delta_A(x, y).$$

By (ii), $\delta(x, y)$ is independent of the choice of A. If $\delta(x, y) = r_f$ for some $f \in M_I$, then by 2.5 of [37], there exists a gallery of type f in A from x to y.

By 29.10, it will suffice to show that for all $x, y \in \Delta$ and all reduced words $f \in M_I$ (as defined in 29.8), $\delta(x, y) = r_f$ whenever there is a gallery from x to y of type f. We prove this assertion by induction with respect to the length of the reduced word f. If f is a reduced word of length at most 1 and

[13] We are using the convention in 29.35 that "A is a subgraph" means that two chambers of A are i-adjacent in Δ for some $i \in I$ if and only if they are i-adjacent in A.

γ is a gallery of type f from x to y, then an element A of $\mathcal{A}$ that contains x and y contains γ (by our convention about the meaning of "subgraph") and hence $\delta(x,y) = \delta_A(x,y) = r_f$. Now let $f \in M_I$ be a reduced word of length n greater than 1. We assume that our assertion is true for all reduced words in M_I whose length is less than n and thus, in particular, for all proper subwords of f.

Let $f = ihj$ for $i, j \in I$ and $h \in M_I$ (which might be empty) and let

$$\gamma = (x, u, \ldots, v, y)$$

be a gallery in Δ of type f. Let A be an element of $\mathcal{A}$ containing both x and y and let A' be an element of $\mathcal{A}$ containing x and v. By induction, we have

$$\delta(x,v) = r_{ih} \tag{29.36}$$

and

$$\delta(u,y) = r_{hj}. \tag{29.37}$$

By 29.36, there is a unique gallery γ_1 in A' from x to v of type ih. Let P be the unique j-panel of Δ containing v and y. By (iii), there exists a special isomorphism π from A' to A fixing x and mapping $P \cap A'$ to $P \cap A$. Thus γ_1^π is a gallery of type ih from x to v^π and $v^\pi \in P \cap A$.

Suppose that $v^\pi = y$. Both u and u^π are i-adjacent to x. Thus if $u \neq u^\pi$, then u and u^π are i-adjacent, so

$$(u, u^\pi, \ldots, v^\pi)$$

is a gallery of type ih from u to y. If $u = u^\pi$, then

$$(u^\pi, \ldots, v^\pi)$$

is a gallery of type h from u to y. By induction, it follows that $\delta(u,y) = r_{ih}$ or r_h. By 29.37, therefore $r_{hj} = r_{ih}$ or r_h. By 29.8.iii, it follows that $u \neq u^\pi$ and that hj is homotopic to ih as defined in 4.1 of [37]. This implies, however, that $f = ihj$ is homotopic to hjj. This conclusion contradicts the assumption that f is reduced. Therefore $v^\pi \neq y$.

Since $v^\pi \neq y$, it follows that

$$(\gamma_1^\pi, y)$$

is a gallery of type $f = ihj$ in A. Therefore $\delta(x,y) = r_f$. □

Definition 29.38. Let P and Q be two panels of a building Δ. We say that P and Q are *comural* if they are contained in the same wall, i.e. if there is a root α in some apartment such that both P and Q are contained in $\mu(\alpha)$ (as defined in 29.32).

Proposition 29.39. *Let Δ be a building, let E be a residue of Δ, let α be a root of an apartment Σ such that $E \cap \alpha \neq \emptyset$ and let $u \in \alpha$. Then* $\text{proj}_E\, u \in \alpha$ *(where* proj_R *is as defined in 29.10.v.)*

Proof. Let $v = \text{proj}_E u$ and $x \in E \cap \alpha$. By 29.10.v, there is a minimal gallery from x to u that passes through v. Therefore $v \in \alpha$ since α is convex (by 29.6.ii and 29.13.iii). □

Definition 29.40. Let α denote a root. We denote by $\partial\alpha$ the intersection of α with the chamber set of the wall $\mu(\alpha)$ (as defined in 29.32). Thus a chamber x is contained in the set $\partial\alpha$ if and only if x is the unique chamber in $\alpha \cap P$ for some panel P. We call $\partial\alpha$ the *border* of α.

Proposition 29.41. *Let A be an apartment of a building Δ, let α be a root of A and let $x \in \partial\alpha$ (where $\partial\alpha$ is as defined in 29.40). Then*

$$\partial\alpha = \{\text{proj}_P x \mid P \in \mu(\alpha)\}.$$

Proof. This holds by 29.39 applied to both α and to its opposite $-\alpha$ in A. □

Proposition 29.42. *Let A and A' be two apartments of a building Δ, let α be a root of A, let α' be a root of A' and suppose that $\mu(\alpha) = \mu(\alpha')$. Then the borders $\partial\alpha$ and $\partial\alpha'$ are either equal or disjoint.*

Proof. This holds by 29.41. □

Proposition 29.43. *Let P and Q be two panels of a building Δ. Then P and Q are comural if and only if* proj_P *restricted to Q and* proj_Q *restricted to P are inverses of each other.*

Proof. Suppose first that P and Q are comural. Then P and Q are both contained in $\mu(\alpha)$ for some root α. Let x be the unique chamber in $P \cap \alpha$ and let u be the unique chamber in $Q \cap \alpha$. By 29.39, $u = \text{proj}_Q x$ and $x = \text{proj}_P u$. Choose $y \in P \backslash \{x\}$. By 29.33, there exists an apartment A containing $\alpha \cup \{y\}$. Then A contains a unique chamber v in $Q \backslash \{u\}$. By 29.13.v, α is a root of A. Let β be the root of A opposite α. Then y is the unique chamber in $P \cap \beta$ and v is the unique chamber in $Q \cap \beta$. By 29.39 again, we have $v = \text{proj}_Q y$ and $y = \text{proj}_P v$. Thus the composition $\text{proj}_P \circ \text{proj}_Q$ restricted to P is the identity. By a similar argument the composition $\text{proj}_Q \circ \text{proj}_P$ restricted to Q is the identity.

Suppose, conversely, that $\text{proj}_P \circ \text{proj}_Q$ restricted to P is the identity map and $\text{proj}_Q \circ \text{proj}_P$ restricted to Q is the identity map. By 29.13.ii, there exists an apartment A containing chambers of both P and Q. Let $A \cap P = \{x, y\}$, let α be the unique root of A containing x but not y and let β be the root opposite α in A. Thus $P \in \mu(\alpha)$. Let $u \in A \cap Q$. By 29.39, $\text{proj}_P u = x$ if $u \in \alpha$ and $\text{proj}_P u = y$ if $u \in \beta$. Since $|A \cap Q| = 2$ and the restriction of proj_P to $A \cap Q$ is injective, it follows that also $Q \in \mu(\alpha)$. □

From now on, we focus on spherical buildings.

Proposition 29.44. *Let Δ be a spherical building and let P and Q be opposite panels (as defined in 9.8 of [37]). Then P and Q are comural.*

Proof. By 9.8 of [37], there exists an apartment Σ cutting both P and Q such that $\text{op}_\Sigma(P \cap \Sigma) = Q \cap \Sigma$, where op_Σ is as defined in 29.9.iv. Let $P \cap \Sigma = \{x_1, x_2\}$ and let α be the unique root of Σ that contains x_1 but not x_2. Then $Q \cap \Sigma = \{y_1, y_2\}$, where $y_i = \text{op}_\Sigma x_i$ for $i = 1$ and 2. By 29.9.i, α contains y_2 but not y_1. Thus the wall $\mu(\alpha)$ contains both P and Q. □

Proposition 29.45. *If Δ is a spherical building, then the number of chambers in a root does not depend on the choice of the root.*

Proof. This holds by 29.18 since every two apartments in a building are isomorphic (by 29.13). □

Proposition 29.46. *Let Δ be a spherical building, let $n = \text{diam}(\Delta)$ and let x and y be two chambers of Δ such that $\text{dist}(x, y) = n - 1$. Then there is a unique root of Δ containing both x and y, and this root is the convex hull of the set $\{x, y\}$*

Proof. By 29.13.ii, there is an apartment Σ containing $\{x, y\}$. By 29.6.iii, there are exactly $n-1$ roots of Σ containing x but not y and an equal number containing y but not x. By 29.9, Σ has $2n$ roots altogether. If a root of Σ contains both x and y, then its opposite in Σ contains neither. We conclude that there is exactly one root of Σ containing both x and y. The claim holds, therefore, by 29.13.iii and 29.20. □

Proposition 29.47. *Let Δ be a spherical building, let m be a wall of Δ and let u be a chamber contained in m. Then there is a unique root containing u whose wall is m.*

Proof. There exists a root α such that $m = \mu(\alpha)$. Suppose that $u \notin \alpha$. By 29.14.iii, there exists an apartment of Δ containing α and u. Replacing α by its opposite in this apartment, we can thus assume that $u \in \alpha$. It remains only to show that α is unique.

Let Σ be an arbitrary apartment containing α, let P be the panel in $\mu(\alpha)$ containing u, let v be the other chamber in $P \cap \Sigma$ and let $w = \text{op}_\Sigma v$. Thus v is the unique chamber of Σ opposite w. By 29.13.iii, it follows that $\text{dist}(w, u) = n-1$, where n is the diameter of Σ. By 29.9.i, $w \in \alpha$, and hence by 29.46, α is the unique root of Δ containing w and u.

Let Q be the panel in m that contains w and let β be a second root containing u whose wall is m. Then $\text{proj}_Q u \in \alpha \cap \beta$ by 29.39. Since w is the unique chamber in $Q \cap \alpha$ and $|Q \cap \beta| = 1$, it follows that $w \in \beta$. Thus $\beta = \alpha$ by the conclusion of the previous paragraph. □

Proposition 29.48. *Let Δ be a spherical building and let α and β be two roots having the same wall. Then either $\alpha = \beta$ or there is a unique apartment containing α and β.*

Proof. This holds by 29.14.iii and 29.47. □

Proposition 29.49. *Let Δ be a spherical building, let m be a wall of Δ and let u, v be two chambers in the same panel of m. Then there exists a unique apartment containing u, v and a root whose wall is m.*

Proof. By 29.47, there exists a unique root α containing u whose wall is m and a unique root β containing v whose wall is m. Since u is not contained in β, we have $\alpha \neq \beta$. By 29.48, therefore, there exists a unique apartment containing α and β. □

Proposition 29.50. *Let Δ be a thick spherical building and let P and Q be two panels of the same type. Then there exists a panel R opposite both P and Q (as defined in 9.8 of [37]).*

Proof. Let σ be the automorphism of the Coxeter diagram Π of Δ defined in 29.9.v. By 9.10 of [37], it will suffice to show that for every two chambers of Δ, there exists a third chamber opposite them both. Let u, v be two chambers of Δ and let $m = \text{dist}(u, v)$. We will proceed by induction with respect to m. Let $m = 1$, let i be the type of the panel containing u and v, let w be a chamber opposite u and let E be the panel of type $\sigma(i)$ containing w. By 9.9 and 9.10 of [37], E contains chambers opposite v. By 29.10.v, it follows that u is opposite every chamber of E except $\text{proj}_E\, u$ and v is opposite every chamber of E except $\text{proj}_E\, v$. Since Δ is thick, there exist chambers in E opposite both u and v.

Now let $m > 1$ and let $(u, \ldots, x, v)$ be a minimal gallery from u to v. By induction, there exists a chamber z opposite both u and x. Let i be the type of the panel containing x and v and let D be the panel of type $\sigma(i)$ containing z. By 9.9 and 9.10 of [37], D contains chambers opposite v. By 29.10.v, it follows that u is opposite every chamber in D except $\text{proj}_D\, u$ and v is opposite every chamber in D except $\text{proj}_D\, v$. Since Δ is thick, there exist chambers in D opposite both u and v. □

Proposition 29.51. *Let Δ be a thick irreducible spherical building, let Π be the Coxeter diagram of Δ and let Ω be the graph with vertex set the set of panels of Δ, where two panels are adjacent in Ω whenever they are comural. Then two panels of Δ are in the same connected component of Ω if and only if their types i and j are joined by a path in Π going only through edges with label 3. In particular, Ω is connected if Π has only single bonds and Ω has two components otherwise. In any case, for each chamber d of Δ, each connected component of Ω contains panels containing d.*

Proof. Let P and Q be two panels and let d be a chamber of Δ. If P and Q have the same type, then by 29.50, there is a panel R opposite both P and Q, and by 29.44, R is comural to both P and Q. Hence panels of the same type lie in the same connected component of Ω. It follows, in particular, that each connected component of Ω contains panels containing d.

If i and j are elements of the vertex set I of Π connected by an edge with label 3, then every $\{i, j\}$-residue E contains i-panels and j-panels that are comural. By the conclusion of the previous paragraph, it follows that two panels of Δ are in the same connected component of Ω if their types i and j are joined by a path in Π going only through edges with label 3. Since Π is spherical, 3 is the only odd label on an edge of Π. The converse holds, therefore, by 29.27. □

Proposition 29.52. *Let Δ be an irreducible spherical building satisfying the Moufang condition, let Σ be an apartment of Δ, let $d \in \Sigma$ and let i, j be two vertices of Π. Then the roots a_i and a_j (as defined in 3.5) are in the same $G^\dagger$-orbit (where $G^\dagger$ is as in 29.15.iii) if and only if i and j are connected by a path in Π passing only through edges with label 3.*

Proof. Let P_i be the i-panel Σ containing d for each vertex i of Π. The elements of $G^\dagger$ are type-preserving. Thus if $a_i = a_j^g$ for some $g \in G^\dagger$, then the wall of a_i contains the i-panel P_i and the j-panel P_i^g. By 29.27, it follows that i and j are connected by a path in Π passing only through edges with label 3 (since Π is spherical).

To prove the converse, it suffices to assume that $\{i, j\}$ is an edge of Π with label 3. Let E be the unique $\{i, j\}$-residue of Σ containing d. By 29.7.iii, E is a circuit of length 6. Let d_1 be the chamber opposite d in E, let Q be the panel of Σ opposite P_j in E and let e be the unique chamber in $Q \cap \Sigma$ distinct from d_1. Then Q is of type i and a_j is the unique root of Σ containing e but not Q. By 29.15.iii, the stabilizer $G_\Sigma^\dagger$ of Σ in $G^\dagger$ acts transitively on the set of chambers of Σ. In particular, there is an element $g \in G_\Sigma^\dagger$ mapping d to e and hence P_i to Q (since $G^\dagger$ is type-preserving). Thus a_i^g is a root of Σ containing e but not Q. Therefore $a_j = a_i^g$. □

Proposition 29.53. *Let Δ and Δ' be two spherical buildings of the same type Π and let ϕ be a σ-homomorphism from Δ to Δ' for some automorphism σ of Π (as defined in 29.2). Suppose that ϕ is surjective. Then every panel of Δ' is of the form $\phi(P)$ for some panel P of Δ.*

Proof. Let x' be a chamber of Δ'. Suppose that y' is a chamber of Δ' that is opposite x'. Since ϕ is surjective, we can choose pre-images x and y of x' and y'. Since ϕ is a homomorphism, we have

$$\text{dist}_{\Delta'}(x', y') \leq \text{dist}_{\Delta}(x, y).$$

Since Δ and Δ' are both of type Π, they have the same diameter (by 29.14.i). Hence, in fact,

$$\text{dist}_{\Delta'}(x', y') = \text{dist}_{\Delta}(x, y)$$

and every minimal gallery from x to y in Δ is mapped to a minimal gallery in Δ' from x' to y'. If f is the type of a minimal gallery γ in Δ from x to y, then $\sigma(f)$ is the type of $\phi(\gamma)$. By 7.7.iii in [37], two distinct minimal galleries with the same first and last chambers in a building cannot have the same type. We conclude that ϕ induces a bijection from the set of minimal galleries in Δ from x to y to the set of minimal galleries in Δ' from x' to y'.

Now let γ' be a minimal gallery in Δ' that begins at x'. Let u' be its last chamber. By 29.13.ii, there is an apartment Σ' of Δ' containing x' and u'. By 29.9.iv, there is a unique chamber v' opposite x' in Σ'. By 29.9.iii and 29.13.iii,

$$\text{dist}_{\Delta'}(x', v') = \text{dist}_{\Delta'}(x', u') + \text{dist}_{\Delta'}(u', v').$$

Hence there is a minimal gallery γ_1' from x' to v' that extends γ'. By the conclusion of the previous paragraph, there is a gallery γ in Δ beginning at x such that $\phi(\gamma) = \gamma'$.

Now let P be a panel of Δ and let i denote its type. Since ϕ is a homomorphism, $\phi(P)$ is contained in a $\sigma(i)$-panel P' of Δ'. Let $x \in P$, let $x' = \phi(x)$ and let $u' \in P' \backslash \{x'\}$. By the conclusion of the previous paragraph, there exists $u \in P$ such that $\phi(x, u) = (x', u')$. Thus $\phi(P) = P'$. □

Proposition 29.54. *Let Δ be a spherical building, let Σ_1 and Σ_2 be two apartments whose intersection is a root α and let Q be a panel of Δ such that $|\alpha \cap Q| = 2$. For $i = 1$ and 2, let P_i be the unique panel of Δ such that $P_i \cap \Sigma_i$ is the panel opposite $\alpha \cap Q$ in Σ_i and let α_i be the root opposite α in Σ_i. Let $\Sigma = \alpha_1 \cup \alpha_2$. Then Σ is an apartment, α_1 and α_2 are roots of Σ and the unique reflection of this apartment that interchanges α_1 and α_2 also interchanges $P_1 \cap \Sigma$ and $P_2 \cap \Sigma$.*

Proof. Let P be a panel in $\mu(\alpha)$ and let u_i be the unique chamber in $P \cap \alpha_i$ for $i = 1$ and 2. By 29.14.iii, there is a unique apartment Σ_0 containing $\alpha_1 \cup \{u_2\}$. By 29.13.v, α_1 is a root of Σ_0. Let α_0 be the root opposite α_1 in Σ_0. Then $\mu(\alpha_0) = \mu(\alpha_1) = \mu(\alpha_2)$ and $u_2 \in \alpha_0 \cap \alpha_2$. By 29.47, it follows that $\alpha_0 = \alpha_2$. Thus $\Sigma_0 = \Sigma$. Hence Σ is an apartment and α_1 and α_2 are roots of Σ.

By 29.34, there are isomorphisms τ_1 from Σ_1 to Σ_2 and τ_2 from Σ_2 to Σ_1 that both act trivially on α. Let τ be the map from Σ to itself that restricts to τ_i on α_i for $i = 1$ and 2. Then τ is a special automorphism of Σ that maps each panel in the wall $\mu(\alpha)$ to itself. Thus τ is the unique reflection of Σ that interchanges α_1 and α_2. The map τ_1 acts trivially on α and hence maps $Q \cap \alpha$ to itself. Therefore τ maps $P_1 \cap \Sigma$ to $P_2 \cap \Sigma$. Similarly, τ maps $P_2 \cap \Sigma$ to $P_1 \cap \Sigma$. □

Proposition 29.55. *Let Δ be a spherical building, let k denote the number of chambers in a root of Δ (which is independent of the root by 29.45) and let Σ_1 and Σ_2 be two apartments such that*

$$0 < |\Sigma_1 \cap \Sigma_2| < k$$

(where $|X|$ denotes the number of chambers in a subset X of Δ). Then there exists an apartment Σ_3 and adjacent chambers u and w such that

(i) *$\Sigma_1 \cap \Sigma_3$ is a root;*
(ii) *$u \in \Sigma_1 \cap \Sigma_2 \subset \Sigma_3$; and*
(iii) *$w \in (\Sigma_2 \cap \Sigma_3) \backslash \Sigma_1$.*

In particular, $|\Sigma_1 \cap \Sigma_2| < |\Sigma_1 \cap \Sigma_3|$ and $|\Sigma_1 \cap \Sigma_2| < |\Sigma_2 \cap \Sigma_3|$.

Proof. By 29.13.iii, $\Sigma_1 \cap \Sigma_2$ is convex. By 29.20, therefore, $\Sigma_1 \cap \Sigma_2$ is an intersection of roots of Σ_1. It follows that we can choose a root α of Σ_1 containing $\Sigma_1 \cap \Sigma_2$ and adjacent chambers u, v such that $u \in \Sigma_1 \cap \Sigma_2$ and $v \in \beta$, where β is the root opposite α in Σ_1. Let w be the unique chamber of Σ_2 adjacent to u such that the subset $\{u, v, w\}$ is contained in a panel.

Then $w \neq v$ since $v \notin \Sigma_1 \cap \Sigma_2$. In particular, $w \notin \Sigma_1$. By 29.14.iii, there exists a unique apartment Σ_3 containing α and w. Thus (ii) and (iii) hold and $\Sigma_1 \cap \Sigma_3$ is a proper convex subset of Σ_1 containing α. By 29.18, no root of Σ_1 contains α properly. By 29.20, it follows that also (i) holds. □

Proposition 29.56. *Let Δ be a spherical building satisfying the Moufang property, let Σ be an apartment of Δ and let u be a non-trivial element of the root group U_α for some root α of Σ. Then u acts trivially on α and $\Sigma \cap \Sigma^u = \alpha$.*

Proof. By 3.14 of [37], every chamber of α is contained in a panel that contains two chambers of α. By 29.15, therefore, u acts trivially on α. In particular, $\alpha \subset \Sigma \cap \Sigma^u$. By 29.13.iii, $\Sigma \cap \Sigma^u$ is a convex subset of Σ. By 29.18 and 29.20, it follows that $|\Sigma \cap \Sigma^u| \leq |\alpha|$. □

Proposition 29.57. *Let Δ be a building of type A_2, let $(x_0, x_1, \ldots, x_6)$ with $x_0 = x_6$ be the chambers in an apartment numbered consecutively and let α_i denote the root $\{x_i, x_{i+1}, x_{i+2}\}$ for each $i \in [1, 3]$. Suppose that the following hold:*

(i) *For all $i \in [1, 3]$, U_i is a subgroup of the pointwise stabilizer of α_i in* $\mathrm{Aut}(\Delta)$ *that acts transitively on the set of apartments of Δ containing α_i.*
(ii) $[U_1, U_2] = [U_2, U_3] = 1$.

Then U_2 equals the root group U_{α_2}.

Proof. Let P_i be the panel of Δ containing x_i and x_{i+1} for $i = 2$ and 3. Let $\rho \in U_2$ and let y be a chamber in $P_2 \backslash \{x_3\}$. By (i) (and 29.14.iii), U_3 contains an element τ mapping x_2 to y. Since ρ fixes x_2, the element ρ^τ fixes $x_2^\tau = y$. By (ii), however, $\rho = \rho^\tau$. Thus ρ fixes y. We conclude that U_2 acts trivially on P_2. By a similar argument, U_2 acts trivially on P_3 (since U_1 acts transitively on $P_1 \backslash \{x_3\}$ and commutes elementwise with U_2). This means, by 29.15, that $U_2 \subset U_{\alpha_2}$. By 29.15.v, it follows that $U_2 = U_{\alpha_2}$. □

Definition 29.58. For each building Δ and each chamber x, let $\Delta_x^{(1)}$ denote the set of chambers of Δ equal to or adjacent to x.

The following result is a just slightly stronger version of 9.7 in [37].[14]

Proposition 29.59. *Let u and v be opposite chambers in a thick spherical building Δ. Then Δ is the convex hull in Δ of*

$$\Delta_u^{(1)} \cup \{v\},$$

where $\Delta_u^{(1)}$ is as in 29.58.

---

[14]Let u and v be opposite chambers in a spherical building Δ, let

$$Z = \Delta_u^{(1)} \cup \{v\}$$

and let g be an element of $\mathrm{Aut}(\Delta)$ that acts trivially on the set Z. Since the chamber u is contained in panels of every color, the element g must be special. By 7.7.iii in [37], it follows that g acts trivially on the convex hull of Z. Therefore 29.59 implies 9.7 in [37].

Proof. Let X denote the convex hull in Δ of

$$\Delta_u^{(1)} \cup \{v\},$$

let x be a chamber adjacent to v and let P be the unique panel containing x and v. Since Δ is connected, it will suffice to show that there exists a chamber s opposite x such that

$$\Delta_s^{(1)} \cup \{x\} \subset X. \tag{29.60}$$

We begin by showing that $x \in X$. By 29.14.ii, there exists a unique apartment Σ containing u and v, and $\Sigma \subset X$. Thus $x \in X$ if $x \in \Sigma$. Suppose that $x \notin \Sigma$. Let z be the unique chamber in $P \cap \Sigma$ that is distinct from v and let $w = \text{op}_\Sigma z$, where op_Σ is as in 29.9.iv. By 29.9.v, u and w are adjacent. Let Q be the unique panel containing them both. Then P and Q are opposite panels as defined in 9.8 of [37]. By 9.9 in [37], every chamber of P is opposite chambers in Q and vice versa. By 29.10.v, it follows that every chamber of P is *not* opposite a unique chamber of Q and vice versa. Let y be the unique chamber of Q not opposite x. Then x is the unique chamber of P not opposite y. Since $x \neq v$, it follows that v is opposite y. There thus exists a minimal gallery from $y \in \Delta_u^{(1)}$ to v that passes through x. Hence $x \in X$ as claimed, whether or not $x \in \Sigma$. Note that we have, in fact, shown that if u_1 and v_1 are any two opposite chambers of Δ, then the convex hull of $\Delta_{u_1}^{(1)} \cup \{v_1\}$ contains $\Delta_{v_1}^{(1)}$.

If x is opposite u, then 29.60 holds with $s = u$ (since $x \in X$). Suppose that x is not opposite u. Then there exists a minimal gallery from u to v that passes through x. By 29.14.ii, therefore, $x \in \Sigma$. Let $y = \text{op}_\Sigma x$. By 29.9.v again, y and u are adjacent. Let Q_1 be the unique panel containing u and y. Thus P and Q_1 are opposite panels. Since Δ is thick, we can choose $w \in Q_1 \backslash \Sigma$. By 29.10.v again, every chamber of P is not opposite a unique chamber of Q_1. By 29.9.iv, y is not opposite v and u is not opposite x. Therefore x and v are both opposite w. By the conclusion of the previous paragraph, the convex hull of $\Delta_v^{(1)} \cup \{w\}$ contains $\Delta_w^{(1)}$ and X contains $\Delta_v^{(1)} \cup \{w\}$. It follows that 29.60 holds with $s = w$. □

Proposition 29.61. *Let Δ' be a thick subbuilding of a spherical building Δ whose rank equals the rank of Δ and suppose that Δ is Moufang. Let G be the setwise stabilizer of Δ' in* Aut(Δ) *and let π be the natural map from G to* Aut(Δ')*. For each root a of Δ, let U_a be the corresponding root group of Δ, and for each root a' of Δ', let $U'_{a'}$ be the corresponding root group of Δ'. Then the following hold:*

(i) *Every apartment of Δ' is also an apartment of Δ.*
(ii) *Every root of Δ' is also a root of Δ.*
(iii) *Δ' is Moufang.*
(iv) *For each root a of Σ, the restriction of the map π to $G \cap U_a$ is injective and its image is the root group U'_a.*

Proof. Assertion (i) holds by 8.8 in [37] and assertion (ii) follows from (i). Let a be a root of Δ'. By the definition of a root group in 29.15,

$$\pi(G \cap U_a) \subset U'_a. \tag{29.62}$$

Let P be a panel in the wall $\mu(a)$ and let Σ_1 and Σ_2 be two apartments of Δ' containing a. By (i) and (ii), Σ_1 and Σ_2 are apartments of Δ and a is a root of Δ. Since Δ is Moufang, there thus exists an element $g \in U_a$ mapping Σ_1 to Σ_2. By 4.12 in [37] and (a) and (b) in 29.15, we can choose a chamber z in a such that every panel containing z contains two chambers in a. Let

$$X_i = \Sigma_i \cup (\Delta')_z^{(1)}$$

for $i = 1$ and 2, where $(\Delta')_z^{(1)}$ is as in 29.58. By the definition of a root group (in 29.15), U_a acts trivially on the set $\Delta_z^{(1)}$. Thus g maps X_1 to X_2. Hence g maps the convex hull of X_1 to the convex hull of X_2. The unique vertex opposite z in Σ_i (for $i = 1$ and 2) is opposite z in Δ'. By 7.19 in [37], Δ' is convex. By 29.59, therefore, the convex hulls of X_1 and of X_2 are both equal to Δ'. Hence $g \in G \cap U_a$. By 29.62, therefore, $\pi(g)$ is an element of U'_a mapping Σ_1 to Σ_2. Thus (iii) holds.

Let $h \in U'_a$. Replacing Σ_2 by Σ_1^h in the previous paragraph, we conclude that there exists $g \in U_a$ such that $\pi(g)$ is also an element of U'_a mapping Σ_1 to Σ_1^h. Thus $\pi(g) = h$ by 29.15.v applied to Δ'. Hence π is surjective. By 29.15.v applied to Δ, the kernel of π restricted to $G \cap U_a$ is trivial. Thus (iv) holds. □

Corollary 29.63. *Let Δ be a Moufang spherical building and let Δ' be a thick irreducible subbuilding of rank at least 2. Then Δ' is also Moufang.*

Proof. Let I be the index set of Δ and let $J \subset I$ be the index set of Δ'. Then Δ' is contained in a J-residue of Δ. The claim holds, therefore, by 11.8 in [37] and 29.61. □

Chapter Thirty

Appendix B

In this appendix, we summarize those aspects of the classification of Moufang spherical buildings that are required in Chapters 16–25. Our description of the classification rests on 3.6.

Notation 30.1. Let Δ be a Moufang spherical building of type Π (as defined in 29.15), let Σ be an apartment of Δ and let d be a chamber of Σ. Let (x, y) be a directed edge[1] of the Coxeter diagram Π, let n be the label on the edge $\{x, y\}$ and let Ξ_d° and $a_x, a_y \in \Xi_d^\circ$ be as in 3.5. Let

$$[a_x, a_y] = (a_1, a_2, \ldots, a_n) \tag{30.2}$$

be as defined in 3.1 (so $a_x = a_1$ and $a_y = a_n$), let

$$U_+^{(e)} = \langle U_{a_1}, U_{a_2}, \ldots, U_{a_n} \rangle,$$

where $e = \{x, y\}$ and U_{a_i} is the root group corresponding to the root a_i for each $i \in [1, n]$, and let

$$\Omega_{xy} = (U_+^{(e)}, U_{a_1}, U_{a_2}, \ldots, U_{a_n}).$$

By 11.27.i in [37], Ω_{xy} is a root group sequence as defined in 12.1 of [37].

Definition 30.3. Let

$$\Omega = (U_+, U_1, U_2, \ldots, U_n)$$

be a root group sequence and let

$$\Omega^{\text{op}} = (U_+, U_n, \ldots, U_2, U_1).$$

By 12.1 of [37], Ω^{op} is also a root group sequence. It is called the *opposite* of Ω. The groups U_i are called the *terms* of Ω, U_1 is its *first term* and U_n is its *last term.* Thus the first term of Ω is the last term of Ω^{op} and vice versa. The number n of terms is called the *length* of Ω. Let $\Omega' = (U'_+, U'_1, U'_2, \ldots, U'_n)$ be a second root group sequence of the same length n. An *isomorphism* from Ω to Ω' is an isomorphism from the group U_+ to the group U'_+ that maps U_i to U'_i for all $i \in [1, n]$. Thus an isomorphism from Ω to Ω' is automatically also an isomorphism from Ω^{op} to $(\Omega')^{\text{op}}$. Thus, in particular, an *automorphism* of Ω (i.e. an isomorphism from Ω to itself) is automatically also an automorphism of Ω^{op}.

Note that $\Omega_{xy} = \Omega_{yx}^{\text{op}}$ for each directed edge (x, y) of Π in 30.1.

[1] By *directed edge* we mean an ordered pair (x, y) such that $\{x, y\}$ is an edge.

Remark 30.4. Let

$$\Omega = (U_+, U_1, U_2, \ldots, U_n)$$

be a root group sequence. By (i) and (ii) in the definition 12.1 in [37] (and the definition of $U_{[i,j]}$ for $1 \le j = i+1 \le n$ in 11.17 in [37]), every element in U_+ can be written canonically in the form $a_1 a_2 \cdots a_n$ with $a_i \in U_i$ for each $i \in [1, n]$, and every element of U_+ can be brought into this canonical form using only commutator relations of the form

$$[a_i, a_j] = a_{i+1} \cdots a_{j-1} \tag{30.5}$$

with $i, j \in [1, n]$, $i + 1 < j$ and $a_k \in U_k$ for all $k \in [i, j]$. Thus if

$$\Omega' = (U'_+, U'_1, U'_2, \ldots, U'_n)$$

is a second root group sequence of length n and for each $i \in [1, n]$, π_i is an isomorphism from U_i to U'_i, then there is an isomorphism π from U_+ to U_+ whose restriction to U_i is π_i for each i if and only if for each relation in U_+ of the form 30.5, the relation

$$[\pi_i(a_i), \pi_j(a_j)] = \pi_{i+1}(a_{i+1}) \cdots \pi_{j-1}(a_{j-1})$$

holds in U'_+. When $\Omega' = \Omega$, we usually express this observation more informally as follows: To show that a collection of automorphisms $\pi_1, \ldots, \pi_n$ of the groups $U_1, \ldots, U_n$ extend to an automorphism of Ω, it suffices to show that they preserve the *commutator relations defining* Ω (by which we mean all relations of the form 30.5).

Definition 30.6. Let

$$\Omega = (U_+, U_1, U_2, \ldots, U_n)$$

be a root group sequence of length n for some $n \ge 3$. Let Π be the Coxeter diagram $\mathsf{I}(n)$ defined in footnote 2 in Chapter 1, let $I = \{x, y\}$ be the vertex set of Π, let Δ be a Moufang building of type Π, let Σ be an apartment of Δ and let d be a chamber of Σ. We will call the triple (Δ, Σ, d) a *realization* of Ω if, after relabeling the two vertices of Π if necessary, Ω is isomorphic to Ω_{xy}, where Ω_{xy} is the root group sequence defined in 30.1 with respect to Σ and d. By 7.5 and 8.11 in [37], Ω has a unique realization.[2] If (Δ, Σ, d) is a realization of Ω, then without loss of generality we can assume that Ω is, in fact, *equal* to Ω_{xy}, in which case we will call (Δ, Σ, d) the *canonical realization* of Ω. Thus if (Δ, Σ, d) is the canonical realization of Ω and

$$[a_x, a_y] = (a_1, \ldots, a_n)$$

for a_x and a_y as in 3.5, then $U_{a_i} = U_i$ for all $i \in [1, n]$.

---

[2]Moufang buildings of type $\mathsf{I}(n)$ for some $n \ge 3$ and Moufang n-gons are essentially the same thing modulo the correspondence between chamber systems of rank 2 and connected bipartite graphs described in 1.9, 7.14 and 7.15 of [37]. This observation allows us to apply the results 7.5 and 8.11 of [36] here. When we say that the realization (Δ, Σ, d) of Ω is unique, we mean that if (Δ', Σ', d') is another realization of Ω, then there exists a special isomorphism from Δ to Δ' mapping the pair (Σ, d) to the pair (Σ', d').

Definition 30.7. Let

$$\Omega = (U_+, U_1, U_2, \ldots, U_n)$$

and

$$\Omega' = (U'_+, U'_1, U'_2, \ldots, U'_n)$$

be two root group sequences, both of length n for some $n \geq 3$. We say that Ω' is a *subsequence* of Ω if U'_+ is a subgroup of U_+ and U'_i is a subgroup of U_i for each $i \in [1, n]$.

By the results summarized in Chapter 17 of [36], every root group sequence is, up to isomorphism or anti-isomorphism, in one of the nine families described in 16.1–16.9 of [36]. The root group sequences of Moufang octagons described in 16.9 of [36] do not play any role here since there are no affine Coxeter diagrams that have an edge with label 8. The first eight families of root group sequences do play an essential role in this book. Here is a summary of some of their salient features:

Notation 30.8. Let X be one of the following symbols:

$$\mathcal{T},\ \mathcal{Q}_{\mathcal{I}},\ \mathcal{Q}_{\mathcal{Q}},\ \mathcal{Q}_{\mathcal{D}},\ \mathcal{Q}_{\mathcal{P}},\ \mathcal{Q}_{\mathcal{E}},\ \mathcal{Q}_{\mathcal{F}},\ \mathcal{H}.$$

We interpret X as an operator that produces a root group sequence

$$X(\Lambda) = (U_+, U_1, \ldots, U_n)$$

from a parameter system Λ of suitable type according to the procedure described in one of the paragraphs 16.1–16.8 in Chapter 16 of [36].[3] In each case, there are groups Λ_i and isomorphisms x_i from Λ_i to U_i for all $i \in [1, n]$,[4] where n, Λ and $\Lambda_1, \ldots, \Lambda_n$ are as follows:

(i) If $X = \mathcal{T}$, then $n = 3$, Λ is an alternative division ring K as defined in 9.1 of [36] and $\Lambda_1 = \Lambda_2 = \Lambda_3 = K$ (i.e. the additive group of K).
(ii) If $X = \mathcal{Q}_{\mathcal{I}}$, then $n = 4$, Λ is an involutory set (K, K_0, σ) as defined in 11.1 of [36], $\Lambda_1 = \Lambda_3 = K_0$ and $\Lambda_2 = \Lambda_4 = K$.
(iii) If $X = \mathcal{Q}_{\mathcal{Q}}$, then $n = 4$, Λ is an anisotropic quadratic space (K, L, q) as defined in 12.2 and 12.4 of [36], $\Lambda_1 = \Lambda_3 = K$ and $\Lambda_2 = \Lambda_4 = L$ (i.e. the additive group of L).
(iv) If $X = \mathcal{Q}_{\mathcal{D}}$, then $n = 4$, Λ is an indifferent set (K, K_0, L_0) as defined in 10.1 of [36], $\Lambda_1 = \Lambda_3 = K_0$ and $\Lambda_2 = \Lambda_4 = L_0$.

---

[3]In Chapter 16 of [36], $X(\Lambda)$ denotes, in fact, the corresponding Moufang n-gon. In this book we let $X(\Lambda)$ denote the root group sequence itself and use the name given in 30.15 for the corresponding building. Thus, for example, if Λ is an anisotropic quadratic space, then $\mathcal{Q}_{\mathcal{Q}}(\Lambda)$ denotes in this book the root group sequence described in 16.3 of [36] and $\mathsf{B}_2^{\mathcal{Q}}(\Lambda)$ the building obtained by applying 8.11 of [36] (and the remarks in footnote 2) to this root group sequence. In other words, if (Δ, Σ, d) is the canonical realization of $\mathcal{Q}_{\mathcal{Q}}(\Lambda)$ as defined in 30.6, then $\mathsf{B}_2^{\mathcal{Q}}(\Lambda) = \Delta$.

[4]Thus

$$U_i = \{x_i(u) \mid u \in \Lambda_i\}$$

for each $i \in [1, n]$.

(v) If $X = \mathcal{Q}_\mathcal{P}$, then $n = 4$, Λ is an anisotropic skew-hermitian pseudo-quadratic space (K, K_0, σ, L_0, q) as defined in 11.17 and 11.18 of [36], $\Lambda_1 = \Lambda_3 = T$, where T is the group defined in 24.5, and $\Lambda_2 = \Lambda_4 = K$.

(vi) If $X = \mathcal{Q}_\mathcal{E}$, then $n = 4$, Λ is a quadratic space (K, L, q) of type E_6, E_7 or E_8 as defined in 12.31 of [36], $\Lambda_1 = \Lambda_3 = S$, where S is the group defined in 16.6 of [36], and $\Lambda_2 = \Lambda_4 = L$.

(vii) If $X = \mathcal{Q}_\mathcal{F}$, then $n = 4$, Λ is a quadratic space (K, L, q) of type F_4 as defined in 14.1 of [36], $\Lambda_1 = \Lambda_3 = W_0 \oplus K$ and $\Lambda_2 = \Lambda_4 = X_0 \oplus F$, where W_0, X_0 and F are as in 16.6 of [36].

(viii) If $X = \mathcal{H}$, then $n = 6$, Λ is an hexagonal system $(J, K, N, \#, T, \times, 1)$ as defined in 15.15 of [36],[5] $\Lambda_1 = \Lambda_3 = \Lambda_5 = J$ (i.e. the additive group of J) and $\Lambda_2 = \Lambda_4 = \Lambda_6 = K$.

Remark 30.9. We observe that the root groups U_i in 30.8 are almost always abelian. The only exceptions are those isomorphic to the group T in case (v) and those isomorphic to the group S in case (vi).

Remark 30.10. By 16.1 in [36] (and 30.3), we have $\mathcal{T}(K) = \mathcal{T}(K)^{\rm op}$ if K is a (commutative) field.

We turn now to Moufang spherical buildings of arbitrary rank.

Notation 30.11. Let Π be one of the Coxeter diagrams in Figure 1.3. We call a directed edge (x, y) *positively oriented* if the vertex x is to the left of the vertex y in Figure 1.3. Let (u, v) be the unique positively oriented directed edge whose label is at least 4 (if there is one).

Definition 30.12. Let Π be a Coxeter diagram. Let ζ be a map from the set of directed edges of Π to the set of root group sequences described in 16.1–16.8 of [36], i.e. in (i)–(viii) of 30.8, together with their opposites as defined in 30.3. We call ζ a *root group labeling of* Π if the following hold:

(i) The root group sequence $\zeta(y, x)$ is the opposite of $\zeta(x, y)$ for all directed edges (x, y) of Π.

(ii) The root group sequence $\zeta(x, y)$ has length equal to the label on the edge $\{x, y\}$ for all directed edges (x, y) of Π.

(iii) For each pair of directed edges (x, y) and (x, z) of Π beginning at the same vertex x, the first term of $\zeta(x, y)$ and the first term of $\zeta(x, z)$ are parametrized by the same group Λ_1.

In the situation described in (iii), we will always use the isomorphisms (both called x_1 in 30.8) from Λ_1 to the first term of $\zeta(x, y)$ and from Λ_1 to the first term of $\zeta(x, z)$ to identify these two first terms (and the two isomorphisms called x_1 as well).

Definition 30.13. Let Π be one of the spherical Coxeter diagrams in Figure 1.3 with $\ell \geq 2$. Let Δ be a Moufang building of type Π, let Σ be an

[5] Hexagonal systems are also known in the literature both as *quadratic Jordan division algebras of degree 3* and as *anisotropic cubic norm structures*.

apartment of Δ, let d be a chamber of Σ and let ζ be a root group labeling of Π as defined in 30.12. For each edge $e = \{x, y\}$ of Π, the group generated by all the terms of $\zeta(x, y)$ is the same as the group generated by all the terms of $\zeta(y, x)$ (by 30.12.i); we denote this group by $U^{(e)}$. We will say that Δ *is linked to* ζ *at* (Σ, d) *via* (π) if (π) is a collection of isomorphisms π_e from $U^{(e)}$ to the group $U_+^{(e)}$ defined in 30.1, one for each edge e of Π, such that the following hold:

(i) For each edge $e = \{x, y\}$ of Π, π_e is an isomorphism from $\zeta(x, y)$ to Ω_{xy} (and hence from $\zeta(y, x)$ to Ω_{yx}).
(ii) Whenever two edges $e = \{x, y\}$ and $f = \{x, z\}$ of Π have a vertex x in common, in which case the first terms of $\zeta(x, y)$ and $\zeta(x, z)$ are the same, the restrictions of π_e and π_f to this first term agree.

Suppose that Δ is linked to a root group labeling ζ of Π at (Σ, d) (with all the notation as in 30.13).[6] The root group labeling ζ is essentially a description of that part of the root group datum of Δ that plays the central role both in 3.6 and in 15.21. In particular, if Δ' is a second Moufang building of type Π that is also linked to ζ, then by 3.6, Δ and Δ' are isomorphic. This means that we can describe the classification of Moufang spherical buildings in terms of root group labelings of Coxeter diagrams. Here is the exact result:

Theorem 30.14. *Suppose that Δ is a Moufang spherical building of rank $\ell \geq 2$ with Coxeter diagram $\Pi = \mathsf{X}_\ell$ with $\mathsf{X} = \mathsf{A}, \mathsf{B}, \ldots, \mathsf{G}$ and let the notion of a root group labeling of Π be as in 30.12. Then Δ is linked (as defined in 30.13) to one of the following root group labelings ζ of Π:*[7]

(i) $\Pi = \mathsf{A}_\ell$ *for some $\ell \geq 2$, and for some field or skew field K, $\zeta(x, y) = \mathcal{T}(K)$ for all positively oriented directed edges (x, y) of Π.*[8]
(ii) $\Pi = \mathsf{A}_2$ *and for some octonion division algebra K, $\zeta(x, y) = \mathcal{T}(K)$ for the one positively oriented directed edge (x, y) of Π.*
(iii) $\Pi = \mathsf{B}_\ell = \mathsf{C}_\ell$ *for some $\ell \geq 2$, and for some anisotropic quadratic space (K, L, q), $\zeta(x, y) = \mathcal{T}(K)$ for all directed edges (x, y) of Π whose label is 3 and $\zeta(u, v) = \mathcal{Q}_\mathcal{Q}(K, L, q)$.*[9]
(iv) $\Pi = \mathsf{B}_\ell = \mathsf{C}_\ell$ *for some $\ell \geq 2$ and for some proper*[10] *involutory set (K, K_0, σ), $\zeta(x, y) = \mathcal{T}(K)$ for all positively oriented directed edges (x, y) of Π whose label is 3 and $\zeta(u, v) = \mathcal{Q}_\mathcal{I}(K, K_0, \sigma)^{\mathrm{op}}$.*[11]

---

[6] By 29.15.iii, it follows that Δ is linked to ζ at *every* pair (Σ', d') such that Σ' is an apartment of Δ and d' a chamber of Σ'. It thus makes sense to say, simply, that Δ is linked to ζ without mentioning the pair (Σ, d).

[7] See 30.8 for references to the definitions in [36] of the various types of algebraic structures that occur in 30.14. See also 30.28.

[8] See 30.11 for the definition of *positively oriented.*

[9] In each case of 30.14 where X_ℓ is not simply laced, (u, v) is as in 30.11.

[10] An involutory set (K, K_0, σ) is *proper* if $\sigma \neq 1$ and the subring $\langle K_0 \rangle$ generated by K_0 equals K. By 23.23 in [36], an involutory set (K, K_0, σ) with $\sigma \neq 1$ that is *not proper* is quadratic of type (iii) or (iv) as defined in 30.21.

[11] In other words, $\zeta(v, u) = \zeta(u, v)^{\mathrm{op}} = \mathcal{Q}_\mathcal{I}(K, K_0, \sigma)$; see 30.3. A similar remark holds in cases (v), (vi), (vii) and (xiii).

(v) $\Pi = \mathsf{B}_\ell = \mathsf{C}_\ell$ *for some* $\ell \geq 3$ *and for some quadratic involutory set* (K, F, σ) *that is either of type (ii) with* $\mathrm{char}(K) \neq 2$ *or of type (iii) or (iv) (in arbitrary characteristic) as defined in 30.21,* $\zeta(x,y) = \mathcal{T}(K)$ *for all positively oriented directed edges* (x,y) *of* Π *whose label is 3 and* $\zeta(u,v) = \mathcal{Q}_{\mathcal{I}}(K,F,\sigma)^{\mathrm{op}}$. [12]

(vi) $\Pi = \mathsf{B}_3 = \mathsf{C}_3$ *for some honorary involutory set* (K, K_0, σ) *as defined in 30.22,* $\zeta(x,y) = \mathcal{T}(K)$ *for the one positively oriented directed edge* (x,y) *of* Π *whose label is 3 and* $\zeta(u,v) = \mathcal{Q}_{\mathcal{I}}(K,K_0,\sigma)^{\mathrm{op}}$.

(vii) $\Pi = \mathsf{B}_\ell = \mathsf{C}_\ell$ *for some* $\ell \geq 2$ *and for some proper anisotropic pseudo-quadratic space*

$$(K, K_0, \sigma, L, q)$$

as defined in 24.4, $\zeta(x,y) = \mathcal{T}(K)$ *for each positively oriented directed edge* (x,y) *of* Π *whose label is 3 and* $\zeta(u,v) = \mathcal{Q}_{\mathcal{P}}(K,K_0,\sigma,L,q)^{\mathrm{op}}$.

(viii) $\Pi = \mathsf{B}_2 = \mathsf{C}_2$ *and* $\zeta(u,v) = \mathcal{Q}_{\mathcal{E}}(K,L,q)$ *for some quadratic space* (K,L,q) *of type* E_6, E_7 *or* E_8.

(ix) $\Pi = \mathsf{B}_2 = \mathsf{C}_2$ *and* $\zeta(u,v) = \mathcal{Q}_{\mathcal{F}}(K,L,q)$ *for some quadratic space* (K,L,q) *of type* F_4.

(x) $\Pi = \mathsf{B}_2 = \mathsf{C}_2$ *and* $\zeta(u,v) = \mathcal{Q}_{\mathcal{D}}(K,K_0,L_0)$ *for some proper*[13] *indifferent set* (K, K_0, L_0).

(xi) $\Pi = \mathsf{D}_\ell$ *for some* $\ell \geq 4$ *and for some field* K, $\zeta(x,y) = \mathcal{T}(K)$ *for all directed edges* (x,y) *of* Π. [14]

(xii) $\Pi = \mathsf{E}_\ell$ *for* $\ell = 6$, 7 *or* 8 *and for some field* K, $\zeta(x,y) = \mathcal{T}(K)$ *for all directed edges* (x,y) *of* Π.

(xiii) $\Pi = \mathsf{F}_4$ *and for some composition algebra* (K,F) *with standard involution* σ *as defined in 30.17 below,* $\zeta(x,y) = \mathcal{T}(K)$ *if* (x,y) *is the unique positively oriented directed edge* (x,y) *of* Π *such that* $y = u$, $\zeta(x,y) = \mathcal{T}(F)$ *if* (x,y) *is the unique positively oriented directed edge of* Π *such that* $v = x$ *and* $\zeta(u,v) = \mathcal{Q}_{\mathcal{I}}(K,F,\sigma)^{\mathrm{op}}$.

(xiv) $\Pi = \mathsf{G}_2$ *and* $\zeta(u,v) = \mathcal{H}(\Lambda)$ *for some hexagonal system*

$$\Lambda = (J, K, N, \#, T, \times, 1).$$

Conversely, if ζ *is one of the root group labelings* ζ *described in (i)–(xiv), then there exists a unique spherical building* Δ *with Coxeter diagram* Π *that is linked to* ζ.

Proof. This holds for $\ell = 2$ by the results formulated in Chapter 17 of [36] as well as 16.12 and 38.5 in [36] and for $\ell \geq 3$ by 40.17, 40.22 and 40.56 in [36]. □

Notation 30.15. Let Δ, $\Pi = \mathsf{X}_\ell$ (where $\mathsf{X} = \mathsf{A}, \mathsf{B}, \dots, \mathsf{G}$) and ζ be as in 30.14. Without loss of generality, we can assume in each case that there

---

[12] See 30.27.

[13] An indifferent set $\Lambda = (K, K_0, L_0)$ is *proper* if neither K_0 nor L_0 is closed under multiplication.

[14] By 30.10, we have $\zeta(x,y) = \zeta(y,x)$, i.e. the orientation of the edge $\{x,y\}$ does not matter. A similar remark holds in case (xii) (and in case (iii) for those edges of Π whose label is 3).

is an apartment Σ of Δ and a chamber d of Σ such that for each directed edge (x, y) of Π, the root group sequence Ω_{xy} defined in 30.1 with respect to (Σ, d) *equals* $\zeta(x, y)$. Under this assumption, we call the triple (Δ, Σ, d) the *canonical realization* of the root group labeling ζ of Π.[15] Let Λ be

- K in cases (i), (ii), (xi) and (xii),
- (K, L, q) in cases (iii), (viii) and (ix),
- (K, K_0, σ) in cases (iv)–(vi),
- (K, K_0, σ, L, q) in case (vii),
- (K, K_0, L_0) in case (x),
- (K, F) in case (xiii) and
- $(J, K, N, \#, T, \times, 1)$ in case (xiv).

We call Λ, respectively K, the *defining parameter system*, respectively, the *defining field* of Δ (see 30.29) and we denote Δ by $\mathsf{X}_\ell(\Lambda)$ except when $\mathsf{X} = \mathsf{B}$ (equivalently, C). In these cases, we denote Δ by:

- $\mathsf{B}_\ell^{\mathcal{Q}}(\Lambda)$ or $\mathsf{C}_\ell^{\mathcal{Q}}(\Lambda)$ in case (iii),
- $\mathsf{B}_\ell^{\mathcal{I}}(\Lambda)$ or $\mathsf{C}_\ell^{\mathcal{I}}(\Lambda)$ in cases (iv)–(vi),
- $\mathsf{B}_\ell^{\mathcal{P}}(\Lambda)$ or $\mathsf{C}_\ell^{\mathcal{P}}(\Lambda)$ in case (vii),
- $\mathsf{B}_2^{\mathcal{E}}(\Lambda)$ or $\mathsf{C}_2^{\mathcal{E}}(\Lambda)$ in case (viii),
- $\mathsf{B}_2^{\mathcal{F}}(\Lambda)$ or $\mathsf{C}_2^{\mathcal{F}}(\Lambda)$ in case (ix),
- $\mathsf{B}_2^{\mathcal{D}}(\Lambda)$ or $\mathsf{C}_2^{\mathcal{D}}(\Lambda)$ in case (x).

Thus in every case we assign Δ a name of the form $\mathsf{X}_\ell^*(\Lambda)$, where X_ℓ is the Coxeter diagram of Δ, Λ is the defining parameter system, $*$ indicates the family of Moufang quadrangles appearing as residues if $\mathsf{X} = \mathsf{B}$ or C and $*$ is simply blank in every other case.

The classification of Moufang spherical buildings as described in 30.8, 30.14 and 30.15 is summarized in Figure 27.1.

Proposition 30.16. *Let Π be one of the spherical diagrams in Figure 1.3 with $\ell \geq 2$, let ζ and ζ' be two root group labelings of Π as defined in 30.12 and let (Δ, Σ, d) and (Δ', Σ', d') be their canonical realizations as defined in 30.15. Suppose that for each directed edge (x, y) of Π, $\zeta'(x, y)$ is a subsequence of $\zeta(x, y)$ as defined in 30.7. Then Δ' is a subbuilding of Δ and $(\Sigma, d) = (\Sigma', d')$.*

Proof. This holds by 40.20 in [36]. □

Notation 30.17. A pair (K, F) is a *composition algebra*[16] if F is a field and one of the following holds:

(i) K is a field containing F and $K^2 \subset F$ but $K \neq F$ (so $\mathrm{char}(K) = 2$).

---

[15]This extends the notion of canonical realization given in 30.6. In general, we will say that a triple (Δ, Σ, d) is a *canonical realization of type 30.14.k* if it is the canonical realization of a pair (Π, ζ) as in case (k) of 30.14, where k is a lowercase Roman numeral in the interval from i to xiv.

[16]In 20.3 of [36], composition algebras are characterized as alternative division rings that are quadratic over a subfield of their center.

(ii) $K = F$.
(iii) K/F is a separable quadratic extension.
(iv) K is a quaternion division algebra with center F.
(v) K is an octonion division algebra with center F.

If (K, F) is in case (k) in this list, we will sometimes say that (K, F) is a composition algebra *of type (k)*. Suppose that (K, F) is a composition algebra. In cases (i) and (ii), let $q(x) = x^2$ for all $x \in K$; in case (iii), let q be the norm of the extension K/F; and in cases (iv) and (v), let q be the reduced norm of K. In each case, the triple (F, K, q) is an anisotropic quadratic space. We call q the *norm* of the composition algebra (K, F) (in all five cases). Let $\sigma: K \to K$ be the identity map in cases (i) and (ii); let σ be the unique non-trivial element of $\mathrm{Gal}(K/F)$ in case (iii); and let σ be the standard involution of K in cases (iv) and (v) as defined in 9.6 and 9.10 of [36]. Thus

$$q(u) = u^\sigma u = uu^\sigma \tag{30.18}$$

for all $u \in K$ and therefore

$$q(uv) = q(u)q(v) \tag{30.19}$$

for all $u, v \in K$ in all five cases. We will call σ the *standard involution* of the composition algebra (K, F) in every case. If K is associative, the triple (K, F, σ) is an involutory set. If f is the bilinear form associated with q, then by 30.18, $f(u, v) = u^\sigma v + v^\sigma u$ for all $u, v \in K$. In particular,

$$f(u, 1) = u + u^\sigma \tag{30.20}$$

for all $u \in K$. The map $T: K \to F$ given by $T(u) = f(u, 1)$ for all $u \in K$ is called the *trace* of (K, F).

Definition 30.21. An involutory set (K, K_0, σ) is *quadratic of type (k)* (or simply *quadratic*) if $F := K_0$ is a subfield of K, (K, F) is a composition algebra of type (k) and σ is the standard involution of this composition algebra as defined in 30.17. Note that if (K, F, σ) is a quadratic involutory set of type (k), then K is associative, so k cannot be v. Note, too, that if (K, F) is an arbitrary composition algebra of type (k) with standard involution σ and k $<$ v, then the triple (K, F, σ) is always an involutory set.

Definition 30.22. An *honorary involutory set* is a triple (K, F, σ) such that (K, F) is a composition algebra of type (v) and σ is its standard involution (as defined in 30.17). If we extend the operator $\mathcal{Q}_\mathcal{I}$ defined in 16.2 of [36] verbatim to honorary involutory sets, we obtain a root group sequence from each honorary involutory set. See, however, 23.5.

Definition 30.23. A composition algebra (K, F) is *inseparable* if K/F is an inseparable extension, equivalently, if it is of type (i). An hexagonal system is *inseparable* if it is isomorphic to $(E/K)^\circ$ (as defined in 15.20 of [36]) for some inseparable extension E/K. All other composition algebras and hexagonal systems are *separable.*

We will now begin to use the notation introduced in 30.15.

Definition 30.24. Let Δ be a Moufang spherical building of rank $\ell \geq 2$ whose Coxeter diagram is different from $\mathsf{I}(8)$. Then Δ is *mixed* (or *of mixed type*) if it is, up to isomorphism, one of the following buildings:

(i) $\mathsf{B}_2^{\mathcal{Q}}(\Lambda)$ for $\Lambda = (K, L, q)$ an anisotropic quadratic space such that the bilinear form associated with q is identically zero and $\dim_K L > 1$.
(ii) $\mathsf{B}_2^{\mathcal{F}}(\Lambda)$ for $\Lambda = (K, L, q)$ a quadratic space of type F_4.
(iii) $\mathsf{B}_2^{\mathcal{D}}(\Lambda)$ for $\Lambda = (K, K_0, L_0)$ an indifferent set.
(iv) $\mathsf{F}_4(\Lambda)$ for Λ an inseparable composition algebra (K, F) as defined in 30.23.
(v) $\mathsf{G}_2(\Lambda)$ for Λ an inseparable hexagonal system $(E/K)^\circ$ as defined in 30.23.

If $\Lambda = (K, L, q)$ is as in (i), then $(K, K, q(L))$ is an improper indifferent set and $\mathsf{B}_2^{\mathcal{Q}}(\Lambda) \cong \mathsf{B}_2^{\mathcal{D}}(K, K, q(L))$ (by 38.4 in [36]). Suppose that Λ is as in (ii)–(iv) and let E be the unique field containing K such that

- $E^2 = F$ in case (ii), where F is as in 14.3 of [36];
- $E^2 = L$ in case (iii), where L is as in 10.2 of [36]; and
- $E^2 = F$ in case (iv).

Then in each case (ii)–(v), E is a field containing K properly such that

$$E^p \subset K \subset E, \tag{30.25}$$

where $p = \text{char}(K)$ and $p = 2$ except in case (v) when $p = 3$. Note that if we identify E with E^p via the isomorphism $x \mapsto x^p$, then

$$K^p \subset E \subset K.$$

In fact, neither K nor E alone is an invariant of the building Δ, but by 35.9, 35.12 and 35.13 in [36], the unordered pair $\{K, E\}$ is an invariant of Δ.

Remark 30.26. By 17.7 in [36], Moufang buildings of type $\mathsf{I}(8)$ (also known as *Moufang octagons*) are parametrized by *octagonal sets*. An octagonal set is a pair (K, σ), where K is a field of characteristic 2 and σ is an endomorphism of K such that $x^{\sigma^2} = x^2$ for all $x \in K$. If (K, σ) is an octagonal set and $F = K^\sigma$, then $K^2 \subset F \subset K$, but it could be that $K^2 = F = K$. This happens, for example, if K is a finite field containing 2^{2k+1} elements for some k and $x^\sigma = x^{2^{k+1}}$ for all $x \in K$. (Note that σ is uniquely determined by $q := |K|$ in this case.) Moufang buildings of type $\mathsf{I}(8)$ are nevertheless considered to be of mixed type. We have excluded them in 30.24 only because they do not play any role in this book.

Remark 30.27. Let (K, F) be a composition algebra with norm q and standard involution σ. Then

$$\mathcal{Q}_{\mathcal{Q}}(F, K, q) = \mathcal{Q}_{\mathcal{I}}(K, F, \sigma)$$

by 16.2 and 16.3 in [36]. Suppose that (K, F) is of type (i) or (ii) and that $\text{char}(K) = 2$ also if the type of (K, F) is (ii). Let $(t, s) \mapsto t * s$ and q_* be the maps from $K \times F$ to F and from F to K given by $t * s = t^2 s$ and $q_*(s) = s$ for all $s \in F$. Then F is a vector space over K with scalar multiplication given by $*$ and (K, F, q_*) is an anisotropic quadratic space. By 16.2 and 16.3 in [36] again, we have

$$\mathcal{Q}_\mathcal{I}(K, F, \sigma)^{\mathrm{op}} = \mathcal{Q}_\mathcal{Q}(K, F, q_*).$$

This explains why we would obtain no new buildings in case (v) of 30.14 were we to allow ℓ to be 2 or the type of (K, K_0, σ) to be (i) or char(K) to be 2 when the type of (K, K_0, σ) is (ii).

Remark 30.28. In cases (iv), (vii) and (x) of 30.14, we assume that the parameter system Λ is proper. In fact, these buildings exist in all three cases also without this assumption. The restriction that Λ be proper is included only to avoid duplications.[17]

Remark 30.29. By 35.6–35.13 in [36], exactly one of the following holds:

(i) Δ is as in case 30.14.i and the pair $\{K, K^{\mathrm{op}}\}$ is an invariant of Δ;
(ii) Δ is of mixed type as defined in 30.24, the pair $\{K, E\}$ defined in 30.24 is an invariant of Δ; or
(iii) Δ is neither as in case 30.14.i nor of mixed type, Λ is an invariant of Δ up to *similarity* [18] and K is an invariant of Δ.

This justifies our referring to Λ and K as *the defining parameter system* and *the defining field* of Δ. Note, too, that a valuation of a skew field K is also a valuation of its opposite K^{op}, and if $\{K, E\}$ is as in (ii), then by 30.25, a valuation ν of K determines a unique valuation of E and vice versa simply through restriction. Thus although the notion of the defining field is not completely well defined, the notion of a valuation of the defining field is.

---

[17]Suppose that $\Lambda = (K, K_0, \sigma)$ is an improper involutory set. If $\sigma = 1$, then it is shown in 38.1 and 38.3 of [36] how to construct an anisotropic quadratic space $\Lambda_\mathcal{Q}$ such that

$$\mathsf{B}_\ell^\mathcal{I}(\Lambda) \cong \mathsf{B}_\ell^\mathcal{Q}(\Lambda_\mathcal{Q}).$$

If $\sigma \neq 1$, then by 23.23 in [36], Λ is quadratic of type (iii) or (iv). Now suppose that

$$\Lambda = (K, K_0, \sigma, L, q)$$

is an improper anisotropic pseudo-quadratic space and let f be the associated skew-hermitian form. If $L = 0$, then

$$\mathsf{B}_\ell^\mathcal{P}(\Lambda) \cong \mathsf{B}_\ell^\mathcal{I}(K, K_0, \sigma).$$

Suppose that $L \neq 0$. Then it is shown in 16.15 and 21.16 of [36] how to obtain an anisotropic quadratic space $\Lambda_\mathcal{Q}$ (if $\sigma = 1$), an involutory set $\Lambda_\mathcal{I}$ (if $\sigma \neq 1$ but f is identically zero) or a proper anisotropic pseudo-quadratic space $\Lambda_\mathcal{P}$ (if $\sigma \neq 1$ and f is not identically zero) such that

$$\mathsf{B}_\ell^\mathcal{P}(\Lambda) \cong \mathsf{B}_\ell^\mathcal{X}(\Lambda_\mathcal{X})$$

for $\mathcal{X} = \mathcal{Q}$, $\mathcal{I}$ or $\mathcal{P}$. (See also 16.18 in [36].) If Λ is an improper indifferent set, then it is shown in 38.4 how to construct an anisotropic quadratic space $\Lambda_\mathcal{Q}$ such that $\mathcal{Q}_\mathcal{D}(\Lambda)$ is isomorphic to $\mathcal{Q}_\mathcal{Q}(\Lambda_\mathcal{Q})$ or $\mathcal{Q}_\mathcal{Q}(\Lambda_\mathcal{Q})^{\mathrm{op}}$.

[18]See the references in 35.6–35.13 of [36] for the various definitions of similarity.

Definition 30.30. A Moufang spherical building of rank $\ell \geq 2$ is *classical* if it is, up to isomorphism, one of the buildings in case (i), (iii)–(v), (vii) or (xi) of 30.14. These are precisely the buildings described in 41.1 of [36].

Definition 30.31. A spherical building Δ is *algebraic* if it is isomorphic to the spherical building $\Delta(G, F)$ associated to the F-points, for some field F, of an absolutely simple algebraic group G as explained in Tits's Boulder notes [31] and in Chapter 5 of [32].[19] All algebraic spherical buildings of rank at least 2 are Moufang. By 5.8 in [32], the pair (G, F) is an invariant of the building $\Delta(G, F)$. A spherical building is *exceptional* if it is the spherical building associated to the F-points, for some field F, of an exceptional absolutely simple algebraic group. The algebraic buildings that are not exceptional are all classical as defined in 30.28. We give a list of all algebraic buildings, exceptional and classical, in 30.32.

Proposition 30.32. *The exceptional spherical buildings of rank at least 2 are, up to isomorphism, the following:*

(i) $\mathsf{A}_2(K)$ *for* K *an octonion division algebra.*
(ii) $\mathsf{B}_3^{\mathcal{I}}(\Lambda)$ *for* Λ *an honorary involutory set (as defined in 30.22).*
(iii) $\mathsf{B}_2^{\mathcal{E}}(\Lambda)$ *for* Λ *a quadratic space of type* E_6, E_7 *or* E_8.
(iv) $\mathsf{E}_6(K)$, $\mathsf{E}_7(K)$ *and* $\mathsf{E}_8(K)$ *for* K *a field.*
(v) $\mathsf{F}_4(\Lambda)$ *for* Λ *a separable composition algebra (as defined in 30.23).*
(vi) $\mathsf{G}_2(\Lambda)$ *for* Λ *a separable hexagonal system (as defined in 30.23).*

The remaining algebraic spherical buildings of rank at least 2 are, up to isomorphism, the following:

(i) $\mathsf{A}_2(K)$ *for* K *a field or a skew field of finite dimension over its center.*
(ii) $\mathsf{B}_\ell^{\mathcal{Q}}(\Lambda)$ *for* Λ *an anisotropic quadratic space* (K, L, q) *such that* $\dim_K L$ *is finite and*
$$\dim_K\{x \in L \mid f(x, L) = 0\} \leq 1,$$
where f *is the bilinear form associated with* q.
(iii) $\mathsf{B}_\ell^{\mathcal{I}}(\Lambda)$ *for* Λ *an involutory set*
$$(K, K_0, \sigma)$$
that is either proper or quadratic of type (ii) (also in characteristic 2), (iii) or (iv) as defined in 30.21 such that K *is finite-dimensional over its center and* $K_0 = K_\sigma$ *or* K^σ *(or both), where* K_σ *and* K^σ *are as in 23.18 and 23.19.*
(iv) $\mathsf{B}_\ell^{\mathcal{P}}(\Lambda)$ *for* Λ *a proper anisotropic quadratic space*
$$(K, K_0, \sigma, L, q)$$
such that (K, K_0, σ) *satisfies all the conditions in (iii) and* $\dim_K L$ *is finite.*

[19] In this case either F is the center of the defining field K of the building Δ (as defined in 30.15) or there is an involution σ of K in the description of Δ in 30.14 and F is the intersection of the center of K with the fixed point set of σ.

(v) $\mathsf{D}_\ell(K)$ *for* K *an arbitrary field.*

Proof. See Chapter 41 in [36]. □

We thus have:

Proposition 30.33. *Every Moufang spherical building is mixed, exceptional or classical. These three classes of buildings are disjoint with one exception: The mixed buildings in case (i) of 30.24 are also classical.*

Proof. This holds by 30.14, 30.24, 30.30 and 30.32. □

The following classification is needed in Chapter 28.

Theorem 30.34. *Let* Δ *be a finite Moufang spherical building whose type* X_ℓ *is different from* $\mathsf{I}(8)$. *Let* K *be the defining field of* Δ *(as defined in 30.15). Then* K *is finite and either* X_ℓ *is simply laced, in which case* $\Delta \cong \mathsf{X}_\ell(K)$, *or* X_ℓ *is not simply laced and* Δ *is, up to isomorphism, one of the following:*[20]

(i) $\mathsf{B}_\ell^{\mathcal{Q}}(K, K, q)$, *where* $q(x) = x^2$ *for all* $x \in K$.
(ii) $\mathsf{B}_\ell^{\mathcal{Q}}(K, E, q)$, *where* E *is the unique field containing* K *such that* E/K *is a quadratic extension and* q *is the norm of this extension.*
(iii) $\mathsf{B}_\ell^{\mathcal{I}}(K, K, \mathrm{id})$.
(iv) $\mathsf{B}_\ell^{\mathcal{I}}(K, F, \sigma)$, *where* F *is the unique subfield of* K *such that* K/F *is a quadratic extension and* σ *is the non-trivial element in* $\mathrm{Gal}(K/F)$.
(v) $\mathsf{B}_\ell^{\mathcal{P}}(K, F, \sigma, K, \alpha N)$, *where* (K, F, σ) *is as in (iv),* N *is the norm of the extension* K/F *and* α *is an element of* K *not in* F.[21]
(vi) $\mathsf{F}_4(K, K)$.
(vii) $\mathsf{F}_4(K, F)$, *where* F *is as in (iv).*
(viii) $\mathsf{G}_2((K/K)^\circ)$, *where* $(K/K)^\circ$ *is as in 15.20 of [36].*
(ix) $\mathsf{G}_2((E/K)^+)$, *where* E *is the unique field containing* K *such that* E/K *is a cubic extension and* $(E/K)^+$ *is as in 15.21 of [36].*

In each case, Δ *is uniquely determined by* K *alone.*

Proof. This holds by 30.14 and the results in Chapter 34 of [36]. □

Notation 30.35. In Chapter 28 we use the following names for the buildings in each case (i)–(ix) of 30.34:

(i) $\mathsf{B}_\ell^{\mathcal{Q}}(K)$ or $\mathsf{B}_\ell^{\mathcal{Q}}(K, 1)$.
(ii) $\mathsf{B}_\ell^{\mathcal{Q}}(K, 2)$.
(iii) $\mathsf{B}_\ell^{\mathcal{I}}(K)$ or $\mathsf{B}_\ell^{\mathcal{I}}(K, 1)$.
(iv) $\mathsf{B}_\ell^{\mathcal{I}}(K, 2)$.
(v) $\mathsf{B}_\ell^{\mathcal{P}}(K)$.

---

[20]If $\ell = 2$ or $\mathrm{char}(K) = 2$ (or both), then the buildings in (i) and (iii) are isomorphic. If $\ell = 2$, then the buildings in (ii) and (iv) are isomorphic. Apart from these exceptions, the buildings in (i)–(ix) are all pairwise non-isomorphic.

[21]Replacing α by another element in $K \backslash F$ has the effect of replacing the anisotropic pseudo-quadratic space $(K, F, \sigma, K, \alpha N)$ by a translate (as defined in 11.26 of [36]) and hence does not produce a different building (by 35.19 in [36]).

(vi) $\mathsf{F}_4(K)$ or $\mathsf{F}_4(K,1)$.
(vii) $\mathsf{F}_4(K,2)$.
(viii) $\mathsf{G}_2(K)$ or $\mathsf{G}_2(K,1)$.
(ix) $\mathsf{G}_2(K,3)$.

In cases (i)–(v), we can replace B by C in our notation without changing the meaning since the Coxeter groups B_ℓ and C_ℓ are the same.

In each case (i)–(ix) of 30.34, the standard name for the group $G^\dagger$ generated by all the root groups is as follows:

(i) $O_{2\ell+1}(q)$, where $q = |K|$.
(ii) $O^-_{2\ell+2}(q)$, where $q = |K|$.
(iii) $PSp_\ell(q)$, where $q = |K|$.
(iv) $U_{2\ell}(q)$, where $q = |F|$.
(v) $U_{2\ell+1}(q)$, where $q = |F|$.
(vi) $F_4(q)$, where $q = |K|$.
(vii) ${}^2E_6(q)$, where $q = |F|$.
(viii) $G_2(q)$, where $q = |K|$.
(ix) ${}^3D_4(q)$, where $q = |K|$.

If $\Delta = \mathsf{A}_\ell(K)$ and $q = |K|$, the standard name for $G^\dagger$ is $PSL_{\ell+1}(q)$ and if $\Delta = \mathsf{D}_\ell(K)$ and $q = |K|$, then the standard name for $G^\dagger$ is $O^+_{2\ell}(q)$. If $\Delta = \mathsf{E}_\ell(K)$ for $\ell = 6$, 7 or 8 and $q = |K|$, then the standard name for $G^\dagger$ is simply $E_\ell(q)$. If (K,σ) is one of the finite octagonal sets described in 30.26 and Δ is the associated Moufang building of type $\mathsf{I}(8)$, then the standard name for $G^\dagger$ is ${}^2F_4(q)$, where $q = |K|$.

By 17.7 in [36], the only finite Moufang buildings of type $\mathsf{I}(8)$ are those parametrized by the finite octagonal sets described in 30.26. Note that by 30.32 and 30.34, all the remaining finite Moufang spherical buildings are algebraic.

* * *

Let a be a root of the Σ. By 3.10, elements of the form $m_\Sigma(u)m_\Sigma(v)$ for $u, v \in U_a^*$ act trivially on Σ and hence normalize U_b for all roots b of Σ. We close this appendix with a set of formulas displaying this action (for certain choices of a and b) and two related collections of formulas (in 30.37 and 30.38) taken from [36]. These results are required in Chapters 19–25.

Proposition 30.36. *Let (Π, ζ) be as in one of the cases (i)–(xiv) of 30.14 and let (Δ, Σ, d) be the canonical realization of the pair (Π, ζ) (as defined in 30.15). Suppose that (K, L, q) is unitary in case (iii).*[22] *Let $\{x, y\}$ be an edge of Π and suppose that*

$$\zeta(x, y) = X(\Lambda)$$

(as opposed to $\zeta(x, y) = X(\Lambda)^{\rm op}$), where Λ and X are as in one of the cases (i)–(viii) of 30.8, and let

$$[a_x, a_y] = (a_1, a_2, \ldots, a_n),$$

[22] See 19.15 and 19.16.

where a_x *and* a_y *are as in 3.5 and* $[a_x, a_y]$ *is as in 3.1. Thus*

$$X(\Lambda) = (U_+, U_1, \ldots, U_n),$$

where $U_i = U_{a_i}$ *for all* $i \in [1, n]$. *Let* x_i *be the isomorphism from* Λ_i *to* U_i *for all* $i \in [1, n]$ *described in 30.8. Let* $e_1 = x_1(\epsilon_1) \in U_1^*$ *and* $e_n = x_n(\epsilon_n) \in U_n^*$, *where* $\epsilon_1 \in \Lambda_1$ *and* $\epsilon_n \in \Lambda_n$ *are as in Figure 5 on page 354 of [36] (and in 30.38 below). Let*

$$h_1(t) = m_\Sigma(e_1)m_\Sigma(x_1(t))$$

for all non-zero $t \in \Lambda_1$ *and let*

$$h_n(t) = m_\Sigma(e_n)m_\Sigma(x_n(t))$$

for all non-zero $t \in \Lambda_n$. *Then (with all the notation in 32.5–32.12 of [36]) the following hold:*

(i) *If* $X = \mathcal{T}$, *then:*
$x_2(u)^{h_1(t)} = x_2(-tu)$ *for all* $u \in K$ *and all* $t \in K^*$;
$x_3(u)^{h_1(t)} = x_3(-t^{-1}u)$ *for all* $u \in K$ *and all* $t \in K^*$;
$x_1(u)^{h_3(t)} = x_1(-ut^{-1})$ *for all* $u \in K$ *and all* $t \in K^*$; *and*
$x_2(u)^{h_3(t)} = x_2(-ut)$ *for all* $u \in K$ *and all* $t \in K^*$.

(ii) *If* $X = \mathcal{Q}_\mathcal{I}$, *then:*
$x_2(u)^{h_1(t)} = x_2(-tu)$ *for all* $u \in K$ *and all* $t \in K_0^*$;
$x_4(u)^{h_1(t)} = x_4(-t^{-1}u)$ *for all* $u \in K$ *and all* $t \in K_0^*$;
$x_1(t)^{h_4(u)} = x_1(u^{-\sigma}tu^{-1})$ *for all* $t \in K_0$ *and all* $u \in K^*$; *and*
$x_2(v)^{h_4(u)} = x_2(u^{-\sigma}vu)$ *for all* $v \in K$ *and all* $u \in K^*$.

(iii) *If* $X = \mathcal{Q}_\mathcal{Q}$, *then:*
$x_3(t)^{h_1(s)} = x_3(t)$ *for all* $t \in K$ *and all* $s \in K^*$;
$x_4(u)^{h_1(s)} = x_4(-s^{-1}u)$ *for all* $u \in L$ *and all* $s \in K^*$;
$x_1(t)^{h_4(u)} = x_1(tq(u)^{-1})$ *for all* $t \in K$ *and all* $u \in L^*$; *and*
$x_3(t)^{h_4(u)} = x_3(tq(u))$ *for all* $t \in K$ *and all* $u \in L^*$.

(iv) *If* $X = \mathcal{Q}_\mathcal{D}$, *then:*
$x_3(s)^{h_1(t)} = x_3(s)$ *for all* $s \in K_0$ *and all* $t \in K_0^*$;
$x_4(u)^{h_1(t)} = x_4(ut^{-2})$ *for all* $u \in L_0$ *and all* $t \in K_0^*$;
$x_1(t)^{h_4(u)} = x_1(tu^{-1})$ *for all* $t \in K_0$ *and all* $u \in L_0^*$; *and*
$x_2(v)^{h_4(u)} = x_2(v)$ *for all* $v \in L_0$ *and all* $u \in L_0^*$.

(v) *If* $X = \mathcal{Q}_\mathcal{P}$, *then:*
$x_2(u)^{h_1(a,t)} = x_2(-tu)$ *for all* $u \in K$ *and all* $(a, t) \in T^*$;
$x_4(u)^{h_1(a,t)} = x_4(-t^{-\sigma}u)$ *for all* $u \in K$ *and all* $(a, t) \in T^*$;
$x_1(a, t)^{h_4(u)} = x_1(-au^{-1}, u^{-\sigma}tu^{-1})$ *for all* $(a, t) \in T$ *and all* $u \in K^*$; *and*
$x_2(v)^{h_4(u)} = x_2(u^{-\sigma}vu)$ *for all* $v \in K$ *and all* $u \in K^*$.

(vi) *If* $X = \mathcal{Q}_\mathcal{E}$, *then:*
$x_2(v)^{h_1(a,t)} = x_2(-\theta(a, v) - tv)$ *for all* $v \in L$ *and all* $(a, t) \in S^*$;
$x_4(v)^{h_1(a,t)} = x_4\big(-q(\pi(a) + t)^{-1}(\theta(a, v) + tv)\big)$ *for all* $v \in L$ *and all* $(a, t) \in S^*$;
$x_1(a, t)^{h_4(v)} = x_1\big(av^{-1}, t/q(v) + \phi(a, v^{-1})\big)$ *for all* $(a, t) \in S$ *and all* $v \in L^*$; *and*

$x_3(a,t)^{h_4(v)} = x_3\big(-av, tq(v)+\phi(a,v)\big)$ *for all* $(a,t)\in S$ *and all* $v\in L^*$.

(vii) *If* $X=\mathcal{Q}_{\mathcal{F}}$, *then:*
$x_2(b,s)^{h_1(a,t)} = x_2\big(\Theta(a,b)+tb, zs+\psi(a,b)\big)$ *for all* $(b,s)\in W_0\oplus F$ *and all* $(a,t)\in (X_0\oplus K)^*$, *where* $z=q_F(a,t)$;
$x_4(b,s)^{h_1(a,t)} = x_4\big((\Theta(a,b)+tb)z^{-1}, sz^{-1}+\psi(a,b)z^{-2}\big)$ *for all* $(b,s)\in W_0\oplus F$ *and all* $(a,t)\in(X_0\oplus K)^*$, *where* $z=q_F(a,t)$;
$x_1(a,t)^{h_4(b,s)} = x_1\big((\Upsilon(a,b)+sa)r^{-2}, tr^{-1}+\omega(a,b)r^{-2}\big)$ *for all* $(a,t)\in X_0\oplus K$ *and all* $(b,s)\in(W_0\oplus F)^*$, *where* $r=q_K(b,s)$;[23] *and*
$x_3(a,t)^{h_4(b,s)} = x_3\big(\Upsilon(a,b)+sa, rt+\omega(a,b)\big)$ *for all* $(a,t)\in X_0\oplus K$ *and all* $(b,s)\in(W_0\oplus F)^*$, *where* $r=q_K(b,s)$.

(viii) *If* $X=\mathcal{H}$, *then:*
$x_4(u)^{h_1(a)} = x_4(u)$ *for all* $u\in K$ *and all* $a\in J^*$;
$x_6(u)^{h_1(a)} = x_6(-N(t)^{-1}a)$ *for all* $u\in K$ *and all* $a\in J^*$;
$x_1(a)^{h_6(t)} = x_1(-t^{-1}a)$ *for all* $a\in J$ *and all* $t\in K^*$;
$x_4(u)^{h_6(t)} = x_4(-tu)$ *for all* $u\in K$ *and all* $t\in K^*$.

Proof. Let E be the $\{x,y\}$-residue of Δ containing d, let a be a root of Σ cutting E and let $u\in U_a^*$. By 29.13.iv, $\Sigma\cap E$ is an apartment of E and $a\cap E$ is a root of this apartment. By 11.10 in [37], U_a acts faithfully on E and the the image of U_a in $\mathrm{Aut}(E)$ is the root group of E corresponding to $a\cap E$. Thus the element $m_\Sigma(u)$ also maps E to itself (by 3.8). By 11.22.ii in [37], in fact, the element of $\mathrm{Aut}(E)$ induced by $\mu_\Sigma(u)$ equals $m_{\Sigma\cap E}(\bar{u})$, where $\bar{u}$ is the image of u in $\mathrm{Aut}(E)$.

Choose $t\in\Lambda_1^*$ and $s\in\Lambda_n^*$. By 3.11, $h_1(t)$ and $h_n(s)$ normalize U_i for all $i\in[1,n]$. By the observations in the previous paragraph, the action of $h_1(t)$ on U_i for all $i\in[2,n]$ as well as the action of $h_n(s)$ on U_i for all $i\in[1,n-1]$ can be read off from 32.5–32.12 of [36]. □

Proposition 30.37. *The following hold (with all the notation of 30.36):*

(i) *If* $X=\mathcal{Q}_{\mathcal{Q}}$, *then*
$$h_1(st)=h_1(s)h_2(t)$$
for all $s,t\in K^*$.

(ii) *If* $X=\mathcal{Q}_{\mathcal{D}}$ *(in which case* $K^2\subset K_0$ *by 10.2 in [36]), then*
$$h_1(st)=h_1(s)h_2(t)$$
for all $s,t\in(K^*)^2$ *and*
$$h_1(s)^2=h_1(s^2)$$
for all $s\in K_0^*$.

(iii) *If* $X=\mathcal{Q}_{\mathcal{E}}$ *or* $\mathcal{Q}_{\mathcal{F}}$, *then*
$$h_1(0,s)h_1(0,t)=h_1(0,st)$$
for all $s,t\in K^*$.

[23] As in Chapter 22, we are using the letter ω here to denote the map called ν in 14.16 of [36]; see 22.1.

Proof. Let

$$H = \langle m_\Sigma(x_1(u))m_\Sigma(x_1(v)) \mid u, v \in \Lambda_1^*\rangle.$$

By 6.7 of [36], the group H acts faithfully on the subgroup U_3U_4. By 30.36.iii, therefore, (i) holds. Similarly, (ii) follows from 30.36.iv. If $X = \mathcal{Q}_\mathcal{E}$ or $\mathcal{Q}_\mathcal{F}$, then by 32.10 and 32.11 of [36], $[x_1(0,t), U_3] = 1$ for all $t \in K^*$ (in both cases). Thus (iii) holds by 30.36.vi and 30.36.vii. □

Proposition 30.38. *Let $X(\Lambda)$, $x_1, \ldots, x_n$, e_1, e_n, ϵ_1 and ϵ_n be as in 30.36 and let $m_i = m_\Sigma(e_i)$ for $i = 1$ and n. Then the following hold:*

(i) *If $X = \mathcal{T}$, then:*
$x_2(u)^{m_1} = x_2(u)$ *for all* $u \in K$;
$x_3(u)^{m_1} = x_2(-u)$ *for all* $u \in K$;
$x_1(u)^{m_3} = x_2(u)$ *for all* $u \in K$;
$x_2(u)^{m_3} = x_1(-u)$ *for all* $u \in K$; *and*
$\epsilon_1 = \epsilon_3 = 1$.

(ii) *If $X = \mathcal{Q}_\mathcal{I}$, then:*
$x_2(u)^{m_1} = x_4(-u)$ *for all* $u \in K$;
$x_3(t)^{m_1} = x_3(t)$ *for all* $t \in K_0$;
$x_4(u)^{m_1} = x_2(u)$ *for all* $u \in K$;
$x_1(t)^{m_4} = x_3(t)$ *for all* $t \in K_0$;
$x_2(u)^{m_4} = x_2(-u^\sigma)$ *for all* $u \in K$;
$x_3(t)^{m_4} = x_1(t)$ *for all* $t \in K_0$; *and*
$\epsilon_1 = \epsilon_4 = 1$.

(iii) *If $X = \mathcal{Q}_\mathcal{Q}$, then:*
$x_2(u)^{m_1} = x_4(-u)$ *for all* $u \in L$;
$x_3(t)^{m_1} = x_3(t)$ *for all* $t \in K$;
$x_4(u)^{m_1} = x_2(u)$ *for all* $u \in L$;
$x_1(t)^{m_4} = x_3(t)$ *for all* $t \in K$;
$x_2(u)^{m_4} = x_2(v)$ *for all* $u \in L$, *where* $v \in L$ *is an element depending on* u *such that* $q(v) = q(u)$;
$x_3(t)^{m_4} = x_1(t)$ *for all* $t \in K$;
$\epsilon_1 = 1$ *and* ϵ_4 *is an element of* L^* *such that* $q(\epsilon_4) = 1$.

(iv) *If $X = \mathcal{Q}_\mathcal{D}$, then:*
$x_2(u)^{m_1} = x_4(u)$ *for all* $u \in L_0$;
$x_3(s)^{m_1} = x_3(s)$ *for all* $s \in K_0$;
$x_4(u)^{m_1} = x_1(u)$ *for all* $u \in L_0$;
$x_1(t)^{m_4} = x_3(t)$ *for all* $t \in K_0$;
$x_2(u)^{m_4} = x_2(u)$ *for all* $v \in L_0$;
$x_3(t)^{m_4} = x_1(t)$ *for all* $t \in K_0$; *and*
$\epsilon_1 = \epsilon_4 = 1$.

(v) *If $X = \mathcal{Q}_\mathcal{P}$, then:*
$x_2(u)^{m_1} = x_4(-u)$ *for all* $u \in K$;
$x_3(a,t)^{m_1} = x_3(a,t)$ *for all* $(a,t) \in T^*$;
$x_4(u)^{m_1} = x_2(u)$ *for all* $u \in K$;
$x_1(a,t)^{m_4} = x_3(-a,t)$ *for all* $(a,t) \in T$;

$x_2(v)^{m_4} = x_2(-v^\sigma)$ *for all* $v \in K$;
$x_3(a,t)^{m_4} = x_1(a,t)$ *for all* $u \in K$;
$\epsilon_1 = (0,1)$ *and* $\epsilon_4 = 1$.

(vi) *If* $X = \mathcal{Q}_{\mathcal{E}}$, *then:*
$x_2(v)^{m_1} = x_4(-v)$ *for all* $v \in L$;
$x_3(a,t)^{m_1} = x_3(a,t)$ *for all* $(a,t) \in S$;
$x_4(v)^{m_1} = x_2(v)$ *for all* $v \in L$;
$x_1(a,t)^{m_4} = x_3(-a,t)$ *for all* $(a,t) \in T$;
$x_2(u)^{m_4} = x_2(v)$ *for all* $u \in L$, *where* $v \in L$ *is an element depending on* u *such that* $q(v) = q(u)$;
$x_3(a,t)^{m_4} = x_1(a,t)$ *for all* $(a,t) \in S$;
$\epsilon_1 = (0,1)$ *and* ϵ_4 *is an element of* L^* *such that* $q(\epsilon_4) = 1$.

(vii) *If* $X = \mathcal{Q}_{\mathcal{F}}$, *then:*
$x_2(b,s)^{m_1} = x_4(b,s)$ *for all* $(b,s) \in W_0 \oplus F$;
$x_3(a,t)^{m_1} = x_3(a,t)$ *for all* $(a,t) \in X_0 \oplus K$;
$x_4(b,s)^{m_1} = x_2(b,s)$ *for all* $(b,s) \in W_0 \oplus F$;
$x_1(a,t)^{m_4} = x_3(a,t)$ *for all* $(a,t) \in X_0 \oplus K$;
$x_2(b,s)^{m_4} = x_2(b,s)$ *for all* $(b,s) \in W_0 \oplus F$;
$x_3(a,t)^{m_4} = x_1(a,t)$ *for all* $(a,t) \in X_0 \oplus K$;
$\epsilon_1 = (0,1) \in X_0 \oplus K$ *and* $\epsilon_4 = (0,1) \in W_0 \oplus F$.

(viii) *If* $X = \mathcal{H}$, *then:*
$x_2(t)^{m_1} = x_6(-t)$ *for all* $t \in K$;
$x_3(a)^{m_1} = x_5(-a)$ *for all* $a \in J$;
$x_4(t)^{m_1} = x_4(t)$ *for all* $t \in K$;
$x_5(a)^{m_1} = x_2(a)$ *for all* $a \in J$;
$x_6(t)^{m_1} = x_2(t)$ *for all* $t \in K$;
$x_1(a)^{m_6} = x_5(-a)$ *for all* $a \in J$;
$x_2(t)^{m_6} = x_4(t)$ *for all* $t \in K$;
$x_3(a)^{m_6} = x_3(a)$ *for all* $a \in J$;
$x_4(t)^{m_6} = x_2(-t)$ *for all* $t \in K^*$;
$x_5(a)^{m_6} = x_1(a)$ *for all* $a \in J$;
$\epsilon_1 = 1 \in J$ *and* $\epsilon_6 = 1 \in K$.

Proof. In light of the observations made in the first paragraph of the proof of 30.36, these facts can be read off from 32.5–32.12 of [36]. □

Bibliography

[1] Abramenko, P. and Brown, K. S. 2008. *Buildings: Theory and Applications.* Springer.

[2] Artin, E. 1967. *Algebraic Numbers and Algebraic Functions.* Gordon and Breach. Basel.

[3] Bourbaki, N. 1968. *Elements of Mathematics: Lie Groups and Lie Algebras,* Chapters 4, 5 and 6. Springer.

[4] Brown, K. S. 1989. *Buildings.* Springer.

[5] Bruhat, F. and Tits, J. 1967. Groupes algébriques simples sur un corps local. In *Proceedings of a Conference on Local Fields* (Driebergen, 1966, T. A. Springer, ed.), pp. 23–36. Springer.

[6] Bruhat, F. and Tits, J. 1972. Groupes réductifs sur un corps local, I. Données radicielles valuées. *Publ. Math. I.H.E.S.* **41**, 5–252.

[7] Bruhat, F. and Tits, J. 1984. Groupes réductifs sur un corps local, II. Schémas en groupes. Existence d'une donnée radicielle valuée. *Publ. Math. I.H.E.S.* **60**, 5–184.

[8] Bruhat, F. and Tits, J. 1984. Schémas en groupes et immeubles des groupes classiques sur un corps local, I. *Bull. Soc. Math. Franc.* **112**, 259–301.

[9] Bruhat, F. and Tits, J. 1987. Schémas en groupes et immeubles des groupes classiques sur un corps local, II. Groupes unitaires. *Bull. Soc. Math. Franc.* **115**, 141–195.

[10] Bruhat, F. and Tits, J. 1987. Groupes algébriques sur un corps local, III. Compléments et applications à la cohomologie galoisienne. *J. Fac. Sci. Univ. Tokyo IA Math.* **34**, 671–698.

[11] Cassels, J. W. S. and Fröhlich, A., ed. 1967. *Algebraic Number Theory.* Thompson Book Company, Inc., Washington.

[12] Cohn, P. M. 1995. *Skew Fields.* Encyclopedia of Mathematics, vol. 57. Cambridge University Press.

[13] Coxeter, H. S. M. 1935. The complete enumeration of finite groups of the form $R_i^2 = (R_iR_j)^{k_{ij}} = 1$. *J. Lond. Math. Soc.* **10**, 21–25.

[14] Garrett, P. 1997. *Buildings and classical groups.* Chapman & Hall, London.

[15] Heydebreck, A. v. 1996. *Klassifikation affiner Gebäude*, Diplom-Arbeit, Fachbereich Mathematik, Universität Frankfurt.

[16] Hocking, J. G. and Young, G. S. 1961. *Topology*, Dover Publications, New York.

[17] Humphreys, J. E. 1990. *Reflection Groups and Coxeter Groups.* Cambridge University Press.

[18] Jacobson, N. 1968. *Structure and Representations of Jordan Algebras.* Amer. Math. Soc. Colloq. Publ. **39**. Amer. Math. Soc., Providence.

[19] Petersson, H. P. 1973. Jordan-Divisionsalgebren und Bewertungen. *Math. Ann.* **202**, 215–243.

[20] Petersson, H. P. 1974. Composition algebras over a field with a discrete valuation. *J. Algebra* **29**, 414–426.

[21] Petersson, H. P. 1975. Exceptional Jordan division algebras over a field with a discrete valuation. *J. Reine Angew. Math.* **274/275**, 1–20.

[22] Petersson, H. P. 1992. Valuations on composition algebras. *Nova J. Algebra Geom.* **1**, 125–131.

[23] Ribenboim, P. 1999. *The Classical Theory of Valuations.* Springer.

[24] Reiner, I. 1975. *Maximal Orders*, Academic Press, London.

[25] Ronan, M. A. 1989. *Lectures on Buildings.* Academic Press, London.

[26] Salzmann, H., Grunhöfer, T., Hähl, H. and Löwen, R. 2007. *The Classical Fields.* Cambridge University Press.

[27] Scharlau, W. 1985. *Quadratic and Hermitian Forms.* Springer.

[28] Serre, J.-P. 1979. *Local Fields.* Springer.

[29] Serre, J.-P. 1980. *Trees.* Springer.

[30] Springer, T. A. 1955. Quadratic forms over fields with a discrete valuation. I. Equivalence classes of definite forms. *Nederl. Akad. Wetensch. Proc. Ser. A* **58** = *Indag. Math.* **17**, 352–362.

[31] Tits, J. 1966. Classification of algebraic semi-simple groups. In *Algebraic Groups and Discontinuous Groups* (Boulder, 1965), pp. 33–62, Proc. Symp. Pure Math. **9**. Amer. Math. Soc., Providence.

[32] Tits, J. 1974. *Buildings of Spherical Type and Finite BN-Pairs*, Lecture Notes in Mathematics, vol. 386. Springer.

[33] Tits, J. 1979. Reductive groups over local fields. In *Proc. Symp. Pure Math.* **33**, Part 1 (*Automorphic Forms, Representations and L-Functions*, Corvallis 1977), pp. 29–69. *Amer. Math. Soc.*, Providence.

[34] Tits, J. 1983. Moufang octagons and Ree groups of type ${}^2F_4$. *Amer. J. Math.* **105**, 539–594.

[35] Tits, J. 1986. Immeubles de type affine. In *Buildings and the Geometry of Diagrams* (Como 1984), pp. 159–190. Lecture Notes in Mathematics **1181**. Springer.

[36] Tits J. and Weiss, R. M. 2002. *Moufang Polygons.* Springer.

[37] Weiss, R. M. 2003. *The Structure of Spherical Buildings.* Princeton University Press.

[38] Weiss, R. M. 2006. *Quadrangular Algebras.* Math. Notes **46**. Princeton University Press.

[39] Wadsworth, A. R. 2002. Valuation theory on finite dimensional division algebras. In *Valuation Theory and its Applications, Vol. I* (Saskatoon, SK, 1999), 385–449. *Field Inst. Commun.* **32**. Amer. Math. Soc., Providence.

Index

www.ingramcontent.com/pod-product-compliance
Ingram Content Group UK Ltd.
Pitfield, Milton Keynes, MK11 3LW, UK
UKHW032106100125
453448UK00005B/309